D0845644

Apoptosis, Cell Signaling, and Human Diseases

Apoptosis, Cell Signaling, and Human Diseases

Molecular Mechanisms, Volume 1

Edited by

Rakesh Srivastava

Department of Biochemistry
University of Texas Health Center at Tyler
Tyler, TX

HUMANA PRESS ✱ TOTOWA, NEW JERSEY

999 Riverview Drive, Suite 208
Totowa, New Jersey 07512

www.humanapress.com

This publication is printed on acid-free paper. ∞
ANSI Z39.48-1984 (American Standards Institute)

Permanence of Paper for Printed Library Materials.

Cover illustration: Figure 3 of Chapter 4, Volume 2, "Cyclin-Dependent Kinase 5: A Target for Neuroprotection?" by Frank Gillardon.

Cover design by Patricia F. Cleary

For additional copies, pricing for bulk purchases, and/or information about other Humana titles, contact Humana at the above address or at any of the following numbers: Tel.: 973-256-1699; Fax: 973-256-8341; E-mail: humana@humanapr.com; or visit our Website: www.humanapress.com

Printed in the United States of America. 10 9 8 7 6 5 4 3 2 1

e-ISBN: 1-59745-200-1 ISBN13 978-1-58829-677-1

Library of Congress Cataloging in Publication Data
Apoptosis, cell signaling, and human diseases : molecular mechanisms / edited by Rakesh Srivastava.
p. ; cm.
Includes bibliographical references and index.
ISBN 1-58829-677-6 (v. 1 : alk. paper) — ISBN 1-58829-882-5 (v. 2 : alk. paper)
1. Apoptosis. 2. Pathology, Molecular. 3. Cellular signal transduction. I. Srivastava, Rakesh, 1956-
[DNLM: 1. Apoptosis—physiology. 2. Cell Transformation, Neoplastic—genetics. 3. Gene Therapy. 4. Signal Transduction. QU 375 A64473 2007]
QH671.A6547 2007
571.9'36—dc22 2006030613

Preface

The aim of *Apoptosis, Cell Signaling, and Human Diseases: Molecular Mechanisms* is to present recent developments in cell survival and apoptotic pathways and their involvement in human diseases, such as cancers and neurodegenerative disorders. This requires an integration of knowledge from several fields of research, including pathology, genetics, virology, cell biology, medicine, immunology, and molecular biology. This edition of the book examines the impact of molecular biology on disease mechanisms. With recent advances in technology such as microarray and proteomics, new biomarkers and molecular targets have been identified. These potential targets will be very useful for the development of novel and more effective drugs for the treatment of human diseases. The challenge now is not only to understand disease mechanisms but also to apply this knowledge to find therapies that are more effective.

Cellular processes play major roles in cell survival and apoptosis. These events are essential for tissue homeostasis and the maintenance of proper growth and development of multicellular organisms. Imbalance in survival and apoptotic pathways may lead to several diseases. Therefore, understanding the molecular mechanisms of cell survival and apoptotic pathways is essential for the treatment and prevention of human diseases. The main focus of *Apoptosis, Cell Signaling, and Human Diseases: Molecular Mechanisms* is to discuss the recent development in cell signaling events, growth, metastasis, and angiogenesis, mechanisms of drug resistance, and targeted therapy for human diseases. Volume 1 contains 15 chapters divided into two sections: "Malignant Transformation and Metastasis" and "Molecular Basis of Disease Therapy"; Volume 2 contains 18 chapters, also divided into two sections: "Kinases and Phosphatases" and "Molecular Basis of Cell Death." Scientists well known in their fields have contributed to this book.

In part I, the pathophysiological processes including the mechanisms by which normal cells are transformed to malignant cells, regulation of cell growth, differentiation and apoptosis by oncogenes and tumor suppressor genes, consequences of DNA damage and the ability of cells to repair damaged DNA in response to stress stimuli, molecular events involved in metastasis and angiogenesis, and roles of transcription factors and cytokines in cell survival and apoptosis, are discussed. The recent development in technology has allowed us to identify new diseases before the appearance of the symptoms. The delay in identification of the disease may be fatal to human life. The incorporation of concepts of engineering to the principles of biology has further revolutionized the field of medicine. Nanotechnology, bioinformatics, microarray and proteomics are powerful tools that are being used in drug discovery and development, and treatment of human diseases.

In part II, biological significance of the kinases, and the cell signaling events that control cell survival and apoptosis are discussed. The identification of over 500 protein kinases encoded by the human genome sequence offers one measure of the importance of protein kinase networks in cell biology. Phosphorylation and dephosphorylation of protein kinases such as protein kinase A (PKA), protein kinase C (PKC), cyclin-dependent kinase (CDK), phosphatidylinositol 3-kinase (PI3K), Akt, and MAP kinase (MAPK) are important for regulating cell cycle, survival, and apoptosis. High-throughput technologies for inactivating genes are producing an inspiring amount of data on the cellular and organismal effects of reducing the levels of individual protein kinases. Despite these technical advances, our understanding of kinase networks remains imprecise. Major challenges include correctly assigning kinases to particular networks, understanding how they are regulated, and identifying the relevant in vivo substrates. Genetic methods provide a way of addressing these questions, but their application requires understanding the mutations and how they affect protein-protein interacts.

Apoptosis is a genetically controlled process that plays important roles in embryogenesis, metamorphosis, cellular homeostasis, and as a defensive mechanism to remove infected, damaged, or mutated cells. Molecules involved in cell death pathways are potential therapeutic targets in immunologic, neurologic, cancer, infectious, and inflammatory diseases. Although a number of stimuli triggers apoptosis, it is mainly mediated through at least three major pathways that are regulated by (i) the death receptors, (ii) the endoplasmic reticulum (ER), and (iii) the mitochondria. Under certain conditions, these pathways may cross talk to enhance apoptosis. Death receptor pathways are involved in immune-mediated neutralization of activated or autoreactive lymphocytes, virus-infected cells, and tumor cells. Consequently, dysregulation of the death receptor pathway has been implicated in the development of autoimmune diseases, immunodeficiency, and cancer. Increasing evidence indicates that the mitochondrial and ER pathways of apoptosis play a critical role in death receptor-mediated apoptosis. Dysregulation of these pathways may contribute to drug resistance.

A lot of progress has been made in understanding the mechanisms of apoptosis. Mitochondria are critical death regulators of the intrinsic apoptotic pathway in response to DNA damage, growth factor withdrawal, hypoxia, or oncogene deregulation. Activation of the mitochondrial pathway results in disruption of mitochondrial homeostasis, and release of mitochondrial proteins. The release of mitochondrial apoptogenic factors is regulated by the pro- and anti-apoptotic Bcl-2 family proteins, which either induce or prevent the permeabilization of the outer mitochondrial membrane. Activation of the death receptor pathway also links the cell-intrinsic pathway through Bid. Mitochondrial membrane permeabilization induces the release of mitochondrial proteins (e.g., cytochrome c, Smac/DIABLO, AIF, Omi/HtrA2, and endonuclease G), which are regulated by proapoptotic and antiapoptotic proteins of Bcl-2 family, and in caspase-dependent and -independent apoptotic pathways. The antiapoptotic members (e.g. Bcl-2 or Bcl-X_L) inhibit the release of mitochondrial apoptogenic factors whereas the proapoptotic members (e.g. Bax, and Bak) trigger the release.

Recent studies suggest that, in addition to mitochondria and death receptors, other organelles, including the endoplasmic reticulum (ER), Golgi bodies, and lysosomes, are also major points of integration of proapoptotic signaling and damage sensing. Each organelle possesses sensors that detect specific alterations, locally activate signal transduction pathways, and emit signals that ensure inter-organellar cross-talk. The genomic responses in intracellular organelles, after DNA damage, are controlled and amplified in the cross-signaling via mitochondria; such signals induce apoptosis, autophagy, and other cell death pathways.

Chromatin remodeling agents modulate gene expression in tumor cells. Acetylation and deacetylation are catalyzed by specific enzyme families, histone acetyltransferases (HATs) and deacetylases (HDACs), respectively. Since aberrant acetylation of histone and nonhistone proteins has been linked to malignant diseases, HDAC inhibitors bear great potential as new drugs due to their ability to modulate transcription, induce differentiation and apoptosis, and inhibit angiogenesis. The preclinical data on HDAC inhibitors are very promising, and several HDAC inhibitors are currently under clinical trials for the treatment of cancers.

Apoptosis, Cell Signaling, and Human Diseases: Molecular Mechanisms will be valuable to graduate students, postdoctoral and medical fellows, and scientists with a working knowledge of biology and pathology who desire to learn about the molecular mechanisms of human diseases and therapy. I hope that individuals of diverse backgrounds will find these volumes very useful.

***Rakesh Srivastava,* PhD**

Contents

Preface v
Contributors xi
Contents of Volume 2 xv

PART I MALIGNANT TRANSFORMATION AND METASTASIS

1 *BCR-ABL* and Human Cancer
María Pérez-Caro and Isidro Sanchez-García ***3***
2 Angiogenesis and Cancer
Yohei Maeshima ***35***
3 Metastasis: *The Evasion of Apoptosis*
Christine E. Horak, Julie L. Bronder, Amina Bouadis, and Patricia S. Steeg ***63***
4 Carcinogenesis: *Balance Between Apoptosis and Survival Pathways*
Dean G. Tang and James P. Kehrer ***97***
5 Aberrations of DNA Damage Checkpoints in Cancer
Marikki Laiho ***119***
6 c-Myc, Apoptosis, and Disordered Tissue Growth
Michael Khan and Stella Pelengaris ***137***
7 Role of Lysophospholipids in Cell Growth and Survival
Xianjun Fang and Sarah Spiegel ***179***
8 Alternative Use of Signaling by the βGBP Cytokine in Cell Growth and Cancer Control: *From Surveillance to Therapy*
Livio Mallucci and Valerie Wells ***203***
9 Control Nodes Linking the Regulatory Networks of the Cell Cycle and Apoptosis
Baltazar D. Aguda, Wee Kheng Yio, and Felicia Ng ***217***

PART II MOLECULAR BASIS OF DISEASE THERAPY

10 Regulation of NF-κB Function: *Target for Drug Development*
Daniel Sliva and Rakesh Srivastava ***239***
11 5-Fluorouracil: *Molecular Mechanisms of Cell Death*
Daniel B. Longley and Patrick G. Johnston ***263***
12 Apoptosis-Inducing Cellular Vehicles for Cancer Gene Therapy: *Endothelial and Neural Progenitors*
Gergely Jarmy, Jiwu Wei, Klaus-Michael Debatin, and Christian Beltinger ***279***
13 Apoptosis and Cancer Therapy
Maurice Reimann and Clemens A. Schmitt ***303***

14 Coupling Apoptosis and Cell Division Control in Cancer: *The Survivin Paradigm*
Dario C. Altieri **321**
15 Clinical Significance of Histone Deacetylase Inhibitors in Cancer
Sharmila Shankar and Rakesh K. Srivastava **335**
Index **363**

Contributors

BALTAZAR D. AGUDA, PhD • *Bioinformatics Institute, Singapore*
DARIO C. ALTIERI, MD • *Department of Cancer Biology , University of Massachusetts Medical School, Worcester, MA*
SHRIKANT ANANT, PhD • *Division of Gastroenterology, Department of Internal Medicine, Washington University School of Medicine, St. Louis, MO*
CHRISTIAN BELTINGER, MD • *University Children's Hospital, Ulm, Germany*
FRED E. BERTRAND • *Department of Microbiology & Immunology, Brody School of Medicine at East Carolina University, Greenville, NC*
JOHN BLENIS, PhD • *Department of Cell Biology, Harvard Medical School, Boston, MA*
AMINA BOUADIS, BS • *Women's Cancers Section, Laboratory of Pathology, Center for Cancer Research, National Cancer Institute, Bethesda, MD*
JULIE L. BRONDER, PhD • *Women's Cancers Section, Laboratory of Pathology, Center for Cancer Research, National Cancer Institute, Bethesda, MD*
ZHEQING CAI, PhD • *Vascular Program, Institute for Cell Engineering; Departments of Pediatrics, Medicine, Oncology, Radiation Oncology; and Institute of Genetic Medicine, The Johns Hopkins University School of Medicine, Baltimore, MD*
FUMIN CHANG • *Department of Microbiology & Immunology, Brody School of Medicine at East Carolina University, Greenville, NC*
Y.S. CHO-CHUNG, MD, PhD • *Cellular Biochemistry Section, Basic Research Laboratory, Center for Cancer Research, National Cancer Institute, National Institutes of Health, Bethesda, MD*
SAMANTHA COORAY, PhD • *Department of Virology, Wright Flemming Institute, Imperial College Faculty of Medicine, St. Mary's Campus, Norfolk Place, London, UK*
TASMAN JAMES DAISH • *Hanson Institute, Adelaide, Australia*
KLAUS-MICHAEL DEBATIN, MD • *University Children's Hospital, Ulm, Germany*
STEIN OVE DØSKELAND, PhD • *Department of Biomedicine, Section of Anatomy and Cell Biology, University of Bergen, Bergen, Norway*
PAUL G. EKERT, MDDS, PhD, FRACP • *The Walter and Eliza Hall Institute of Medical Research, Victoria, Australia*
XIANJUN FANG, PhD • *Department of Biochemistry, Medical College of Virginia , Virginia Commonwealth School of Medicine, Richmond, VA*
RICHARD A. FRANKLIN • *Department of Microbiology & Immunology and Leo Jenkins Cancer Center, Brody School of Medicine at East Carolina University, Greenville, NC*
GRO GAUSDAL, CSci • *Department of Biomedicine, Section of Anatomy and Cell Biology, University of Bergen, Bergen, Norway*
FRANK GILLARDON, PhD • *Boehringer INGELHEIM Pharma GmbH & Co. KG, CNS Research, Biberach an der Riss, Germany*
LARS HERFINDAL, CSci • *Department of Biomedicine, Section of Anatomy and Cell Biology, University of Bergen, Bergen, Norway*

CHRISTINE E. HORAK, PhD • *Women's Cancers Section, Laboratory of Pathology, Center for Cancer Research, National Cancer Institute, Bethesda, MD*
STEVEN IDELL, PhD • *The Texas Lung Injury Institute, The University of Texas Health Center at Tyler, Tyler, TX*
GERGELY JARMY, PhD • *University Children's Hospital, Ulm, Germany*
PATRICK G. JOHNSTON, PhD • *Drug Resistance Laboratory, Centre for Cancer Research and Cell Biology, Queen's University Belfast, Belfast, N. Ireland*
JAMES P. KEHRER, PhD • *College of Pharmacy, Washington State University, Pullman, WA*
MICHAEL KHAN, PhD, FRCP • *Molecular Medicine, Biomedical Research Institute , University of Warwick, Coventry, UK*
CAMILLA KRAKSTAD, PhD • *Department of Biomedicine, Section of Anatomy and Cell Biology, University of Bergen, Bergen, Norway*
SHARAD KUMAR, PhD • *Hanson Institute, Adelaide, Australia*
MARIKKI LAIHO, MD, PhD • *Hartman Institute and Molecular Cancer Biology Research Program, Biomedicum Helsinki, University of Helsinki, Helsinki, Finland*
DANIEL B. LONGLEY, PhD • *Drug Resistance Laboratory, Centre for Cancer Research and Cell Biology, Queen's University Belfast, Belfast, N. Ireland*
PETER LOW, PhD • *Department of General Zoology, Eotvos University, Budapest, Hungary*
YOHEI MAESHIMA, MD, PhD • *Assistant Professor of Medicine, Okayama University Graduate School of Medicine and Dentistry, Department of Medicine and Clinical Science, Okayama, Japan*
LIVIO MALLUCCI, MD • *King's College London, Biomedical and Health Sciences, Pharmaceutical Science Research Division, London, UK*
JAMES A. MCCUBREY, PhD • *Department of Microbiology and Immunology, East Carolina University, Brody School of Medicine, Greenville, NC*
MAURIZIO MEMO, PhD • *Department of Biomedical Sciences and Biotechnologies, University of Brescia Medical School, Brescia, Italy*
MARIA SAVERIA GILARDINI MONTANI, PhD • *Department of Environmental Science, La Tuscia University, Viterbo, Italy*
MARIA V. NESTEROVA • *Cellular Biochemistry Section, Basic Research Laboratory, CCR, National Cancer Institute, Bethesda, MD*
FELICIA NG • *Bioinformatics Institute, Singapore*
STELLA PELENGARIS, Bsc, PhD • *Molecular Medicine , Biomedical Research Institute, University of Warwick, Coventry, UK*
MARÍA PÉREZ-CARO, PhD • *Instituto Biologia Molecular y Celular del Cancer (IBMCC), CSIC/University of Salamanca, Salamanca, Spain*
MAURICE REIMANN, PhD • *Hematology, Oncology, and Tumor Immunology, Charité - Universitätsmedizin Berlin (CVK), Berlin, Germany*
MARY E. REYLAND, PhD • *Department of Craniofacial Biology, University of Colorado Health Sciences Center, Denver, CO*
PHILIPPE ROUX, PhD • *Department of Cell Biology, Harvard Medical School, Boston, MA*

ISIDRO SÁNCHEZ-GARCÍA, MD, PhD • *Instituto Biologia Molecular y Celular del Cancer (IBMCC), CSIC/University of Salamanca, Salamanca, Spain*
CLEMENS A. SCHMITT, MD • *Hematology, Oncology, and Tumor Immunology, Charité - Universitätsmedizin Berlin (CVK), and Max-Delbrück-Center for Molecular Medicine, Berlin, Germany*
GREGG L. SEMENZA, MD, PhD • *Vascular Program, Institute for Cell Engineering; Departments of Pediatrics, Medicine, Oncology, Radiation Oncology; and Institute of Genetic Medicine, The Johns Hopkins University School of Medicine, Baltimore, MD*
SHARMILA SHANKAR, PhD • *Department of Biochemistry, University of Texas Health Center at Tyler, Tyler, TX*
SREERAMA SHETTY, PhD • *The Texas Lung Injury Institute, The University of Texas Health Center at Tyler, Tyler, TX*
JOHN SILKE, PhD • *The Walter and Eliza Hall Institute of Medical Research, Parkville, Australia*
TOMASZ SKORSKI, MD, PhD • *Center for Biotechnology and Molecular Carcinogenesis Section, College of Science and Technology, Temple University, Philadelphia, PA*
DANIEL SLIVA, PhD • *Department of Medicine, Indiana School of Medicine, and Cancer Research Laboratory, Methodist Research Institute, Indianapolis, IN*
SARAH SPIEGEL, PhD • *Department of Biochemistry, Medical College of Virginia , Virginia Commonwealth School of Medicine, Richmond, VA*
RAKESH SRIVASTAVA, PhD • *Department of Biochemistry, University of Texas Health Center at Tyler, Tyler, TX*
PATRICIA S. STEEG, PhD • *Molecular Therapeutics Program, Molecular Targets Faculty, and Women's Cancers Section, Laboratory of Pathology, Center for Cancer Research, National Cancer Institute, Bethesda, MD*
LINDA S. STEELMAN • *Department of Microbiology & Immunology, Brody School of Medicine at East Carolina University, Greenville, NC*
SRIPATHI M. SUREBAN • *Department of Internal Medicine, Washington University School of Medicine, St. Louis, MO*
DEAN G. TANG, PhD • *The University of Texas M. D. Anderson Cancer Center, Department of Carcinogenesis, Science Park-Research Division, Smithville, TX*
DAVID M. TERRIAN • *Leo Jenkins Cancer Center and Department of Anatomy & Cell Biology, Brody School of Medicine at East Carolina University, Greenville, NC*
JIWU WEI, MD • *University Children's Hospital, Ulm, Germany*
VALERIE WELLS, PhD • *King's College London, Biomedical and Health Sciences, Pharmaceutical Science Research Division, London, UK*
WEE KHENG YIO • *Bioinformatics Institute, Singapore*
JAI YU, Bsc • *The Walter and Eliza Hall Institute of Medical Research, Parkville, Australia*

Contents of Volume 2

Part I Kinases and Phosphatases

1 Significance of Protein Kinase A in Cancer
Maria V. Nesterova and Y.S. Cho-Chung
2 Protein Kinase C and Cancer
Mary E. Reyland
3 The Role of PI3K-Akt Signal Transduction in Virus Infection
Samantha Cooray
4 Cyclin-Dependent Kinase 5: *A Target for Neuroprotection?*
Frank Gillardon
5 Critical Roles of the Raf/MEK/ERK Pathway in Apoptosis and Drug Resistance
James A. McCubrey, Fred E. Bertrand, Linda S. Steelman, Fumin Chang, David M. Terrian, and Richard A. Franklin
6 MAP Kinases Signaling in Human Diseases
Philippe Roux and J. Blenis
7 Serine/Threonine Protein Phosphates in Apoptosis
Gro Gausdal, Camilla Krakstad, Lars Herfindal, and Stein Ove Døskeland
8 Urokinase/Urokinase Receptor-Mediated Signaling in Cancer
Sreerama Shetty and Steven Idell

Part II Molecular Basis of Cell Death

9 Signaling Pathways That Protect the Heart Against Apoptosis Induced by Ischemia and Reperfusion
Zheqing Cai and Gregg L. Semenza
10 Cyclooxygenase-2 Gene Expression: *Transcriptional and Posttranscriptional Controls in Intestinal Tumorigenesis*
Shrikant Anant and Sripathi M. Sureban
11 Death Receptors: *Mechanisms, Biology, and Therapeutic Potential*
Sharmila Shankar and Rakesh Srivastava
12 DNA Damage-Dependent Apoptosis
Tomasz Skorski
13 The Role of Proteasome in Apoptosis
Peter Low
14 Apoptosis Induction in T Lymphocytes by HIV
Maria Saveria Gilardini Montani
15 Inhibitors of Apoptosis Proteins and Caspases
Jai Yu, John Silke, and Paul G. Ekert
16 Intracellular Pathways Involved in DNA Damage and Repair to Neuronal apoptosis
Maurizio Memo

17 The Biology of Caspases
Tasman James Daish and Sharad Kumar

18 Oxidative Stress in the Pathogenesis of Diabetic Neuropathy
Mahdieh Sadidi, Ann Marie Sastry, Christian M. Lastoskie, Andrea A. Vincent, Kelli A. Sullivan, and Eva L. Feldman

I

Malignant Transformation and Metastasis

1

BCR-ABL and Human Cancer

María Pérez-Caro and Isidro Sánchez-García

Summary

The *BCR-ABL* oncogene was the first chromosomal abnormality shown to be associated with a specific human malignancy, the chronic myelogenous leukemia (CML), resulting from a reciprocal t(9;22) translocation characterized by the formation of a shortened chromosome, named Philadelphia chromosome (Ph), in which the tyrosine kinase of c-ABL is constitutively activated. This chromosomal translocation generates *BCR-ABL* fusion genes which can be translated in three different oncoproteins, p210, p190, and p230, associated with three different pathologies in humans, CML, acute lymphoblastic leukemia (ALL), and chronic neutrophilic leukemia, respectively. The molecular mechanisms and downstream pathways of *BCR-ABL* are poorly understood mainly as a result of the lack of a good in vivo model that mimics the human disease. Nevertheless, additional in vitro and in vivo models have led to the design of several novel therapeutic approaches. That is the case of Imatinib mesylate (Gleevec, STI571; Novartis Pharma AG), a drug targeting the tyrosine kinase activity of *BCR-ABL* and an effective therapy for chronic phase CML but not advanced stages of CML and patients with Ph^+ ALL. The resistance mutations in the kinase domain of *BCR-ABL* together with the stem cell origin of the chromosomal translocation and the inability of the imatinib to inhibit the *BCR-ABL* activity in these cells have revealed the limitations of Gleevec, providing new lessons for the development of alternative therapies in oncology.

Key Words: *BCR-ABL*; chronic myelogenous leukemia (CML); cancer stem cells; stem cells; mouse models; human cancer; disease progression; tyrosine kinase inhibition; imatinib mesylate.

1. Introduction: The Philadelphia Chromosome

The Philadelphia (Ph) chromosome was the first specific cytogenetic lesion associated with a human malignant disease, namely chronic myeloid leukemia (CML) *(1)*. It took 13 yr before it was appreciated that the Ph chromosome is the result of a t(9;22) reciprocal chromosomal translocation and another decade before this translocation was shown to involve the *ABL* proto-oncogene, normally in chromosome 9, and a previously unknown gene, later termed *BCR* for breakpoint cluster region, on chromosome 22 *(2)*. The regulated activity tyrosine kinase of ABL was then defined as the pathogenic principle of a defined biphasic disease of the hematopoietic system; the initial chronic phase, characterized by an excessive production of myeloid cells that retain a normal differentiation program, and an acute phase termed "blast crisis" which resembles acute leukemia *(3)*. In order to understand the mechanisms of the Ph chromosome and malignancy, it is necessary to know the biology and functions of the translocated genes *ABL* and *BCR*.

The proto-oncogene c-*ABL* belongs to the family of the nonreceptor tyrosine kinases (TKs) and was originally identified for its homology with v-*ABL* (Abelson murine

From: *Apoptosis, Cell Signaling, and Human Diseases: Molecular Mechanisms, Volume 1*
Edited by R. Srivastava © Humana Press Inc., Totowa, NJ

leukemia virus) and ability to induce an acute neoplastic transformation in mice *(4)*. The mammalian c-*ABL* gene is ubiquitously expressed. It encodes two 145 kDa isoforms arising from alternative splicing of two distinct first exons (1a/I and 1b/IV in the human and mouse respectively) *(5)*. The c-ABL proto-oncoprotein is normally distributed both in the nucleus (where it is bound to chromatin) and in the cytoplasm (where it co-localizes with F-actin), but it is principally found in the nucleus *(6)*. The pattern of expression and the endocellular location of the c-ABL protein suggest that this molecule plays an important role in cellular biology and also exerts multiple functions in various cell compartments; however, the exact role of c-ABL still needs further clarification *(7)*. To realize the possible functions of c-ABL at both the cytoplasmic and nuclear level, it is also important to consider the different domains present in the c-ABL protein.

As in nonreceptor TKs, the c-ABL protein possesses three Src homology (SH) regions—1, 2, and 3—at the N-terminal extremity that mediate catalytic (SH1), phosphor-tyrosine, and proline-rich sequence binding proteins (SH2–SH3), respectively *(8)*. The C-terminal extremity of this protein contains a domain of interaction with F-actin that suggests an important role of c-ABL in the mechanisms which regulate the variations of the cellular morphology and the intercellular adhesion *(9)*. Indeed, several diverse functions have been attributed to c-ABL. The normal ABL protein is involved in the regulation of the cell cycle *(10)*, in the cellular response to genotoxic stress *(11)*, and in the transmission of information about cellular environment through integrin signaling; but overall, it appears that this protein serves as a cellular module that integrates signals from various extracellular and intracellular sources and that influences decisions with regard to cell cycle and apoptosis *(12)*.

In the event of translocation, c-*ABL* from chromosome 9 associates with a portion of a gene on chromosome 22 termed *BCR*. Instead of the small portion of the *BCR* gene overlapping the breakpoint region, the entire gene is relatively large and consists of about 25 exons, including two putative alternative first (e1′) and second (e2′) exons *(13)*. At the end, a 160-kDa protein is formed and, as with c-*ABL*, is ubiquitously expressed in the cell. The first N-terminal exon encodes for a region with serine-threonine kinase activity. A coiled-coil domain at the N-terminus of BCR allows dimmer formation in vivo *(14)*. The center of the molecule contains a region with *dbl*-like (from the *DBL* proto-oncogene) and pleckstrin homology (PH) domains that stimulate the exchange of guanidine triphosphate (GTP) for guanidine diphosphate (GDP) on Rho guanidine exchange factors. The C-terminus encodes a GTPase activating protein for p21^{rac} *(15)*, an essential activating component of the neutrofil respiratory burst. In addition, BCR can be phosphorylated on several tyrosine residues, especially tyrosine 177 (which binds Grb-2), involved in the activation of the Ras pathway *(16)*. Although these and other data argue for a role of BCR in signal transduction, their true biological relevance remains to be determined.

1.1. BCR-ABL *Chromosomal Translocation: p210*$^{BCR\text{-}ABL}$ *and Chronic Myeloid Leukemia*

Although the presence of a Ph chromosome translocation always parallels the presence of a *BCR-ABL* rearrangement, there is variability at the molecular level concerning the type of rearrangement between *BCR* and *ABL*. As a consequence, *BCR-ABL* hybrid genes can generate different types of fusion transcripts and proteins showing a

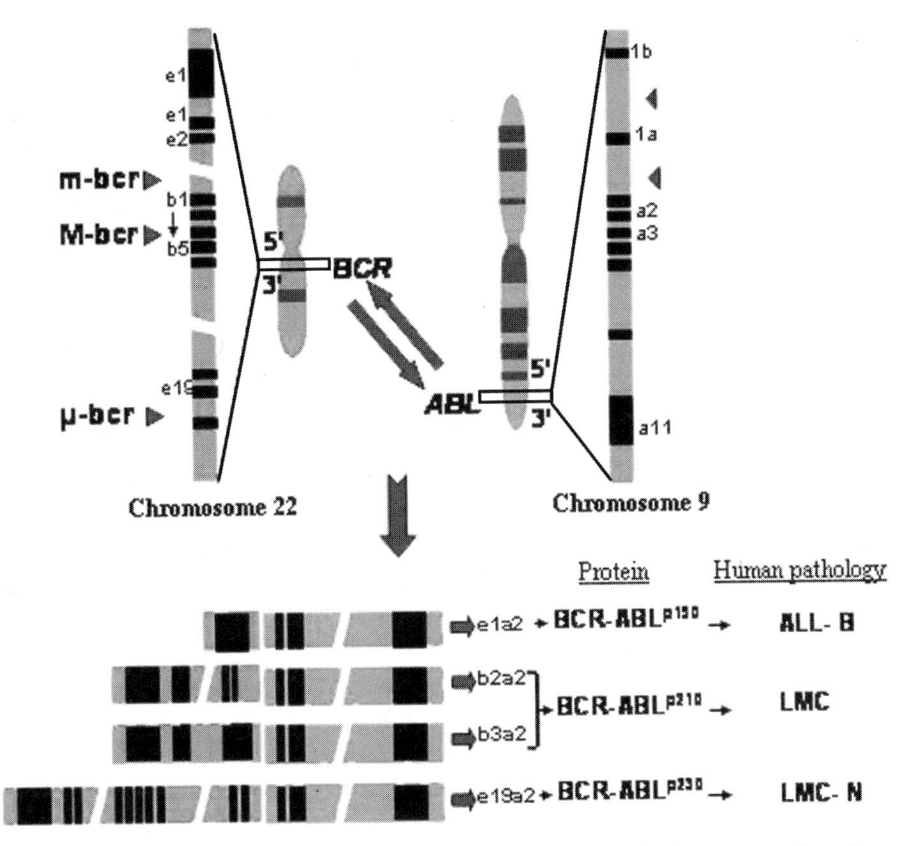

Fig. 1. The chromosomal translocation t(9;22)(q34;q11). The Ph chromosome is a shortened chromosome 22 that results from the translocation of 3′ *ABL* segments on chromosome 9 to 5′ *BCR* segments on chromosome 22. Breakpoints (*arrowheads*) on the *ABL* gene are located 5′ (toward the centromere) of exon a2 in most cases. Various breakpoint locations have been identified along the *BCR* gene on chromosome 22. Depending on which breakpoints are involved, different-sized segments from *BCR* are fused with the 3′-sequences of the *ABL* gene. This results in chimeric protein products (p190, p210, and p230) with variable molecular weights and presumably variable function. The abbreviation m-bcr denotes minor breakpoint cluster region, M-bcr major breakpoint cluster region, and μ-bcr a third breakpoint location in the *BCR* gene that is downstream from the M-bcr region between exons e19 and e20.

preferential but not exclusive association with different leukemia phenotypes *(17,18)*. The breakpoints within the *ABL* gene al 9q34 can occur anywhere over a large (>300 Kb) area at its 5′-end, either upstream of the first alternative exon 1b, downstream of the alternative exon 1a, or, more frequently, between both. Interestingly, exon 1, even if retained in the genomic fusion, is never part of the chimeric mRNA *(17)*. All *BCR-ABL* fusion genes contain a variable 5′ portion derived from *BCR* sequences and a 3′ portion almost invariably of the *ABL* gene sequence (*see* Fig.1).

In CML, the breakpoints on chromosome 22 are restricted to a central region of the *BCR* gene called major breakpoint cluster region (M-*BCR*). This region contains five exons corresponding to *BCR* exons 10 to 14, but originally numbered from 1 to 5. Generally, the break occurs within introns located between exons b2 and b3 or b3 and b4 with the *ABL* exon a2, forming the fusion gene b2a2 or b3a2, respectively. The translocation product of this *BCR-ABL* fusion gene is a novel chimeric 210 kDa protein termed $p210^{BCR\text{-}ABL}$. The presence of this characteristic Ph chromosome $p210^{BCR\text{-}ABL}$ is the hallmark cytogenetic event in about 90% of the cases of CML.

The natural history of CML ends with an acute leukemia phase usually fatal in a few months. CML arises from the clonal expansion of a single pluripotent hematopoietic stem cell. It follows a biphasic course with an initial phase of myeloid hyperplasia in which myeloid and other hematopoietic cell lineages are involved in the process of clonal proliferation and differentiation *(18)*. Nevertheless, the initial expansion results not only from an increased rate of proliferation but also from the occurrence of additional cell divisions during myeloid maturation and the ability of CML clones to reach higher cell densities than normal with enhanced cell survival *(19)*. The BCR-ABL oncoprotein has constitutively expressed tyrosine kinase activity as a result of oligomerization of the coiled coil region of $p210^{BCR\text{-}ABL}$ and deletion of the inhibitory SH3 domain of ABL. This results in phosphorylation of $p210^{BCR\text{-}ABL}$ itself on the Y-177 tyrosine residue and leads to recruitment of GRB2 *(20)*, a small adapter molecule that can activate the RAS pathway, and in the phosphorylation of JAK2 and STAT1/STAT5 *(21,22)*, whose activation contribute to growth factor independence. Furthermore, $p210^{BCR\text{-}ABL}$ activates the PI(3)/Akt pathway *(23,24)*, increases expression of Bcl-2 *(25)*, and phosphorylates STAT5 *(26)* leading to the increased resistance of CML progenitors to apoptosis. Finally, $p210^{BCR\text{-}ABL}$ is localized almost exclusively in the cytosol because of the loss of a nuclear localizing signal and thus has increased binding to actin as compared with c-ABL *(27,28)*. This results in phosphorylation of a number of neighboring cytoskeletal proteins that contribute to the aberrant adhesion receptor function and may explain the premature circulation of progenitors and precursors of the blood *(29)*.

Therefore, the features of chronic phase CML, expansion and premature circulation of the malignant myeloid population, can be explained by activation of mutagenic pathways, antiapoptotic pathways, and abnormal cytoskeletal function. The same characteristics—increased mutagenicity and decreased susceptibility to apoptosis—may also be responsible for disease progression *(30)*.

After variable periods, CML may enter an advanced blast phase characterized by the appearance of poorly differentiated Ph^{+} myeloid or lymphoid blast cells in the peripheral blood. *BCR-ABL* activity promotes the accumulation of molecular and chromosomal alterations directly or indirectly responsible for the reduced apoptosis susceptibility and the enhanced proliferative potential and differentiation arrest of CML blast-crisis cells. This phenotype is normally accompanied by nonrandom clonal chromosomal defects, the duplication of the Ph chromosome being the most frequent defect that increases the levels of both *BCR-ABL* mRNA and the protein associated with disease progression, trisomy 8, trisomy 19, and an isochromosome 17q replacing the normal 17 *(31)*. The abnormalities in other proto-oncogenes or tumor suppressor genes during the blast crisis include point mutations in the coding sequences of *RAS (32)*, *p53 (33)*, gene amplification of *MYC* that occurs as a result of increased transcription or trisomy 8 *(34)*, or rearrangements of *Rb (35)* and *p16 (36)*. The suppressor gene 16^{INK4a}, located on chromosome 9, is an inhibitor of cyclin D that phosphorylates Rb, thus preventing cell cycle arrest in G1 *(37)*. Sequential studies of CML patients demonstrate homozygous deletions acquired in association with the progression to lymphoid blast crisis in approx 50% of cases *(38)*. Moreover, epigenetic alterations, such as the methylation of several promoters (including the proximal promoter of c-*ABL* oncogene), appear to be specifically and consistently associated with the progression of CML *(39,40)*. Silencing of the c-*ABL* promoter by methylation will further decrease nuclear c-*ABL* levels, thereby further decreasing apoptosis and, perhaps, enhancing genomic instability, features that are all related to disease progression.

BCR-ABL utilizes a number of alternative pathways or multiple signaling pathways to transform cells and protect them from cell death. Although the relative importance of these pathways in the natural history of the disease remains to be elucidated, the identification and cloning of molecules that are part of the signaling cascade in leukemic cells creates new possibilities for the development of more effective therapeutic compounds. One of these molecules was recently identified because of its potentially important role in cancer dissemination is *SLUG (41)*. *SLUG* (*SNAI2*), a member of the snail family of zinc-finger transcription factors, was first identified because of its critical role for the normal development of neural crest-derived cells and development of the mesoderm in chick embryos *(42)*. Much of the knowledge regarding the function of this gene in humans and mice was derived from analysis of loss-of-function mutations *(43–46)*; however, little was known about how the aberrant expression of SLUG contributed to malignancy. The generation of a Slug-overexpressing mouse model *(41)* aided in the understanding of *SLUG*'s relevance to human cancer development *(41,46–48)*. Slug-expressing mice developed mesenchymal cancers, mainly leukemias, facilitated cell migration, and promoted survival of these migrating cells *(49)*. In addition, *BCR-ABL* oncogenes induced *SLUG* expression, showing that the leukemogeneic capacity of the $p190^{BCR\text{-}ABL}$ oncogene is dependent on the presence of *SLUG (46)*. All these results were consistent with a model in which tumor cells harboring the BCR-ABL fusion protein constitutively express *SLUG*, promoting both aberrant survival of tumor cells and migration of defective target cells (*see* Fig. 2). *SLUG* expression might, therefore, be defined as a common pathway of cell invasion for other leukemias and sarcomas, representing an important mechanism of tumor invasion as well as an attractive target for therapeutic modulation of invasiveness in the treatment of human cancer *(41)*.

1.2. Other Forms of BCR-ABL *Preferentially Associated With Specific Human Leukemia*

As it was introduced previously, the presence of a Ph chromosome translocation always parallels the presence of a *BCR-ABL* rearrangement whose variability at the molecular level depends only on the type of rearrangement between both implicated genes. All *BCR-ABL* fusions genes contain a 5′ portion derived from *BCR* sequences and a 3′ portion that, with very few exceptions, includes the entire *ABL* gene sequence with the exclusion of the first two alternative exons, 1a and 1b, which never form part of the chimeric mRNA. Instead of the $p210^{BCR\text{-}ABL}$ fusion transcript resulting from the alternative splicing between the M-*BCR* and the *ABL* exon 2a; there are two other breakpoint cluster regions located in *BCR* that are characterized as minor-*BCR* (m-*BCR*) and micro-*BCR* (μ-*BCR*). The m-*BCR* is located upstream of M-*BCR* in the long 54.4 kb intron between the two alternative exons e2′ and e2. The translation product of this fusion gene, which is transcribed into an e1a2 7.5 kb mRNA, is a 190-kDa protein, designated $p190^{BCR\text{-}ABL}$, and is associated with Ph-positive ALL in 70% of

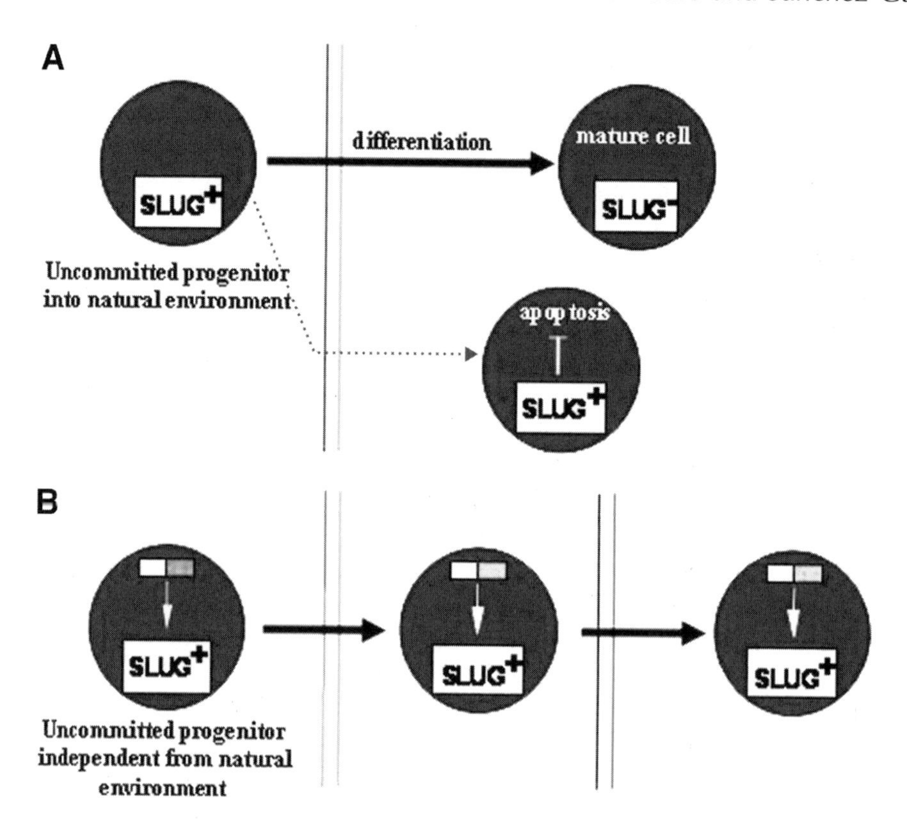

Fig. 2. Model of the role of Slug in BCR-ABL-dependent cancer. **(A)** In the hematopoietic system, normal uncommitted progenitor cells differentiate to mature cells. During this transition, the expression of Slug is downregulated *(179,180)*. These normal uncommitted progenitor cells are responsive to environmental cues, which regulate the number of mature cells produced and limit the self-renewal of immature cells. When in physiological situations, these normal uncommitted progenitor cells migrate, Slug would promote survival to allow them to carry out their function. If this is not achieved in a specific period of time, they would undergo apoptosis as they have been deprived of required external signals *(41)*. **(B)** In cancer development, the differentiation capacity of the target cell is blocked, but inhibition of differentiation is not sufficient for transformation because survival and proliferation of target cells would be restricted to a particular microenvironment. Thus, other genetic changes that allow cells to grow outside their normal environment in addition to mutations that block differentiation must exist. *BCR-ABL* fusion oncogenes created as a result of chromosomal abnormalities associated with leukemias block differentiation and have the capacity to activate target genes such as Slug, which promotes survival (with independence of the required external signals) and migration of the defective target cells into different environments *(41)*.

cases *(50)*. Both in vivo and in vitro studies suggest that P190 is characterized by a higher transforming activity than P210, and this could explain why P190 is preferentially associated with an acute leukemia phenotype. However, the relationship between the BCR-ABL fusion protein and the leukemia phenotype is an intriguing question. Although the P190 fusion protein is almost exclusively associated with an acute leukemia phenotype, mainly lymphoid, there are also rare chronic-phase CML cases which exclusively express P190 instead of P210 *(18)*. These P190 CMLs frequently appear to be associated with the chronic myelomonocytic leukemia like (CMML) phenotype *(51)*. It appears that the presence of p190$^{BCR\text{-}ABL}$ in a committed early myeloid cell could result in a myeloproliferative defect that includes the monocytic lineage, whereas p210$^{BCR\text{-}ABL}$ in the same type of progenitor restricts the excessive proliferation to the granulocytic pathway. Finally, a longer type of *BCR-ABL* transcript in which the breakpoint takes place at the very 3′ end of the *BCR* gene (m-*BCR*) and joins *BCR* exon 19 with *ABL* exon 2 (e19a2) was originally described 15 yr ago *(52)*; this transcript includes the same portion of the *ABL* sequence as the other more common types but contains almost all of the *BCR* coding sequences, resulting in a fusion chimeric protein of 230 kDa (p230$^{BCR\text{-}ABL}$) being present in a subset of patients with Ph+ chronic neutrophilic leukemia (CNL). Thus, p230$^{BCR\text{-}ABL}$ contains additional sequences that are not found in the p190 or p210 variants, specifically the calcium phospholipid binding (CalB) domain and two-thirds of the domain associated with the GTPase activating activity for p21rac (GAP). CNL is a rare disorder marked by a sustained mature neutrophilic expansion typically more indolent as compared with classical CML; progression to blast crisis is uncommon *(53)*.

Based on the observation that the ABL part in the chimeric protein is almost invariably constant whereas the BCR portion varies greatly, one may deduce that ABL is likely to carry the transforming principle and the different sizes of the BCR sequence may dictate the phenotype of the disease.

1.3. TEL (ETV6)-ABL and Rare Human Leukemias

Instead of the classical rearrangement between *BCR* and *ABL* genes, several patients with clinical features of CML were found to have a different fusion of the *TEL* (*ETV6*) gene with *ABL* expressing a chimeric TEL-ABL protein that contains the same portion of the ABL tyrosine kinase fused to TEL rather than to BCR. However, this fusion in human acute or chronic leukemia appears to be a rare event *(54,55)*.

The *TEL* gene (also known as *ETS*-variant gene 6 [*ETV6*]), is located on chromosome 12q13 and was originally identified at the breakpoint of a (5;12) translocation in a patient with CMML, where it was fused to the platelet-derived growth factor β (*PDGF-βR*) receptor gene *(56)*. *TEL* encodes a ubiquitously expressed 452 amino acid protein with two regions of homology to the Ets family of transcription factors: the pointed (PNT) homology domain and the DNA binding domain. Subsequently, *TEL* was recognized to be rearranged in many different chromosomal translocations in human leukemia such as t(12;21) in pre-B ALL in which *TEL* is fused to *AML1* gene *(57)*.

The *TEL-ABL* fusion oncogenes produce two different TEL-ABL proteins; patients with B-lymphoblastic leukemia (B-ALL) and atypical CML; the first 4 exons of *TEL* are fused to *ABL* exon 2; the other patients had *TEL* exons 1 through 5 fused to *ABL* exon 2. The result is two fusion proteins with increased tyrosine kinase activity that are localized to the cytoplasm and F-actin cytoskeleton *(58,59)*, similar to BCR-ABL.

In *TEL*, the PNT domain mediates oligomerization *(58)* and is required for activation of the TEL-PDGF-βR *(60)*, TEL-JAK2 *(61)*, and TEL-ABL *(58)* fusion tyrosine kinases. The fact that BCR contains a coiled-coil oligomerization domain also requires for activation of BCR-ABL kinase activity and transformation *(62)* has led to the suggestion that oligomerization mediated for these partners of ABL may be the critical event in the pathogenesis of these leukemias.

1.4. BCR-ABL *in Healthy Individuals*

With rare and questionable exceptions, all patients with bona fide CML have a *BCR-ABL* rearrangement in their malignant cell clone. In the same way, the finding of a *BCR-ABL* fusion gene in hematopoietic cells is a sign of impending or overt CML (or Ph^+ ALL, and AML). Nevertheless, there is some indirect evidence that the formation of the *BCR-ABL* hybrid gene may be detected at low frequency in the blood of many healthy individuals *(63,64)*. These data have important biological implications because they suggest that in order to successfully produce a leukemic phenotype it is necessary that the fusion gene structure allow for the production of a functional protein with direct or indirect oncogenic properties and that the chromosomal translocation occur in a relatively early precursor cell with self renewal capacity. In other words, only the combination of a correct fusion gene in the correct primitive hematopoietic progenitor has the potential selective advantage and can successfully become an expanding clone *(64–66)*.

It is therefore likely that the *BCR-ABL* genes detected in the circulating leukocytes from healthy individuals *(67)* do not reflect incipient leukemia, because they were generated by relatively mature, harmless progenitors from which the derived clones are eventually lost through normal cell differentiation and death.

2. The In Vitro BCR-ABL Induced Transformation is Influenced by the Cellular Context

Despite the large body of data accumulated in recent years, the molecular pathways by which BCR-ABL proteins induce transformation remain, in part, obscure. To elucidate the molecular pathogenesis of CML and to identify targets for therapeutic intervention for CML, the roles and relative importance of the domains of *BCR-ABL*, of the *BCR-ABL*–activated signaling pathways and of the micro-environment of BCR-ABL–targeted cells in neoplastic transformation by *BCR-ABL* must be examined in biological model systems.

A variety of model systems for *BCR-ABL* transformation with different advantages and disadvantages have been developed and used.

Fibroblast lines have been used extensively in CML research because of their easy manipulation. Fibroblast transformation, anchor-independent growth in soft agar, is the standard in vitro test for tumorigenicity *(68)*. Conversely, it became clear that the introduction of *BCR-ABL* into fibroblasts has diverse effects depending of the type of fibroblast used. $p210^{BCR\text{-}ABL}$, for example, is able to transform Rat-1 fibroblasts *(69)* but not NIH3T3 *(70)*. Indeed, NIH3T3 fibroblasts are not transformed by either of the two *BCR-ABL* oncogenes or by *v-ABL* itself *(69)*. However, it has been shown that $p210^{BCR\text{-}ABL}$ and v-*ABL* convert the factor-dependent Ba/F3 cell line into factor-independent and make it tumorogenic *(71)*, preventing apoptotic death by inducing a *BCL-2* expression pathway *(72)*. These different results were obtained from studies in fibroblasts and must consequently be interpreted carefully.

Hematopoietic cell lines as in vitro CML models, including cell lines with myeloid differentiation such as the well-known K562, allow for fairly good study of the blast-crisis phase but are insufficient as models of chronic-phase CML. Although all these lines are derived from blast crisis, they contain genetic lesions in addition to *BCR-ABL* and their immortalization is not successful because of their limited life span *(73)*, in contrast with their Ph– counterparts.

Transformation of factor-dependent cells lines to growth factor independence is an important feature of BCR-ABL and other oncoproteins that contain an activity tyrosine kinase *(58,74)*. Murine cell lines such as Ba/F3 and 32D and human cell lines such as MO7 have been used to study the effects of *BCR-ABL*. A particular advantage of murine lines is the fact that they are derived from nonmalignant hematopoietic cells; this, unfortunately, does not rule out the development of additional mutations *(75)* that confer a selective growth advantage.

The advance of using cell lines is the relative easy of obtaining a large number of clonally derived cells for biochemical analysis, genetic manipulation, and biological examination. It is particularly important to use the same cellular context for studies that compare cellular and molecular events affected by various *ABL* oncogenes or their mutants where the oncogenes may have different oncogenic potentials in different blood lineages in vivo. Moreover, because the tyrosine kinase activity of BCR-ABL, as well as other oncoproteins with an intrinsic tyrosine kinase activity, is essential for it's oncogenic potential, the factor-dependent hematopoietic cell lines are very useful for searching and testing specific pharmacological inhibitors which should restore the factor-dependence of these cell lines *(76–78)*. However, development of leukemia is a complex process that involves both the effects of *BCR–ABL* within its target cells and the interactions of *BCR–ABL* target cells with the rest of the in vivo environment. Cell lines are limited in representing the physiologically relevant target cells of BCR–ABL in vivo.

To overcome the limitations of established cell lines, primary cells isolated from mice or humans have also been used to study *BCR–ABL* transformation. It has been shown that *BCR–ABL* is able to stimulate cytokine-independent growth of primary bone marrow cells and to stimulate growth of hematopoietic cells differentiated from embryonic stem (ES) cells; adapting an ES cell differentiation system on a macrophage colony stimulating factor (M-CSF)$^{-/-}$ stromal cell line (OP9) preferentially supports differentiation of myeloid and erythroid cells without overproduction of macrophages, favoring hematopoietic development *(79)* and inducing their differentiation into hemangioblasts capable of forming hematopoietic and endothelial cells.

In these studies, the oncogene expression increased the percentage and absolute number of immature progenitor elements and the normal balance of erythroid to myeloid colonies was reversed to a dominance of myeloid over erythroid *(80)*, reproducing one cardinal feature of the clinical disease. In vitro ES-differentiated progenitors can contribute to long-term, multilineage hematopoiesis in vivo *(81)*. and *BCR-ABL*-expressing ES-derived hematopoietic progenitor clones generate an acute leukemia in vivo *(82)*, suggesting that ES-derived hematopoietic progenitors may be a potential cell population that can be used to study *BCR-ABL* effects in vivo.

Taking all these assays together, there is no doubt that the study of cell lines contributed significantly to the understanding of CML. Although these systems can overcome some disadvantages of established cell lines, they are still limited in representing in vivo biology of the disease.

3. Animal Models

The deficiencies of the in vitro assays and the lack of a functional assay using human cells have driven the search for a murine CML model that accurately recapitulates this disease. Because CML is a stem cell disease, it is critical to model the disease by targeting the *BCR-ABL* oncogene into multipotential hematopoietic stem and progenitor cells and express it in these cells and their progenies, similar to the way *BCR-ABL* is generated and expressed in CML patients. Instead of engrafting human CML cells in immunodeficient mice, there are three methods that are generally used for targeting oncogenes into mice transgenic, knock-in, and retroviral transduction. All of these approaches, as well as some others, have been used for investigating BCR-ABL leukemogenesis.

3.1. Syngeneic Mice

Murine factor-dependent cell lines, such as 32D or BaF3, transduced with the translocation *BCR-ABL* oncogene give rise to an aggressive leukemia when transplanted into mice with the same genetic background; these are known as syngeneic mice. Infection with a p210$^{BCR\text{-}ABL}$-expressing retrovirus leads to the outgrowth of clonal cell lines with the phenotypic properties of immature lymphoid cells. Inoculation of these cells into syngeneic mice allows tumor progress to be followed from localized masses at the site of intraperitoneal injection to the spleen and other lymphoid organs *(83)*. After 2 wk, tumors are widely disseminated and simulate a form of acute leukemia more than chronic phase CML. Although it is not a good model for the study of initial chronic phase CML, it constitutes an excellent in vivo model to test the efficacy of new drugs in vivo.

3.2. Engraftment of Immunodeficient Mice With Human BCR-ABL Positive Cells

Cell lines derived from human blast crisis are propagated relatively easily in severe combined immunodeficiency (SCID) mice *(84)*. In addition to the SCID defect in V(D)J recombination, these animals lack functional natural killer (NK) cells. Chronic-phase CML cells and, even more so, cells from accelerated phase or blast crisis readily engraft in these mice, showing a significant correlation between engraftment and disease state.

Early attempts by many groups to establish an in vivo transplantation model of chronic phase CML by using SCID mice as recipients of human cells were not successful *(85,86)*. Even blast crisis CML cells, when injected intraperitoneally into SCID mice, under the renal capsule, or into subcutaneously implanted human fetal bone fragments, were found to disseminate poorly to the bone marrow of the mice *(86,87)*. Whereas the two growth-factor independent cell lines K562 and U937 grew aggressively and induced leukemia in these animals, three other myeloid cell lines, which require interleukin 3 or granulocyte-macrophage colony-stimulating factor (GM-CSF) for continuous growth in vitro, failed to induce disease.

The discovery that nonobese diabetic NOD/SCID mice have additional defects in NK cell activity as well as defective macrophage and complement function *(88)* has allowed superior engraftment of normal *(89,90)* and leukemic *(91)* human hematopoietic cells, proving that they are better recipients for CML cells. Through comparison with the SCID model, engraftment of NOD/SCID mice with multiple types of normal and leukemic human cells was higher and could be achieved with lower numbers of CD34$^+$ cell enriched populations. These experiments provide a foundation for both the future characterization of the phenotype and properties of normal and Ph$^+$ cells that have long-term in vivo repopulating activity as well as for the development of strategies to selectively manipulate these populations in vivo.

3.3. Transduction of Murine Bone Marrow Cells With BCR-ABL *Retroviruses*

In 1990, several groups reported that a CML-like myeloproliferative syndrome could be induced when p210*BCR-ABL*-infected marrow was transplanted into syngeneic recipients *(92–94)*. However, in these first bone marrow transplantation experiments, mice that received bone marrow cells transduced with p210*BCR-ABL* also developed other hematopoietic neoplasms such as B-ALL and macrophage tumors. More importantly, the low efficiency of inducing a myeloproliferative disorder (MPD) combined with poor reproducibility made them useful in studying the biology of *BCR-ABL* in CML. The improvement of BCR-ABL retroviral stocks, with refined culture conditions and more efficient expression of *BCR-ABL* in desired targets cells, has allowed for more accurate reproduction of human CML in mice *(95,96)*. Common features of the disease include increased numbers of peripheral blood cells (PBCs) (with a predominance of granulocytes), splenomegaly, and extramedullary hematopoiesis in liver and pulmonary haemorrhages resulting from extensive granulocyte infiltration in the lung. Additionally, the disease is mostly polyclonal and can be transplanted into secondary recipient mice. The ability of *BCR-ABL* to induce a much more efficient and reproducible CML-like MPD in mice in this model likely resulted from proficient expression of the transgene in the correct cell type, the hematopoietic stem cell and progenitor cell *(97,98)*. Nevertheless, pulmonary hemorrhage, a complication not found in human CML, was a frequent cause of death in these studies, indicating that these novel models may have their own distinct problems even though they have allowed for the identification of targets both within and downstream of BCR-ABL.

3.4. Transgenic Mouse Models

Until quite recently, the design of transgenic mouse models of human CML had been unsuccessful, probably because of the lack of an appropriate promoter for the direct expression of the translocated *BCR-ABL* gene. In human disease, p210*BCR-ABL* is expressed under the control of the *BCR* promoter. However, *BCR* promoter directed expression of p210*BCR-ABL* results in embryonic lethality in mice *(99)*; thus, in order to create a mouse model for p210*BCR-ABL*+ leukemias, it was necessary to employ different promoters. Many groups have made use of alternative promoters such as metallothionein, *TEC* (hematopoietic specific) *(100)*, immunoglobulin heavy chain enhancer (Eμ), a tetracycline-repressible promoter *(101)*, or the part of the long-terminal repeat (LTR) of the myeloproliferative sarcoma virus (MPSV), in order to generate transgenic mice (Table 1). Although these models developed hematological malignancies, all of them were exclusively diagnosed as ALL rather than granulocyte hyperplasia; much less frequently, a myeloid phenotype after a latency of 8 to 44 wk *(100)* was diagnosed.

Additional mouse models of *BCR-ABL* leukemogenesis have been generated based on homologous recombination. This approach has been used to fuse p190*BCR-ABL* coding sequences into the endogenous *BCR* gene *(102)*. After 4 mo, 95% of chimeric mice expressing one p190*BCR-ABL* allele developed pre-B-ALL. This approach of producing transgenic *BCR-ABL* mice was able to replicate acute leukemia but no information is available on the use of this knock-in strategy to target p210*BCR-ABL* at the *BCR* locus.

The most recent development in the hallmark of CML knock-in model mice consists of an inducible genetic model based on the specific expression, in hematopoietic stem cells, of *BCR-ABL* regulated by a mouse stem cell leukemia (SCL) enhancer *(103)*.

Table 1
Genotype-Phenotype Correlations in Men and Mice When Human Cancer-Gene Defect Targets Primitive Stem Cells

Human genotype	Human phenotype	Mouse genotype (knock-in)	Mouse phenotype	Ref.
t(9;11)(p22;q23) (*MLL-AF9*)	AML	Mll (exon 8)-AF9	Chimeric mice develop AML	*182*
t(9;22)(q34;q11) (*BCR-ABLp190*)	B-cell ALL	Bcr (exon 1)-ABL	Chimeric mice develop B-cell ALL Embryonic lethal (Het)	*102*
t(8;21)(q22;q22) (*AML1-ETO*)	AML	AML1 (exon 5)-ETO AML1 (exon 4)-ETO	Embryonic lethal (Het)	*183,184*
inv(16)(q13;q22) (*CBFB-MYH11*)	AML(M4E_0)	Cbfb (exon 5)-MYH11	Embryonic lethal (Het)	*181*
t(7;22)(p22;q12) (Ews-ERG)	Ewing's sarcoma	Ews (exon 7)-ERG*	T-cell leukemias	*190*

*This rearrangement is induced by *Cre*-recombinase expression under the control of a lymphoid specific cassette.

These animals are able to develop a chronic-phase CML-like disease, making it easier to study the leukemogenic mechanism during disease initiation and progression.

3.5. Knock-Out Animal Models With CML Phenotypes

Consistent with the importance of RAS signaling in promoting growth of myeloid cells, JUNB, an antagonist of the RAS downstream target JUN and negative regulator of the cell proliferation and survival, was shown to act as a tumor suppressor in myeloid cells *(104)*. *Junb*-null mice have severe vascular defects in the placenta, leading to early embryonic lethality. However, inactivation of JUNB, specifically in the long-term self-renewing hematopoietic stem cells *(105)*, led to the development of a CML-like disease in mice, supporting the idea that CML originates from hematopoietic stem cells and that inactivation of JUNB could be important for CML development.

Gene-knockout studies, as it has been shown, have revealed key negative regulators of myelopoiesis, (e.g., JUNB). Moreover, mice with disruption of the gene encoding interferon consensus sequence binding protein (ICSBP or interferon regulatory factor 8) develop a chronic phase-CML like disease *(106)*. Expression of this ICSBP protein is significantly decreased in mice with *BCR-ABL*-induced CML disease *(107)*. These data indicate that ICSBP is a tumor suppressor and its downregulation is important for the pathogenesis of CML.

Many questions remain unanswered. One of these that is yet unresolved is the requirement of BCR-ABL expression for the development and maintenance of a *BCR-ABL*-induced leukemia. It is widely accepted that translocation of *BCR-ABL* is the primary oncogenic event shown to be essential for the development of Ph^+ leukemias, but its maintenance function in the diesase remains unsolved. In vitro experiments in *BCR-ABL* tetracycline-regulatable ES cell differentiation system have revealed that continual expression of *BCR-ABL* is required for the expansion of progenitors and myeloid cells *(108)*; nevertheless, in vivo mouse models are not fully determinant in this final conclusion. Although a transgenic BCR-ABL tetracycline-regulatable B-ALL mouse model in which administration of tetracycline induces complete remission of the phenotype does exist *(109)*, not all of the studied animal lines have confirm this pattern; some succumbed to a rapidly progressing B-cell leukemia that was independent of *BCR-ABL* expression after 2 to 4 wk of remission.

Altogether, it is clear that the available mouse models of *BCR-ABL* induced diseases do not closely mimic CML in humans and that a better understanding of the mechanism of disease progression will require generation of other transgenic lines in which *BCR-ABL* expression is under the control of promoters active only in primitive hematopoietic progenitors. The establishment of disease models of *BCR-ABL*-induced leukemia with a long myeloproliferative phase would be necessary for assessing the mechanisms of disease progression.

4. The Cellular Origin of the Human Disease

A complete understanding of the cancer process requires more detailed knowledge of the origins of neoplastic growth. Identifying the initial target cell origin of cancer and the respective contributions of different genetic events to cell-fate determination and disruption of local homeostasis would be crucial for the development of novel nontoxic therapies that influence tumor-cell behavior.

Two main hypotheses have been put forward to explain the origin of cancer. One view is that all tumors come from multipotent stem cells, as these have the long lifespan required for the accumulation of many genetic alterations that are found in most tumors. In this scenario, tumor-cell fate and behavior are determined by specific combinations of genetic or gene-expression changes that have occurred during tumor development *(110)*. An alternative view is that there are a variety of target cells that range from stem cells to committed cells that have begun the process of cell determination and differentiation, the nature of the target cell being the important influence on malignant potential *(111)* (*see* Fig. 3). The parallelism described in past years between the mechanistic similarities of normal stem cells and cancer cells has led to the notion of "cancer stem cells" *(110,112,113)* as the original cells in which cancer begins.

Leukemia provides the best evidence that normal stem cells are the targets of transforming mutations *(66,114)*. The cells capable of initiating human acute myeloid leukemia (AML) in NOD/SCID mice have a $CD34^+CD38^-$ phenotype (a population enriched for hematopoietic stem cells [HSCs]), phenotype similar to normal HSCs *(115,116)*. Conversely, $CD34^+CD38^+$ leukemia cells can not transfer disease to mice in the vast majority of cases (*see* Fig. 4). In CML, leukemia-associated chromosomal rearrangements have been also found in $CD34^+CD38^-$ cells *(117–119)*. All these suggest that HSCs, rather than committed progenitors, are the target for leukemic transformation.

Aside from the studies that point to a stem cell origin, other authors have suggested that the Ph^+ rearrangement might happen in stem cells more primitive than $CD34^+CD38^-$ cells (i.e., it may occur at the level of putative hemangioblasts *[120]*). The identification of the *BCR-ABL* fusion gene in various proportions of endothelial cells generated from CML patients' bone marrow cells in vitro has led to theorization that CML might originate from bone marrow derived hemangioblastic precursor cells. Indeed, recent studies have provided direct evidence that cells capable of initiating human CML have a differentiation potential similar to hemangioblasts *(121)*, which suggests that hemangioblasts, rather than HSCs, are the target for leukemic transformation in CML.

These advances, which have recently extended to the study of breast cancer *(122)* and glioblastoma *(123)*, indicate that many types of cancer cells can be organized into hierarchies, ranging from malignant cancer stem cells, which have extensive proliferative potential, to differentiated cancer cells, which have limited proliferative potential. In conclusion, all results indicate that cancer can be considered a disease of unregulated self-renewal in which genome alterations convert normal stem cell self-renewal pathway into engines for neoplastic proliferation *(124)*.

5. CML Treatment Points to *BCR-ABL* Targeted Therapy

CML is a clonal proliferative malignancy originating from a pluripotent HSC. Until recently, the therapeutic armamentarium for CML was limited to allogeneic stem cell transplantation (SCT), conventional chemotherapy with agents such as busulfan (BUS) and hydroxyurea (HU), and treatment with interferon (IFN)-α based regimens *(125)*. All of these options have major drawbacks with respect to efficacy and tolerability. Currently, allogeneic SCT is the only treatment with known curative potential in CML. However, most patients are not eligible for this therapy because of advanced age or lack of a suitable stem-cell donor *(126)*, there is a significant risk of treatment-related morbidity

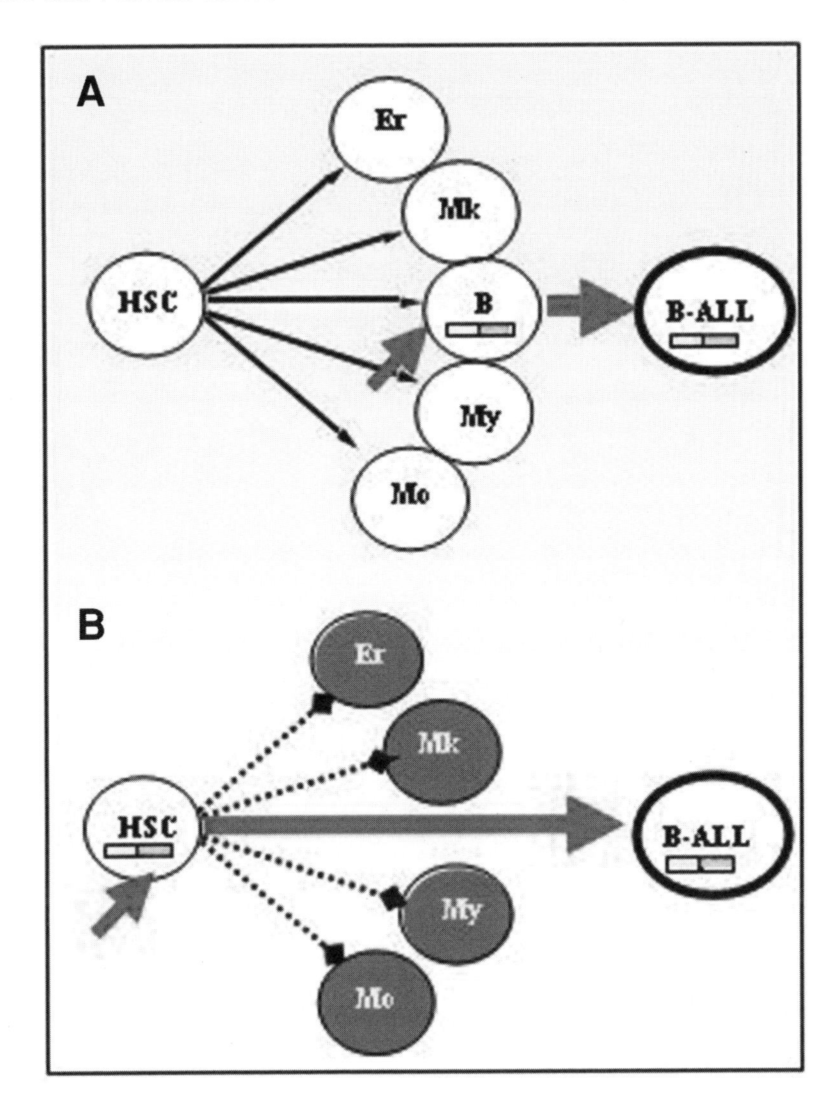

Fig. 3. Proposed model for the role of *BCR-ABL*^{p190}, in particular, and chromosomal abnormalities, in general, in cell lineage specification. **(A)** Adopting a lineage among two or more options is a fundamental developmental decision in multicellular organisms. In the context of the hematopoietic system, a small pool of multipotent stem cells maintains the multiple cell lineages that constitute blood. Ph-ALL has been thought to affect progenitor cells committed to B-cell lymphoid differentiation. **(B)** The demonstration that normal uncommitted progenitor cells are the target for leukemic transformation in Ph-ALL, indicate that genes created by chromosomal abnormalities into the target stem cell are candidate instigators of lineage choice decisions. Consistent with this, forced expression of these genes can, in certain cellular environments, select or impose a lineage outcome *(66,114)*.

and mortality. IFN-α treatment leads to both hematological and cytogenetic response in many chronic phase CML patients *(127)* and its effectiveness at treating a significant portion of CML patients suggests that IFN-α therapy may have some selectivity toward destroying Ph^+ leukemic cells or inhibiting the oncogenic activity of BCR-ABL. On the

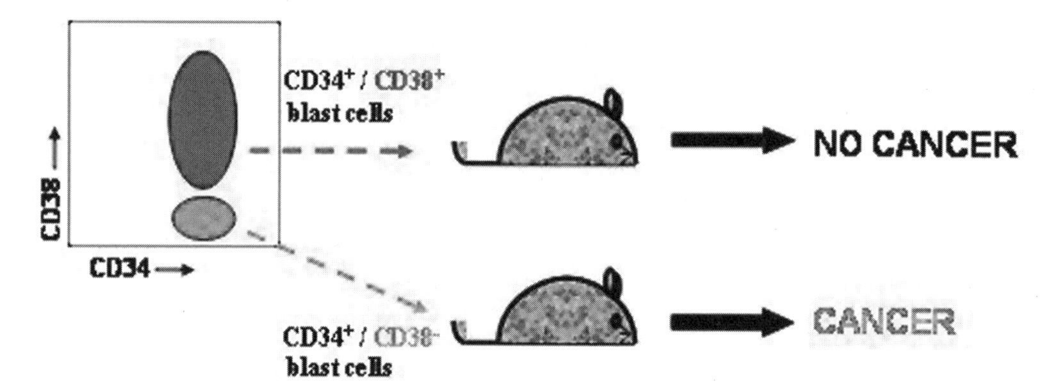

Fig. 4. Nature of the leukemic target cell. Leukemia provides the best evidence that normal stem cells are the targets of transforming mutations. Several assays in AML *(115,116)* and CML *(117–119)* have shown that cells capable of initiating leukemia in NOD/SCID mice have a $CD34^+CD38^-$ phenotype, similar to normal HSC. Nevertheless, $CD34^+CD38^+$ cells are not able to transfer disease to mice, suggesting that HSCs are the target for leukemic transformation.

other hand, up to 20% of CML patients do not tolerate IFN-α therapy because of toxicity problems *(128)*.

Until recently, the use of nonselective cytotoxic drugs that affect normal and malignant cells has been the habitual treatment for many cancers, including Ph^+ leukemias. Because most available chemotherapeutic agents show some degree of *S*-phase specificity, cells that are not actively dividing, as stem and progenitor cells, have shown resistant to such drugs. This raises the possibility that the quiescent leukemic cells identified in CML patients are likely to survive standard chemotherapy regimens. For this and other reasons, novel approaches in the Ph^+ leukemias treatment are being targeted to the BCR-ABL tyrosine kinase. BCR-ABL fusion oncoprotein is present in almost all patients with CML and is expressed at high levels only in the leukemic cells. In addition, because the mitogenic and antiapoptotic effects are dependent on the tyrosine kinase activity of BCR-ABL, inhibition of its activity triggers growth arrest and apoptosis of the leukemic cells.

Efforts to inhibit the function of BCR-ABL have been performed since the early 1990s. Because BCR-ABL is a tyrosine kinase protein, development of molecular targeted therapy has been directed toward inhibiting this tyrosine kinase activity. More than 100 oncogenes have been defined that contribute to carcinogenesis *(129,130)*; tyrosine kinases represent a large fraction of these known dominant oncogenic proteins *(130)*. However, the high degree of commonality among kinase ATP-binding regions *(131)* suggests that it would be difficult to develop compounds that specifically inhibited a single or limited set of tyrosine kinases without having cross-reactivity toward others.

The identification of naturally occurring compounds with specific inhibitory activity toward tyrosine kinases such as Herbimycin A *(132)*, Genistein *(133)*, or Erbstatin *(134)*, and the capacity of these to inhibit *BCR-ABL* cellular transformation *(135,136)*, suggest that inhibition of deregulated ABL tyrosine kinase activity may be a viable therapeutic approach toward ABL-induced leukemias. However, the effect of decreased tyrosine kinase activity was due more to activation of the proteasome 20S than to a specific inhibition of the kinase domain itself. More selective inhibition was obtained by drugs named tyrphostins, from tyrosine phosphorylation inhibitors *(137,138)*, which are more specific for individual tyrosine kinases but with a limited specificity and potency at the cellular level.

Novel treatments strategies for BCR-ABL+ leukemias can be divided, currently, in two groups—treatments centered in downregulation of intracellular BCR-ABL protein levels and those focused in the inhibition of BCR-ABL tyrosine kinase activity. The introduction, as part of this last group, of imatinib mesylate (Gleevec, Glivec formerly STI571; Novartis Pharma AG; Basel, Switzerland), a potent and specific inhibitor of the BCR-ABL tyrosine kinase *(139)*, has caused a rapid change in the management of patients with CML and, although it is presently indicated as first-line therapy for CML in all phases, interest in identifying new BCR-ABL inhibitors is increasing. In 2000, the novel pyrido (2,3-*d*) pyrimidine derivative, PD180970, originally identified as a Scr tyrosine kinase inhibitor *(140)*, was also shown to potently inhibit BCR-ABL tyrosine kinase activity and induce apoptosis of Ph^+ leukemic cells *(141)*. Treatment with PD180970 inhibits the phosphorylation of Gab2 and Crkl, affecting neither the viability nor growth of BCR-ABL negative human leukemic cells *(141)*. These results highlight PD180970 as a promising therapeutic agent against BCR-ABL Ph^+ leukemia. Finally, and consistent with in vitro and pharmacokinetic profile, a novel selective inhibitor of BCR-ABL, AMN107, has been described that is significantly more potent than imatinib and active against a number of imatinib resistant BCR-ABL mutants *(142)*. Crystallographic analysis of ABL-AMN107 complexes has provided a structural explanation for the differential activity of this molecule and imatinib against imatinib-resistance BCR-ABL, standing out as a promising new inhibitor for the therapy of CML and Ph^+ ALL.

6. Clinical Trials With Gleevec

Searching molecules with tyrosine kinase inhibitory activity led to the identification of imatinib mesylate (STI571, Gleevec). Initially, Gleevec was developed as a specific inhibitor of the platelet derived growth factor receptor tyrosine kinase (PDGF-R) *(143,144)* and a suppressed tumor growth of PDGF-R activated cell lines in vivo. Surprisingly, this compound not only inhibited PDGF-R tyrosine kinase activity but quantitatively inhibited not only all ABL tyrosine kinases (including the 210-kD BCR-ABL and the 190-kD BCR-ABL *[144]*), but also the c-kit tyrosine kinase (the receptor for stem cell factor) *(143,145)*, and ARG *(146)*.

The selective inhibition of BCR-ABL by imatinib was shown to be mediated via interaction between the inhibitor and the aminoacids constituting the ATP binding cleft of the tyrosine kinase. Imatinib competitively binds to the ATP binding site *(147)*, inhibiting the tyrosine phosphorylation of its substrates and the growth of the affected leukemic cells. Imatinib acts by downregulating antiapoptotic XIAP, cIAP1, and Bcl-xL, without affecting Bcl-2, Bax, Apaf-1, Fas (CD95), Fas ligand, and BCR-ABL levels *(148)*. Imatinib also inhibits STAT5, Akt kinase, and NF-κB activities in the BCR-ABL positive cells *(149)*.

The high selectivity of imatinib mesylate at suppressing BCR-ABL-induced cell expansion in vitro and in vivo along with imatinib mesylate's pharmacological properties *(150)* have led to the testing of this in CML patients refractory to INF-α therapy.

6.1. Gleevec Treatment in CML Patients

In 1998, a phase I study of imatinib mesylate treatment on chronic phase CML patients refractory or resistant to interferon-based therapy began. When these patients became nonresponsive to IFN-α therapy, no treatment options proven to be effective against CML were available. Druker, Talpaz, Sawyers, and colleagues reported that 53 of 54 patients had complete hematologic responses after 4 wk of imatinib mesylate treatment *(151)*, most of them maintaining normal blood counts and complete hematological responses for over 1 yr *(151)*.

The success of imatinib mesylate treatment on chronic phase CML patients in phase I studies led to large-scale phase II and III studies, reinforcing the positive results from phase I. Even more impressive, in a randomized phase III study of 553 patients not previously treated with IFN-α, only 1.5% of imatinib mesylate–treated patients had disease progression *(152)*.

6.2. Gleevec Treatment in Acute Leukemia and Blast Crisis

The high rate of remission in imatinib mesylate-treated chronic phase CML patients treated led to expansion of the study to determine the effect of imatinib mesylate therapy on blast crisis CML patients and Ph^+ B-ALL patients. However, only 18% of patients had responses and the remaining 20 imatinib mesylate treated patients with lymphoid blast crisis CML or Ph^+ B-ALL had a similar lower level of disease suppression compared with imatinib mesylate treated chronic phase CML patients *(153)*. During the aggressive phase of CML blast crisis, numerous secondary genetic aberrations, including duplication of the Ph chromosome, were common *(154)*. These findings suggested that failure of imatinib mesylate treatment in blast phase CML patients could result from either the presence of added oncogenic events beyond the Ph chromosome or a heightened level of BCR-ABL tyrosine kinase activity ineffectively suppressed by the dose of imatinib mesylate. If duplication of the Ph chromosome was the reason for this low response, then higher doses of imatinib mesylate could be effective to suppress the ABL tyrosine kinase activity. Treatment showed that imatinib mesylate at a higher dose of 800 mg/d is toxic, limiting the therapeutic applicability of imatinib mesylate in treating CML patients expressing higher amounts of BCR-ABL.

6.3. Mechanisms of Resistance to Gleevec

The presence of BCR-ABL mutants resistant to or with reduced sensitivity to Imatinib mesylate inhibition appears to be the most frequent mechanism by which CML patients relapse from Imatinib mesylate therapy *(155–157)*. More than 80% of blast phase CML cases have definable additional genetic aberrations including trisomy 8, i(17q) *(31)*, loss of p53 function *(33)*, MYC amplification *(34)*, RB deletion/rearrangement *(35)*, and p16INK4A rearrangement/ deletion *(36)*. These findings suggested that the high rate of relapse in blast crisis CML patients treated with imatinib mesylate could result from the acquisition of additional oncogenic events that render Ph^+ leukemic cell growth independent of BCR-ABL tyrosine kinase activity. Surprisingly, the large majority of accelerated and blast crisis Ph^+ leukemic cells from imatinib mesylate–resistant CML patients maintained high levels of BCR-ABL tyrosine kinase activity.

To determine how imatinib mesylate achieves its high specificity at inhibiting BCR-ABL transformation, the crystal structure of the kinase domain of ABL complexed with imatinib mesylate was resolved. A critical component in determining the inactive vs active conformation of tyrosine kinases is the orientation and phosphorylation status of residues in the activation loop of the kinase. The initial search for BCR-ABL mutants resistant to imatinib mesylate focused on the identification of mutants in the ABL kinase domain predicted to obstruct direct imatinib mesylate binding. The ABL kinase T334I mutant was described as the most frequent BCR-ABL mutant in CML blast crisis patients resistant to imatinib mesylate therapy *(155)*. Numerous additional BCR-ABL mutants resistant to imatinib mesylate have subsequently been identified in CML patients *(156,157)*. Using a random mutagenesis approach, a range of BCR-ABL mutants outside of the ABL kinase domain, in the cap, SH3, and SH2 domains of ABL were identified as resistant to imatinib mesylate inhibition *(158)*. These domains play pivotal roles in the transition from an inactive to active conformation of ABL. Because imatinib mesylate binds to the inactive conformation of ABL, mutations in these ABL domains could be predicted to abrogate imatinib mesylate binding.

The BCR portion of BCR-ABL has not been sequenced in BCR-ABL clones resistant to imatinib mesylate treatment and different regions of BCR are critical for BCR-ABL activation. So mutations in BCR that up-regulate BCR-ABL tyrosine kinase activity may also be relatively resistant to imatinib mesylate. Some imatinib mesylate resistant patient samples with an activated BCR-ABL tyrosine kinase domain contain no mutations in the ABL kinase domain *(159)*. Sequencing BCR-ABL in its entirety in these samples may identify novel mutations in unexpected regions of BCR-ABL that render the protein insensitive to imatinib mesylate treatment.

BCR-ABL expression has been shown to down-regulate proteins involved in DNA repair. Mutation in the xeroderma pigmentosum group B protein results in a human DNA repair disorder, and BCR-ABL expression inactivates this enzyme by binding to and phosphorylating the enzyme *(160)*. BRCA1 deficiency in mice leads to heightened sensitivity to DNA double-strand breaks *(161)*, and this protein is down-regulated upon BCR-ABL induction in multiple cell types *(162–164)*. Determining which DNA repair mechanism is largely responsible for the mutator phenotype present in BCR-ABL expressing cells may identify therapeutic approaches that could suppress the generation of imatinib mesylate–resistant BCR-ABL mutants in CML patients.

6.4. Ph^+ Cell Populations that are not Elimated by Gleevec

The origin of CML begins in a HSC, a target population that is largely quiescent *(165)*. In vitro and in vivo studies in the last years have shown that these quiescent Ph^+ HSCs, origin of CML, are insensitive to imatinib mesylate treatment *(166,167)*, not been this cells eliminated in CML patients.

Imatinib mesylate inhibits malignant primitive progenitor growth primarily through inhibition of their abnormally increased proliferation rather than selective induction of apoptosis *(168)*. This suggests that BCR/ABL kinase activity may be required for abnormal proliferation and expansion of Ph^+ cells in CML but may not be essential for preservation of primitive malignant cells. BCR-ABL kinase inhibition by imatinib mesylate may therefore remove the proliferative advantage of Ph^+ progenitors and their progeny cells allowing regrowth of coexisting Ph^- cells without eliminate all Ph^+ primitive progenitors.

In conclusion, although imatinib mesylate therapy is able to suppress the oncogenic activity of wild-type *BCR-ABL* expression, the treatment is unable to destroy all Ph$^+$ cell populations. As in Ph$^+$ HSCs, the inability of imatinib mesylate to eliminate Ph$^+$ progenitors suggests that CML patients will require long-term imatinib mesylate monotherapy, which may lead to the generation and selection of *BCR-ABL* escape mutants resistant to the inhibitor.

7. New Therapeutic Options

Although imatinib mesylate produces high rates of complete clinical and cytogenetic responses in the chronic phase, resistance is universal and clinical relapse develops rapidly in the advanced phases of the disease. Because of alternative mechanisms driving the growth and survival of the malignant clones could be responsible for imatinib resistance, novel tyrosine kinase inhibitors that target BCR-ABL as well as agents that downregulate BCR-ABL levels, regardless of wild-type or mutant status, may need to be developed for the future therapy of CML.

In recent years, different therapeutic options have been developed in order to offer other possibilities to imatinib resistance BCR-ABL leukemias.

7.1. Alternative Small Molecules Inhibitors

Small molecules inhibitors belonging to the class of pyrido(2,3-d)pyrimidine compounds (PD 166326, PD 173995 and PD 180970) have been shown to selectively suppress the tyrosine kinase activity of small group of proteins including BCR-ABL *(169–171)*. These compounds required about 10-fold less drug compared with imatinib to inhibit BCR-ABL tyrosine kinase at similar levels suggesting to be useful at treating Ph$^+$ patients resistant to the highest doses of imatinib because of high levels of BCR-ABL expression.

7.2. Inhibiting BCR-ABL *mRNA Production*

Inhibiton of *BCR-ABL* production at the mRNA level via antisense oligonucleotide, small interfering RNA, peptide nucleic acid, or ribozyme expression of BCR-ABL junction sequences effectively suppresses BCR-ABL in vitro transformation *(77,172–178)*. However, these strategies are limited by the lack of proven technologies to introduce sufficient amount of such agents into a patient and thereby suppress the leukemia.

7.3. Suppressing BCR-ABL *Protein Levels*

Heat shock proteins (HSP) help proteins avoid misfolded, inactive, or aggregated states *(179)*. Inactivation of Hsp90 using benzoquinone ansamycins such as geldanamycin or 17-AAG results in a rapid degradation of the client proteins of Hsp90, BCR-ABL for example. Consistently with this, geldanamycin selectively increases apoptosis of BCR-ABL expressing cells *(180,181)* without severely affecting normal protein translation.

Arsenic trioxide, in the same way, is able to decreased BCR-ABL protein levels but, in this case, inhibiting *BCR-ABL* translation *(182)*.

7.4. Obstruction of Signaling Components Critical for BCR-ABL *Oncogenesis*

Imatinib has been shown to elicit MAPK activation in imatinib-resistant, Ph$^+$ leukemic cells. Interruption of this pathway in conjunction with imatinib is associated with a highly synergistic induction of mitochondrial damage and apoptosis *(183)*. Farnesil transferase inhibitors (SCH66336), which are known to block Ras mediated signaling, have been shown to inhibit cell growth of BCR-ABL positive leukemic cells and, in combination with imatinib, inhibit synergistically colony growth of primary CML, causing apoptosis of imatinib resistant Ph^+-expressing cells.

In summary, accumulating experience with treatment of advanced stages of CML with imatinib has indicated that resistance to this agent is common. Novel BCR-ABL targeted agents that inhibit its tyrosine kinase activity or reduce its intracellular levels hold great promise in preventing and treating imatinib-resistance disease.

8. Conclusions

The origin of the translocation *BCR-ABL* begins in a stem cell. The BCR-ABL story has demonstrated that identifying and selectively targeting the primary oncogenic event essential for the development of Ph^+ leukemias is sufficient to suppress the cancer. BCR-ABL expression has been shown to regulate a plethora of signaling molecules, and it is unclear whether inhibition of any single signal transduction pathway would be sufficient to block leukemogenesis. The discovering of imatinib as a potent inhibitor of the tyrosine kinase activity BCR-ABL and its powerlessness in the inhibition of the Ph^+ HSC, has led, out of necessity, to the identification of other important factors in the pathogenesis of CML and to design alternative treatments strategies. Therapeutic approaches aimed at suppressing the leukemic activity of a Ph^+ cell by imatinib mesylate-like mechanisms are unlikely to destroy all cancerous cells in a patient and would most likely require life-long treatment. Eliminating the cellular origin of leukemia is the only curative approach for the disease; in CML, this would require destroying the Ph^+ HSC that is largely quiescent. The well-understood pathogenesis of Ph^+ leukemias is turning out to be fertile ground for the testing of novel therapies. Defining the effectiveness of these treatment approaches against imatinib mesylate resistant patients may help guide the next wave of cancer therapeutics.

Acknowledgments

We thank all the members of lab 13 for their helpful comments and constructive discussions on the manuscript. Research in our group was supported by MEyC (BIO2000-0453-P4-02, SAF2003-01103, and FIT-010000-2004-157), Junta de Castilla y León (CSI06/03), ADE de Castilla y León (04/04/SA/0001), FIS (PI020138, G03/179, and G03/136), and USAL-CIBASA project. MPC is a recipient of a MCyT fellowship.

References

1. Nowell PC, Hungerford DA. A minute chromosome in human chronic granulocytic leukemia. Science 1960;132:1497.
2. Groffen J, Stephenson JR, Heisterkamp N, de Klein A, Bartram CR, Grosveld G. Philadelphia chromosomal breakpoints are clustered within a limited region, bcr, on chromosome 22. Cell 1984;36(1):93–99.
3. Klucher KM, Lopez DV, Daley GQ. Secondary mutation maintains the transformed state in BaF3 cells with inducible BCR/ABL expression. Blood 1998;91(10):3927–3934.

4. Abelson HT, Rabstein LS. Influence of prednisolone on Moloney leukemogenic virus in BALB-c mice. Cancer Res 1970;30(8):2208–2212.
5. Ben-Neriah Y, Bernards A, Paskind M, Daley GQ, Baltimore D. Alternative 5′ exons in c-abl mRNA. Cell 1986;44(4):577–586.
6. Van Etten RA, Jackson P, Baltimore D. The mouse type IV c-abl gene product is a nuclear protein, and activation of transforming ability is associated with cytoplasmic localization. Cell 1989;58(4):669–678.
7. Saglio G, Cilloni D. Abl: the prototype of oncogenic fusion proteins. Cell Mol Life Sci 2004;61(23):2897–2911.
8. Laneuville P. Abl tyrosine protein kinase. Semin Immunol 1995;7(4):255–266.
9. Woodring PJ, Hunter T, Wang JY. Regulation of F-actin-dependent processes by the Abl family of tyrosine kinases. J Cell Sci 2003;116(Pt 13):2613–2626.
10. Sawyers CL, McLaughlin J, Goga A, Havlik M, Witte O. The nuclear tyrosine kinase c-Abl negatively regulates cell growth. Cell 1994;77(1):121–131.
11. Yuan ZM, Shioya H, Ishiko T, et al. p73 is regulated by tyrosine kinase c-Abl in the apoptotic response to DNA damage. Nature 1999;399(6738):814–817.
12. Deininger MW, Goldman JM, Melo JV. The molecular biology of chronic myeloid leukaemia. Blood 2000;96:3343–3356.
13. Chissoe SL, Bodenteich A, Wang YF, et al. Sequence and analysis of the human ABL gene, the BCR gene, and regions involved in the Philadelphia chromosomal translocation. Genomics 1995;27(1):67–82.
14. McWhirter JR, Galasso DL, Wang JY. A coiled-coil oligomerization domain of Bcr is essential for the transforming function of Bcr-Abl oncoproteins. Mol Cell Biol 1993;13(12):7587–7595.
15. Diekmann D, Brill S, Garrett MD, et al. Bcr encodes a GTPase-activating protein for p21rac. Nature 1991;351(6325):400–402.
16. Ma G, Lu D, Wu Y, Liu J, Arlinghaus RB. Bcr phosphorylated on tyrosine 177 binds Grb2. Oncogene 1997;14(19):2367–2372.
17. Melo JV. The diversity of BCR-ABL fusion proteins and their relationship to leukemia phenotype. Blood 1996;88(7):2375–2384.
18. Saglio G, Pane F, Martinelli G, Guerrasio A. BCR/ABL transcripts and leukemia phenotype: an unsolved puzzle. Leuk Lymphoma 1997;26(3–4):281–286.
19. Clarkson B, Strife A. Linkage of proliferative and maturational abnormalities in chronic myelogenous leukemia and relevance to treatment. Leukemia 1993;7(11):1683–1721.
20. Ma G, Lu D, Wu Y, Liu J, Arlinghaus RB. Bcr phosphorylated on tyrosine 177 binds Grb2. Oncogene 1997;14(19):2367–2372.
21. Shuai K, Halpern J, ten Hoeve J, Rao X, Sawyers CL. Constitutive activation of STAT5 by the BCR-ABL oncogene in chronic myelogenous leukemia. Oncogene 1996;13(2):247–254.
22. Ilaria RL, Jr, Van Etten RA. P210 and P190(BCR/ABL) induce the tyrosine phosphorylation and DNA binding activity of multiple specific STAT family members. J Biol Chem 1996;271(49):31,704–31,710.
23. Skorski T, Kanakaraj P, Nieborowska-Skorska M, et al. Phosphatidylinositol-3 kinase activity is regulated by BCR/ABL and is required for the growth of Philadelphia chromosome-positive cells. Blood 1995;86(2):726–736.
24. Franke TF, Kaplan DR, Cantley LC. PI3K: downstream AKTion blocks apoptosis. Cell 1997;88(4):435–437.
25. Sanchez-Garcia I, Grutz G. Tumorigenic activity of the BCR-ABL oncogenes is mediated by BCL2. Proc Natl Acad Sci USA 1995;92(12):5287–5291.
26. de Groot RP, Raaijmakers JA, Lammers JW, Koenderman L. STAT5-Dependent CyclinD1 and Bcl-xL expression in Bcr-Abl-transformed cells. Mol Cell Biol Res Commun 2000;3(5):299–305.

27. Van Etten RA, Jackson PK, Baltimore D, Sanders MC, Matsudaira PT, Janmey PA. The COOH terminus of the c-Abl tyrosine kinase contains distinct F- and G-actin binding domains with bundling activity. J Cell Biol 1994;124(3):325–340. Erratum in J Cell Biol 1994;124(5):865.
28. Salgia R, Li JL, Ewaniuk DS, et al. BCR/ABL induces multiple abnormalities of cytoskeletal function. J Clin Invest 1997;100(1):46–57.
29. Gordon MY, Dowding CR, Riley GP, Goldman JM, Greaves MF. Altered adhesive interactions with marrow stroma of haematopoietic progenitor cells in chronic myeloid leukaemia. Nature 1987;328(6128):342–344.
30. Cortez D, Reuther G, Pendergast AM. The Bcr-Abl tyrosine kinase activates mitogenic signaling pathways and stimulates G1-to-S phase transition in hematopoietic cells. Oncogene 1997;15(19):2333–2342.
31. Chopra R, Pu QQ, Elefanty AG. Biology of BCR-ABL. Blood Rev 1999;13(4):211–229.
32. Cogswell PC, Morgan R, Dunn M, et al. Mutations of the ras protooncogenes in chronic myelogenous leukemia: a high frequency of ras mutations in bcr/abl rearrangement-negative chronic myelogenous leukemia. Blood 1989;74(8):2629–2633.
33. Feinstein E, Cimino G, Gale RP, et al. p53 in chronic myelogenous leukemia in acute phase. Proc Natl Acad Sci USA 1991;88(14):6293–6297.
34. Sawyers CL. The role of myc in transformation by BCR-ABL. Leuk Lymphoma 1993; 11(Suppl 1):45–46.
35. Ahuja H, Bar-Eli M, Arlin Z, et al. The spectrum of molecular alterations in the evolution of chronic myelocytic leukemia. J Clin Invest 1991;87(6):2042–2047.
36. Towatari M, Adachi K, Kato H, Saito H. Absence of the human retinoblastoma gene product in the megakaryoblastic crisis of chronic myelogenous leukemia. Blood 1991;78(9): 2178–2181.
37. Quesnel B, Preudhomme C, Fenaux Pp16ink4a gene and hematological malignancies. Leuk Lymphoma 1996;22(1-2):11–24.
38. Sill H, Goldman JM, Cross NC. Homozygous deletions of the p16 tumor-suppressor gene are associated with lymphoid transformation of chronic myeloid leukemia. Blood 1995;85(8):2013–2016.
39. Zion M, Ben-Yehuda D, Avraham A, et al. Progressive de novo DNA methylation at the bcr-abl locus in the course of chronic myelogenous leukemia. Proc Natl Acad Sci USA 1994;91(22):10,722–10,726.
40. Asimakopoulos FA, Shteper PJ, Krichevsky S, et al. ABL1 methylation is a distinct molecular event associated with clonal evolution of chronic myeloid leukemia. Blood 1999; 94(7):2452–2460.
41. Perez-Mancera PA, Gonzalez-Herrero I, Perez-Caro M, et al. SLUG in cancer development. Oncogene 2005;24(19):3073–3082.
42. Nieto MA, Sargent MG, Wilkinson DG, Cooke J. Control of cell behavior during vertebrate development by Slug, a zinc finger gene. Science 1994;264(5160):835–839.
43. Perez-Losada J, Sanchez-Martin M, Perez-Caro M, Perez-Mancera PA, Sanchez-Garcia I. The radioresistance biological function of the SCF/kit signaling pathway is mediated by the zinc-finger transcription factor Slug. Oncogene 2003;22(27):4205–4211.
44. Jiang R, Lan Y, Norton CR, Sundberg JP, Gridley T. The Slug gene is not essential for mesoderm or neural crest development in mice. Dev Biol 1998;98(2):277–285.
45. Sanchez-Martin M, Rodriguez-Garcia A, Perez-Losada J, Sagrera A, Read AP, Sanchez-Garcia I. SLUG (SNAI2) deletions in patients with Waardenburg disease. Hum Mol Genet 2002;11(25):3231–3236.
46. Sanchez-Martin M, Perez-Losada J, Rodriguez-Garcia A, et al. Deletion of the SLUG (SNAI2) gene results in human piebaldism. Am J Med Genet A 2003;122(2):125–132.

47. Inukai T, Inoue A, Kurosawa H, et al. SLUG, a ces-1-related zinc finger transcription factor gene with antiapoptotic activity, is a downstream target of the E2A-HLF oncoprotein. Mol Cell 1999;4(3):343–352.
48. Khan J, Bittner ML, Saal LH, et al. cDNA microarrays detect activation of a myogenic transcription program by the PAX3-FKHR fusion oncogene. Proc Natl Acad Sci USA 1999;96(23):13,264–13,269.
49. Perez-Losada J, Sanchez-Martin M, Rodriguez-Garcia A, et al. Zinc-finger transcription factor Slug contributes to the function of the stem cell factor c-kit signaling pathway. Blood 2002;100(4):1274–1286.
50. Chan LC, Karhi KK, Rayter SI, et al. A novel abl protein expressed in Philadelphia chromosome positive acute lymphoblastic leukaemia. Nature 1987;325(6105):635–637.
51. Melo JV, Myint H, Galton DA, Goldman JM. P190BCR-ABL chronic myeloid leukaemia: the missing link with chronic myelomonocytic leukaemia? Leukemia 1994;8(1):208–211.
52. Saglio G, Guerrasio A, Rosso C, et al. New type of Bcr/Abl junction in Philadelphia chromosome-positive chronic myelogenous leukemia. Blood 1990;76(9):1819–1824.
53. Pane F, Frigeri F, Sindona M, et al. Neutrophilic-chronic myeloid leukemia: a distinct disease with a specific molecular marker (BCR/ABL with C3/A2 junction). Blood 1996; 88(7):2410–2414. Erratum in Blood 1997;89(11):4244.
54. Janssen JW, Ridge SA, Papadopoulos P, et al. The fusion of TEL and ABL in human acute lymphoblastic leukaemia is a rare event. Br J Haematol 1995;90(1):222–224.
55. Nilsson T, Andreasson P, Hoglund M, et al. ETV6/ABL fusion is rare in Ph-negative chronic myeloid disorders. Leukemia 1998;12(7):1167–1168.
56. Golub TR, Barker GF, Lovett M, Gilliland DG. Fusion of PDGF receptor beta to a novel ets-like gene, tel, in chronic myelomonocytic leukemia with t(5;12) chromosomal translocation. Cell 1994;77(2):307–316.
57. Golub TR, Barker GF, Bohlander SK, et al. Fusion of the TEL gene on 12p13 to the AML1 gene on 21q22 in acute lymphoblastic leukemia. Proc Natl Acad Sci USA 1995;92(11): 4917–4921.
58. Golub TR, Goga A, Barker GF, et al. Oligomerization of the ABL tyrosine kinase by the Ets protein TEL in human leukemia. Mol Cell Biol 1996;16(8):4107–4116.
59. Hannemann JR, McManus DM, Kabarowski JH, Wiedemann LM. Haemopoietic transformation by the TEL/ABL oncogene. Br J Haematol 1998;102(2):475–485.
60. Jousset C, Carron C, Boureux A, et al. A domain of TEL conserved in a subset of ETS proteins defines a specific oligomerization interface essential to the mitogenic properties of the TEL-PDGFR beta oncoprotein. EMBO J 1997;16(1):69–82.
61. Schwaller J, Frantsve J, Aster J, et al. Transformation of hematopoietic cell lines to growth-factor independence and induction of a fatal myelo- and lymphoproliferative disease in mice by retrovirally transduced TEL/JAK2 fusion genes. EMBO J 1998;17(18):5321–5333.
62. McWhirter JR, Galasso DL, Wang JY. A coiled-coil oligomerization domain of Bcr is essential for the transforming function of Bcr-Abl oncoproteins. Mol Cell Biol 1993;13(12):7587–7595.
63. Biernaux C, Sels A, Huez G, Stryckmans P. Very low level of major BCR-ABL expression in blood of some healthy individuals. Bone Marrow Transplant 1996;17(Suppl 3):S45–S47.
64. Sanchez-Garcia I. Consequences of chromosomal abnormalities in tumor development. Annu Rev Genet 1997;31:429–453.
65. Sanchez-Garcia I. Chromosomal Abnormalities, Cancer and Mouse Models: The critical role of translocation-associated genes in human cancer. Curr Genomics 2000;1:71–80.
66. Perez-Losada J, Gutierrez-Cianca N, Sanchez-Garcia I. Philadelphia-positive B-cell acute lymphoblastic leukemia is initiated in an uncommitted progenitor cell. Leuk Lymphoma 2001;42(4):569–576.

67. Bose S, Deininger M, Gora-Tybor J, Goldman JM, Melo JV. The presence of typical and atypical BCR-ABL fusion genes in leukocytes of normal individuals: biologic significance and implications for the assessment of minimal residual disease. Blood 1998;92(9):3362–3367.
68. Tordaro GJ, Green H. An assay for cellular transformation by SV40. Virology 1964; 23(1):117–119.
69. Lugo TG, Witte ON. The BCR-ABL oncogene transforms Rat-1 cells and cooperates with v-myc. Mol Cell Biol 1989;9(3):1263–1270.
70. Daley GQ, McLaughlin J, Witte ON, Baltimore D. The CML-specific P210 bcr/abl protein, unlike v-abl, does not transform NIH/3T3 fibroblasts. Science 1987;237(4814):532–535.
71. Daley GQ, Baltimore D. Transformation of an interleukin 3-dependent hematopoietic cell line by the chronic myelogenous leukemia-specific P210bcr/abl protein. Proc Natl Acad Sci USA 1988;85(23):9312–9316.
72. Sanchez-Garcia I, Grutz G. Tumorigenic activity of the BCR-ABL oncogenes is mediated by BCL2. Proc Natl Acad Sci USA 1995;92(12):5287–5291.
73. Spencer A, Yan XH, Chase A, Goldman JM, Melo JV. BCR-ABL-positive lymphoblastoid cells display limited proliferative capacity under in vitro culture conditions. Br J Haematol 1996;94(4):654–658.
74. Carroll M, Tomasson MH, Barker GF, Golub TR, Gilliland DG. The TEL/platelet-derived growth factor beta receptor (PDGF beta R) fusion in chronic myelomonocytic leukemia is a transforming protein that self-associates and activates PDGF beta R kinase-dependent signaling pathways. Proc Natl Acad Sci USA 1996;93(25):14,845–14,850.
75. Klucher KM, Lopez DV, Daley GQ. Secondary mutation maintains the transformed state in BaF3 cells with inducible BCR/ABL expression. Blood 1998;91(10):3927–3934.
76. Ruibao Ren. The molecular mechanism of chronic myelogenous leukemia and its therapeutic implications: studies in a murine model. Oncogene 2002;21:8629–8642.
77. Cobaleda C, Sanchez-Garcia I. In vivo inhibition by a site-specific catalytic RNA subunit of RNase P designed against the BCR-ABL oncogenic products: a novel approach for cancer treatment. Blood 2000;95(3):731–737.
78. Cobaleda C, Perez-Losada J, Sanchez-Garcia I. Chromosomal abnormalities and tumor development: from genes to therapeutic mechanisms. Bioessays 1998;20(11):922–930.
79. Nakano T, Kodama H, Honjo T. Generation of lymphohematopoietic cells from embryonic stem cells in culture. Science 1994;265(5175):1098–1101.
80. Era T, Witte ON. Regulated expression of P210 Bcr-Abl during embryonic stem cell differentiation stimulates multipotential progenitor expansion and myeloid cell fate. Proc Natl Acad Sci USA 2000;97(4):1737–1742.
81. Kyba M, Perlingeiro RC, Daley GQ. HoxB4 confers definitive lymphoid-myeloid engraftment potential on embryonic stem cell and yolk sac hematopoietic progenitors. Cell 2002;109(1):29–37.
82. Peters DG, Klucher KM, Perlingeiro RC, Dessain SK, Koh EY, Daley GQ. Autocrine and paracrine effects of an ES-cell derived, BCR/ABL-transformed hematopoietic cell line that induces leukemia in mice. Oncogene 2001;20(21):2636–2646.
83. McLaughlin J, Chianese E, Witte ON. In vitro transformation of immature hematopoietic cells by the P210 BCR/ABL oncogene product of the Philadelphia chromosome. Proc Natl Acad Sci USA 1987;84(18):6558–6562.
84. Sawyers CL, Gishizky ML, Quan S, Golde DW, Witte ON. Propagation of human blastic myeloid leukemias in the SCID mouse. Blood 1992;79(8):2089–2098.
85. Sawyers CL, Gishizky ML, Quan S, Golde DW, Witte ON. Propagation of human blastic leukemias in the SCID mouse. Blood 1992;79:2089–2098.
86. Cesano A, Hoxie JA, Lange B, Nowell PC, Bishop J, Santoli D. The severe combined immunodeficient (SCID) mouse as a model for myeloid leukemias. Oncogene 1992;7:827–836.

87. Namikawa R, Ueda R, Kyoizumi S. Growth of human myeloid leukemia in the human marrow environment of SCID-hu mice. Blood 1993;82:2526–2536.
88. Shultz LD, Schweitzer PA, Christianson SW, et al. Multiple defects in innate and adaptive immunologic function in NOD/LtSz-*scid* mice. J Immunol 1995;154:180–191.
89. Larochelle A, Vormoor J, Lapidot T, et al. Engraftment of immune-deficient mice with primitive hematopoietic cells from β-thalassemia and sickle cell anemia patients: Implications for evaluating human gene therapy protocols. Hum Mol Genet 1995;4:163–172.
90. Cashman JD, Lapidot T, Wang JCY, et al. Kinetic evidence of the regeneration of multilineage hematopoiesis from primitive cells in normal human bone marrow transplanted into immunodeficient mice. Blood 1997;89:4307–4316.
91. Bonnet D, Dick JE. Human acute myeloid leukemia is organized as a hierarchy that originates from a primitive hematopoietic cell. Nat Med 1997;3:730–737.
92. Daley GQ, Van Etten RA, Baltimore D. Induction of chronic myelogenous leukemia in mice by the P210bcr/abl gene of the Philadelphia chromosome. Science 1990;247(4944): 824–830.
93. Elefanty AG, Hariharan IK, Cory S. bcr-abl, the hallmark of chronic myeloid leukaemia in man, induces multiple haemopoietic neoplasms in mice. EMBO J 1990;9(4):1069–1078.
94. Kelliher MA, McLaughlin J, Witte ON, Rosenberg N. Induction of a chronic myelogenous leukemia-like syndrome in mice with v-abl and BCR/ABL. Proc Natl Acad Sci USA 1990;87(17):6649–6653.
95. Zhang X, Ren R. Bcr-Abl efficiently induces a myeloproliferative disease and production of excess interleukin-3 and granulocyte-macrophage colony-stimulating factor in mice: a novel model for chronic myelogenous leukemia. Blood 1998;92(10):3829–3840.
96. Pear WS, Miller JP, Xu L, et al. Efficient and rapid induction of a chronic myelogenous leukemia-like myeloproliferative disease in mice receiving P210 bcr/abl-transduced bone marrow. Blood 1998;92(10):3780–3792.
97. Hawley RG. High-titer retroviral vectors for efficient transduction of functional genes into murine hematopoietic stem cells. Ann NY Acad Sci 1994;716:327–330.
98. Cherry SR, Biniszkiewicz D, van Parijs L, Baltimore D, Jaenisch R. Retroviral expression in embryonic stem cells and hematopoietic stem cells. Mol Cell Biol 2000;20(20): 7419–7426.
99. Heisterkamp N, Jenster G, Kioussis D, Pattengale PK, Groffen J. Human bcr-abl gene has a lethal effect on embryogenesis. Transgenic Res 1991;1(1):45–53.
100. Honda H, Hirai H. Model mice for BCR/ABL-positive leukemias. Blood Cell Mol Dis 2001;27(1):265–278.
101. Huettner CS, Zhang P, Van Etten RA, Tenen DG. Reversibility of acute B-cell leukaemia induced by BCR-ABL1. Nat Genet 2000;24(1):57–60.
102. Castellanos A, Pintado B, Weruaga E, et al. A BCR-ABL(p190) fusion gene made by homologous recombination causes B-cell acute lymphoblastic leukemias in chimeric mice with independence of the endogenous bcr product. Blood 1997;90(6):2168–2174.
103. Koschmieder S, Gottgens B, Zhang P, et al. Inducible chronic phase of myeloid leukemia with expansion of hematopoietic stem cells in a transgenic model of BCR-ABL leukemogenesis. Blood 2005;105(1):324–334.
104. Schorpp-Kistner M, Wang ZQ, Angel P, Wagner EF. JunB is essential for mammalian placentation. EMBO J 1999;18(4):934–948.
105. Passegue E, Wagner EF, Weissman IL. JunB deficiency leads to a myeloproliferative disorder arising from hematopoietic stem cells. Cell 2004;119(3):431–443.
106. Holtschke T, Lohler J, Kanno Y, et al. Immunodeficiency and chronic myelogenous leukemia-like syndrome in mice with a targeted mutation of the ICSBP gene. Cell 1996;87(2):307–317.

107. Hao SX, Ren R. Expression of interferon consensus sequence binding protein (ICSBP) is downregulated in Bcr-Abl-induced murine chronic myelogenous leukemia-like disease, and forced coexpression of ICSBP inhibits Bcr-Abl-induced myeloproliferative disorder. Mol Cell Biol 2000;20(4):1149–1161.
108. Hoover RR, Gerlach MJ, Koh EY, Daley GQ. Cooperative and redundant effects of STAT5 and Ras signaling in BCR/ABL transformed hematopoietic cells. Oncogene 2001;20(41): 5826–5835.
109. Huettner CS, Zhang P, Van Etten RA, Tenen DG. Reversibility of acute B-cell leukaemia induced by BCR-ABL1. Nat Genet 2000;24(1):57–60.
110. McCulloch EA. Stem cells in normal and leukemic hemopoiesis (Henry Stratton Lecture, 1982). Blood 1983;62(1):1–13.
111. Cline MJ. The molecular basis of leukemia. N Engl J Med 1994;330(5):328–336.
112. Trott, KR. Tumour stem cells: the biological concept and its application in cancer treatment. Radiother Oncol 1994;30:1–5.
113. Kummermehr, J, Trott, KR. In: Potten CS, ed., *Stem Cells*. 363–399 New York: Academic Press; 1997:363–399.
114. Cobaleda C, Gutierrez-Cianca N, Perez-Losada J, et al. A primitive hematopoietic cell is the target for the leukemic transformation in human philadelphia-positive acute lymphoblastic leukemia. Blood 2000;95(3):1007–1013.
115. Bonnet D, Dick JE. Human acute myeloid leukemia is organized as a hierarchy that originates from a primitive hematopoietic cell. Nat Med 1997;3:730–737.
116. Hope KJ, Jin L, Dick JE. Human acute myeloid leukemia stem cells. Arch Med Res 2003;34(6):507–514.
117. Mauro MJ, Druker BJ. Chronic myelogenous leukemia. Curr Opin Oncol 2001;13:3–7.
118. Jamieson CH, Ailles LE, Dylla SJ, et al. Granulocyte-macrophage progenitors as candidate leukemic stem cells in blast-crisis CML. N Engl J Med 2004;351(7):657–667.
119. Cozzio A, Passegue E, Ayton PM, Karsunky H, Cleary ML, Weissman IL. Similar MLL-associated leukemias arising from self-renewing stem cells and short-lived myeloid progenitors. Genes Dev 2003;17(24):3029–3035.
120. Gunsilius E, Duba HC, Petzer AL, et al. Evidence from a leukaemia model for maintenance of vascular endothelium by bone-marrow-derived endothelial cells. Lancet 2000;355:1688–1691.
121. Fang B, Zheng C, Liao L, et al. Identification of human chronic myelogenous leukemia progenitor cells with hemangioblastic characteristics. Blood 2005;105(7):2733–2740.
122. Al-Hajj M, Wicha MS, Benito-Hernandez A, Morrison SJ, Clarke MF. Prospective identification of tumorogenic breast cancer cells. Proc Natl Acad Sci USA 2003;100:3983–3988.
123. Singh SK. Identification of a cancer stem cell in human brain tumours. Cancer Res 2003;63:5821–5828.
124. Pardal R, Clarke MF, Morrison SJ. Applying the principles of stem-cell biology to cancer. Nat Rev Cancer 2003;3(12):895–902.
125. Silver RT, Woolf SH, Hehlmann R, et al. An evidence-based analysis of the effect of busulfan, hydroxyurea, interferon, and allogeneic bone marrow transplantation in treating the chronic phase of chronic myeloid leukemia: developed for the American Society of Hematology. Blood 1999;94:1517–1536.
126. Goldman JM, Druker BJ. Chronic myeloid leukemia: current treatment options. Blood 2001;98(7):2039–2042.
127. Talpaz M, Kantarjian HM, McCredie KB, et al. Hematologic remission and cytogenetic improvement induced by recombinant human interferon alpha A in chronic myelogenous leukemia. N Engl J Med 1986;314:1065–1069.
128. Salesse S, Verfaillie CM. BCR/ABL: from molecular mechanisms of leukemia induction to treatment of chronic myelogenous leukemia. Oncogene 2002;21(56):8547–8559.

129. Blume-Jensen P, Hunter T. Oncogenic kinase signalling. Nature 2001;411:355–365.
130. Levitzki A, Gazit A. Tyrosine kinase inhibition: an approach to drug development. Science 1995;267:1782–1788.
131. Venter JC, Adams MD, Myers EW, Li PW, Mural RJ. The sequence of the human genome. Science 2001;291:1304–1351.
132. Uehara Y, Hori M, Takeuchi T, Umezawa H. Screening of agents which convert 'transformed morphology' of Rous sarcoma virus-infected rat kidney cells to 'normal morphology': identification of an active agent as herbimycin and its inhibition of intracellular src kinase. Jpn. J. Cancer Res 1985;76:672–675.
133. Akiyama T, Ishida J, Nakagawa S, Ogawara H, Watanabe S. Genistein, a specific inhibitor of tyrosine specific protein kinases. J Biol Chem 1987;262:5592–5595.
134. Umezawa H, Imoto M, Sawa T, Isshiki K, Matsuda N. Studies on a new epidermal growth factor-receptor kinase inhibitor, erbstatin, produced by MH435-hF3. J Antibiot 1986;39: 170–173.
135. Carlo-Stella C, Dotti G, Mangoni L, Regazzi E, Garau D. Selection of myeloid progenitors lackingBCR/ABL mRNA in chronic myelogenous leukemia patients after in vitro treatment with the tyrosine kinase inhibitor genistein. Blood 1996;88:3091–3100.
136. Kawada M, Tawara J, Tsuji T, Honma Y, Hozumi M. Inhibition of Abelson oncogene function by erbstatin analogues. Drugs Exp Clin Res 1993;19:235–241.
137. Yaish P, Gazit A, Gilon C, Levitzki A. Blocking of EGF-dependent cell proliferation by EGF receptor kinase inhibitors. Science 1988;242:933–935.
138. Levitzki A. Tyrosine kinases as targets for cancer therapy. Eur J Cancer 2002;38 (Suppl 5):S11–S18.
139. Buchdunger E, Zimmermann J, Mett H, et al. Inhibition of the Abl protein-tyrosine kinase in vitro and in vivo by a 2-phenylaminopyrimidine derivative. Cancer Res 1996;56(1): 100–104.
140. Kraker AJ, Hartl BG, Amar AM, Barvian MR, Showalter HD, Moore CW. Biochemical and cellular effects of c-Src kinase-selective pyrido[2, 3-d]pyrimidine tyrosine kinase inhibitors. Biochem Pharmacol 2000;60(7):885–898.
141. Dorsey JF, Jove R, Kraker AJ, Wu J. The pyrido[2,3-d]pyrimidine derivative PD180970 inhibits p210Bcr-Abl tyrosine kinase and induces apoptosis of K562 leukemic cells. Cancer Res 2000;60(12):3127–3131.
142. Weisberg E, Manley PW, Breitenstein W, et al. Characterization of AMN107, a selective inhibitor of native and mutant Bcr-Abl. Cancer Cell 2005;7(2):129–141.
143. Savage DG, Antman KH. Imatinib mesylate—a new oral targeted therapy. N Engl J Med 2002;346(9):683–693
144. Druker BJ, Sawyers CL, Capdeville R, Ford JM, Baccarani M, Goldman JM. Chronic myelogenous leukemia. Hematology (Am Soc Hematol Educ Program). 2001;87–112. Review.
145. Mauro MJ, O'Dwyer M, Heinrich MC, Druker BJ. STI571: a paradigm of new agents for cancer therapeutics. J Clin Oncol 2002;20(1):325–334.
146. Okuda K,Weisberg E, Gilliland DG, Griffin JD. ARG tyrosine kinase activity is inhibited by STI571. Blood 2001;97:2440–2448.
147. Schindler T, Bornmann W, Pellicena P, Miller WT, Clarkson B, Kuriyan J. Structural mechanism for STI-571 inhibition of abelson tyrosine kinase. Science 2000;289(5486):1938–1942.
148. Fang G, Kim CN, Perkins CL, Ramadevi N, Winton E, Wittmann S, Bhalla KN. CGP57148B (STI-571) induces differentiation and apoptosis and sensitizes Bcr-Abl-positive human leukemia cells to apoptosis due to antileukemic drugs. Blood 2000;96(6):2246–2253.
149. Deininger MW, Vieira S, Mendiola R, Schultheis B, Goldman JM, Melo JV. BCR-ABL tyrosine kinase activity regulates the expression of multiple genes implicated in the pathogenesis of chronic myeloid leukemia. Cancer Res 2000;60(7):2049–2055.

150. Buchdunger E, O'Reilly T, Wood J. Pharmacology of imatinib (STI571). Eur J Cancer 2002;38(Suppl 5):S28–S36.
151. Druker BJ, Talpaz M, Resta DJ, Peng B, Buchdunger E. Efficacy and safety of a specific inhibitor of the BCR-ABL tyrosine kinase in chronic myeloid leukemia. N Engl J Med 2001;344:1031–1037.
152. Druker BJ, O'Brien SG, Cortes J, Radich J. Chronic myelogenous leukemia. Hematology 2002;111–135. Review.
153. Druker BJ, Sawyers CL, Kantarjian H, Resta DJ, Reese SF. Activity of a specific inhibitor of the BCR-ABL tyrosine kinase in the blast crisis of chronic myeloid leukemia and acute lymphoblastic leukemia with the Philadelphia chromosome. N Engl J Med 2001;344: 1038–1042.
154. Mitelman F. The cytogenetic scenario of chronic myeloid leukemia. Leuk Lymphoma 1993;11:11–15.
155. Gorre ME, Mohammed M, Ellwood K, Hsu N, Paquette R. Clinical resistance to STI-571 cancer therapy caused by BCR-ABL gene mutation or amplification. Science 2001;293: 876–880.
156. Shah NP, Nicoll JM, Nagar B, Gorre ME, Paquette RL. Multiple BCR-ABL kinase domain mutations confer polyclonal resistance to the tyrosine kinase inhibitor imatinib (STI571) in chronic phase and blast crisis chronic myeloid leukemia. Cancer Cell 2002;2:117–125.
157. Branford S, Rudzki Z,Walsh S, Parkinson I, Grigg A. Detection of BCRABL mutations in imatinib-treated CML patients is virtually always accompanied by clinical resistance and mutations in the ATP phosphate-binding loop (P-loop) are associated with a poor prognosis. Blood 2003;102:276–283.
158. Shah NP, Nicoll JM, Nagar B, et al. Multiple BCR-ABL kinase domain mutations confer polyclonal resistance to the tyrosine kinase inhibitor imatinib (STI571) in chronic phase and blast crisis chronic myeloid leukemia. Cancer Cell 2002;2(2):117–125.
159. Schindler T, Bornmann W, Pellicena P, Miller WT, Clarkson B, Kuriyan J. Structural mechanism for STI-571 inhibition of abelson tyrosine kinase. Science 2000;289(5486):1938–1942.
160. Azam M, Latek RR, Daley GQ. Mechanisms of autoinhibition and STI-571/imatinib resistance revealed by mutagenesis of BCR-ABL. Cell 2003;112:831–843.
161. Hochhaus A, Kreil S, Corbin AS, La Rosee P, Muller MC. Molecular and chromosomal mechanisms of resistance to imatinib (STI571) therapy. Leukemia 2002;16:2190–2196.
162. Takeda N, Shibuya M, Maru Y. The BCR-ABL oncoprotein potentially interacts with the xeroderma pigmentosum group B protein. Proc Natl Acad Sci USA 1999;96:203–207.
163. Tomlinson GE, Chen TT, StastnyVA,Virmani AK, Spillman MA. Characterization of a breast cancer cell line derived from a germ-line BRCA1 mutation carrier. Cancer Res 1998;58:3237–3242
164. Deutsch E, Jarrousse S, Buet D, Dugray A, Bonnet ML. Down-regulation of BRCA1 in BCR-ABL-expressing hematopoietic cells. Blood 2003;101:4583–4588.
165. Spangrude GJ, Heimfeld S, Weissman IL. Purification and characterization of mouse hematopoietic cells. Science 1988;241:58–62.
166. Graham SM, Jorgensen HG, Allan E, Pearson C, Alcorn MJ. Primitive, quiescent, Philadelphia-positive stem cells from patients with chronic myeloid leukemia are insensitive to STI571 in vitro. Blood 2002;99:319–325.
167. Chu S, Xu H, Shah NP, et al. Detection of BCR-ABL kinase mutations in CD34+ cells from chronic myelogenous leukemia patients in complete cytogenetic remission on imatinib mesylate treatment. Blood 2005;05(5):2093–2098.
168. Bhatia R, Holtz M, Niu N, et al. Persistence of malignant hematopoietic progenitors in chronic myelogenous leukemia patients in complete cytogenetic remission following imatinib mesylate treatment. Blood 2003;101(12):4701–4707.

169. Wisniewski D, Lambek CL, Liu C, Strife A, Veach DR. Characterization of potent inhibitors of the Bcr-Abl and the c-kit receptor tyrosine kinases. Cancer Res 2002;62: 4244–4255.
170. Huang M, Dorsey JF, Epling-Burnette PK, Nimmanapalli R, Landowski TH. Inhibition of Bcr-Abl kinase activity by PD180970 blocks constitutive activation of Stat5 and growth of CML cells. Oncogene 2002;21:8804–8816.
171. Huron DR, Gorre ME, Kraker AJ, Sawyers CL, Rosen N, Moasser MM. A novel pyridopyrimidine inhibitor of Abl kinase is a picomolar inhibitor of Bcr-abl-driven K562 cells and is effective against STI571-resistant Bcr-abl mutants. Clin Cancer Res 2003;9:1267–1273.
172. Rapozzi V, Burm BE, Cogoi S, van derMarel GA, van Boom JH. Antiproliferative effect in chronic myeloid leukaemia cells by antisense peptide nucleic acids. Nucleic Acids Res 2002;30:3712–3721.
173. Wilda M, Fuchs U, Wossmann W, Borkhardt A. Killing of leukemic cells with a BCR/ABL fusion gene by RNA interference (RNAi). Oncogene 2002;21:5716–5724.
174. Szczylik C, Skorski T, Nicolaides NC, Manzella L, Malaguarnera L. Selective inhibition of leukemia cell proliferation by BCR-ABL antisense oligodeoxynucleotides. Science 1991;253:562–565.
175. Lange W, Cantin EM, Finke J, Dolken G. In vitro in vivo effects of synthetic ribozymes targeted against BCR/ABL mRNA. Leukemia 1993;7:1786–1794.
176. Wright L, Wilson SB, Milliken S, Biggs J, Kearney P. Ribozyme-mediated cleavage of the bcr/abl transcript expressed in chronic myeloid leukemia. Exp Hematol 1993;21:1714–1718.
177. Choo Y, Sanchez-Garcia I, Klug A. In vivo repression by a site-specific DNA-binding protein designed against an oncogenic sequence. Nature 1994;372(6507):642–645.
178. Cobaleda C, Sanchez-Garcia I. RNase P: from biological function to biotechnological applications. Trends Biotechnol 2001;19(10):406–411.
179. Neckers L. Hsp90 inhibitors as novel cancer chemotherapeutic agents. Trends Mol Med 2002;8(4 Suppl):S55–S61.
180. Maloney A, Workman P. HSP90 as a new therapeutic target for cancer therapy: the story unfolds. Expert Opin Biol Ther 2002;2(1):3–24.
181. Nimmanapalli R, O'Bryan E, Bhalla K. Geldanamycin and its analogue 17-allylamino-17-demethoxygeldanamycin lowers Bcr-Abl levels and induces apoptosis and differentiation of Bcr-Abl-positive human leukemic blasts. Cancer Res 2001;61(5):1799–1804.
182. Perkins C, Kim CN, Fang G, Bhalla, KN. Arsenic induces apoptosis of multidrug-resistant human myeloid leukemia cells that express Bcr-Abl or overexpress MDR, MRP, Bcl-2, or Bclx(L). Blood 2000;95:1014–1022.
183. Yu C, Krystal G, Varticovksi L, McKinstry R, Rahmani M, Dent P, Grant S. Pharmacologic mitogen-activated protein/extracellular signal-regulated kinase kinase/mitogen-activated protein kinase inhibitors interact synergistically with STI571 to induce apoptosis in Bcr/Abl-expressing human leukemia cells. Cancer Res 2002;62(1):188–199.
184. Becker MW, Clarke MF. SLUGging away at cell death. Cancer Cell 2002;2(4):249–251.
185. Inoue A, Seidel MG, Wu W, et al. Slug, a highly conserved zinc finger transcriptional repressor, protects hematopoietic progenitor cells from radiation-induced apoptosis in vivo. Cancer Cell 2002;2(4):279–288.
186. Castilla LH, Wijmenga C, Wang Q, et al. Failure of embryonic hematopoiesis and lethal hemorrhages in mouse embryos heterozygous for a knocked-in leukemia gene CBFB-MYH11. Cell 1996;87:687–696.
187. Corral J, Lavenir I, Impey H, et al. An Mll-AF9 fusion gene made by homologous recombination causes acute leukemia in chimeric mice: a method to create fusion oncogenes. Cell.1996;85:853–861.

188. Yergeau DA, Hetherington CJ, Wang Q, et al. Embryonic lethality and impairment of haematopoiesis in mice heterozygous for an AML1-ETO fusion gene. Nature Genet 1997;15:303–306.
189. Okuda T, Cai Z, Yang S, et al. Expression of a knocked-in AML1-ETO leukemia gene inhibits the establishment of normal definitive hematopoiesis and directly generates dysplastic hematopoietic progenitors. Blood 1998;91:3134–3143.
190. Forster A, Pannell R, Drynan LF, et al. The invertor knock-in conditional chromosomal translocation mimic. Nat Methods 2005;2:27–30.

2

Angiogenesis and Cancer

Yohei Maeshima

Summary

Angiogenesis, the creation of neovasculatures from pre-existing ones, is required for various physiological processes. However, pathological angiogenesis is a hallmark of malignant tumors, metastasis and various ischemic as well as inflammatory disorders. Angiogenesis is regulated by the balance between proangiogenic factors and antiangiogenic factors, and concentrated effort in this area of research has led to the discovery of a growing number of angiogenesis-associated factors and the complex interactions among these factors. Understanding of the regulatory mechanisms of these factors in mediating the angiogenic process involved in tumor growth prompted the application of antiangiogenic factors on experimental tumor models with successful outcomes. Based on these experimental results, some antiangiogenic agents have been tested in clinical trials. In this review, the process and regulators of angiogenesis, the involvement of angiogenesis in cancer development and the application of antiangiogenic therapies on established tumors would be discussed. Among various antiangiogenic reagents, special emphasis will be given to antiangiogenic reagents derived from vascular basement membranes, a crucial regulator of angiogenesis, rather than a structural tissue component.

Key Words: Angiogenesis; endothelial cell; blood vessels; tumor; vascular basement membrane; endostatin; tumstatin; integrin; VEGF; angiopoietins.

1. Introduction

Angiogenesis, the formation of new capillaries from preexisting blood vessels, is required for several physiological processes as well as pathological conditions. In recent years, the involvement of angiogenesis in the development of malignant tumors and metastasis has been extensively studied. Therapeutic strategies to suppress angiogenesis has been examined on experimental tumor models with successful outcomes. Based on these experimental data, some antiangiogenic agents have been tested in clinical trials. In this review, the process and regulators of angiogenesis, the involvement of angiogenesis in cancer development, and the application of antiangiogenic therapies on established tumors will be discussed. Among various antiangiogenic reagents, special emphasis will be given to antiangiogenic reagents derived from vascular basement membranes.

2. The Process of Angiogenesis

Angiogenesis, the formation of new blood vessels from preexisting ones, is composed of several steps (*see* Fig. 1): (1) the degradation of vascular basement membrane matrix by protease, (2) migration and proliferation of endothelial cells into interstitium, (3) endothelial tube formation, (4) recruitment and attachment of mesenchymal cells to

From: *Apoptosis, Cell Signaling, and Human Diseases: Molecular Mechanisms, Volume 1*
Edited by R. Srivastava © Humana Press Inc., Totowa, NJ

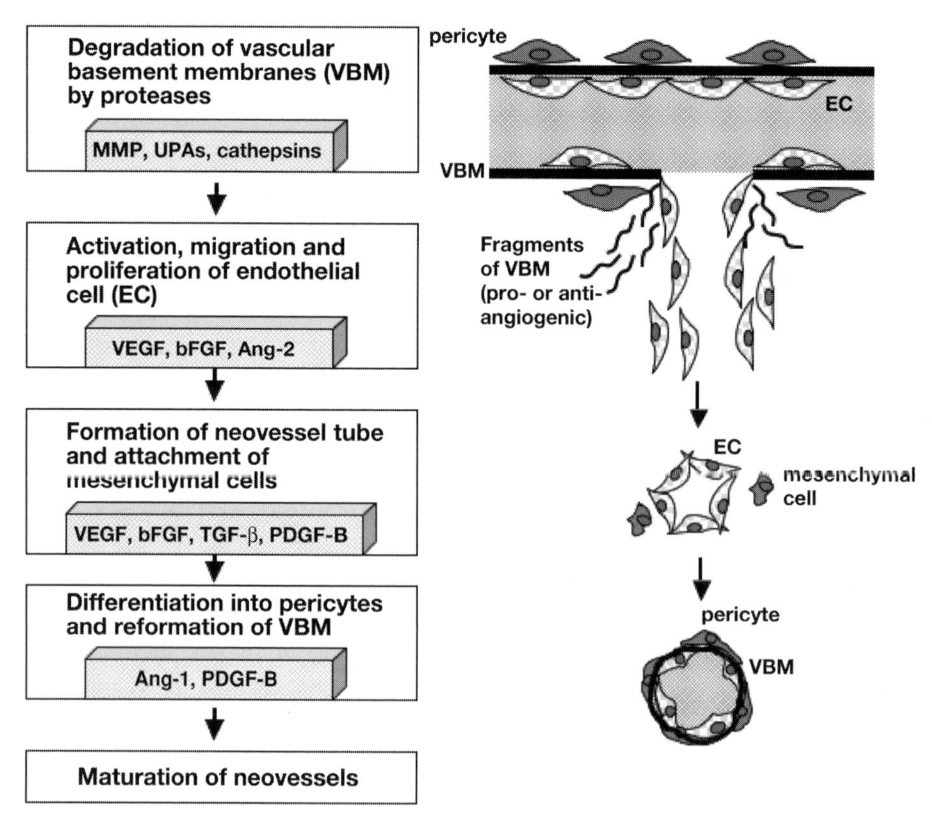

Fig. 1. The vascular microenvironment in the process of angiogenesis. Vascular basement membranes (VBM) are degraded by proteases such as MMPs in activated blood vessels. Proangiogenic growth factors (e.g.,VEGF, bFGF, PDGF) are released from VBM, produced by surrounding tumor cells, fibroblasts, and immune cells, and induce proliferation and migration of endothelial cells (EC). Fragments of VBM have pro- or antiangiogenic activities and regulate angiogenesis. Formation of EC tube with the attachment of pericytes leads to the maturation of neovessels with VBM.

the endothelial cell tube, and (5) maturation of blood vessels with the formation of vascular basement membrane *(1)*. In these angiogenic steps, several angiogenic factors and extracellular matrix (ECM) proteins are involved. Although small blood vessels consist only of endothelial cells, larger vessels are surrounded by mural cells such as pericytes and smooth muscle cells. In the quiescent stage, endothelial cells reside on vascular basement membranes mainly composed of type IV collagen, laminin, nidogen, and heparan sulfate proteoglycan (HSPG) *(2)*. Upon angiogenic stimuli, proteases represented by matrix metalloprotease (MMP) are activated and, in turn, degrade vascular basement membranes, thus leading to the migration of activated endothelial cells into interstitium. Proteinases also expose new cryptic epitopes and produce fragments of ECM proteins such as type IV collagen, potentially leading to the regulation of migratory capacity of endothelial cells as well as smooth muscle cells *(3)*. A provisional matrix of fibronectin, fibrin, and other components provides a supporting scaffold, guiding endothelial cells to their targets and thus facilitating neovessel formation. Proangiogenic growth factors, as well as these provisional matrix proteins, affect the migration, proliferation, and tube formation of endothelial cells via specific cell surface receptors for those growth factors as well as integrin families. Integrin receptors are composed of heterodimers of α and β subunit and display a variety of biological functions upon binding to various matrix proteins and peptides, assisting vascular cells to build new vessels in coordination with their surrounding matrix environments *(4,5)*. Among various integrin families, the αvβ3 and αvβ5 integrins have been extensively investigated for possible involvement in angiogenic process, and pharmacological blockade of these integrins has been reported to suppress pathological angiogenesis *(6,7)*. However, genetic deletion studies of the αvβ3 and αvβ5 integrins suggest that integrins expressed on vascular endothelial cells inhibit angiogenesis by suppressing vascular endothelial growth factor (VEGF)-mediated endothelial cell survival, through transdominant effect on other proangiogenic integrins or by mediating the antiangiogenic activity of thrombospondin-1 and other angiogenesis inhibitors *(4,8,9)*. The possible positive or negative roles of integrins on angiogenesis depending on specific conditions thus require further clarification.

Recent advances in molecular biology have further revealed the involvement of antiangiogenic factors in regulating angiogenesis *(10,11)*. The balance between proangiogenic factors and antiangiogenic factors is considered to be critical in regulating angiogenic process. Thus, excess amount of proangiogenic stimuli leads to the progression of angiogenesis, but excess antiangiogenic milieu results in the resolution of angiogenic process. Various factors are known to induce angiogenesis (Table 1), but among those, proangiogenic capacity of VEGF, basic fibroblast growth factor (bFGF) and angiopoietins are well investigated and established to date. At present, there are many known antiangiogenic factors (Table 2) but these factors were already investigated extensively following the discovery of endostatin, a C-terminal domain of type XVIII collagen, exhibiting a potent antiangiogenic antitumor effect *(12)*. Subsequently various intrinsic antiangiogenic factors have been reported including angiostatin, restin, arresten, canstatin, and tumstatin *(13–18)*.

Endothelial cells differentiate from angioblasts in the embryo and from endothelial progenitor cells (EPC), multipotent adult progenitor cells or side-population cells in the adult bone marrow *(19,20)*. A portion of angiogenic factors are also known to recruit endothelial progenitor cells and facilitate the formation of neovessels *(21)*. Endothelial and hematopoietic progenitors share common markers and are affected by common signaling pathways. In fact, hematopoietic stem cells (HSCs) and leukocytes stimulate angiogenesis via releasing angiogenic factors or by transdifferentiation to endothelial cells *(22,23)*. The involvement of VEGF, placental growth factor (PlGF), angiopoietin-1, and inhibitor of differentiation (Id) in this process had been reported *(24,25)*. Angiogenesis is involved in a variety of physiological and pathological conditions, but initially it plays a critical role in the development of blood vessels and the cardiovascular system. Vasculogenesis, neovessel formation originating from EPCs, is critically involved in the development and is tightly linked to angiogenesis. Recent evidence indicates that EPCs contribute to angiogenic process also in the adult with ischemic, malignant, or inflammatory disorders. Therapeutic efficacy of EPC to stimulate vessel growth in ischemic disorders has been reported *(26,27)*. Angiogenesis and arteriogenesis refer to the sprouting and subsequent stabilization of these sprouts by mural cells. Angiogenesis is involved in physiological settings such as ovarian cycle, endometrium

Table 1
Proangiogenic Factors

VEGF	PDGF-BB
bFGF	Midkine
PlGF	Pleiotrophin
G-CSF	TGF-α, β
HGF	Follistatin
Angiopoietin-1/2	IL-3, 8
PD-ECGF	TNF-α
Angiogenin	Proliferin

Table 2
Antiangiogenic Factors

IFN-α,β,γ	Kringle 5
PF4	PEDF
Prolactin 16 kD frag.	IL-4, 10, 12
Angiostatin	Troponin-1
Endostatin	IP-10
Vasostatin	MIG
Arresten	Thrombospondin-1
Canstatin	Kininostastin
Tumstatin	TIMPs
Vasohibin	PEX
Angioarrestin	PAI
Endorepellin	Meth-1, 2

remodeling, and wound healing. Excessive angiogenesis is known to cause various disorders, and the best known are cancer, rheumatoid arthritis, psoriasis, and visual disturbance represented by diabetic retinopathy and retinopathy of prematurity *(11)*. Recent findings further added common disorders such as atherosclerosis, obesity, bronchial asthma, infectious disease, and endometriosis to the growing list of diseases associated with the involvement of angiogenesis. Several congenital and inherited diseases, such as DiGeorge syndrome, are also caused by abnormal vascular remodeling and angiogenesis. Insufficient angiogenesis and abnormal vessel regression lead to hypertension, preeclampsia, neurodegenaration as well as heart and brain ischemia.

3. Factors Involved in Regulating Angiogenesis

3.1. VEGF

The role of VEGF (also termed as VEGF-A) in regulating angiogenesis has been intensely investigated, and VEGF-signaling is critically involved in physiological as well as pathological angiogenesis represented by tumor growth *(28)*. A gene family of VEGF consists of VEGF-A, VEGF-B, VEGF-C, and PlGF *(28)*. VEGF-A is a key regulator of blood vessel growth, and VEGF-C and VEGF-D are involved in regulating lymphatic angiogenesis *(29)*. Inactivation of a single *Vegf* allele in mice resulted in embryonic lethality between days 11 and 12 as a result of numerous developmental anomalies, defective vascularization in several organs, and reduction of nucleated red blood cells within the blood islands in the yolk sac, suggesting an essential role of VEGF in embryonic vasculogenesis and angiogenesis *(30,31)*. Mice deficient in VEGF-B or PlGF *(32)* did not exhibit any evident developmental abnormalities. VEGF induces the proliferation of endothelial cells in vitro *(28)*, and induces angiogenic response in vivo *(33)*. VEGF also induces survival of endothelial cells partly mediated via activating phosphatidylinositol (PI)-3 kinase-Akt pathway and by the increased production of antiapoptotic proteins Bcl-2 and A1 *(34,35)*. VEGF also promotes monocyte chemotaxis and induces vascular permeability, leading to inflammation and other pathological circumstances *(36,37)*. Consistent with its role in regulating vascular permeability, VEGF also induces fenestration of endothelial cells in some vascular beds *(38)*.

Four different splice variant isoforms of VEGF ($VEGF_{121}$, $VEGF_{165}$, $VEGF_{189}$, $VEGF_{206}$) have been reported *(39,40)*. The properties of native VEGF (45 kDa) closely correspond with those of predominant isoform, $VEGF_{165}$, which is secreted but a significant fraction remains bound to the cell surface and ECM *(41)*. Expression of VEGF mRNA is induced under hypoxic condition in endothelial cells, mediated via hypoxia-inducible factor (HIF)-1 *(42)*. In addition, growth factors such as transforming growth factor (TGF)-α, TGF-β, insulin-like growth factor (IGF)-1, basic fibroblast growth factor (FGF), and platelet-derived growth factor (PDGF) upregulates the expression of VEGF mRNA *(28)*. Inflammatory cytokines such as interleukin (IL)-1α and IL-6 induce the expression of VEGF mRNA in synovial fibroblasts, leading to the development of joint lesions in rheumatoid arthritis.

VEGF binds to tyrosine kinase receptors VEGFR-1 (Flt-1) and VEGFR-2 (KDR/Flk-1) *(28)*, and VEGF-C and VEGF-D bind to VEGFR-3 *(29)*. VEGF also bind to a family of co-receptors, the neuropilins.

3.1.1. VEGFR-1 (Flt-1)

Not only VEGF-A, but also VEGF-B and PlGF, which do not bind to Flk-1, bind to Flt-1. Upon binding to VEGF, Flt-1 undergoes weak tyrosine autophosphorylation *(43)*. A soluble extracellular domain of Flt-1 (sFlt-1) serves as a VEGF inhibitor *(44)*. Recent studies have revealed a synergistic effect of VEGF and PlGF in vivo as evidenced by impaired tumorigenesis in PlGF deficient mice *(32)*. Flt-$1^{-/-}$ mice die *in utero* between days 8.5 and 9.5 resulting primarily from excessive proliferation of angioblasts, suggesting that Flt-1 negatively regulates the activity of VEGF at least during early development *(45)*. The role of Flt-1 in hematopoiesis and recruitment of endothelial progenitors have also been demonstrated *(23)*. PlGF promoted collateral vessel growth in a model of myocardial infarction and treatment with anti-Flt-1 inhibited pathological conditions such as tumor, rheumatoid arthritis, and atherosclerosis suggest the involvement of Flt-1 in mediating angiogenic response *(25)*.

3.1.2. VEGFR-2 (KDR/Flk-1)

VEGFR-2, a high-affinity receptor of VEGF, serves as a major mediator of VEGF in inducing mitogenesis, angiogenesis, and vascular permeability *(46)*. Flk-1-null mice die *in utero* between days 8.5 and 9.5 resulting primarily from a lack of vasculogenesis and failure to develop blood islands and organized blood vessels, suggesting the pivotal role of Flk-1 in developmental angiogenesis and hematopoiesis *(47)*. Upon binding of the ligand, Flk-1 undergoes dimerization, tyrosine phosphorylation, and phosphorylates several proteins in endothelial cells such as phospholipase C-γ, PI-3 kinase, Ras GTPase-activating protein, and the Src family of tyrosine kinases *(48,49)*. VEGF induces endothelial cell growth by activating the Raf-Mek-Erk pathway via protein kinase C, but not via Ras *(50)*. Activation of Flk-1 is required for the effect of VEGF in preventing apoptosis of endothelial cells through activating PI-3 kinase-Akt signaling pathway *(34)*.

3.1.3. Neuropilin

Neuropilin binds the semaphorin family and is implicated in neuronal guidance, and enhances $VEGF_{165}$-mediated chemotaxis *(51)*. It binds to $VEGF_{165}$ in an isoform-specific manner, presents $VEGF_{165}$ to VEGFR-2 enhancing the signaling mediated via VEGFR 2 *(51)*. Neuropilin-1-null mice exhibit embryonic lethality demonstrating the role of neuropilin-1 in the development of vascular system *(52)*.

3.2. Angiopoietins

The angiopoietin family comprises three structurally related proteins, Angiopoietin (Ang)-1, Ang-2, and Ang-3/4 *(53)*. Angiopoietin-1 consistently activates and phosphorylates Tie-2 receptor, but Ang-2 inhibits activation of Tie-2 and can even specifically block Ang-1-dependent phosphorylation *(54)*. Mice deficient in Ang-1 die at embryonic day 12.5, exhibiting some defects in vascular maturation, suggesting the important role of Ang-1/Tie-2 signaling in inducing endothelial-matrix interactions *(55)*. Mice deficient in Ang-2 die 2 wk after birth, exhibiting some defects in retinal vascularization and lymphatic function *(56)*. The expression of Ang-2 has been reported to be up-regulated by hypoxia, hypoxia inducible factor (HIF)-1α, VEGF, angiotensin-II, leptin, and other factors *(57,58)*. In contrast, very little is known about the regulatory mechanism of Ang-1. Ang-1 tightens vessels by affecting junctional molecules *(59)*, and is involved in the attachment of mesenchymal cells to endothelial-tube and differentiation to "pericytes," resulting in mature, "nonleaky" blood vessels *(55)*. Ang-1 binds to Tie-2 and promotes the firm attachment of pericytes *(55)*. The biological roles of angiopoietins are, however, pleiotropic and context-dependent. Ang-1 induces vessel growth in skin, ischemic limbs. and in some tumors *(60)*, possibly because of its effect on endothelial cell survival and mobilization of EPCs and HSCs *(61)*. However, Ang-1 also suppresses angiogenesis in tumors and the heart *(62,63)*. Ang-2, a natural antagonist of Ang-1, loosens the attachment of pericytes resulting in promoting sprouting angiogenesis in the presence of VEGF *(54)*. However, when insufficient angiogenic signals are present, Ang-2 causes endothelial cell death and vessel regression *(54)*.

3.3. Proteases

In the process of ECM remodeling during vessel sprouting, proteases such as urokinase plasminogen activator (uPA) and its inhibitor PAI-1, matrix metalloproteinases (MMPs) and tissue inhibitor of MMP, TIMPs, heparinases, and cathepsins play important roles *(64,65)*. HSPGs present in the basement membranes sequester proangiogenic growth factors such as VEGF and bFGF. Proteinases liberate matrix-bound proangiogenic factors, thus facilitating sprouting angiogenesis. MMP-9 and MMP-2 are known to be required for the mobilization of the sequestered VEGF and thus the initiation of tumor angiogenesis *(66)*. Although MMP-9 does not effectively degrade perlecans or HSPGs, it degrades type IV collagen effectively, possibly disrupting the organization of basement membranes including HSPGs and perlecans, thus leading to the release of proangiogenic VEGF. Stromal and immune cells are known to produce MMPs *(67)*. Following matrix degradation and the release of VEGF, angiogenesis is initiated and tumors begin to grow resulting in the further recruitment of immune cells and fibroblasts. The effect of MMP-9 in releasing matrix-bound VEGF is potentially important at the early stage of tumor progression associated with angiogenic switch, it is not likely, however, to be required after the angiogenesis has been initiated *(68)*. Although insufficient breakdown of ECM prevents vascular endothelial cells from migrating into the interstitium, excessive degradation removes essential support and guidance for migrating endothelial cells, thus leading to sustained angiogenic response *(64)*. Following degradation of basement membranes by MMPs, cryptic domains of partially degraded collagens with proangiogenic activity become exposed *(69)*. On the other hand, proteinases are known to liberate matrix-bound or matrix comprising angiogenesis inhibitors such as endostatin *(12)*, canstatin *(16)*, arresten *(15)*, tumstatin *(17,70)*, restin *(14)*, endorepellin *(71)*, and thrombospondin (TSP)-1, resulting in the resolution of angiogenesis. Expression of MMPs increases as the tumors grow larger, and the levels of MMPs in serum and urine are known to be useful indicators of tumor progression and metastasis. Degradation of basement membranes by MMPs therefore acts as both a positive (at the early stage) and a negative (at the middle to late stage) regulator of tumor angiogenesis *(68,72)*. Interestingly, a domain of MMP-2, PEX, can inhibit the interaction of MMP-2 and $\alpha v\beta 3$ integrin, thus leading to the suppression of tumor growth *(73)*.

3.4. Ephrins

The Eph family of receptor tyrosine kinases and corresponding ligands ephrins had originally been identified to determine embryonic patterning and neuronal targeting *(74)*. EphA receptors bind the ephrinA ligands; EphB receptors bind ephrinB ligands, with the exception of EphA4 binding to both ephrins *(75)*. Mice deficient in ephrinB2 or EphB4 die during embryogenesis with severe cardiovascular defects, suggesting the involvement in the primary capillary network remodeling and patterning in embryonic vasculatures *(76,77)*. Reciprocal expression pattern of ephrinB2 and EphB4 in arterial and venous endothelial cells, suggests that they might interact at the arterial-venous interface *(78)*. EphA2 was observed in tumor-associated vascular endothelial cells, ephrinA1 was detected in tumor as well as endothelial cells and soluble EhpA2-Fc exhibited anti-angiogenic and anti-tumor effect, suggesting the involvement in tumor angiogenesis *(79)*.

3.5. Integrins

Integrins, cell surface receptors for extracellular matrix, exist as heterodimers of α and β subunits, forming at least 25 different combinations of receptors with distinct and overlapping specificity for ECM proteins *(80)*. Integrins transmit bidirectional "outside-in" or "inside-out" signals *(4)*. The $\alpha v\beta 3$ integrin is predominantly expressed on endothelial cells in tumor vasculature *(81)*, and antibodies against $\alpha v\beta 3$ or $\alpha v\beta 5$ integrin or RGD peptide recognized by these receptors efficiently blocked angiogenesis in tumors and retina *(7,82,83)*. Antibodies against $\alpha v\beta 3$ interfered with bFGF-induced angiogenic effect, and antibodies against $\alpha v\beta 5$ interfered with VEGF-induced angiogenic effect *(83)*. Genetically engineered mice lacking $\beta 3$ or $\beta 5$-integrin are fertile and viable, and αv-integrin deficient mice showed extensive angiogenesis *(8,84,85)*. Mice lacking one or both of these integrins ($\beta 3$ or $\beta 5$-integrin) exhibited enhanced tumor growth and angiogenesis *(85)*, suggesting the possibility that these integrins serve as negative regulators of angiogenesis. Previous reports have demonstrated the *trans*-dominant negative regulation of both the $\alpha 5\beta 1$ and $\alpha 2\beta 1$ integrin by $\alpha_{IIb}\beta 3$ integrin *(86)*. Apparent discrepancy on the function of $\alpha v\beta 3$ and $\alpha v\beta 5$ integrins might be mediated via *trans*-dominant inhibition of other proangiogenic integrins ($\alpha 5\beta 1$, $\alpha 1\beta 1$, or $\alpha 2\beta 1$ integrin) in endothelial cells *(4)*. Thus, reagents directed at $\alpha v\beta 3$ or $\alpha v\beta 5$ integrin might indirectly inhibit proangiogenic function of other integrins such as $\alpha 5\beta 1$ integrin, instead of directly inhibiting proangiogenic activity of $\alpha v\beta 3$ or $\alpha v\beta 5$ integrin. Inhibitory action of $\alpha v\beta 3$ integrin on the level of Flk1, resulting in reduced responses to VEGF *(87)*, further support the function of $\alpha v\beta 3$ integrin as a negative regulator of angiogenesis.

3.6. Cadherin

Vascular endothelial cadherin (VE-cadherin), a transmembrane glycoprotein located at adherent junctions of endothelial cells, plays a pivotal role in maintaining vascular integrity. Cytoplasmic tail of VE-cadherin interacts with β- and γ-catenin, which in turn promote the anchorage to actin cytoskeleton. Targeted inactivation or truncation of the VE-cadherin gene resulted in embryonic lethality resulting from altered vascular remodeling and impairment of VEGF-mediated endothelial cell survival and angiogenesis *(88)*, suggesting that VE-cadherin is required for proper angiogenic process. In fact, antibodies against VE-cadherin blocked angiogenesis and tumor growth *(89)*.

4. Mechanisms of Tumor Angiogenesis

Primary tumors and metastases initially exist as small avascular masses, and avascular *in situ* carcinomas may exist for months or years, consequently remaining as a small mass of a few cubic millimeters. Subsequently, some tumor cells switch to the angiogenic phenotype and induce the ingrowth of new capillary blood vessels to allow further tumor growth *(10,90)*. Many primary epithelial tumors, originally separated from underlying blood vessels by basement membranes, develop in this manner, requiring the degradation of basement membranes to access the vasculatures. Expansion of tumor mass occurs not only by increased perfusion of blood through the tumor, but also by paracrine stimulation of growth factors and matrix proteins produced by the newly formed capillary endothelium *(11)*. The switch to the angiogenic phenotype depends on a net balance of proangiogenic factors and antiangiogenic factors. Proangiogenic factors include bFGF, VEGF, and other factors secreted from tumor cells or infiltrating macrophages or mobilized from the ECM. Increased levels of proangiogenic factors accompanied by decreased levels of antiangiogenic factors are required for the 'angiogenic switch'.

In nonangiogenic *in situ* tumors or in dormant micrometastases, proliferation of tumor cells continues, but is balanced by high rates of apoptosis of tumor cells *(91)*. The angiogenic switch in human tumors are driven by: (1) angiogenic oncogenes which up-regulate proangiogenic factors and/or down-regulate antiangiogenic factors; (2) hypoxia leading to the activation of HIF-1 resulting in the activation of VEGF; (3) fibroblasts in the tumor bed producing proangiogenic factors; and (4) endothelial progenitor cells derived from bone marrow which migrate into tumor vasculatures *(92)*. Once the angiogenic switch is turned on, neovessels converge on the dormant *in situ* tumors and tumor cells cluster around each microvessel. The angiogenic switch is associated with a marked decrease in tumor cell apoptosis.

The interaction of tumor cells with host blood vessels has been described in association with VEGF and Ang-2 *(53)*; tumor cells initially home in on and grow by co-opting existing blood vessels, and start off as well-vascularized small tumors *(93)*. The host blood vessels regress in response to the inappropriate co-option as a result of elevated autocrine expression of Ang-2, starve the tumor, and render tumors avascular and hypoxic. The co-opted blood vessels undergo apoptosis and the tumor becomes avascular and hypoxic, leading to marked induction of VEGF from tumor cells. In the presence of elevated levels of VEGF, regression of destabilized co-opted blood vessels will stop, and subsequent sprouting angiogenesis from these blood vessels will allow for the survival and growth of the tumor. The microenvironment with high levels of Ang-2 may explain why tumor vasculatures fail to mature, and tend to be leaky and hemorrhagic *(53)*.

Antiangiogenic therapy can inhibit endothelial cell proliferation in a tumor bed or increase endothelial cell apoptosis. Endothelial cells in the tumor bed have significantly higher proliferation rates than quiescent endothelium and they should be more susceptible to cytotoxic agents. Several chemotherapeutic agents, in fact, inhibit angiogenesis in addition to inducing tumor cell death. For example, paclitaxel, which inhibits microtubule polymerization, inhibits proliferation and migration of endothelial cells, and inhibits tumor growth *(94)*. Continuous administration of chemotherapeutic agents such as cyclophosphamide or vinblastine at a lower dosage than the usual maximum tolerated dose (antiangiogenic schedule) resulted in complete tumor regression and long-term survival in experimental animal models *(95)*.

5. Inhibitors of Angiogenesis (*see* Table 3)

5.1. Endostatin

Type XVIII collagen is a triple helical molecule harboring several heparan sulfate side-chains, thus classified as a major proteoglycans of epithelial and vascular basement membranes *(96,97)*. The C-terminal globular noncollagenous (NC) domain contains the antiangiogenic fragment, endostatin *(12)*. This antiangiogenic domain is separated from an upstream trimerization region by a protease-sensitive hinge *(96)*. Upon cleavage within this hinge region, endostatin will be released into circulation and tissues *(98)*. Type XVIII is known to be involved in the normal development of vasculature in the retina, and a splice mutation in human type XVIII collagen has been associated with Knobloch syndrome, a disease with insufficient development of retinal vasculatures leading to retinal degeneration and blindness *(99)*.

Type XV collagen, sharing homology with type XVIII collagen *(100)*, is a disulfide-bonded proteoglycan containing chondroitin sulfate side-chains. These collagens are

Table 3
Endogenous Basement Membrane Derived Angiogenesis Inhibitors

Angiogenesis inhibitor	Parent protein	Domain	Receptors	Inhibitory activities on EC
Endostatin	α1 chain of type XVIII collagen	NC1	Glypicans, flk-1, α5β1-integrin	Proliferation, migration, tube formation, survival, tumor growth
Restin	α1 chain of type XV collagen	NC10	Unknown	Proliferation, migration, tumor growth
Tumstatin	α3 chain of type IV collagen	NC1	αvβ3-integrin	Proliferation, tube formation, survival, protein translation, tumor growth
Arresten	α1 chain of type IV collagen	NC1	α1β1-integrin	Proliferation, migration, tube formation, survival, tumor growth
Canstatin	α2 chain of type IV collagen	NC1	α3β1-integrin	Proliferation, migration, tube formation, survival, tumor growth
α6(IV)NC1	α6 chain of type IV collagen	NC1	Unknown	Proliferation
Endorepellin	Perlecan	Domain V	α2β1-integrin	Migration, tube formation, survival, blood vessel growth *in vivo*

multiplexin subclass of nonfibrillar collagens containing an N-terminal NC domain, triple-helical regions with multiple interruptions by NC domains, and C-terminal NC1. Type XV collagen is widely distributed in several basement membranes of various tissues, including vascular basement membranes *(101)*. Studies from mice deficient in type XV collagen revealed the important structural role of this collagen in stabilizing skeletal muscle cells and microvessels *(102)*. Similar to type XVIII collagen, the C-terminal NC domain, termed as restin, was shown to possess anti-angiogenic activity *(14,103)*.

Endostatin, a 20 kDa C-terminal NC1 domain of type XVIII collagen, possesses potent antiangiogenic activity *(12)*. Endostatin was first isolated from the conditioned medium of murine hemangioendothelioma cells. Endostatin inhibits endothelial cell proliferation, migration and tube formation in vitro *(104–106)* and possesses potent inhibitory effect on tumor growth in vivo *(12,107)*. The inhibitory effects of endostatin on the expression of VEGF in tumor cells and on vascular permeability had been reported *(108)*. Endostatin also blocks VEGF-mediated proangiogenic signaling via direct interaction with flk-1 *(109)*. Recent reports have demonstrated the therapeutic potential of endostatin in nonneoplastic disorders with involvement of angiogenic process such as rheumatoid arthritis and proliferative diabetic retinopathy *(110,111)*. These reports have also demonstrated the antiangiogenic activity of synthetic peptide derived from N-terminal domain of human endostatin in vitro and in vivo *(112)*.

Cell surface receptors for endostatin had been extensively studied recently. Rehn et al. reported that recombinant human endostatin interacted with α5- and αv-integrins on the surface of human endothelial cells, and demonstrated the functional significance of this interaction *(113)*. More recently, interaction of endostatin with α5β1 integrin in association with cell surface HSPGs and caveolin have been reported *(114,115)*. The interaction of endostatin with α5β1 integrin led to the activation of caveolin-associated Src tyrosine kinase, p190RhoGAP phosphorylation, followed by inactivation of RhoA, and disassembly of actin stress fibers and focal adhesion, the mechanism potentially involved in the inhibitory effect of endostatin on endothelial cell migration *(115)*. Sudhakar et al. reported that human endostatin binding to α5β1 integrin led to the inhibition of the activation of focal adhesion kinase and subsequent inhibition of mitogen-activated protein kinases (MAPKs) including extracellular signal-regulated kinase (ERK)-1/2 and p38MAPK, resulting in the inhibition of endothelial cell migration *(116)*. In addition, heparan sulfate glycosaminoglycans, or glypicans, were reported to bind to endostatin, and this interaction was important for mediating endostatin's antiangiogenic activities in vitro *(117)*. The requirement of E-selectin, an inducible leukocyte adhesion molecule specifically expressed on endothelial cells, for the antiangiogenic activity of endostatin has also been reported *(118)*. Suppressive effect of endostatin on the expression of Id1 and Id3, essential proteins in regulating cell growth and differentiation, and the transcription factor Ets-1 which promotes angiogenesis via inducing several target genes such as MMP-1, MMP-3, MMP-9 and β3 integrin, had been recently reported *(119)*.

5.2. Tumstatin

Type IV collagen, the most abundant constituent of the basement membranes, is expressed as six distinct α-chains, namely, α1–α6 *(120)*, assembles into triple helices, and forms organized networks. Network assembly of type IV collagen is essential for structural integrity and biological functions of basement membranes. In general, type IV collagen possesses the ability to promote cell adhesion, migration, differentiation, and cell growth *(121)*. Although type IV collagen is found only in basement membranes in normal conditions, it is associated with organ fibrosis and accumulates in tumor interstitium. The α-chains consist of N-terminal cysteine-rich 7S domain, the middle collagenous triple helical domain with Gly-X-Y repeats interrupted by short NC sequences, and a C-terminal globular NC1 *(2)*. Superstructures self-associate from triple helical monomers to form either dimers (via interactions of NC1 domains) or tetramers (via interaction of 7S domains), with 56 possible combinations *(122)*. The α1(IV) and α2(IV) chains are ubiquitously distributed in human basement membranes and heterotrimers composed of 2 α1(IV) and 1 α2(IV) chains are predominant *(123)*. In contrast, the localization of the other four isoforms is tissue- and organ-specific *(124,125)*. The distribution of the α3(IV) chain is limited to certain basement membranes, such as glomerular basement membrane, several basement membranes of the cochlea, ocular basement membrane of the anterior lens capsule, Descemet's membrane, ovarian and testicular basement membrane *(126)*, and alveolar capillary basement membrane *(125,127)*. This chain is absent from epidermal basement membranes of the skin and the vascular basement membrane of liver *(125)*.

The NC1 domain of type IV collagen is considered to play a crucial role in the assembly of type IV collagen to form trimers, and thus influence basement membrane organization and modulation of cell behavior *(2,121)*. Petitclerc et al. demonstrated the potent biological activity of recombinant α2(IV), α3(IV), and α6(IV) NC1 domains to inhibit angiogenesis and tumor growth *(18)*.

We identified the pivotal role of recombinant human α3(IV)NC1 in inhibiting the proliferation of capillary endothelial cells and blood vessel tube formation, and also in inducing endothelial cell specific apoptosis *(17)*. We named this domain as "umstatin"(for its unique property of causing "tumor-stasis"), to add another member to the family of endogenous inhibitors of angiogenesis derived from larger proteins, such as angiostatin, endostatin and restin *(12–14)*. Tumstatin further inhibited in vivo neovascularization in matrigel plug assays and suppressed tumor growth of human renal cell carcinoma (786-O) and prostate carcinoma (PC-3) in xenograft mice models *(17)*. Endothelial cell apoptosis of tumor vasculatures was induced in tumstatin-treated mice with PC-3 tumors.

Goodpasture syndrome is an autoimmune disease characterized by pulmonary hemorrhage and/or rapidly progressive glomerulonephritis *(127)*. These symptoms are caused by the disruption of glomerular and pulmonary alveolar basement membrane through immune injury associated with autoantibody against α3(IV) NC1 *(127)*. The most probable disease-related pathogenic autoepitope was identified in the N-terminal portion *(128,129)* and was further confined to be within the N-terminal 40 amino acids *(130,131)*. Truncated tumstatin (tum-1) lacking N-terminal 53 amino acids encompassing the pathogenic Goodpasture autoepitopes, was fully active in inhibiting tumor growth and angiogenesis, and another deletion mutant tum-2, consisting of 124 amino acids in the N-terminal half portion of tumstatin possessed equivalent antiangiogenic activity specific to endothelial cells *(17)*. In contrast, deletion mutant tum-3 consisting of 120 amino acids in the C-terminal half portion of tumstatin, failed to inhibit

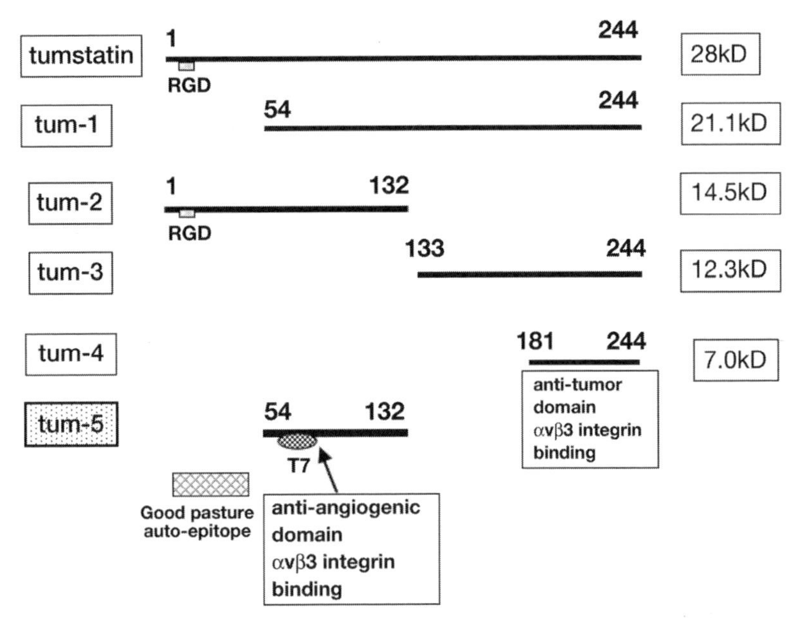

Fig. 2. Anti-angiogenic domain of tumstatin. Tumstatin, a 244-kDa domain derived from C-terminal NC1 domain of α3 chain of type IV collagen, possesses antiangiogenic tum-5 domain that binds to αvβ3 integrin in an RGD-independent manner. This domain does not include Goodpasture autoepitope and the core antiangiogenic domain is located at T7 peptide domain. C-terminal tum-4 domain possesses antitumor activity with binding to αvβ3 integrin in an RGD-independent manner.

angiogenic responses in endothelial cells. These results collectively suggested the localization of antiangiogenic domain within the amino acids 54–124 of tumstatin, which differs from the Goodpasture auto-epitope (*see* Fig. 2).

Previously, the α3(IV)NC1 domain had been shown to bind and inhibit the growth of melanoma cell lines in vitro via amino acids 185-203 of α3(IV)NC1 domain *(132)*. Deletion mutant tum-4 consists of 64 amino acids in the C-terminus of tumstatin, which includes the 185–203 peptide region. Although tum-4 did not inhibit the proliferation of endothelial cells, it inhibited the proliferation of melanoma cells (WM-164). The antitumor activities of tumstatin were thus considered to localize in distinct region of tumstatin, different from antiangiogenic domain *(17)*.

The NC1 domain of α1 and α2 chain of type IV collagen (arresten and canstatin, respectively) also possesses antiangiogenic activity. Arresten, the 26-kDa NC1 domain of the α1 chain of type IV collagen, inhibits endothelial cell proliferation, migration, tube formation, and in vivo neovascularization using matrigel plug assay. Arresten also inhibits the growth of two human xenograft tumor models in nude mice and the development of tumor metastases. The antiangiogenic activity of arresten was mediated via interaction with cell surface proteoglycans and the α1β1 integrin on endothelial cells *(15)*. Canstatin, an endogenous 24-kDa fragment of human α2 chain of type IV collagen inhibits endothelial cell proliferation, migration and murine endothelial cell tube formation. In addition, canstatin potently induced apoptosis of endothelial cells, without exhibiting any effects on nonendothelial cells. Treatment with canstatin did not inhibit the activation of ERK1/2. The proapoptotic effect of canstatin on endothelial cells was associated with a down-regulation of the antiapoptotic protein, FLIP. Canstatin also suppressed in vivo growth of tumors in two human xenograft mouse models with decreased CD31-positive tumor vasculatures *(16)*. The cell surface receptors of canstatin are considered as αvβ3 and α3β1 integrin. Canstatin inhibited the activation of PI3-kinase/Akt resulting in the induction of apoptosis dependent upon signaling events transduced through Fas membrane death receptors *(133)*.

5.2.1. Tumstatin Binds to αvβ3 Integrin

Within the N-terminal portion of tumstatin, there is a RGD sequence (amino acids 7–9). In general, the RGD sequence is considered as an important binding site for αvβ3 integrin. Previous studies have identified the 185–203 amino acid sequence of α3(IV)NC1 as a ligand for αvβ3 integrin and responsible for the associated antitumor cell property *(134)*. We found distinct additional RGD independent αvβ3 integrin binding site within 54–132 amino acids of tumstatin *(135)*. This site does not include the RGD sequence, and, as described above, this site is not essential for inhibition of tumor cell proliferation but necessary for the antiangiogenic activity. A fragment of tumstatin containing 54–132 amino acids could bind both endothelial cells and melanoma cells, but could only inhibit proliferation of endothelial cells, with no effect on tumor cell proliferation. Another fragment of tumstatin containing the 185–203 amino acid (tum-4) could bind both endothelial cells and melanoma cells, but could only inhibit the proliferation of melanoma cells. These results may suggest the involvement of additional cell-specific ligand receptor interactions for exerting cell-specific antiproliferative effects. The presence of cyclic RGD peptides did not affect the αvβ3 integrin-mediated antiangiogenic activity of tumstatin, although significant inhibition of endothelial cell binding to vitronectin was observed. Thus, the two distinct RGD-independent αvβ3 binding sites on tumstatin mediated two separate antitumor activities, suggesting unique αvβ3 integrin mediated mechanisms governing the two distinct anti-tumor properties of tumstatin *(135)*.

5.2.2. Tum-5 Domain

In order to directly demonstrate the antiangiogenic activity of the putative 54–132 amino acids of tumstatin (tum-5 domain), the recombinant tum-5 was produced in *Escherichia coli* and *Pichia pastoris*. Recombinant tum-5 proteins were not recognized by antisera obtained from patients with Goodpasture syndrome, excluding the possibility that these recombinant proteins might induce an autoimmune disorder. Tum-5 specifically inhibited proliferation and caused apoptosis of endothelial cells with no significant effect on nonendothelial cells such as melanoma cell line (WM-164) or prostate carcinoma cell line (PC-3) *(136)*. Tum-5 also inhibited tube formation of endothelial cells on matrigel and induced G1 cell cycle arrest of endothelial cells. In addition, antiangiogenic effect of tum-5 was demonstrated in vivo using both a Matrigel plug assay in C57BL/6 mice, and human prostate cancer (PC-3) xenografts in nude mice. Tum-5 at 1 mg/kg significantly inhibited growth of PC-3 tumors in association with a decrease in CD31 positive tumor vasculatures. Cell attachment assays revealed that tum-5 could bind to αvβ3 integrin in an RGD-independent manner, and was thus independent of vitronectin binding. Competition proliferation assays utilizing soluble αvβ3 integrin protein to compete αvβ3 integrin receptors on the cell surface revealed that the antiproliferative effect of tum-5 was reversed dose-dependently with an increasing amount of αvβ3 soluble protein, suggesting the requirement of tum-5-αvβ3 integrin interaction in antiangiogenic effect of tum-5 *(136)*. These results suggested the potential role of αvβ3 integrin in negative regulation of angiogenesis by binding to tum-5.

5.2.3. Tumstatin Peptide

Reduction and alkylation of tumstatin and tum-5 failed to alter the antiangiogenic activity in vitro and in vivo, suggesting this activity was independent of disulfide bond requirement *(137)*. Therefore, five overlapping synthetic peptides were designed so that they would cover the tum-5 domain. Among these peptides, only the T3 (69–88 amino acids) and T7 peptide (74–98 amino acids) inhibited proliferation and induced apoptosis specifically in endothelial cells. T3 peptide, similar to tumstatin and the tum-5 domain, could bind and exert antiangiogenic effects via the αvβ3 integrin in an RGD-independent manner. Restoration of a disulfide bond between two cystines within the T3 peptide did not affect on the antiangiogenic activity of T3 peptide. Antiangiogenic effect of the peptides was further confirmed in vivo using a Matrigel plug assay in C57BL/6 mice. These results suggested that the antiangiogenic activity of tumstatin was localized to a 25 amino acid region of tumstatin (T7 peptide) *(137)*.

5.2.4. Tumstatin Inhibits Protein Synthesis in Endothelial Cells

Apoptosis, programmed cell death, is regulated in part at the level of protein synthesis, and is generally associated with inhibition of cap-dependent protein translation *(138–140)*. Because tumstatin selectively stimulates apoptosis of endothelial cells via binding to αvβ3 integrin, the potential effects of tumstatin peptides on protein synthesis have been examined. Tum-5, T3, and T7 peptide inhibited protein synthesis in endothelial cells, but not in other nonendothelial cells. In contrast, endostatin did not exhibit similar effects on protein synthesis. Tumstatin peptides exhibited endothelial cell-specific inhibitory effect on cap-dependent translation, and this inhibitory effect was not observed in β3-integrin-deficient endothelial cells *(141)*.

In many cell types, ligand binding to integrin induced phosphorylation of focal adhesion kinase (FAK) and subsequent activation of various signaling molecules *(142,143)*. For instance, phosphorylated FAK interacts with and activates phosphatidylinositol 3-kinase (PI3-knase) and protein kinase B (PKB/Akt), leading to cell survival *(142,144)*. In fact, inhibition of PI3-kinase in endothelial cells has been shown to repress protein synthesis *(145)*. Tumstatin peptides inhibited phosphorylation and activation of PI3-kinase and phosphorylation of Akt in endothelial cells *(141)*. Akt activates Rapamycin/FKBP-target 1 protein (RAFT1), also known as mammalian target of rapamycin (mTOR), which in turn phosphorylates eukaryotic initiation factor 4E (eIF4E)-binding protein (4E-BP1) *(146,147)*. Unphosphorylated 4E-BP1 forms complex with eIF4E and inhibits cap-dependent translation *(148)*. In fact, stimulation of cells with growth factors induces phosphorylation of 4E-BP1, dissociation of 4E-BP1 from eIF4E, leading to the induction of protein translation *(147,148)*. Tumstatin peptides suppressed mTOR kinase activity, phosphorylation of 4E-BP1, and dissociation of 4E-BP1 from eIF4E, thus resulting in inhibition of cap-dependent translation. In contrast, tumstatin peptides failed to inhibit phosphorylation of ERK1/2 upon vitronectin attachment or stimulation with VEGF in endothelial cells. Overexpression of constitutive active Akt resulted in reversal of tumstatin peptide-induced inhibition of cap-dependent translation. These results suggested the possible inhibitory effects of tumstatin peptides on protein synthesis of endothelial cells through negative regulation of mTOR signaling. Alternatively, interaction of tumstatin peptide with $\alpha v\beta 3$ integrin may induce negative regulatory signals counteracting growth factor-initiated cell survival signals (*see* Fig. 3).

5.2.5. Endogenous Tumstatin Inhibits Angiogenesis and Tumor Growth

In addition to the antiangiogenic effect of exogenously added tumstatin, the endogenous functions of tumstatin have been reported by Hamano et al. *(70)*. Mice deficient in $\alpha 3$ chain of type IV collagen, a precursor of tumstatin, exhibited normal pregnancy, development, and wound healing process, but had accelerated pathological angiogenesis and tumor growth. Although tumstatin inhibited angiogenesis in mice with the physiological expression levels of $\beta 3$ integrin, inhibitory effect of tumstatin was not observed in $\beta 3$ integrin-null mice, suggesting the requirement of $\beta 3$ integrin in tumstatin's antiangiogenic effect *(70)*. Treatment of basement membranes with MMP-9 generated tumstatin fragments in vitro. Other MMPs—MMP-2, 3, and 13—could release tumstatin fragment from basement membranes, although the efficiency was significantly less than MMP-9. In fact, mice deficient in MMP-9 exhibit significantly decreased circulating blood concentrations of tumstatin compared with wild-type mice. The growth of Lewis lung carcinoma tumors was similar in both MMP-9(+/+) and MMP-9(−/−) mice until the tumors reached 500 mm^3, but the tumors on MMP-9(−/−) mice exhibited accelerated tumor growth afterwards. Intravenous administration of tumstatin to raise the concentration of tumstatin to normal levels resulted in the retardation of tumor growth similar to the level of wild-type mice, suggesting the importance of physiological levels of tumstatin in regulating tumor growth *(70)*. MMP-9 has been implicated as a positive regulator of angiogenic switch, leading to reduced tumor growth and tumor vasculatures upon genetic ablation of MMP-9 in the early stage of tumor growth *(66)*. However, pharmacological inhibition of MMP-9 resulted in acceleration of tumor growth *(72)*. Whereas MMP-9 may mediate the angiogenic switch leading to the initial burst of tumor growth, it also suppresses tumor growth by generating endogenous inhibitors of angiogenesis such as tumstatin *(70)*. The physiological endogenous anti-angiogenic role of TSP-1 and endostatin via binding to CD36 and $\alpha 5\beta 1$ integrin, respectively, had also been recently demonstrated *(149)*.

5.2.6. Application of Tumstatin Peptide on Non-neoplastic Disorder

Diabetic nephropathy is complicated in 30–40% of patients with type 2 diabetes and is the most common pathological disorders predisposing end-stage renal diseases (ESRD) in Japan and in the Western World *(150)*. Hyperglycemia is involved in the progression of diabetic nephropathy and early alterations in diabetic nephropathy include glomerular hyperfiltration, glomerular and tubular epithelial hypertrophy, and the development of microalbuminuria *(151)*. These early alterations are followed by the development of glomerular basement membrane thickening, the accumulation of extracellular matrix components in the mesangium as well as in the interstitium, and the increase of urinary albumin excretion, eventually leading to glomerulosclerosis and

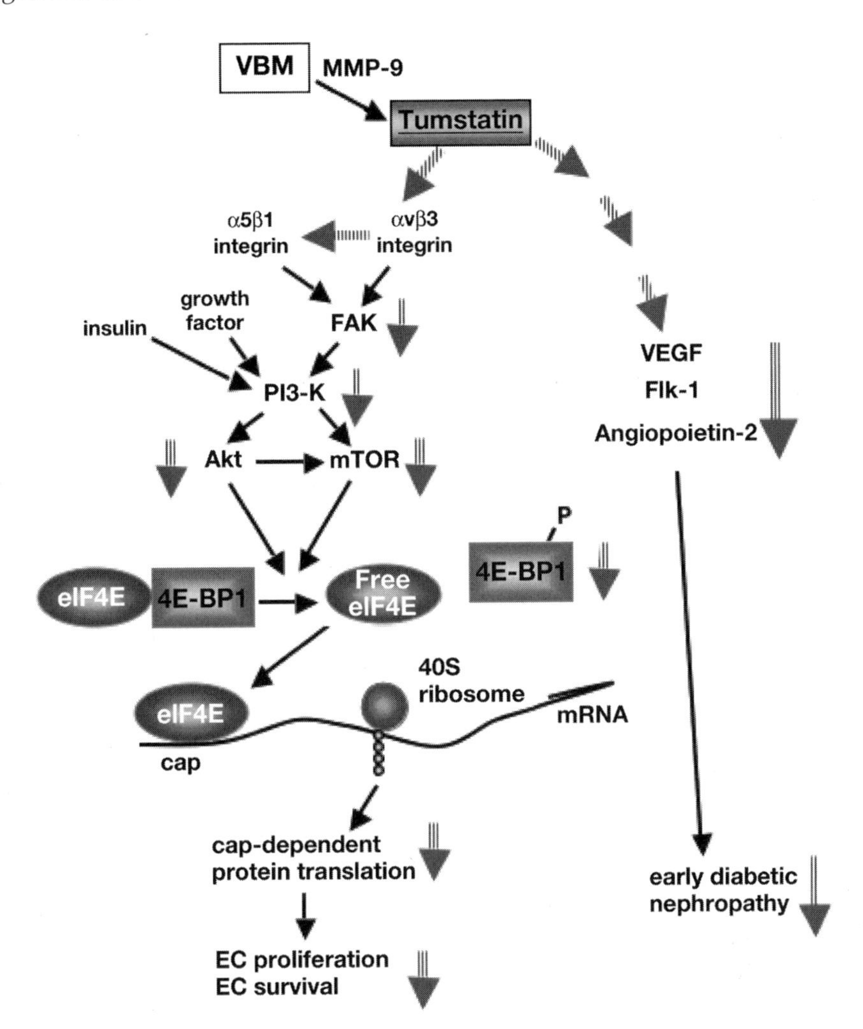

Fig. 3. Schematic of the mechanism of action of tumstatin in inhibiting angiogenesis. Proteases (e.g., MMP-9) degrade vascular basement membrane (VBM) to generate tumstatin, which subsequently binds to αvβ3 integrin on endothelial cells. Tumstatin negatively regulates the pathway that includes focal adhesion kinase (FAK), phosphatidylinositol 3-kinase (PI3-K), Akt, mammalian target of rapamycin (mTOR), eukaryotic translation initiation factor (eIF4E), and eIF4E-binding protein 1 (4E-BP1), resulting in the inhibition of protein synthesis and proliferation of endothelial cells (EC). Upon binding to αvβ3 integrin, tumstatin may also cause transdominant inhibition of angiogenic signals from α5β1 and α2β1 integrin. Tumstatin derived peptide suppresses the renal expression of VEGF, angiopoietin-2 and flk-1, leading to the suppression of early alterations in diabetic nephropathy. The effects of tumstatin are shown by striped arrows.

progressive loss of renal function *(152,153)*. Recent reports by Yamamoto et al. implicated the therapeutic efficacy of the tumstatin peptide in the early stage of diabetic nephropathy via down-regulating renal expression of VEGF, flk-1, and Ang-2 *(154)*, further emphasizing the potential biological functions of tumstatin on various disorders involving angiogenic process (*see* Fig. 3).

5.3. Endorepelin

5.3.1. Perlecan

Perlecan is a major HSPG of basement membranes and vascular and avascular ECM, involved in regulating cell growth, differentiation, cell adhesion, and the development of blood vessels, cartilage, and the nervous system *(155)*. Perlecan-null mutations result in early embryonic lethality accompanied by severe cephalic and cartilage abnormalities *(156)*. Embryos that survive initially usually develop later malformations of the cardiovascular system *(157)*. In humans, two rare skeletal disorders, dyssegmental dysplasia silver-handmaker type (DSSH) and Schwartz-Jampel syndrome (SJS) are caused by mutations of genes encoding perlecan *(158,159)*. Perlecan is considered to exert proangiogenic effects, because it binds to and protect growth factors from degradation and it interacts with adhesion molecules *(160)*.

5.3.2. Endorepellin, an Antiangiogenic C-Terminal Fragment of Perlecan

Perlecan, a C-terminal 85-kDa domain V of perlecan inhibits endothelial cell migration, tube formation, and growth of blood vessels *(71)*. Endorepellin is suspected of inhibiting angiogenesis in a dominant-negative fashion, endothelial cells secrete perlecan. Endorepellin reversibly alters the actin cytoskeleton of endothelial cells, potentially associated with its effects on cell motility and other alterations in cell morphology *(161)*. The interaction of endorepellin with $\alpha 2\beta 1$ integrin triggers a unique signaling pathway that causes an increase in the second messenger camp;activation of protein kinase A and FAK; transient activation of p38 mitogen-activated protein kinase and heat shock protein 27, followed by a rapid down-regulation of the latter two proteins; and ultimately disassembly of actin stress fibers and focal adhesions associated with the inhibitory effect on endothelial cell migration and angiogenesis *(161)*. Collectively, a family of C-terminal fragment of endogenous matrix proteins (endostatin, tumstatin and endorepellin) harbors potent antiangiogenic effects potentially generated via processing by proteases.

5.4. Anti-VEGF Reagents

Based on the well-known role of VEGF in promoting angiogenesis and tumor growth, therapeutic strategies to block the biological effect of VEGF had been developed and, to date, clinically tested on patients with refractory tumors. The therapeutic effect of neutralizing anti-VEGF or VEGF receptor antibody, small molecule inhibitor of tyrosine kinase receptor of VEGF in experimental tumor models has been reported *(79)*.

Observation of proteinuria upon treatment with anti-VEGF antibody or soluble flt-1 (an antagonist of VEGF) has been reported in experimental animal models *(162)* and clinical trials, suggesting the possible side effect on glomerular filtration barrier.

6. Conclusion

Here, we described on the biological roles of angiogenic response in developing tumors and the mediators and inhibitors of angiogenesis, emphasizing the endogenous inhibitors of angiogenesis derived from basement membrane proteins. Further publications on the role of angiogenesis in developing cancer implicate the significance of this research field and its potential to lead to the development of novel therapeutic approaches to halt and regress tumor growth. Investigation on the biological function of endogenous angiogenesis inhibitors may provide insight into the regulatory mechanism of endothelial cells and other vascular cell components in the setting of various pathological disorders.

Acknowledgments

A portion of this study was supported by research grant from a grant-in-aid for Scientific Research from the Ministry of Education, Science and Culture of Japan (YM), Grant-in-Aid from the Tokyo Biochemical Research Foundation (YM), and from the Japan Diabetes Foundation (YM). YM is a recipient of the 2001 Research Award from the Okayama Medical Foundation, the 2002 Research Award from the KANAE Foundation for Life & Socio-Medical Science, the 2002 Young Investigator Award from the Japan Society of Cardiovascular Endocrinology and Metabolism, the 2003 Research Award from the Kobayashi Magobei Memorial Foundation for Medical Science, the 2003 Research Award from the Ryobi Teien Memorial Foundation, the 2004 Research award from the Inamori Foundation, the 2004 Research award and from the SUZUKEN memorial foundation, the 2004 Research Award from the Sanyo Broadcast Academic and Cultural Foundation, and the 2005 Oshima Award (Young Investigator Award) from the Japanese Society of Nephrology.

I appreciate Dr. Raghu Kalluri (Beth Israel Deaconess Medical Center and Harvard Medical School, Boston) and Dr. Hirofumi Makino (Okayama University Graduate School of Medicine, Dentistry and Pharmaceutical Sciences, Okayama, Japan) for their mentorship in my research career.

References

1. Kalluri R, Sukhatme VP. Fibrosis and angiogenesis. Curr Opin Nephrol Hypertens 2000;9(4):413–418.
2. Timpl R. Macromolecular organization of basement membranes. Curr Opin Cell Biol 1996;8(5):618–624.
3. Hangai M, Kitaya N, Xu J, et al. Matrix metalloproteinase-9-dependent exposure of a cryptic migratory control site in collagen is required before retinal angiogenesis. Am J Pathol 2002;161(4):1429–1437.
4. Hynes RO. A reevaluation of integrins as regulators of angiogenesis. Nat Med 2002;8(9):918–921.
5. Hood JD, Cheresh DA. Role of integrins in cell invasion and migration. Nat Rev Cancer 2002;2(2):91–100.
6. Brooks PC, Clark RA, Cheresh DA. Requirement of vascular integrin alpha v beta 3 for angiogenesis. Science 1994;264(5158):569–571.
7. Brooks PC, Montgomery AM, Rosenfeld M, et al. Integrin alpha v beta 3 antagonists promote tumor regression by inducing apoptosis of angiogenic blood vessels. Cell 1994;79(7):1157–1164.
8. Bader BL, Rayburn H, Crowley D, Hynes RO. Extensive vasculogenesis, angiogenesis, and organogenesis precede lethality in mice lacking all alpha v integrins. Cell 1998;95(4): 507–519.
9. Hodivala-Dilke KM, McHugh KP, Tsakiris DA, et al. Beta3-integrin-deficient mice are a model for Glanzmann thrombasthenia showing placental defects and reduced survival. J Clin Invest 1999;103(2):229–238.
10. Folkman J. Anti-angiogenesis: new concept for therapy of solid tumors. Ann Surg 1972;175(3):409–416.

11. Folkman J. Angiogenesis in cancer, vascular, rheumatoid and other disease. Nat Med 1995;1(1):27–31.
12. O'Reilly MS, Boehm T, Shing Y, et al. Endostatin: an endogenous inhibitor of angiogenesis and tumor growth. Cell 1997;88(2):277–285.
13. O'Reilly MS, Holmgren L, Shing Y, et al. Angiostatin: a novel angiogenesis inhibitor that mediates the suppression of metastases by a Lewis lung carcinoma [see comments]. Cell 1994;79(2):315–328.
14. Ramchandran R, Dhanabal M, Volk R, et al. Antiangiogenic activity of restin, NC10 domain of human collagen XV: comparison to endostatin. Biochem Biophys Res Commun 1999;255(3):735–739.
15. Colorado PC, Torre A, Kamphaus G, et al. Anti-angiogenic cues from vascular basement membrane collagen. Cancer Res 2000;60(9):2520–2526.
16. Kamphaus GD, Colorado PC, Panka DJ, et al. Canstatin, a novel matrix-derived inhibitor of angiogenesis and tumor growth. J Biol Chem 2000;275(2):1209–1215.
17. Maeshima Y, Colorado PC, Torre A, et al. Distinct antitumor properties of a type IV collagen domain derived from basement membrane. J Biol Chem 2000;275(28):21,340–21,348.
18. Petitclerc E, Boutaud A, Prestayko A, et al. New Functions for Non-collagenous Domains of Human Collagen Type IV. Novel integrin ligands inhibiting angiogenesis and tumor growth in vivo. J Biol Chem 2000;275(11):8051–8061.
19. Luttun A, Carmeliet G, Carmeliet P. Vascular progenitors: from biology to treatment. Trends Cardiovasc Med 2002;12(2):88–96.
20. Reyes M, Dudek A, Jahagirdar B, Koodie L, Marker PH, Verfaillie CM. Origin of endothelial progenitors in human postnatal bone marrow. J Clin Invest 2002;109(3):337–346.
21. Rehman J, Li J, Orschell CM, March KL. Peripheral blood "endothelial progenitor cells" are derived from monocyte/macrophages and secrete angiogenic growth factors. Circulation 2003;107(8):1164–1169.
22. Takakura N, Watanabe T, Suenobu S, et al. A role for hematopoietic stem cells in promoting angiogenesis. Cell 2000;102(2):199–209.
23. Hattori K, Heissig B, Wu Y, et al. Placental growth factor reconstitutes hematopoiesis by recruiting VEGFR1(+) stem cells from bone-marrow microenvironment. Nat Med 2002;8(8):841–849.
24. Lyden D, Hattori K, Dias S, et al. Impaired recruitment of bone-marrow-derived endothelial and hematopoietic precursor cells blocks tumor angiogenesis and growth. Nat Med 2001;7(11):1194–1201.
25. Luttun A, Tjwa M, Moons L, et al. Revascularization of ischemic tissues by PlGF treatment, and inhibition of tumor angiogenesis, arthritis and atherosclerosis by anti-Flt1. Nat Med 2002;8(8):831–840.
26. Asahara T, Isner JM. Endothelial progenitor cells for vascular regeneration. J Hematother Stem Cell Res 2002;11(2):171–178.
27. Rafii S, Lyden D. Therapeutic stem and progenitor cell transplantation for organ vascularization and regeneration. Nat Med 2003;9(6):702–712.
28. Ferrara N, Davis-Smyth T. The biology of vascular endothelial growth factor. Endocr Rev 1997;18(1):4–25.
29. Karkkainen MJ, Makinen T, Alitalo K. Lymphatic endothelium: a new frontier of metastasis research. Nat Cell Biol 2002;4(1):E2–E5.
30. Ferrara N, Carver-Moore K, Chen H, et al. Heterozygous embryonic lethality induced by targeted inactivation of the VEGF gene. Nature 1996;380(6573):439–442.
31. Carmeliet P, Ferreira V, Breier G, et al. Abnormal blood vessel development and lethality in embryos lacking a single VEGF allele. Nature 1996;380(6573):435–439.

32. Carmeliet P, Moons L, Luttun A, et al. Synergism between vascular endothelial growth factor and placental growth factor contributes to angiogenesis and plasma extravasation in pathological conditions. Nat Med 2001;7(5):575–583.
33. Leung DW, Cachianes G, Kuang WJ, Goeddel DV, Ferrara N. Vascular endothelial growth factor is a secreted angiogenic mitogen. Science 1989;246(4935):1306–1309.
34. Gerber HP, McMurtrey A, Kowalski J, et al. Vascular endothelial growth factor regulates endothelial cell survival through the phosphatidylinositol 3′-kinase/Akt signal transduction pathway. Requirement for Flk-1/KDR activation. J Biol Chem 1998;273(46): 30,336–30,343.
35. Gerber HP, Dixit V, Ferrara N. Vascular endothelial growth factor induces expression of the antiapoptotic proteins Bcl-2 and A1 in vascular endothelial cells. J Biol Chem 1998;273(21):13,313–13,316.
36. Senger DR, Galli SJ, Dvorak AM, Perruzzi CA, Harvey VS, Dvorak HF. Tumor cells secrete a vascular permeability factor that promotes accumulation of ascites fluid. Science 1983;219(4587):983–985.
37. Dvorak HF, Brown LF, Detmar M, Dvorak AM. Vascular permeability factor/vascular endothelial growth factor, microvascular hyperpermeability, and angiogenesis. Am J Pathol 1995;146(5):1029–1039.
38. Roberts WG, Palade GE. Increased microvascular permeability and endothelial fenestration induced by vascular endothelial growth factor. J Cell Sci 1995;108(Pt 6):2369–2379.
39. Houck KA, Ferrara N, Winer J, Cachianes G, Li B, Leung DW. The vascular endothelial growth factor family: identification of a fourth molecular species and characterization of alternative splicing of RNA. Mol Endocrinol 1991;5(12):1806–1814.
40. Tischer E, Mitchell R, Hartman T, et al. The human gene for vascular endothelial growth factor. Multiple protein forms are encoded through alternative exon splicing. J Biol Chem 1991;266(18):11,947–11,954.
41. Park JE, Keller GA, Ferrara N. The vascular endothelial growth factor (VEGF) isoforms: differential deposition into the subepithelial extracellular matrix and bioactivity of extracellular matrix-bound VEGF. Mol Biol Cell 1993;4(12):1317–1326.
42. Semenza G. Signal transduction to hypoxia-inducible factor 1. Biochem Pharmacol 2002; 64(5-6):993–998.
43. de Vries C, Escobedo JA, Ueno H, Houck K, Ferrara N, Williams LT. The fms-like tyrosine kinase, a receptor for vascular endothelial growth factor. Science 1992;255(5047):989–991.
44. Kendall RL, Thomas KA. Inhibition of vascular endothelial cell growth factor activity by an endogenously encoded soluble receptor. Proc Natl Acad Sci USA 1993;90(22): 10,705–10,709.
45. Fong GH, Zhang L, Bryce DM, Peng J. Increased hemangioblast commitment, not vascular disorganization, is the primary defect in flt-1 knock-out mice. Development 1999;126(13): 3015–3025.
46. Ferrara N, Gerber HP, LeCouter J. The biology of VEGF and its receptors. Nat Med 2003;9(6):669–676.
47. Shalaby F, Rossant J, Yamaguchi TP, et al. Failure of blood-island formation and vasculogenesis in Flk-1-deficient mice. Nature 1995;376(6535):62–66.
48. Guo D, Jia Q, Song HY, Warren RS, Donner DB. Vascular endothelial cell growth factor promotes tyrosine phosphorylation of mediators of signal transduction that contain SH2 domains. Association with endothelial cell proliferation. J Biol Chem 1995;270(12):6729–6733.
49. Eliceiri BP, Paul R, Schwartzberg PL, Hood JD, Leng J, Cheresh DA. Selective requirement for Src kinases during VEGF-induced angiogenesis and vascular permeability. Mol Cell 1999;4(6):915–924.

50. Takahashi T, Ueno H, Shibuya M. VEGF activates protein kinase C-dependent, but Ras-independent Raf-MEK-MAP kinase pathway for DNA synthesis in primary endothelial cells. Oncogene 1999;18(13):2221–2230.
51. Soker S, Takashima S, Miao HQ, Neufeld G, Klagsbrun M. Neuropilin-1 is expressed by endothelial and tumor cells as an isoform- specific receptor for vascular endothelial growth factor. Cell 1998;92(6):735–745.
52. Kawasaki T, Kitsukawa T, Bekku Y, et al. A requirement for neuropilin-1 in embryonic vessel formation. Development 1999;126(21):4895–4902.
53. Yancopoulos GD, Davis S, Gale NW, Rudge JS, Wiegand SJ, Holash J. Vascular-specific growth factors and blood vessel formation. Nature 2000;407(6801):242–248.
54. Maisonpierre PC, Suri C, Jones PF, et al. Angiopoietin-2, a natural antagonist for Tie2 that disrupts in vivo angiogenesis. Science 1997;277(5322):55–60.
55. Suri C, Jones PF, Patan S, et al. Requisite role of angiopoietin-1, a ligand for the TIE2 receptor, during embryonic angiogenesis. Cell 1996;87(7):1171–1180.
56. Gale NW, Thurston G, Hackett SF, et al. Angiopoietin-2 is required for postnatal angiogenesis and lymphatic patterning, and only the latter role is rescued by Angiopoietin-1. Dev Cell 2002;3(3):411–423.
57. Oh H, Takagi H, Suzuma K, Otani A, Matsumura M, Honda Y. Hypoxia and vascular endothelial growth factor selectively up-regulate angiopoietin-2 in bovine microvascular endothelial cells. J Biol Chem 1999;274(22):15,732–15,739.
58. Mandriota SJ, Pepper MS. Regulation of angiopoietin-2 mRNA levels in bovine microvascular endothelial cells by cytokines and hypoxia. Circ Res 1998;83(8):852–859.
59. Thurston G, Rudge JS, Ioffe E, et al. Angiopoietin-1 protects the adult vasculature against plasma leakage. Nat Med 2000;6(4):460–463.
60. Shim WS, Teh M, Bapna A, et al. Angiopoietin 1 promotes tumor angiogenesis and tumor vessel plasticity of human cervical cancer in mice. Exp Cell Res 2002;279(2):299–309.
61. Hattori K, Dias S, Heissig B, et al. Vascular endothelial growth factor and angiopoietin-1 stimulate postnatal hematopoiesis by recruitment of vasculogenic and hematopoietic stem cells. J Exp Med 2001;193(9):1005–1014.
62. Visconti RP, Richardson CD, Sato TN. Orchestration of angiogenesis and arteriovenous contribution by angiopoietins and vascular endothelial growth factor (VEGF). Proc Natl Acad Sci USA 2002;99(12):8219–8224.
63. Ahmad SA, Liu W, Jung YD, et al. The effects of angiopoietin-1 and -2 on tumor growth and angiogenesis in human colon cancer. Cancer Res 2001;61(4):1255–1259.
64. Luttun A, Dewerchin M, Collen D, Carmeliet P. The role of proteinases in angiogenesis, heart development, restenosis, atherosclerosis, myocardial ischemia, and stroke: insights from genetic studies. Curr Atheroscler Rep 2000;2(5):407–416.
65. Jackson C. Matrix metalloproteinases and angiogenesis. Curr Opin Nephrol Hypertens 2002;11(3):295–299.
66. Bergers G, Brekken R, McMahon G, et al. Matrix metalloproteinase-9 triggers the angiogenic switch during carcinogenesis. Nat Cell Biol 2000;2(10):737–744.
67. Coussens LM, Werb Z. Inflammation and cancer. Nature 2002;420(6917):860–867.
68. Kalluri R. Basement membranes: structure, assembly and role in tumour angiogenesis. Nat Rev Cancer 2003;3(6):422–433.
69. Xu J, Rodriguez D, Petitclerc E, et al. Proteolytic exposure of a cryptic site within collagen type IV is required for angiogenesis and tumor growth in vivo. J Cell Biol 2001;154(5):1069–1079.
70. Hamano Y, Zeisberg M, Sugimoto H, et al. Physiological levels of tumstatin, a fragment of collagen IV alpha3 chain, are generated by MMP-9 proteolysis and suppress angiogenesis via alphaV beta3 integrin. Cancer Cell 2003;3(6):589–601.

71. Mongiat M, Sweeney SM, San Antonio JD, Fu J, Iozzo RV. Endorepellin, a novel inhibitor of angiogenesis derived from the C terminus of perlecan. J Biol Chem 2003;278(6): 4238–4249.
72. Pozzi A, Moberg PE, Miles LA, Wagner S, Soloway P, Gardner HA. Elevated matrix metalloprotease and angiostatin levels in integrin alpha 1 knockout mice cause reduced tumor vascularization. Proc Natl Acad Sci USA 2000;97(5):2202–2207.
73. Brooks PC, Silletti S, von Schalscha TL, Friedlander M, Cheresh DA. Disruption of angiogenesis by PEX, a noncatalytic metalloproteinase fragment with integrin binding activity. Cell 1998;92(3):391–400.
74. Holder N, Klein R. Eph receptors and ephrins: effectors of morphogenesis. Development 1999;126(10):2033–2044.
75. Kullander K, Klein R. Mechanisms and functions of Eph and ephrin signalling. Nat Rev Mol Cell Biol 2002;3(7):475–486.
76. Gerety SS, Wang HU, Chen ZF, Anderson DJ. Symmetrical mutant phenotypes of the receptor EphB4 and its specific transmembrane ligand ephrin-B2 in cardiovascular development. Mol Cell 1999;4(3):403–414.
77. Wang HU, Chen ZF, Anderson DJ. Molecular distinction and angiogenic interaction between embryonic arteries and veins revealed by ephrin-B2 and its receptor Eph-B4. Cell 1998;93(5):741–753.
78. Adams RH, Wilkinson GA, Weiss C, et al. Roles of ephrinB ligands and EphB receptors in cardiovascular development: demarcation of arterial/venous domains, vascular morphogenesis, and sprouting angiogenesis. Genes Dev 1999;13(3):295–306.
79. Marme D. The impact of anti-angiogenic agents on cancer therapy. J Cancer Res Clin Oncol 2003;129(11):607–620.
80. Giancotti FG, Ruoslahti E. Integrin signaling. Science 1999;285(5430):1028–1032.
81. Varner JA, Brooks PC, Cheresh DA. REVIEW: the integrin alpha V beta 3: angiogenesis and apoptosis. Cell Adhes Commun 1995;3(4):367–374.
82. Brooks PC, Stromblad S, Klemke R, Visscher D, Sarkar FH, Cheresh DA. Antiintegrin alpha v beta 3 blocks human breast cancer growth and angiogenesis in human skin. J Clin Invest 1995;96(4):1815–1822.
83. Friedlander M, Theesfeld CL, Sugita M, et al. Involvement of integrins alpha v beta 3 and alpha v beta 5 in ocular neovascular diseases. Proc Natl Acad Sci USA 1996;93(18): 9764–9769.
84. Huang X, Griffiths M, Wu J, Farese RV, Jr, Sheppard D. Normal development, wound healing, and adenovirus susceptibility in beta5-deficient mice. Mol Cell Biol 2000;20(3):755–759.
85. Reynolds LE, Wyder L, Lively JC, et al. Enhanced pathological angiogenesis in mice lacking beta3 integrin or beta3 and beta5 integrins. Nat Med 2002;8(1):27–34.
86. Diaz-Gonzalez F, Forsyth J, Steiner B, Ginsberg MH. Trans-dominant inhibition of integrin function. Mol Biol Cell 1996;7(12):1939–1951.
87. Reynolds AR, Reynolds LE, Nagel TE, et al. Elevated Flk1 (vascular endothelial growth factor receptor 2) signaling mediates enhanced angiogenesis in beta3-integrin-deficient mice. Cancer Res 2004;64(23):8643–8650.
88. Carmeliet P, Lampugnani MG, Moons L, et al. Targeted deficiency or cytosolic truncation of the VE-cadherin gene in mice impairs VEGF-mediated endothelial survival and angiogenesis. Cell 1999;98(2):147–157.
89. Liao F, Li Y, O'Connor W, et al. Monoclonal antibody to vascular endothelial-cadherin is a potent inhibitor of angiogenesis, tumor growth, and metastasis. Cancer Res 2000;60(24): 6805–6810.
90. Hanahan D, Folkman J. Patterns and emerging mechanisms of the angiogenic switch during tumorigenesis. Cell 1996;86(3):353–364.

91. Holmgren L, O'Reilly MS, Folkman J. Dormancy of micrometastases: balanced proliferation and apoptosis in the presence of angiogenesis suppression. Nat Med 1995;1(2):149–153.
92. Folkman J. Angiogenesis and apoptosis. Semin Cancer Biol 2003;13(2):159–167.
93. Holash J, Maisonpierre PC, Compton D, et al. Vessel cooption, regression, and growth in tumors mediated by angiopoietins and VEGF. Science 1999;284(5422):1994–1998.
94. Belotti D, Vergani V, Drudis T, et al. The microtubule-affecting drug paclitaxel has antiangiogenic activity. Clin Cancer Res 1996;2(11):1843–1849.
95. Browder T, Butterfield CE, Kraling BM, et al. Antiangiogenic scheduling of chemotherapy improves efficacy against experimental drug-resistant cancer. Cancer Res 2000;60(7): 1878–1886.
96. Rehn M, Pihlajaniemi T. Alpha 1(XVIII), a collagen chain with frequent interruptions in the collagenous sequence, a distinct tissue distribution, and homology with type XV collagen. Proc Natl Acad Sci USA 1994;91(10):4234–4238.
97. Halfter W, Dong S, Schurer B, Cole GJ. Collagen XVIII is a basement membrane heparan sulfate proteoglycan. J Biol Chem 1998;273(39):25,404–25,412.
98. Sasaki T, Fukai N, Mann K, Gohring W, Olsen BR, Timpl R. Structure, function and tissue forms of the C-terminal globular domain of collagen XVIII containing the angiogenesis inhibitor endostatin. Embo J 1998;17(15):4249–4256.
99. Suzuki OT, Sertie AL, Der Kaloustian VM, et al. Molecular analysis of collagen XVIII reveals novel mutations, presence of a third isoform, and possible genetic heterogeneity in Knobloch syndrome. Am J Hum Genet 2002;71(6):1320–1329.
100. Muragaki Y, Abe N, Ninomiya Y, Olsen BR, Ooshima A. The human alpha 1(XV) collagen chain contains a large amino-terminal non-triple helical domain with a tandem repeat structure and homology to alpha 1(XVIII) collagen. J Biol Chem 1994;269(6):4042–4046.
101. Myers JC, Kivirikko S, Gordon MK, Dion AS, Pihlajaniemi T. Identification of a previously unknown human collagen chain, alpha 1(XV), characterized by extensive interruptions in the triple-helical region. Proc Natl Acad Sci USA 1992;89(21):10,144–10,148.
102. Eklund L, Piuhola J, Komulainen J, et al. Lack of type XV collagen causes a skeletal myopathy and cardiovascular defects in mice. Proc Natl Acad Sci USA 2001;98(3): 1194–1199.
103. Sasaki T, Larsson H, Tisi D, Claesson-Welsh L, Hohenester E, Timpl R. Endostatins derived from collagens XV and XVIII differ in structural and binding properties, tissue distribution and anti-angiogenic activity. J Mol Biol 2000;301(5):1179–1190.
104. Yamaguchi N, Anand-Apte B, Lee M, et al. Endostatin inhibits VEGF-induced endothelial cell migration and tumor growth independently of zinc binding. Embo J 1999;18(16): 4414–4423.
105. Dhanabal M, Volk R, Ramchandran R, Simons M, Sukhatme VP. Cloning, expression, and in vitro activity of human endostatin. Biochem Biophys Res Commun 1999;258(2):345–352.
106. Hanai J, Dhanabal M, Karumanchi SA, et al. Endostatin causes G1 arrest of endothelial cells through inhibition of cyclin D1. J Biol Chem 2002;277(19):16,464–16,469.
107. Dhanabal M, Ramchandran R, Volk R, et al. Endostatin: yeast production, mutants, and antitumor effect in renal cell carcinoma. Cancer Res 1999;59(1):189–197.
108. Hajitou A, Grignet C, Devy L, et al. The antitumoral effect of endostatin and angiostatin is associated with a down-regulation of vascular endothelial growth factor expression in tumor cells. Faseb J 2002;16(13):1802–1804.
109. Kim YM, Hwang S, Pyun BJ, et al. Endostatin blocks vascular endothelial growth factor-mediated signaling via direct interaction with KDR/Flk-1. J Biol Chem 2002;277(31): 27,872–27,879.
110. Matsuno H, Yudoh K, Uzuki M, et al. Treatment with the angiogenesis inhibitor endostatin: a novel therapy in rheumatoid arthritis. J Rheumatol 2002;29(5):890–895.

111. Takahashi K, Saishin Y, Silva RL, et al. Intraocular expression of endostatin reduces VEGF-induced retinal vascular permeability, neovascularization, and retinal detachment. Faseb J 2003;17(8):896–898.
112. Cattaneo MG, Pola S, Francescato P, Chillemi F, Vicentini LM. Human endostatin-derived synthetic peptides possess potent antiangiogenic properties in vitro and in vivo. Exp Cell Res 2003;283(2):230–236.
113. Rehn M, Veikkola T, Kukk-Valdre E, et al. Interaction of endostatin with integrins implicated in angiogenesis. Proc Natl Acad Sci USA 2001;98(3):1024–1029.
114. Wickstrom SA, Alitalo K, Keski-Oja J. Endostatin associates with integrin alpha5beta1 and caveolin-1, and activates Src via a tyrosyl phosphatase-dependent pathway in human endothelial cells. Cancer Res 2002;62(19):5580–5589.
115. Wickstrom SA, Alitalo K, Keski-Oja J. Endostatin associates with lipid rafts and induces reorganization of the actin cytoskeleton via down-regulation of RhoA activity. J Biol Chem 2003;278(39):37,895–37,901.
116. Sudhakar A, Sugimoto H, Yang C, Lively J, Zeisberg M, Kalluri R. Human tumstatin and human endostatin exhibit distinct antiangiogenic activities mediated by alpha v beta 3 and alpha 5 beta 1 integrins. Proc Natl Acad Sci USA 2003;100(8):4766–4771.
117. Karumanchi SA, Jha V, Ramchandran R, et al. Cell surface glypicans are low-affinity endostatin receptors. Mol Cell 2001;7(4):811–822.
118. Yu Y, Moulton KS, Khan MK, et al. E-selectin is required for the antiangiogenic activity of endostatin. Proc Natl Acad Sci USA 2004;101(21):8005–8010.
119. Abdollahi A, Hahnfeldt P, Maercker C, et al. Endostatin's antiangiogenic signaling network. Mol Cell 2004;13(5):649–663.
120. Prockop DJ, Kivirikko KI. Collagens: molecular biology, diseases, and potentials for therapy. Annu Rev Biochem 1995;64:403–434.
121. Madri JA. Extracellular matrix modulation of vascular cell behaviour. Transpl Immunol 1997;5(3):179–183.
122. Myllyharju J, Kivirikko KI. Collagens, modifying enzymes and their mutations in humans, flies and worms. Trends Genet 2004;20(1):33–43.
123. Paulsson M. Basement membrane proteins: structure, assembly, and cellular interactions. Crit Rev Biochem Mol Biol 1992;27(1-2):93–127.
124. Kalluri R, Shield CF, Todd P, Hudson BG, Neilson EG. Isoform switching of type IV collagen is developmentally arrested in X- linked Alport syndrome leading to increased susceptibility of renal basement membranes to endoproteolysis. J Clin Invest 1997;99(10):2470–2478.
125. Kashtan CE. Alport syndrome and thin glomerular basement membrane disease. J Am Soc Nephrol 1998;9(9):1736–1750.
126. Frojdman K, Pelliniemi LJ, Virtanen I. Differential distribution of type IV collagen chains in the developing rat testis and ovary. Differentiation 1998;63(3):125–130.
127. Hudson BG, Reeders ST, Tryggvason K. Type IV collagen: structure, gene organization, and role in human diseases. Molecular basis of Goodpasture and Alport syndromes and diffuse leiomyomatosis. J Biol Chem 1993;268(35):26,033–26,036.
128. Kalluri R, Sun MJ, Hudson BG, Neilson EG. The Goodpasture autoantigen. Structural delineation of two immunologically privileged epitopes on alpha3(IV) chain of type IV collagen. J Biol Chem 1996;271(15):9062–9068.
129. Hellmark T, Segelmark M, Unger C, Burkhardt H, Saus J, Wieslander J. Identification of a clinically relevant immunodominant region of collagen IV in Goodpasture disease [see comments]. Kidney Int 1999;55(3):936–944.
130. Hellmark T, Burkhardt H, Wieslander J. Goodpasture disease. Characterization of a single conformational epitope as the target of pathogenic autoantibodies. J Biol Chem 1999;274 (36):25,862–25,868.

131. Netzer KO, Leinonen A, Boutaud A, et al. The goodpasture autoantigen. Mapping the major conformational epitope(s) of alpha3(iv) collagen to residues 17-31 and 127-141 of the nc1 domain [In Process Citation]. J Biol Chem 1999;274(16):11,267–11,274.
132. Han J, Ohno N, Pasco S, Monboisse JC, Borel JP, Kefalides NA. A cell binding domain from the alpha3 chain of type IV collagen inhibits proliferation of melanoma cells. J Biol Chem 1997;272(33):20,395–20,401.
133. Panka DJ, Mier JW. Canstatin inhibits Akt activation and induces Fas-dependent apoptosis in endothelial cells. J Biol Chem 2003;278(39):37,632–37,636.
134. Shahan TA, Ziaie Z, Pasco S, et al. Identification of CD47/integrin-associated protein and alpha(v)beta3 as two receptors for the alpha3(IV) chain of type IV collagen on tumor cells. Cancer Res 1999;59(18):4584–4590.
135. Maeshima Y, Colorado PC, Kalluri R. Two RGD-independent alpha vbeta 3 integrin binding sites on tumstatin regulate distinct anti-tumor properties. J Biol Chem 2000;275(31): 23,745–23,750.
136. Maeshima Y, Manfredi M, Reimer C, et al. Identification of the anti-angiogenic site within vascular basement membrane-derived tumstatin. J Biol Chem 2001;276(18):15,240–15,248.
137. Maeshima Y, Yerramalla UL, Dhanabal M, et al. Extracellular matrix-derived peptide binds to alpha(v)beta(3) integrin and inhibits angiogenesis. J Biol Chem 2001;276(34):31,959–31,968.
138. Brown EJ, Schreiber SL. A signaling pathway to translational control. Cell 1996;86(4): 517–520.
139. Bushell M, McKendrick L, Janicke RU, Clemens MJ, Morley SJ. Caspase-3 is necessary and sufficient for cleavage of protein synthesis eukaryotic initiation factor 4G during apoptosis. FEBS Lett 1999;451(3):332–336.
140. Gingras AC, Raught B, Sonenberg N. Regulation of translation initiation by FRAP/mTOR. Genes Dev 2001;15(7):807–826.
141. Maeshima Y, Sudhakar A, Lively JC, et al. Tumstatin, an endothelial cell-specific inhibitor of protein synthesis. Science 2002;295(5552):140–143.
142. Vuori K. Integrin signaling: tyrosine phosphorylation events in focal adhesions. J Membr Biol 1998;165(3):191–199.
143. Ruoslahti E. Fibronectin and its integrin receptors in cancer. Adv Cancer Res 1999;76:1–20.
144. Chen HC, Guan JL. Association of focal adhesion kinase with its potential substrate phosphatidylinositol 3-kinase. Proc Natl Acad Sci USA 1994;91(21):10,148–10,152.
145. Vinals F, Chambard JC, Pouyssegur J. p70 S6 kinase-mediated protein synthesis is a critical step for vascular endothelial cell proliferation. J Biol Chem 1999;274(38):26,776–26,782.
146. Brunn GJ, Hudson CC, Sekulic A, et al. Phosphorylation of the translational repressor PHAS-I by the mammalian target of rapamycin. Science 1997;277(5322):99–101.
147. Gingras AC, Gygi SP, Raught B, et al. Regulation of 4E-BP1 phosphorylation: a novel two-step mechanism. Genes Dev 1999;13(11):1422–1437.
148. Pause A, Belsham GJ, Gingras AC, et al. Insulin-dependent stimulation of protein synthesis by phosphorylation of a regulator of 5′-cap function. Nature 1994;371(6500):762–767.
149. Sund M, Hamano Y, Sugimoto H, et al. Function of endogenous inhibitors of angiogenesis as endothelium-specific tumor suppressors. Proc Natl Acad Sci USA 2005;102(8):2934–2939.
150. Ritz E, Rychlik I, Locatelli F, Halimi S. End-stage renal failure in type 2 diabetes: A medical catastrophe of worldwide dimensions. Am J Kidney Dis 1999;34(5):795–808.
151. Osterby R, Parving HH, Nyberg G, et al. A strong correlation between glomerular filtration rate and filtration surface in diabetic nephropathy. Diabetologia 1988;31(5):265–270.
152. Makino H, Yamasaki Y, Haramoto T, et al. Ultrastructural changes of extracellular matrices in diabetic nephropathy revealed by high resolution scanning and immunoelectron microscopy. Lab Invest 1993;68(1):45–55.

153. Makino H, Kashihara N, Sugiyama H, et al. Phenotypic modulation of the mesangium reflected by contractile proteins in diabetes. Diabetes 1996;45(4):488–495.
154. Yamamoto Y, Maeshima Y, Kitayama H, et al. Tumstatin Peptide, an inhibitor of angiogenesis, prevents glomerular hypertrophy in the early stage of diabetic nephropathy. Diabetes 2004;53(7):1831–1840.
155. Iozzo RV. Matrix proteoglycans: from molecular design to cellular function. Annu Rev Biochem 1998;67:609–652.
156. Arikawa-Hirasawa E, Watanabe H, Takami H, Hassell JR, Yamada Y. Perlecan is essential for cartilage and cephalic development. Nat Genet 1999;23(3):354–358.
157. Costell M, Carmona R, Gustafsson E, Gonzalez-Iriarte M, Fassler R, Munoz-Chapuli R. Hyperplastic conotruncal endocardial cushions and transposition of great arteries in perlecan-null mice. Circ Res 2002;91(2):158–164.
158. Arikawa-Hirasawa E, Wilcox WR, Le AH, et al. Dyssegmental dysplasia, Silverman-Handmaker type, is caused by functional null mutations of the perlecan gene. Nat Genet 2001;27(4):431–434.
159. Nicole S, Davoine CS, Topaloglu H, et al. Perlecan, the major proteoglycan of basement membranes, is altered in patients with Schwartz-Jampel syndrome (chondrodystrophic myotonia). Nat Genet 2000;26(4):480–483.
160. Bix G, Iozzo RV. Matrix revolutions: 'tails' of basement-membrane components with angiostatic functions. Trends Cell Biol 2005;15(1):52–60.
161. Bix G, Fu J, Gonzalez EM, et al. Endorepellin causes endothelial cell disassembly of actin cytoskeleton and focal adhesions through alpha2beta1 integrin. J Cell Biol 2004; 166(1):97–109.
162. Sugimoto H, Hamano Y, Charytan D, et al. Neutralization of circulating vascular endothelial growth factor (VEGF) by anti-VEGF antibodies and soluble VEGF receptor 1 (sFlt-1) induces proteinuria. J Biol Chem 2003;278(15):12,605–12,608.

3

Metastasis

The Evasion of Apoptosis

Christine E. Horak, Julie L. Bronder, Amina Bouadis, and Patricia S. Steeg

Summary

Metastasis, the spread of tumor cells from the primary site of growth to a secondary site, is the leading cause of cancer related deaths. The process of metastasis can be conceptually broken down into a series of sequential steps: acquisition of an invasive phenotype, intravasation, travel through the circulatory system, arrest at a secondary site, and extravasation and colonization at that site. The metastatic cell must be able to escape cell death at several steps in this process. Thus, it is not surprising that metastatic cells can be molecularly characterized by altered expression of several apoptosis genes and altered signaling through cell survival and death pathways. In addition, this same acquired resistance to cell death that has allowed metastatic cells to survive and colonize in a secondary organ has also permitted resistance to courses of chemotherapy. Herein, we describe the aberrant apoptotic signaling that metastatic cells have evolved, how this aberrant signaling impedes traditional chemotherapeutic approaches that induce cell death, and potential therapeutic approaches to overcome apoptotic resistance of metastasis.

Key Words: Metastasis; apoptosis; Bcl-2 family; TNF-α family; DAP kinase; maspin; transcription factors; anoikis; molecularly targeted therapy; chemotherapeutic resistance.

1. Introduction to Metastasis

Tumor metastasis is a significant contributor to death in cancer patients. More than 90% of cancer deaths are caused by the development of metastases, because the primary tumor can often be treated successfully if detected early *(1)*. Simply defined, metastasis is the spread of malignant tumor cells from the primary tumor site to a secondary site within the body.

Metastasis is a complex, multistep process that can be divided into five stages as illustrated in Fig. 1. In the first stage, the primary tumor cells acquire an invasive phenotype. They exhibit alterations in cell-to-cell interactions, as well as cell-to-extracellular matrix (ECM) interactions leading to initiation of motility. The second stage, termed intravasation, culminates in the escape of tumor cells from the primary site and penetration into circulation, possibly via the lymphatic vessels. During the third stage, malignant cells passively and efficiently travel through the circulation until they reach a capillary bed and either adhere to the vessel walls or arrest as a result of size constraints. Extravasation is the fourth step of the metastatic process wherein tumor cells exit the circulatory vessels at the destined site. Following extravasation, metastatic tumor cells proliferate, forming micrometastatic or macrometastatic lesions within the organ in a process called colonization.

From: *Apoptosis, Cell Signaling, and Human Diseases: Molecular Mechanisms, Volume 1*
Edited by R. Srivastava © Humana Press Inc., Totowa, NJ

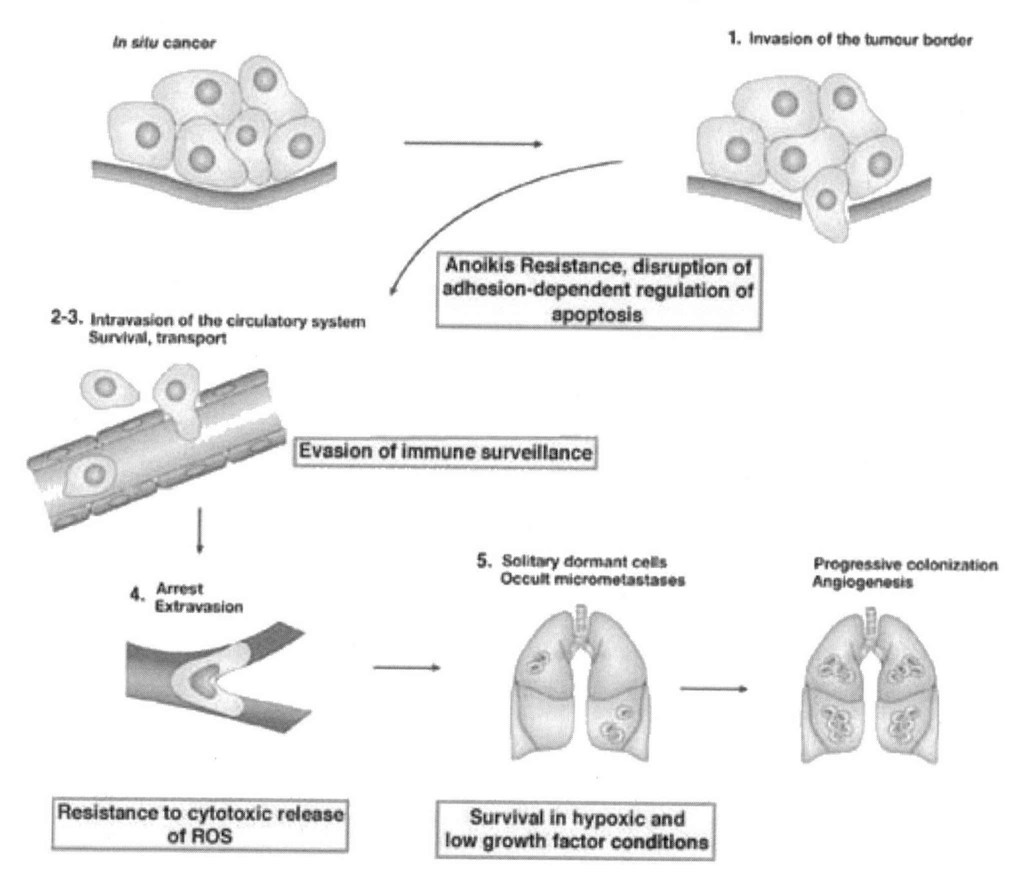

Fig. 1. Metastasis formation through evasion of apoptosis. The stages of metastasis (1–5) are shown: (1) Destruction and invasion of the basement membrane by *in situ* cancer through changes in cell–cell and cell–ECM interaction, and initiation of motility. (2–3) Intravasation, survival and transport of the metastatic cells via the lymphatic or the circulatory system. (4) Arrest in capillary bed and extravasation into the destined organ. (5) Proliferation of metastatic cells at the new location and colonization. Boxed text highlights apoptotic evasion events within each of these stages.

Metastasis is inherently an inefficient process *(2)*. The initial steps of metastasis are performed with very high efficiency, whereas the final stage—initiation of colonization and persistence—is considerably less efficient. Elegant in vivo microscopy studies following the fate of malignant cells through the metastatic process have established that most cells die during extravasation and colonization *(3)*. It is estimated that only 0.01 to 0.1% of malignant cells in circulation actually give rise to macrometastases *(3)*. This inefficiency is caused primarily by cell death as a result of the many stresses that are introduced before and after the cells reach the new environment. Such stresses include the loss of cell–cell contact, recognition and destruction of the tumor cells by the immune system, and lack of necessary growth factors, all of which trigger apoptosis *(4)*.

2. Apoptotic Resistance in Metastatic Progression

Apoptosis is a genetically determined process by which a cell self-destructs and commits suicide in order to maintain cellular homeostasis. It is hypothesized to block metastatic dissemination by eliminating misplaced cells. The process of apoptosis is the primary contributor to metastatic inefficiency. The success of the metastatic process relies on the ability of malignant cells to evade apoptosis. A strong correlation between the metastatic aggressiveness of cell lines and resistance to apoptosis has been established from countless studies. In addition, there is evidence correlating disease prognosis and the apoptotic index, a measure of the amount of programmed cell death within tumor tissue specimens *(5–11)*.

Acquired apoptotic resistance is critical at several steps in the metastatic cascade, but perhaps the most important stage is resistance to cell death induced by loss of cell–cell and cell–ECM contact. Matrix-independent survival of metastatic cancer cells before intravasation and during the passage through blood and/or lymph compartments is an essential component of the metastatic cascade. Detachment of cells from the ECM results in a type of programmed cell death referred to as anoikis. Some of the earliest evidence of anoikis resistance in metastatic cells was reported by Glinsky et al., who showed that higher survival of metastatic cancer cells compared with nonmetastatic or poorly metastatic cells was associated with the loss of homotypic and heterotypic aggregation and disruption of adhesion-dependent regulation of apoptosis *(12)*. Since then, there have been numerous reports of anoikis resistance in metastatic cell lines and lesions. The molecular basis of this resistance will be discussed later in this chapter.

En route to metastatic colonization, malignant cells must also evade immune surveillance in circulation and at the site of colonization. They must develop a mechanism to escape cell death induced by cytotoxic lymphocytes and natural killer (NK) cells that recognize and destroy misplaced cells. Further, malignant cells must escape cytotoxicity induced by reactive oxygen species (ROS) release from endothelial cells while crossing the blood/tissue barrier during the process of extravasation *(13)*. Finally, malignant cells must be able to successfully colonize at the secondary site. They must conquer hypoxic conditions and proliferate in an environment lacking the necessary growth factors.

It has been shown that even after successful formation of micrometastases, macroscopic tumors may not develop because of dormancy *(3)*. Folkman and colleagues first described metastatic dormancy as the process in which tumor cells are metabolically active, but the rate of proliferation is balanced by the rate of apoptosis such that no net tumor growth occurs *(14)*. A second type of dormancy involves significant numbers of nondividing solitary tumor cells that persist for long periods of time. These cells do not proliferate, as measured by retention of fluorescent nanospheres (these would be diluted upon cell division), nor do they undergo apoptosis *(15)*. The nature of dormancy remains elusive, but genetic and/or environmental influences may shunt cells out of dormancy into an antiapoptotic, highly proliferative state allowing metastatic colonization to ensue.

As we shall see in this chapter, the process of apoptotic evasion is a genetically and molecularly regulated process. It not only creates a permissive environment for metastasis, but it also allows tumor cells to withstand the action of therapy *(16)*.

3. Metastasis Assays

To discern the effects of apoptotic mediators on metastasis, an understanding of standard metastasis assays is in order. Measurements of metastasis discussed in this chapter range from in vitro and in vivo metastasis assays to clinical correlates in human tumor cohorts. Hallmarks of metastasis may be measured by cell culture-based assays such as soft agar colonization, wound scratch assays, and Boyden chamber chemotaxis assays, which assess anchorage-independent growth, motility, and invasion. The process of metastasis can only truly be assayed in vivo with mouse modeling experiments. There are genetically engineered models of metastasis (transgenics) as well as transplantable murine models—syngeneic (host and tumor cells derived from similar genetic background) and human–mouse xenograft (host must be immunocompromised to prevent tissue rejection) (reviewed in ref. *17*). Both syngeneic and xenograft models can be employed in either of the two classical metastasis assays—experimental and spontaneous. In the experimental metastasis assay, tumor cells are directly injected into the circulation, commonly via the tail vein, portal vein, or arterial circulation (intracardiac). A significant outcome of the experimental metastasis assay has been generation of clonal variants with differing metastatic potential, resulting from successive injections and explants *(17)*. However, in the experimental metastasis assay, early steps of the metastatic process, invasion and intravasation, are bypassed. In the spontaneous metastasis assay, orthotopic transplantation (injection of tumor cells into the tissue from which the tumor was derived) can lead to spontaneous metastasis to distant sites, which may more accurately reflect the metastatic process in its entirety. Strategies to image in vivo metastases typically include bioluminescence (luciferase or green fluorescent protein-labeled tumor cells), magnetic resonance imaging, or positron enhanced tomography *(17)*.

Human tumor cohort studies are by far the most relevant to human disease, but provide merely indirect, correlative information. Immunohistochemistry (IHC), reverse-transcription (RT)-PCR, and immunoblotting analysis are employed to measure gene and protein expression in tumor cohorts. This expression data can then be correlated to a number of clinicopathological factors including tumor grade, stage, survival, and risk for local or distant metastases. The significance of these associations thus dictates whether the protein of interest is a reliable prognostic indicator.

4. Molecular Mechanisms of Apoptotic Resistance

Tumor cells go through multiple molecular and genetic changes en route to achieving a metastatic state. Colonization at a secondary site requires the acquisition of an antiapoptotic, prosurvival phenotype. Apoptotic resistance may be achieved through one or more of the following general mechanisms:

1. Disruption of caspase function by mutation of caspase genes, altered expression of inhibitors of apoptosis (IAPs) and/or death-associated protein kinases (DAP kinases).
2. Altered activity and/or expression of components of the intrinsic/mitochondrial pathway of apoptosis, which includes Bcl-2 family members, Apaf-1 and maspin.
3. Altered activity and/or expression of components of the extrinsic pathway of apoptosis, which includes the tumor necrosis factor (TNF)-α family of ligands and the death receptors.
4. Altered activity and/or expression of transcription factors, such as NF-κB, and p53, signaling pathways, such as the TGF-β, Akt, and stress-activated protein kinase (SAPK) cascade, and focal adhesion complex molecules, all of which ultimately modulate components of the intrinsic and extrinsic apoptotic pathways.

In this section, we will describe these mechanisms of apoptotic resistance and will define their significance in metastatic progression. The specific molecular pathways that contribute to these processes are schematically represented in Fig. 2.

4.1. Disruption of Caspase Function in Metastasis

Both the intrinsic and extrinsic apoptotic pathways lead to the activation of caspases, a family of cysteine proteases. Therefore, altering the activity of these genes

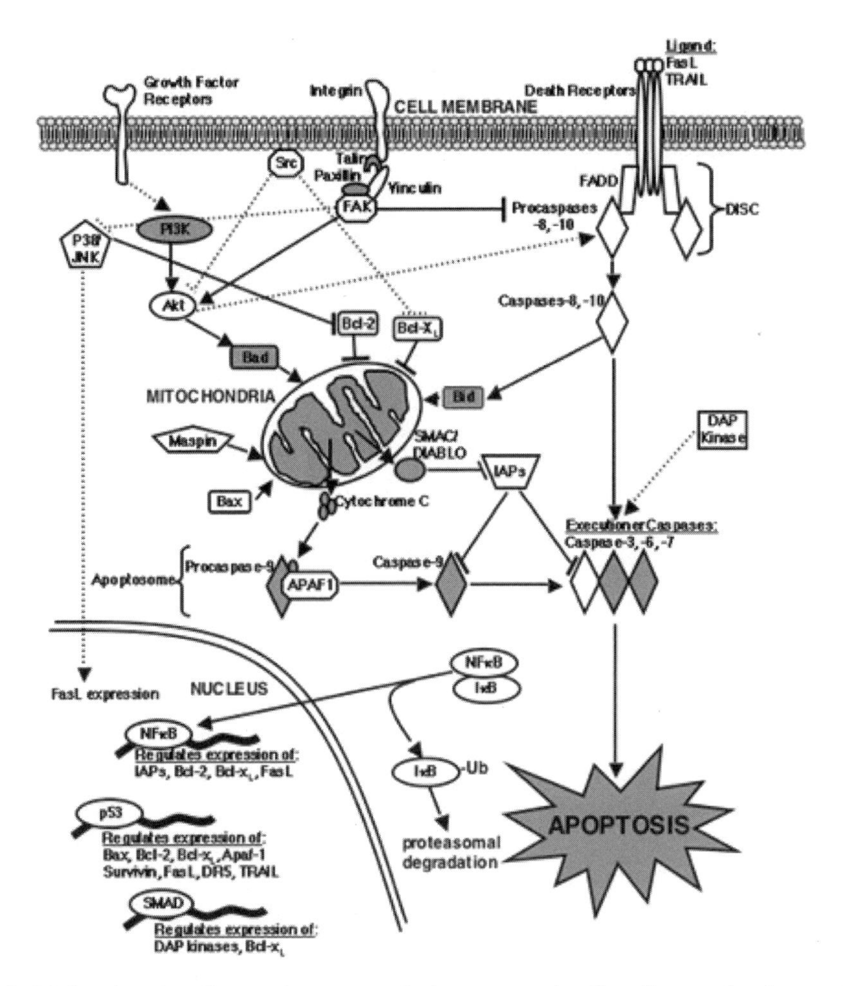

Fig. 2. Molecular signaling pathways regulating apoptosis. Signaling molecules are labeled and represented by shaded and unshaded shapes. An unshaded shape indicates that altered expression or activation of the protein is involved in metastatic spread, whereas a shaded shape indicates that there is no evidence for the protein's role in metastasis as of yet. Solid lines represent the positive (lines ending in arrowheads) or negative (lines ending in perpendicular line) regulatory interaction between proteins. Dashed lines represent a putative and/or indirect relationship between signaling molecules.

could significantly contribute to the process of apoptotic evasion. Caspase function may be disrupted in metastasis by mutation or diminished expression of caspase genes, overexpression of the IAPs, or reduced expression of DAP kinases.

4.1.1. Mutation and Altered Expression of Caspase Genes

The caspase family can be subdivided into two classes, the initiators and the effectors. The initiators include caspase-9, a key element of the protein complex central to the intrinsic/mitochondrial pathway of apoptosis called the apoptosome (Fig. 2), and the highly homologous caspases-8 and -10, components of the death-induced signal complex (DISC) complex downstream of the death receptors (Fig. 2) described later in this section. These initiators cleave and activate the primary effectors, caspases-3, -6, and -7.

As both apoptotic pathways converge on caspases, altered expression and/or activation of these proteases could impact signaling through both of these pathways. Caspases, however, are highly redundant, therefore disruption of a single caspase may be compensated for by other family members.

Mutation and altered expression of caspases have been associated with various forms of cancer, but most of these instances correlated with early events in tumor formation. Nonetheless, there are a few examples of caspase expression and mutation correlating with metastatic progression. In particular, apoptotic resistance of medulloblastoma cell lines may be attributed to loss of *caspase-8* and/or *caspase-10* mRNA expression resulting from aberrant promoter methylation *(18,19)*. Analysis of medulloblastoma specimens also revealed that reduced *caspase-8* expression is a significant prognostic indicator of progression-free survival *(20)*. *Caspase-10* mutations, which render the protein functionally defective, have been identified specifically in non-small-cell lung cancers (NSCLC) with lymph node metastases, when compared with NSCLC without metastases *(21)*.

There is conflicting evidence for the role of caspase-3 in metastasis. Two retrospective studies comparing caspase-3 expression with clinical outcome of NSCLC indicated that caspase-3 protein levels, negatively correlated with lymph node metastasis *(22,23)*. No correlation was found between caspase-3 transcript levels, apoptotic index, and tumor progression of hepatocellular carcinomas (HCC) *(24)*, however; for gastric cancer, the data is conflicting. One group found that lower caspase-3 protein levels correlated with lymph node metastasis of gastric lymphoma *(25)*. Another group found the opposite relationship—positive caspase-3 expression correlated with lymph node status—but they also showed that most of the caspase-3 protein in gastric carcinomas was not activated *(26)*. Although caspases are clearly critical in manifesting apoptosis, most of the data linking reduced expression and mutation of caspases to apoptotic resistance and metastatic progression is indirect and correlative. Therefore, a causal relationship is speculative at best.

4.1.2. Altered Expression of Caspase Inhibitors of Apoptosis

IAPs are a family of caspase inhibitors that target both the intrinsic and extrinsic apoptotic pathways through specific inhibition of caspases-3, -7, and -9 (Fig. 2) *(27)*. Eight members of the IAP family have been identified, including XIAP, cIAP1, cIAP2, and survivin. The best-characterized IAP, survivin, is overexpressed in malignancies of the lung, prostate, breast, colon, esophagus, pancreas, stomach, and central nervous system (CNS) *(27–31)*. *Survivin* expression was significantly higher in prostate cancers with lymph node metastases *(32)*; however, there was no significant relationship between survivin expression in colorectal cancer and risk for lymph node or distant metastases *(30)*. In fact, decreased *survivin* expression was found in hepatic metastases in comparison with primary colon tumors *(33)*. Thus, although a clear link between elevated *survivin* expression and transformation has been established, the contribution of survivin to metastasis is undefined.

4.1.3. Loss of Death-Associated Protein Kinase Expression is Related to Apoptotic Resistance

The DAP kinases are a family of calcium/calmodulin, serine/threonine kinases, which contain death-domains and are associated with the cytoskeleton (reviewed in refs. *34,35*). The founding member of the family was identified in a functional genomics screen, which involved selection for genes that confer resistance to interferon (IFN)-γ-stimulated apoptosis when expression was suppressed by antisense cDNA *(36)*. DAP kinases promote apoptosis by acting downstream of the initiator caspase-8, but upstream of the effector caspases (Fig. 2) *(37)*. Therefore DAP kinases, like IAPs, have the potential to affect both the extrinsic and intrinsic pathways of apoptosis. The best evidence for the role of DAP kinase in metastatic progression comes from Lewis lung carcinoma mouse model experiments *(36)*. Loss of DAP kinase expression in murine Lewis lung cancer cells corresponded to enhanced metastatic behavior compared with syngeneic cells that express DAP kinase. When DAP kinase expression was restored in these highly aggressive lung cancer cells, their metastatic capacity was reduced in mice. The presumed cause of this reduced metastatic capacity was enhanced apoptosis of cancer cells as indicated by diminished *in situ* TdT-dUTP terminal nick-end labeling (TUNEL), a measure of DNA cleavage, of tumor sections *(36)*.

Loss of DAP kinase expression also correlates with patient prognosis according to clinical data. IHC analysis of DAP kinase protein in 128 breast cancer specimens indicated that higher DAP kinase expression correlated with patient survival *(38)*. Loss of DAP kinase expression may be linked to hypermethylation of its promoter *(39)*. In NSCLC, methylation of the DAP kinase promoter was associated with an increase in patient tumor size and lymph node involvement *(40)*. Furthermore, methylation-specific PCR was used to analyze ten brain metastases derived from different types of cancer. In nine of ten metastases, the DAP kinase promoter was hypermethylated, whereas normal brain tissue samples did not exhibit such hypermethylation *(41)*. Although the DAP kinase family has only recently been identified, these studies suggest that it may be an important therapeutic target for conquering metastasis in the future.

4.2. Altered Signaling Through the Intrinsic Apoptotic Pathway in Metastasis

The intrinsic or mitochondrial apoptotic pathway relies on formation of a protein complex dubbed the apoptosome, composed of caspase-9, Apaf-1 and mitochondrial-derived cytochrome *c* (Fig. 2). Bcl-2 family members, as well as the serine protease inhibitor maspin, regulate cytochrome *c* release. Several components of this intrinsic pathway are differentially expressed in metastatic lesions.

4.2.1. Loss of Apaf-1 Expression in Metastasis

The reduced expression of Apaf-1 (apoptosis protease activating factor) has been linked to apoptotic resistance and metastatic spread of melanoma. Its down-regulation has been observed in metastatic melanoma cell lines compared with lines derived from primary melanoma tumors *(42)*. Several metastatic melanoma cell lines exhibited *Apaf-1* promoter hypermethylation and epigenetic silencing of the gene *(43)*. Loss of Apaf-1 protein in tumor tissue samples also appeared to correlate with metastases of melanoma *(42)*. Recently, it has been demonstrated that *Apaf-1* haploinsufficiency, and therefore its diminished expression, increased with colorectal carcinoma progression and was associated with liver metastasis *(44)*.

4.2.2. Altered Expression of the Bcl-2 Family

The Bcl-2 family consists of 25 pro- (Bax, Bak, Bid) and antiapoptotic (Bcl-2, Bcl-x_L) members that heterodimerize to form permeability pores in the mitochondrial membrane.

Proapoptotic family members promote cytochrome *c* release into the cytoplasm, prompting caspase activation and execution of the apoptotic program while antiapoptotic members prevent this cytochrome *c* release (Fig. 2) *(45)*. Dysregulation of both pro-and antiapoptotic Bcl-2 family members occurs during cellular transformation, and upregulation of *Bcl-2* is linked to metastatic progression in vitro.

4.2.2.1. Bcl-2

Examination of *Bcl-2* expression in a pancreatic tumor cell line and sublines of increasing metastatic potential revealed that progression from the low to highly metastatic state was accompanied by increasing levels of *Bcl-2 (46)*. Moreover, the metastatic variants developed resistance to apoptosis induced by a variety of biochemical stimuli. Furthermore, *Bcl-2* transfected parental cell lines produced a higher incidence of hepatic metastases in vivo as compared with the low *Bcl-2*-expressing parental cell line *(46)*. Bcl-2 levels also correlated with prostate cancer progression. Relative to a slowly growing parental Dunning R-3327 rat prostate cancer cell line, rapidly proliferating, androgen-independent variants expressed higher levels of Bcl-2 protein *(47)*. *Bcl-2* overexpressing murine melanoma cells experienced a 40% reduction in interleukin converting enzyme (ICE) and caspase protease activities and were more capable of surviving limited dilution cloning, indicating a decreased requirement for cell–cell contact *(48)*. Related findings were reported in *Bcl-2*-transfected mitogen activated protein kinase kinase (MEK)-transformed mammary epithelial cells (EpH4) that were treated with the actin depolymerizing agent latrunculin A, which induces apoptosis *(49)*. *Bcl-2* overexpression rescued parental cells from latrunculin-induced apoptosis, but was insufficient to enhance in vitro migration or invasion, hallmarks of metastasis. However, in mice inoculated with *Bcl-2* transfectants, the pulmonary metastatic burden was dramatically augmented, despite no differences in the kinetics of primary tumor development *(49)*. These studies suggest that there are distinct mechanisms driving primary tumor formation and metastasis and that Bcl-2-mediated inhibition of apoptosis is one such pathway.

It is hypothesized that Bcl-2 dysregulation and inhibition of apoptosis may expose breast tumor cells to further genetic damage, leading to accumulation of genetic alterations that increase the risk of metastases *(50)*. In an analysis of T_1 primary breast tumors, intense Bcl-2 staining predicted ninefold excess odds for lymph node metastasis when compared with Bcl-2 negative tumors *(50)*. Bcl-2 expression was inversely correlated to histological grade and Bcl-2 positive tumors had a lower apoptotic index, supporting a relationship between high Bcl-2 expression, resistance to apoptosis, and concomitant metastasis. In direct contrast, Le et al. *(51)* reported that Bcl-2 expression was associated with a low risk of distant metastases of breast cancer. Confounding issues were discrepancies in staining methods, duration of follow-up, and consideration of overall relapse versus specific types of relapse *(51)*. Thus, the role of Bcl-2 expression in breast cancer metastasis has yet to be clearly defined.

High Bcl-2 levels are associated with prostate cancer progression. By tissue microarray analysis, significant Bcl-2 overexpression was found in hormone refractory prostate cancer and prostate cancer metastases in comparison with localized prostate cancer *(52)*. Likewise, lymph node and bone metastases from prostate cancer exhibited statistically significant increases in Bcl-2 staining compared with prostatic intraepithelial neoplasia and localized prostate cancer *(47)*.

4.2.2.2. BCL-X_L

Another member of the Bcl-2 family with an emerging role in metastatic progression is Bcl-x_L. *Bcl-x_L* overexpression and apoptotic resistance are often linked to *Bcl-2* expression levels, but it is evident that these proteins have nonredundant functions in apoptosis and metastasis. For example, only *Bcl-x_L* has been linked to amorphosis (programmed cell death that occurs upon disruption of the actin cytoskeleton) resistance. In a cytoskeletal-based functional screen to identify genes bestowing resistance to amorphosis, the *Bcl-x_L* coding sequence alone conferred amorphosis resistance to sensitive cells *(53)*.

Numerous in vitro and in vivo transfection studies confirm the significance of *Bcl-x_L* in tumor metastasis and highlight Bcl-x_L's role in several distinct aspects of metastasis. *Bcl-x_L*-overexpressing breast carcinoma cells were less adherent to laminin, fibronectin, and collagen IV than vector-transfected controls *(16,54)*. Cell viability in suspension, a measurement of anoikis sensitivity, was drastically enhanced in *Bcl-x_L* overexpressing MDA-MB-435 cells *(54)*. Proficient survival in suspension also translates to improved viability in the circulation. In a competition assay in which a mix of control and *Bcl-x_L*-transfected MDA-MB-435 cells were injected into the tail vein of nude mice, there was marked enrichment of viable *Bcl-x_L* overexpressing cells in the vasculature over a period of 4 h *(54)*. MDA-MB-435 cells overexpressing *Bcl-x_L* were also several-fold more resistant to transforming growth factor (TGF)-β- and TNF-α induced apoptosis than control transfectants, and this phenotype was abrogated by *Bcl-x_L* antisense *(55)*.

Intramammary fat pad implantation of *Bcl-xL* transfected MDA-MB-435 cells resulted in a discernable rise in the incidence and size of lung metastases *(16)*. Furthermore, lymph node metastases were present in mice implanted with *Bcl-x_L* overexpressing MDA-MB-435 cells, but not in control mice *(54)*. Quite striking is the demonstration that breast cancer cells overexpressing Bcl-x_L exhibited heightened genomic instability as measured by genomic fingerprinting. This genetic instability sequentially increased from the *Bcl-x_L*-overexpressing primary tumor to the metastases found in bone, lymph nodes, and lungs of nude mice *(56)*. Moreover, *Bcl-x_L* overexpression in primary breast cancer has been associated with high tumor grade and increased risk for nodal metastases *(57)*. Together, these studies suggest that dysregulation of the antiapoptotic molecule Bcl-x_L has consequences on the genomic level that can affect the apoptotic and cellular processes contributing to metastatic tumor cell survival.

4.2.2.3. BAX

Bax is a proapoptotic Bcl-2 family member that translocates to the mitochondria and promotes the release of cytochrome *c* and Smac/Diablo, a negative regulator of IAPs (Fig. 2). It has been proposed that high expression of *Bax* results in a greater propensity for tumor cells to undergo apoptosis, thereby curbing metastasis and providing a better prognosis for the patient.

Acquisition of a prosurvival phenotype characterized by low-*Bax* expression accompanies tumor cells in their progression to metastasis. Metastatic variants derived from lymph node metastases of prostate cancer were more resistant than parental or non-metastatic cells to pharmacologically induced apoptosis, and displayed delayed kinetics of the classic hallmarks of apoptosis, DNA fragmentation, and PARP cleavage *(58)*. Notably, the highly metastatic variants expressed low-steady-state *Bax* levels when compared with the parental and non-metastatic variants.

Given the above in vitro data, it is not surprising then that several studies predicted Bax to be a useful clinical prognostic indicator. IHC performed in 119 primary tumors from women with metastatic breast adenocarcinoma revealed heterogeneous Bax negativity compared with the almost uniform Bax positivity found in non-neoplastic epithelium and carcinoma *in situ* *(59)*. Patients with Bax negative tumors had significantly shorter time to progression (6.6 mo compared with 10.6 mo in Bax positive patients) and reduced overall survival (8.1 mo vs 15.7 mo). In patients with colorectal carcinoma, Bax expression was significantly correlated with less lymph vessel invasion, less depth of invasion, and better prognosis (5-yr survival rate) *(60)*. IHC of primary tumors from patients with oral and oropharyngeal squamous cell carcinoma indicated that Bax positivity was greater in the lower disease stages T_1/T_2 than in the more advanced T_3/T_4 stages *(61)*. Even more convincing, Bax positivity was significantly elevated in patients without lymph node metastases than in patients with lymph node metastases, and these Bax-positive patients experienced improved overall 5-yr survival *(61)*.

Expression of other Bcl-2 family members in conjunction with Bax can also be valuable potential prognostic markers. Together, low Bax and high Bcl-2 staining were associated with low apoptotic index, high tumor grade, axillary lymph node involvement, and metastasis in patients with primary breast carcinoma *(62)*. Whether these analyses will result in the development of Bax-directed therapies for metastatic cancer remains to be seen.

4.2.3. Maspin

The serpin (serine protease inhibitor) maspin has garnered increasing interest as a molecular marker and potential therapeutic target in several types of cancer, but most especially in breast cancer (reviewed in ref. *63*). Mounting evidence suggests that maspin sensitizes cells to apoptosis via the mitochondrial pathway. Overexpression of *maspin* increased *Bax* and activated Caspases-8 and -9 in breast and prostate cell lines *(64,65)*. Conversely, *Bax* and *Caspase-8* and *-9* repression eliminated sensitization to apoptosis-inducing drugs via maspin in these cell lines *(65)*. Additionally, *Bcl-2* overexpression protected endothelial cells from maspin-mediated apoptosis in vitro *(66)*. A recent report suggested that maspin regulation of mitochondrial apoptosis might be direct. Latha et al. showed that maspin translocates to the mitochondrial membrane, similar to Bcl-2 family members, altering the mitochondrial membrane potential and leading directly to cytochrome *c* release *(67)*. Thus far, maspin is the only serpin shown to induce apoptosis and actually, its extracellular serine protease inhibitor activity is not involved in its apoptosis-promoting activity *(64)*.

Several mouse models implicate maspin-mediated apoptosis in metastatic progression *(68)*. Transgenic mice expressing *maspin* in mammary epithelial cells were crossed to an oncogenic SV40 T mouse strain, which always develop mammary tumors. Progeny showed reduced primary tumor growth and incidence of pulmonary metastases. The observed reduction can be attributed to enhanced apoptosis of cancerous mammary epithelial cells, at least in part (decreased angiogenesis and motility of malignant cells may also contribute) *(69)*. A xenograft model using the highly metastatic human breast cancer cell line TM40D transfected with *maspin*, again showed that *maspin*-expressing tumor cells had reduced primary tumor growth and were incapable of supporting pulmonary metastasis formation *(70)*. Finally, polyoma middle T oncogenic mice were allowed to develop primary mammary tumors and were then administered maspin protein via liposome carriers. Primary tumor growth and lung metastases were inhibited in maspin-treated mice *(71)*.

In clinical correlates, maspin expression has also been linked to cancer aggressiveness. Downregulation of maspin has been reported in invasive and metastatic breast carcinoma samples. IHC analysis of a large cohort of breast cancer tissue specimens revealed a stepwise diminuition of maspin expression as tumors progress from *in situ* to invasive to lymph node metastasis-positive *(72,73)*. Maspin expression was also negatively associated with invasive depth and metastasis in gastric carcinomas *(74)*.

Several lines of evidence suggest that the diminished *maspin* expression observed in metastatic lesions may be attributed to hypermethylayion of its promoter. Breast and gastric cancer cell lines treated with the demethylating agent 5-aza-2′-deoxycytidine (5-aza C) had enhanced *maspin* expression *(75–77)*. Novel microarray and chromatin immunoprecipitation techniques provided direct evidence for *maspin* promoter methylation in pancreatic cancer *(78,79)*. There is a growing body of evidence for the role of maspin in the intrinsic pathway of apoptosis and in metastasis suppression, suggesting that targeted re-expression with of this serpin with demethylating agents may be a logical approach for the treatment of apoptotic resistance of metastases.

4.3. Alteration of the Extrinsic Apoptotic Pathway in Metastasis

The extrinsic pathway of apoptosis is mediated by death receptors of the TNF-α receptor family. This pathway is activated by ligand binding to one of the death receptors, followed by formation of the DISC, which is composed of Fas-associated death domain containing protein (FADD), and caspase-8 or -10. DISC then either signals downstream effector caspases (3, 6, and 7) leading directly to cell demise or signals through the Bcl-2 family member Bid and the intrinsic, mitochondrial pathway of cellular apoptosis (Fig. 2). Two ligands of the TNF-α family have been implicated in progression of cancer cells toward a metastatic phenotype—FasL and TNF-related apoptosis-inducing ligand (TRAIL).

Resistance to FasL and TRAIL-mediated apoptosis can be conferred by multiple mechanisms. The mechanisms specific to these ligands will be explained below, but disruption of the DISC complex can potentiate both FasL and TRAIL resistance. Somatic mutations in *FADD* that functionally disrupt the protein have been observed at a higher frequency in metastatic NSCLC compared with primary tumors *(21)*. *Caspase-8* and *-10* mutations have been implicated in TRAIL resistance in medulloblastoma and NSCLC *(18–21)*. Some general apoptotic and survival pathways that have previously been discussed or will be discussed later in this chapter can mediate FasL and TRAIL resistance, as well. The transcription factor p53, which is mutated in 50% of cancers, regulates expression of both *TRAIL* and *FasL*, as well as the Fas receptor *(80)*. Several lines of evidence also suggest that mutation of the Bcl-2 family member *Bax*, which occurs frequently in mismatched-repair deficient tumors, also contributes to TRAIL-induced apoptotic resistance *(81,82)*. Other signaling pathways such as TGF-β, protein kinase C (PKC), Akt cell survival, and mitogen-activated protein kinase (MAPK) pathways affect sensitivity of cancer cells to TRAIL and FasL induced apoptosis *(83–86)*.

4.3.1. Fas/FasL-Specific Mechanisms of Apoptotic Resistance

The death receptor ligand FasL is normally expressed by cytoxic T lymphocytes and NK cells and is involved in immunosurveillance of tumor cells that are expressing the Fas receptor. Fas/FasL interactions have been shown to be involved in suppression of melanoma metastasis *(87)*. *FasL*-deficient mice had a higher incidence and number of spontaneous metastases compared with wild-type syngeneic mice injected with K1735 murine melanoma cells.

Whereas Fas/FasL signaling is important in suppressing metastasis, metastatic tumor cells are frequently resistant to Fas-induced apoptosis. Hedlund et al. studied seven prostate cancer cell lines for sensitivity to Fas-mediated apoptosis. Cell lines derived from primary tumors were more sensitive to apoptosis and the four most resistant lines were derived from metastatic lesions *(88)*. Similarly, Fas resistance has been observed in pancreatic ductal carcinoma cell lines and osteosarcoma cell lines with differing metastatic potentials *(89,90)*.

Multiple mechanisms have been implicated in FasL-resistance of the metastatic cell, including reduced expression of the Fas receptor. IHC, and RT-PCR analysis of colorectal cancer specimens revealed that Fas expression was reduced in colorectal adenomas, colorectal carcinomas, and liver metastases compared with matched primary colorectal carcinomas *(91)*. Similarly, Wang et al. found that Fas levels were significantly lower in metastatic liver lesions compared with matched primary colorectal tumor specimens *(92)*. In HCCs, Fas levels correlated with disease-free survival. Fas-positive HCCs exhibited reduced incidence of intrahepatic metastatic foci and enhanced incidence of apoptotic tumor cells *(93)*. Furthermore, poor patient prognosis has been correlated with Fas:FasL ratios in hepatic and ovarian carcinomas *(94,95)*.

Another mechanism of Fas-mediated apoptosis resistance is enhanced expression of soluble Fas (sFas). The sFas receptor is released extracellularly and will compete for binding to FasL expressed by cytotoxic T cells and NK cells, thus blocking ligand engagement of Fas-expressing tumor cells. One study indicated that it is not tumor cells that have altered their sFas levels, but it is the surrounding noncancerous hepatocytes and mononuclear liver cells that express sFas to permit metastatic growth in the liver *(93)*. Circulating sFas concentration in breast cancer patients' serum was measured by enzyme-linked immunosorbant assay (ELISA). High-sFas patients led to poorer prognosis for both overall and disease-free survival *(96)*. Similarly, another study found sFas plasma levels to be a significant independent predictor of poor overall survival of metastatic breast cancer *(97)*. In addition, serum levels of sFas were higher in patients with advanced melanoma that were refractory to chemotherapy *(98)*. Expression of the FasL decoy receptor DRc3 similarly affects FasL-initiated apoptosis by antagonizing ligand binding to Fas. Wu et al. determined that DcR3 positively correlated with malignancy in tumor patients and is associated with tumor differentiation, lymph node, and metastatic status *(99)*.

In perhaps the most aggressive mechanism of tumor self-defense, metastatic cells have acquired a way to use the host's Fas/FasL immune defense against it in what has been termed the "tumor counterattack phenomenom." Tumor cells can express FasL on both their cell surfaces and in a soluble, extracellular form, which can then induce programmed cell death in neighboring tumor-infiltrating lymphocytes (TILs). High FasL expression levels are associated with lymph node metastasis grade in breast carcinoma *(100)*. Mann et al. provided evidence for FasL involvement in colorectal carcinoma evasion of TILs *(101)*. They determined FasL expression and CD25 levels (a measure of lymphocyte activation) in matched primary tumor and liver metastases from 21 colorectal carcinoma patients. FasL was detected in 30 to 40% of tumors, but in 60 to 100% of metastases. The level of lymphocyte activation was lower in 90% of the metastases *(101)*. As further evidence of the FasL-mediated mechanism of immune evasion in colorectal cancer, Okada et al. found that the *in situ* nick translation labeling index (a measure of apoptosis) of TILs from 41 colorectal patients was higher in colorectal carcinomas with lymph node metastases versus those without metastases *(102)*. In addition, increased apoptosis of Fas-positive TILs was observed in metastatic gastric carcinoma *(103)*.

4.3.2. TRAIL Specific Mechanisms of Apoptotic Resistance

TRAIL is highly homologous to FasL, but unlike FasL, it is more widely expressed in a variety of tissues. TRAIL has five known receptors, two of which are complete with death domains (DR4 and DR5), whereas the other three are decoy receptors missing functional death domains (DcR1, DcR2 and OPG). TRAIL has garnered much interest as a potential therapeutic cancer target as tumor cells are more susceptible to TRAIL-induce cell death than normal epithelial cells, presumably because tumor cells express DR4 and DR5 at higher levels. Normal cells may express higher levels of the decoy receptors, thus protecting them from TRAIL-mediated apoptosis *(104)*.

Although its function is not completely understood, TRAIL expression on liver NK cells appears to play a role in tumor cell surveillance. Mice depleted of NK cells or treated with a TRAIL-neutralizing antibody showed a significant increase in spontaneous liver metastases *(105)*. Cretney et al. developed a mouse model of *TRAIL* deficiency *(106)*. *TRAIL* gene-targeted mice have an absence of *TRAIL* expression in liver and spleen mononuclear cells and an absence of TRAIL cytotoxicity. These mice were more prone to spontaneous metastases, showed decreased response to the antimetastatic effects of therapeutic doses of interleukin-12, and α-galactosylceramide, and were more sensitive to the carcinogen methylcholanthrene *(106)*. In ovarian cancer tissue specimens, *TRAIL* expression correlated with patient survival *(107)*. As we shall see in the latter part of this chapter, *TRAIL* replacement by gene therapy in mice can effectively suppress tumor growth and experimental metastases *(108–110)*.

TRAIL is clearly an important and an effective molecule for eliminating cancerous, even metastatic cells; however, TRAIL resistance has been detected. Several melanoma cell lines exhibit resistance to TRAIL and interestingly, primary oral cancer cell lines are more susceptible to TRAIL-induced apoptosis than their metastatic counterparts *(111,112)*. There are several mechanisms of apoptotic resistance that are unique to the TRAIL pathway. Mutation of TRAIL receptors *DR4* and *DR5* and a higher incidence of allelic loss of the TRAIL receptors have been detected in breast cancers with metastases. Transfection of the mutated receptors found in these breast tumors suppressed TRAIL-induced apoptosis *(113)*. Loss of receptor expression has also been associated with TRAIL resistance in some melanoma cell lines *(111)*.

Tumor cell expression of decoy receptors is an alternate mechanism of resistance to TRAIL apoptosis. Zhang et al. found that TRAIL-resistant melanoma cell lines express high levels of DcR1 and DcR2, which may antagonize TRAIL binding to the complete TRAIL receptors *(111)*. DcR2-positive metastatic gastric carcinoma cells were also identified in patient ascites. These cells were resistant to TRAIL-induced apoptosis *(114)*. Osteoprotegerin (OPG), the other decoy receptor, is expressed in bone stroma cells of prostate cancer patients with detectable bone metastases. A prostate cancer cell line treated with media conditioned by the patient derived bone stroma exhibited reduced sensitivity to TRAIL-induced apoptosis, suggesting that expression of OPG by bone stroma may create a permissive environment for bone metastasis colonization *(115)*.

4.4. Transcriptional Regulators of Apoptotic Resistance in Metastasis

Whereas apoptotic mediators contribute to tumor aggressiveness and metastasis by virtue of their direct effects on the intrinsic and extrinsic death pathways, transcription factors are indirect but pivotal metastatic regulators through their pleiotropic effects on gene regulation (Fig. 2). Modulation of genes involved in apoptosis, together with genes associated with invasion, angiogenesis, proliferation, and survival, identify transcription factors as attractive targets for therapeutic intervention. Several such factors are outlined below.

4.4.1. NF-κB

In its inactive state, NF-κB is bound in the cytoplasm by IκB. IκB phosphorylation leads to its ubiquitination and proteasomal degradation, thereby releasing free, activated NF-κB. Activated NF-κB then translocates to the nucleus where it promotes transcription of a myriad of antiapoptotic genes including *Bcl-2*, *Bcl-*x_L, *Traf-1*, *Traf-2*, *survivin*, *cIAP-1*, *cIAP-2*, *c-FLIP*, *FasL*, and *p21*, as well as genes implicated in angiogenesis and invasion such as *VEGF*, *iNOS*, *MMP2*, *MMP9*, and *uPA* *(116,117)* (Fig. 2).

As cells progress to a metastatic state, there is constitutive activation of NF-κB and target genes, resulting in an apoptotic resistant phenotype. Microarray analysis of primary murine keratinocytes, transformed PAM212 cells, and lymph- or lung-metastatic sublines indicated that approximately half of the genes differentially enriched in the metastatic variants represented NF-κB-related genes—either genes implicated in NF-κB regulation (ubiquitin activating enzyme E1 which degrades IκB), or genes that are NF-κB targets (*cIAP1*, *FasL*) *(118)*. Similarly, enhanced activation of the NF-κB pathway was found in mammary cells undergoing an in vitro correlate of breast cancer progression, the epithelial–mesenchymal transition (EMT). In a comparison between transformed mammary epithelial cells (EpRas) and those that undergo EMT in response to TGF-β (EpRasXT), enriched expression of NF-κB target genes was demonstrated in EpRasXT cells *(119)*.

In human breast tumors, there is a substantial and striking correlation between constitutive NF-κB activation and progression to a more aggressive state, characterized by hormone-independent growth. NF-κB was constitutively activated in 50% of primary breast tumors *(120)*. Furthermore, those tumors with enhanced NF-κB binding activity were estrogen receptor (ER) negative. In an independent analysis, high NF-κB activity was observed in ER^-, $ErbB2^+$ human breast tumors. Blockade of NF-κB signaling, and addition of heregulin to the ER^-, $ErbB2^+$ SKBR3 cell line dramatically increased apoptosis *(121)*. The connection between constitutive NF-κB activity and the poorly differentiated, highly metastatic ER^-, $ErbB2^+$ phenotype emphasizes the significance of NF-κB signaling in the pathway to aggressive disease. These studies also highlight NF-κB as a prospective molecular candidate for inhibition.

Suppression of NF-κB activity can inhibit metastasis without affecting tumorigencity in vivo. In an orthotopic nude mouse model, all of the mice injected with either the murine pancreatic cancer cell line AsPc-1 or stable transfectants of this cell line expressing a hyperstable IκB mutant developed pancreatic tumors *(122)*. In the parental AsPc-1-tumor bearing mice, 8/11 mice developed liver metastases and ascites, and all mice had peritoneal metastases. Remarkably, only one-tenth of AsPc-1/IκB-mutant bearing mice developed peritoneal metastases and ascites, and none of these mice developed hepatic metastases.

Pharmacological inhibition of NF-κB activity can be achieved through proteasome inhibitors that block IκB degradation and other compounds that inhibit IκB phosphorylation, such as BAY 11-7082 and BAY 11-7085. The specific 20S proteasome inhibitor PS-341 impeded NF-κB activation, slowed the proliferation of human small cell lung cancer cell lines in vitro and in BALB/c mice, and limited angiogenesis *(123)*. NF-κB inhibition renders cells susceptible to apoptosis during tumor cell readherence and implantation. The human colorectal cancer cell lines HT-29 and DLD-1 were transiently suspended and allowed to readhere in vitro while in the presence of BAY 11-7085 *(124)*. Augmentation of apoptosis occurred in the presence of drug, and an even more impressive reduction of the apoptotic threshold was found when cells were transduced with an IκB super repressor construct specifically targeting NF-κB. Moreover, BAY 11-7085 reduced colon cancer cell implantation in an in vivo model of metastasis *(124)*, highlighting the potential therapeutic utility of NF-κB inhibition.

4.4.2. p53

Another transcription factor fundamentally important to apoptosis and metastasis is the tumor suppressor p53, which regulates hundreds of genes through transactivation or repression. Activation of p53 can trigger a plethora of biological outcomes including cell-cycle arrest, apoptosis or anoikis, senescence, and differentiation. These outcomes are influenced by cell context, target gene selectivity, covalent posttranslational modifications of p53 itself, and activation of the proteins with which p53 interacts *(125)*. Proapoptotic genes positively regulated by p53 include members of the death receptor families *Fas* and *DR5* as well as mitochondrial associated proteins *Bax*, *Bak*, *Noxa*, *Puma*, *Pig3*, and *Apaf-1 (126,127)*. p53 also represses the promoters of *Bcl-2*, $Bclx_L$, and *survivin*, all antiapoptotic molecules (Fig. 2).

Although the contribution of p53 to apoptosis and cell-cycle control is apparent, the mechanisms whereby p53 affects metastatic growth are less understood. In ras^+ myc^+ initiated mouse prostate tumors, loss of p53 function lead to widespread metastases *(128)*. In mouse embryo fibroblasts transformed with E1A and ras (C8), p53 wild-type cells underwent apoptosis upon detachment from the substratum; however, those cells with dominant negative or null p53 were conferred a survival advantage *(129)*. In an elegant in vivo experiment, athymic nude mice were injected with C8 fibroblasts containing either wild-type or mutant p53, lungs were either isolated within 2 to 4 hs or after several wk. Cells isolated from the lungs 2 to 4 h following injection were predominantly p53 mutant, and accordingly, the majority of lung macrometastases carried mutant *p53* alleles. Thus, loss of p53-mediated apoptosis and improved tumor cell survival in the vasculature provide a foundation for metastatic growth. Loss of the proapoptotic function of p53 also renders cells more genetically unstable. There is a substantial correlation between human tumors bearing null or mutant p53, lymph node metastasis, and mutations in *Her2*, *EGFR*, and *myc (130)*. Hence, stimulation or restoration of wild-type p53 function has potential therapeutic implications, and these will be addressed later in this chapter.

4.4.3. TGF-b Signaling

TGF-β signaling affects numerous biological processes such as cell-cycle progression, developmental regulation, chemotaxis, adhesion, angiogenesis, growth suppression, and apoptosis *(131)*. Although clearly a master regulator of metastasis, the definitive role of TGF-β in tumor cell progression and metastasis is quite complex. In the early stages of carcinogenesis, TGF-β is thought to function as a tumor suppressor, especially in normal, nontransformed tissues. However, during progression to a more invasive metastatic state, TGF-β signaling sensitivity is attenuated and accompanied by enhanced TGF-β stromal secretion, resulting in a switch from tumor suppressor to metastasis promoter. The mechanisms underlying when and how this complex transition occurs are uncertain, but may involve gene dosage effects in the TGF-β signaling pathway and aberrant activation of the Ras/MAPK or PI3K pathways *(132)*.

TGF-β binding induces complex formation between the TGF-β type II receptor (TβRII) and the type I receptor (TβRI). TβRII transphosphorylates TβRI, which subsequently phosphorylates transcriptional activator SMAD proteins. SMADs form heterodimeric complexes and translocate to the nucleus where they bind DNA directly or associate with other transcription factors and modulate expression of genes such as *DAPK*, *c-myc*, *c-Jun*, *p21*, and *VEGF* *(133)*. What largely determines the TGF-β-mediated transcriptional response are environmental cues, cell context, and developmental stage *(133)*. TGF-β signaling is implicated in initiation, integration, and execution of apoptosis by cooperating with Fas and TNF-α, activating the SAPK/JNK cascade, and down-regulating $Bclx_L$/Bcl-2 while activating caspases- 1, 3, 8, and 9 *(131)*.

Gene expression profiling of normal human renal epithelium, non-metastatic renal cell carcinoma (RCC), and primary and disseminated lesions showed a reduction in both TβRII and TGF-β-regulated genes in RCC and to a greater degree in metastatic lesions *(134)*. TβRII expression was reduced in 1 out of 16 primary RCC tumors that did not produce metastases, but was reduced in 8 out of 9 primary tumors in which metastases were present *(134)*. Similarly, microarray analysis of a NSCLC cell line and two highly metastatic sublines revealed reduced *SMAD4* expression along with alterations in genes involved in invasion and apoptosis in the metastatic sublines *(135)*. These data suggest that escape from apoptosis and attenuation of TGF-β-mediated signaling are potential mechanisms influencing metastatic progression in cancers such as RCC and lung carcinoma.

Both in vitro and in vivo evidence support TGF-β signaling desensitization as a driving metastatic force. This can occur through loss of TβRII function itself, or through loss of downstream signaling. Transgenic mice bearing the dominant negative TβRII mutant that disrupts TGF-β signaling developed larger and more frequent lung and liver metastases than parental mice despite no change in primary tumor size *(136)*. Defects in TGF-β signaling also contribute to the metastatic phenotype by altering anoikis sensitivity. Upon restoration of functional TGF-β signaling in the human breast carcinoma cell line MDA-MB-468, cells lost several phenotypic markers characteristic of invasion and metastasis. They underwent rapid apoptosis marked by caspase activation, became sensitized to anoikis, and lost the ability to survive in soft agar *(137)*. Disturbance of the apoptotic machinery via disruption of TGF-β signaling fosters an environment apt for matrix-independent survival, invasion, and subsequent metastasis.

There are translational implications for loss of TGF-β signaling and metastatic progression. A somatic missense mutation in the catalytic core of TβRI that inactivates TGF-β signaling was identified in lymph node metastases from primary breast cancer *(138)*. In prostate cancer, poor clinical outcome was associated with overproduction of TGF-β and loss of TβRII *(139)*. However, increased expression of TGF-β was widespread in colorectal and gastric carcinoma, highlighting the dual functions of TGF-β signaling and its resultant complexities in human cancer *(140)*. Many other examples exist in which genetic or epigenetic inactivation of *TβRI*, *TβRII*, or the *SMADs* was found in human tumors *(140)*, exemplifying the concept that inactivation of the TGF-β signaling pathway is a fundamental route to conferring survival advantages upon tumor cells.

4.5. Resistance to Anoikis

Epithelial cells normally adhere to a basement membrane. Loss of this cell adherence results in a type of programmed cell death referred to as anoikis. As described above, surviving this loss of cell–cell adherence is critical for the spread of metastasis, as tumor cells must disengage themselves from the primary site of tumor growth to enter circulation. Therefore resistance to anoikis is key to metastatic progression. Many highly metastatic cell lines, including melanoma, gastric carcinoma, and oral squamous cell carcinomas exhibit anoikis-resistance in vitro *(141–143)*.

Known mechanisms of anoikis resistance involve activation of the PI3K-Akt pathway, a cell survival pathway that culminates by upregulating many of the proteins involved in cell death previously described, including Bcl-2 family members and components of the extrinsic apoptotic pathway (Fig. 2). Anoikis resistance of Ras-transformed cells can be completely explained by activation of the PI3K/Akt pathway *(144)*. Additionally, in human colorectal cancer tumor specimens, Akt activating-phosphorylation has been found to be negatively associated with apoptosis and positively correlated with stages of malignant progression, such as invasive depth and lymph node status *(145)*.

Anoikis may also be manifested by coordinated downregulation of the SAPK pathway, incorporating either MAP kinases p38 or JNK (Fig. 2). As evidence of a functional link to anoikis resistance, SAPK inhibitors have been shown to abolish anoikis and enhance cell adherence *(146)*. These SAPKs also converge on many of the apoptotic regulators that we have already addressed. Evidence suggests that JNK phosphorylation inhibits Bcl-2 function thus promoting apoptosis *(147)*. Additionally, the p38 pathway has been shown to upregulate *FasL* expression *(148)*. In this section, we will briefly describe the upstream components that regulate the Akt and SAPK pathways to avoid anoikis.

Integrins are cell surface proteins that interact with components of the extracellular matrix and upregulate Akt and downregulate SAP kinases. The α4 integrin is expressed in a number of metastatic melanomas and osteosarcomas. Engagement of α4 integrin with a monoclonal antibody (MAb) increased anoikis of a human osteosarcoma cell line in vitro *(149)*. In another example of integrin signaling involvement in metastatic progression, periostin, which binds α5β3 integrins thus enhancing Akt activation and cell survival, was overexpressed in 80% of colorectal cancers, with the metastatic tumors exhibiting the highest expression *(150)*.

Integrins interact with the actin cytoskeleton through a multiprotein complex called the focal adhesion complex consisting of talin, vinculin and paxillin (Fig. 2). Cancer cells lacking *vinculin* have been shown to be highly metastatic *(151,152)* and murine embryonic carcinoma cells devoid of vinculin are resistant to anoikis. The observed increase in survival in vinculin null cells may be attributed to enhanced recruitment of focal adhesion kinase (FAK) to the focal adhesion complex *(153)*. The nonreceptor tyrosine kinase FAK is an important regulator of anoikis. Its expression positively correlates with metastasis in a wide variety of cancers. In a mouse xenograft model, *FAK* expression was silenced by RNA interference in pancreatic adenocarcinoma cells. *FAK* gene silencing promoted anoikis and resulted in decreased metastasis in mice *(154)*. Anoikis resistance in FAK-activated malignancies may be mediated by downregulation of p38 signaling

(146). However, FAK may also potentially promote anoikis resistance by sequestering a component of DISC, thus blocking the extrinsic apoptotic pathway *(155)*. A definitive mechanism of FAK-mediated anoikis resistance has yet to be resolved.

Anoikis resistance can also be attributed to hyperactivation of another nonreceptor tyrosine kinase, Src. In human colon cancer cell lines, Src expression and activity corresponded with increased Akt phosphorylation and anoikis resistance *(156)*. Ectopic expression of kinase-defective mutants of Src in human metastatic colon cancer cells resulted in enhanced sensitivity to apoptosis and appeared to be associated with a decrease in *Bcl-x*$_L$ *(157)*. The Src substrate p130 (Cas), which is normally dephosphorylated and inactivated upon cell detachment in anoikis-sensitive cells, is not dephoshorylated in anoikis-resistant lung adenocarcinoma cells. This constitutive p130 (Cas) phosphorylation is dependent on Src activity *(158)*.

Growth factor receptors can also mediate anoikis resistance by a variety of molecular mechanisms. Anoikis-sensitive head and neck squamous cell carcinoma (HNSCC) could be protected from cell death upon loss of cell adherence by the hepatocyte growth factor. The hepatocyte growth factor receptor (c-met) has a well-established role in metastasis progression. The observed anoikis resistance is dependent on Akt and on the MAPK Erk, which regulates cell proliferation *(159)*. EphrinA2 is another receptor tyrosine kinase that can suppress anoikis. Overexpression of *Ephrin A2* in pancreatic adenocarcinoma cell lines resulted in anoikis resistance and enhanced cellular invasiveness in vitro, whereas *EphrinA2* silencing allowed for anoikis in vitro and reduced tumor growth and metastasis in vivo *(160)*. EphrinA2 anoikis resistance may be FAK-mediated as FAK phosphorylation corresponds to EphrinA2 activity. The neurotrophin receptors, or Trk receptors, also promote anoikis resistance. TrkB was identified in a functional screen for genes that suppress anoikis when they are overexpressed *(161)*. *TrkB* ectopically expressed in nonmalignant epithelial cells led to anoikis-resistance in vitro and rapidly growing and metastasizing tumors when injected in mice. This resistance is mediated by constitutive activation of the PI3K-Akt survival signaling pathway *(161)* and *TrkB* is frequently overexpressed in a variety of highly aggressive human malignancies *(162–164)*. As further evidence of the connection between Trk receptors and metastasis, a Pan-trk inhibitor CEP-701 inhibited metastatic colonization and enhanced overall survival in a rat prostatic cancer model *(165)*.

We have not exhaustively cataloged the molecular mechanisms of anoikis resistance, but we have identified many of the key players. Other growth factors and cell-adhesion molecules that affect Akt, p38 and/or FAK or Src activity might also contribute to anoikis resistance.

5. Resistance to Pharmacologically Induced Apoptosis

As described earlier in this chapter, the ability of cancer cells to evade apoptosis during dissemination and colonization contributes to the development of metastases. In a similar vein, the ability of these same cells to resist pharmacologically induced apoptosis is a conduit for metastatic propagation. Often the cells that are highly resistant to chemotherapy are those with the highest metastatic propensity. Outlined below are three agents well characterized for their role in apoptosis resistance and metastasis. This discussion is by no means exhaustive, but rather, highlights specific instances in which highly metastatic cells are also those that are highly resistant to exogenously or pharmacologically induced apoptosis. A greater understanding of how metastatic cells develop resistance to pharmacologically induced apoptosis may improve our strategies for the prevention or treatment of metastasis.

5.1. 5-Fluorouracil and Derivatives

The fluoropyrimidine 5-fluorouracil (5-FU) is the most widely used chemotherapeutic agent for the treatment of colorectal cancer *(166)*. 5-Fluorouracil and its derivatives UFT (composed of a 5-fluorouracil analog, tegafur, and uracil) and S-1 are cytotoxic agents that inhibit thymidylate synthase, a key enzyme in the production of thymidylate. DNA synthesis in the absence of thymidylate leads to a futile cycle of deoxyuridine incorporation into and excision from DNA, leading to an apoptotic "thymineless" death *(167)*.

Several mechanisms can be attributed to fluoropyrimidine resistance, among which are increased thymidylate synthase expression and intrinsic pharmacokinetic resistance. Significantly higher levels of thymidylate synthase mRNA were found in hepatic metastases from patients resistant to fluoropyrimidine-based therapy *(168)*. Another potential mechanism responsible for 5-FU resistance is apoptotic resistance. Although the biochemistry of the fluoropyrimidines is understood, the molecular pathway leading to fluoropyrimidine-induced apoptosis is not well characterized; however, p53 and Bax are putative molecules at the apex of this apoptotic cascade *(169)*.

Two 5-FU resistant derivatives of the human colorectal carcinoma cell line HCT116 were examined and characterized by their genotypic and phenotypic alterations in comparison with the parental cell line *(170)*. The apoptotic fractions of the resistant variants were approx 50 and 33% lower than the parental cell line. Microarray analysis revealed differential regulation of genes involved in apoptosis, metastasis, proliferation, and DNA repair. Thus, drug resistance may be acquired not only by deregulating a drug's target enzyme, but also, by interfering with a complex network of apoptotic and metastatic pathways.

In the Dunning rat prostate metastatic cancer model (R-3327 AT3), highly metastatic Bcl-2-transfected AT3 cells were more resistant than vector control AT3 cells to 5-FU-induced apoptosis, and were 40-fold more resistant to 5-FU cytotoxicity *(47)*. S-1, a more recently developed oral fluoropyrimidine with a therapeutic index higher than 5-FU, caused persistent apoptosis in vivo and reduced tumor burden *(171)*. Treatment with S-1 statistically reduced the number of peritoneal micro and macrometastases of gastric cancer and prolonged survival until sacrifice *(172)*. UFT, an oral fluoropyrimidine, was used in combination with the bisphosphonate zolendronic acid in the 4T1/luc murine breast cancer model, which spontaneously disseminates to bone following orthotopic mammary fat pad inoculation *(173)*. In vivo, this drug regimen suppressed bone metastases without affecting primary tumor growth. In vitro, the combination of 5-FU and zolendronic acid significantly increased the cellular apoptotic fraction *(173)*. Hence, fluoropyrimidine sensitivity is related to inhibition of metastatic growth, at least in part, through the induction of apoptosis.

5.2. Taxanes

Primary tumors and metastases are differentially sensitive to the taxanes, a class of agents that promote microtubule assembly and stabilize microtubules against depolymerization. The parental taxane paclitaxel (Taxol) and its second generation analog docetaxel have a broad spectrum of antitumor activity which may be attributed to the continuum of signaling pathways altered in response to these agents. Disruption of microtubule dynamics alters gene expression, apoptotic signaling, angiogenesis, and the EGFR and Ras proliferation signaling pathways (reviewed in ref. *174*). In a pharmacodynamic analysis of Taxol in primary and metastatic rat prostate MAT-Lylu tumors, accumulation and retention of Taxol was identical in the primary and metastatic tumors *(175)*. However, primary tumors were consistently more sensitive than metastases to Taxol-induced growth inhibition *(175)*. Thus, differential sensitivity to the taxanes results not from innate pharmacokinetic differences between the primary tumor and its metastases, but rather, from intrinsic differences in the responses of the molecular signaling pathways.

One study, investigating the mechanisms behind Taxol's antimetastatic effects, revealed Taxol-induced alterations in gene expression that were involved in angiogenesis and metastasis. In the C57BL/6 mouse melanoma model, Taxol failed to affect primary tumor size but inhibited spontaneous lung metastases *(176)*. Taxol induced expression of *E-cadherin* and the metastasis suppressor *Nm23*, while downregulating *VEGF* expression. Not surprisingly, Taxol-treated cells underwent significant apoptosis. These data describe an intricate interplay between microtubule disruption and pathways of angiogenesis, metastasis, and apoptosis.

Where taxanes are incapable of inducing apoptosis, resistance and metastatic outgrowth emerge. Treatment with antimicrotubule agents like the taxanes leads to phosphorylation of Bcl-2 and Bcl-x_L, which inhibits their mitochondrial membrane channel function and subsequent antiapoptotic activity (Table 1) *(174)*. When nonmetastatic breast carcinoma MCF-7 or MDA-MB-468 cells were transfected with either *Bcl-2* or *Bcl-*x_L, Taxol resistance was elevated approximately 100-fold *(16)*. Similarly, primary cultures of *Bcl-*x_L-transfected MDA-MB-435 breast cancer cells established from lung, lymph node, or bone metastases showed almost twofold greater resistance to docetaxel *(56)*. Therefore high *Bcl-2* and *Bcl-*x_L expression, which support apoptosis resistance, anoikis resistance, and genomic instability (as outlined earlier in this chapter), is not only conducive to metastatic growth, but also challenges the pro-apoptotic effects of the taxanes.

5.3. Nitric Oxide

There is considerable controversy regarding the effects of nitric oxide (NO) and nitric oxide inhibitors on tumor cell metastasis. NO has pleiotropic effects on pathways known to contribute to metastasis; namely angiogenesis, migration, and apoptosis. When expressed at low levels, NO mediates vasodilation, smooth muscle relaxation, neurotransmission, and inhibition of platelet aggregation; when induced to higher levels, NO can contribute to inflammation and carcinogenesis *(177)*. Production of NO is catalyzed by a family of enzymes, the nitric oxide synthases (NOS), of which there are three characterized isoforms: NOSI (nNOS) and NOSIII (eNOS) are constitutively expressed neuronal and endothelial isoforms, respectively, whereas NOSII (iNOS) can be induced upon exposure to inflammatory cytokines.

NO production triggers p53-mediated apoptosis via induction of DNA damage and PARP cleavage *(177)*. Although this would imply a straightforward correlation between high NO expression, apoptosis, and therefore, inhibition of metastasis, the actual role of

Table 1
Potential Therapeutic Molecular Targets in Apoptosis and Metastasis

Gene	Function	Therapeutic intervention	Resultant effect to stimulate apoptosis
Bcl-2; Bcl-x_L	Mitochondrial membrane permeability	Bcl-2 antisense, Taxanes	Phosphorylation of Bcl-2/ Bcl-x_L and subsequent inhibition of mitochondrial membrane channel function
Bax	Mitochondrial membrane permeability	5-fluorouracil, gene therapy	Elevation of expression
NF-κB	Transcription factor	Proteasome inhibitors (PS-341)	Blockade of IκB degradation, which maintains NF-κB in an inactive state
p53	Transcription factor	DNA damaging agents (ionizing radiation, adriamycin), antimetabolites (5-fluorouracil), proteasome inhibitors (PS-341), cyclin dependent kinase inhibitors (flavopiridol), gene therapy	Activation of p53 by increasing its stability and expression
DAPK	Calcium/calmodulin serine/threonine kinase	De-methylating agents (5-Aza C)	Reversal of promoter methylation and subsequent gene activation
Maspin	Serine protease inhibitor		
TRAIL	TNF family ligand involved in extrinsic apoptotic pathway	Gene therapy, DR monoclonal antibodies, TRAIL agonists	Restoration of expression; tumor cell-specific apoptosis

NO in metastasis is unclear. NO sensitivity to apoptosis and metastasis is dependent on several factors: NO source (host tissue versus tumor cell), cell context (genetic constitution, cell type, p53 status), and NOS isoform expression *(178)*.

Murine melanoma K-1735 cells and their high, low, and nonmetastatic clones were exposed to various NO-inducing cytokines and cytotoxicity was examined *(179)*. The low and nonmetastatic clones were highly sensitive and produced significant levels of NO upon cytokine induction. In contrast, the highly metastatic clones did not produce NO to any discernible degree and, accordingly, had lower NOS activity than the nonmetastatic cells. These in vitro experiments suggested an inverse correlation between NOSII activity and metastatic potential. Similar conclusions were reached in clonal populations of murine pancreatic adenocarcinoma cells (Panc02), in which high *NOSII* expressing clones failed to produce primary tumors and metastases, whereas low-*NOSII* expressers were tumorigenic and produced liver metastases and ascites *(180)*. Confounding these two studies are the results from Jadeski et al. *(181)* who observed the converse, that is, highly metastatic clones of the murine mammary carcinoma C3H/HeJ model expressed high levels of *NOSIII* and *NOSII*, whereas the weakly metastatic clones expressed low levels of *NOSIII*. Whether these differences were cell context specific or caused by specificity of NO-producing isoforms remains unclear.

Another significant factor underlying NO sensitivity and metastasis is the effect of host-derived NO production. When the $NOSII^{-/-}$ KX-dw cell line was injected into syngeneic $NOSII^{+/+}$ and $NOSII^{-/-}$ C57BL/6 mice, tumors from $NOSII^{+/+}$ mice had high *NOSII* expression and activity and produced few lung metastases *(182)*. On the contrary, tumors in $NOSII^{-/-}$ mice had undetectable *NOSII* expression and a significant number of pulmonary metastatic nodules. The authors concluded that host physiological *NOSII* expression directly inhibited tumor growth and metastasis. However, although host derived *NOSII* may differentially modulate metastatic progression, tumor cell NO sensitivity and production play an equally important role. In $NOSII^{+/+}$ or $NOSII^{-/-}$ syngeneic mice, B16 melanoma cells were highly resistant to NO-induced cytotoxicity whereas the murine ovarian carcinoma M5076 cells remained sensitive *(183)*. Despite no differences in primary tumor size, B16 cells produced more metastases in $NOS^{+/+}$ mice than in $NOS^{-/-}$ mice, and M5076 cells produced fewer metastases in $NOS^{+/+}$ mice than $NOS^{-/-}$ mice.

Just as the in vitro and in vivo evidence regarding the role played by NO in tumor cell metastasis is contradictory, so too is the clinical data. High total NOS expression corresponded to improved survival in lung carcinoma patients *(184)*. When analysis was conducted separately on the three NOS isoforms, high-NOSII staining was associated with lower grade lung carcinoma *(184)*. However, compelling but directly contrasting evidence suggested that high-*NOSII* mRNA or protein levels were positively correlated with lymph node metastases in oral squamous cell carcinoma, breast cancer, and head and neck cancer (reviewed ref. *177*). Here again, differences may result from tumor cell type specificity. What is questionable then, is the efficacy of NO inhibitors as therapeutic agents, given the complexity of tumor–host interactions. Based upon the above evidence, it seems a better understanding of NO's promotion or inhibition of metastasis is required prior to launching NO intervention into the clinical setting.

6. Overcoming Apoptotic Resistance

Conventional therapeutic approaches in the treatment of cancer typically induce apoptosis. We have established that metastatic tumor cells have evolved a multitude of mechanisms to evade apoptosis by either directly disrupting apoptotic pathways, downstream targets of these pathways, or the actual apoptotic machinery. Many metastases are thus refractory to traditional therapeutic approaches and this explains why metastasis is so fatal. Effective treatment strategies will thus rely on overcoming resistance to apoptosis.

6.1. Paradigm for Diagnosing and Treating Apoptotic Resistance

Bold et al. *(185)* offer the following paradigm for tackling this resistance issue: upon biopsy, the apoptotic responsiveness of the tumor should be assessed by determining its apoptotic index and the expression levels of p53, Bcl-2 family members and other molecular components of programmed cell death. If the tumor is responsive to apoptotic stimuli, then a traditional course of chemotherapy and ionizing radiation may be followed. However, if the tumor has a resistant phenotype, additional steps must be taken to restore sensitivity to apoptotic stimuli. These additional steps include gene therapy and/or a combination of gene therapy/chemotherapy, which will be discussed in brief below.

6.2. Restoring Sensitivity

Restoring sensitivity to apoptotic stimuli involves either restoring expression of key regulators, artificially supplying cancer cells with the critical gene involved in apoptosis, or inhibiting prosurvival pathways active in metastatic cells. Of course, in order to restore function, the aberrant gene/protein must first be identified. Table 1 summarizes the therapeutic approaches described in this section that are currently being used or may potentially be used to conquer apoptotic resistance in cancer.

Effective drugs have been identified to stimulate re-expression of particular apoptotic regulators. Several genes involved in apoptosis have reduced expression in metastatic cells as a result of hypermethylation of their promoters, including *Apaf-1*, *maspin* and *DAP* kinases *(40,41,43,75–77)*. Therefore, combination chemotherapy with such de-methylating agents as 5-aza C may restore cytotoxicity of apoptosis-inducing compounds. In metastatic lesions with functional, wild-type p53, induction of *p53* expression with such DNA-damaging agents as adriamycin may repress cell survival pathways thereby prompting cell death when administered in addition to apoptosis-inducing drugs. For cancers harboring mutations in p53's DNA-binding region, pharmacological compounds that stabilize the DNA-binding domain may also hold promise in a combination chemotherapeutic course with cytotoxic drugs *(186)*.

Gene therapy and/or administration of recombinant proteins for apoptotic regulators that are not functional in metastatic cells are additional approaches to restore sensitivity to cytoxic drugs. Several groups, including Introgen Therapeutics and Onyx Pharmaceuticals, are attempting *p53* gene therapeutic approaches with adenoviral-based vectors. In lung cancer, viral delivery of wild-type *p53* substantially enhanced the apoptotic index, which may be further augmented by cytotoxic drug therapy and/or radiotherapy *(187)*. *TRAIL* is a particularly intriguing candidate for gene therapy given its propensity to induce apoptosis specifically in tumor cells. Administration of adenoviral vectors expressing *TRAIL* or recombinant TRAIL protein delivered by liposomes has shown much potential in experimental animal models *(108,109,188)*. In a mouse model of breast cancer, *TRAIL* gene therapy in combination with the chemothera peutic agent paclitaxel synergistically reduced lung metastases *(110)*. Other potential apoptotic regulators that may be appropriate for this kind of gene replacement therapy include *Bax* (currently in phase I clinical trial) *(189)*, *maspin* and the *DAP* kinases. It should be added, however, that the efficacy of any of these gene therapeutic approaches depends on targeted gene delivery specifically to sites of tumor growth.

TRAIL-induced apoptosis might also be triggered in cancers with a monoclonal agonist antibody to the TRAIL receptors. One such agonist antibody against DR4 is currently in a phase I trial (Human Genome Sciences) and a similar recombinant antibody exhibited potent antitumor effects in a mouse model *(190)*. Of course, this therapeutic approach relies on DR expression. Tumor cells that have developed TRAIL resistance by diminished DR expression or decoy receptor expression will not be amenable to agonist-based therapy.

The latest approach in treating apoptotic resistance involves antisense technology. Short synthetic RNA or DNA oligonucleotides with sequences complementing the mRNA sequence of specific genes are delivered to tumor cells to interfere with the translation of the coded protein. This approach would be utilized to disrupt expression of genes that function to suppress apoptosis, such as *Bcl-2*, *Bcl-x_L* and *survivin*. In fact, antisense *Bcl-2* therapy (Genasense®, Genta, Inc.) in combination with chemotherapeutic agents such as paclitaxel, cyclophosphamide, and irinotecan is in phase II and III clinical trials for the treatment of several types of cancer including prostate and melanoma (reviewed by ref. *189*). But again, the efficacy of this therapeutic approach depends on targeted delivery of the disruptive oligos to the tumor.

Drug-mediated inhibition of apoptotic inhibitors may also enhance the cytotoxic effects of traditional chemotherapy. Proteasome inhibitors, such as PS-341, have potential in overcoming apoptotic resistance by increasing the stability of a number of apoptotic promoters, including p53 and the NF-κB inhibitor IκB, and triggering the SAP kinase JNK *(191)*. Suppression of signaling through the neurotrophic receptors, which block anoikis, with the pan-Trk inhibitor CEP-701 has some potential in relieving cell survival signals and reducing metastases in prostate cancer models *(162,165)*. Finally, several potential small-molecule antagonists of Bcl-2 are under investigation, including GX-01 from Gemin-X and the natural product Tetrocarcin-A (Kyowa-Hakko Kogyo) (reviewed ref. *189*). As technology improves and drug discovery progresses, additional more specific and efficacious modes of therapy will become available for the treatment of apoptotic resistance and metastatic disease.

7. Conclusion

We are moving into the era of tailor-designed therapeutics based on molecular diagnosis. Much progress has been made elucidating the molecular mechanisms of cell death resistance in metastatic progression. Given the sheer number of potential molecular players that may contribute to this resistance, we need to develop more sensitive and comprehensive diagnostics for determining the faulty gene/genes responsible for apoptotic resistance, in order to best design a course of therapy to overcome it.

In light of the emerging field of cancer stem cell research, we must also consider what population of tumor cells should assessed for faulty gene function in a clinical diagnosis. Much of the data regarding apoptotic resistance in metastasis is derived from tumor cohort studies, which generally consider tumors as clonal entities, in particular those studies that use immunoblotting and RT-PCR to measure gene expression. However, it should be stressed that tumors consist of heterogeneous populations of cells. Recently, cancer stem cells have been isolated from solid tumors and it is these cells that are believed to be responsible for tumor generation and/or regeneration, in the case of metastasis *(192–195)*. Cancer stem cells, like normal stem cells have the capacity for self-renewal and the ability to undergo a broad range of differentiation events. Unlike normal tissue stem cells, however, cancer stem cells do not have restricted proliferation (reviewed in refs. *194,196*). Thus far, information pertaining to the behavior of cancer stem cells is limited, but some evidence implies that they are both resistant to anoikis and resistant to some cytotoxic drugs *(193,197)*. This newly emerging model of cancer development via cancer stem cells would suggest that both diagnostic and therapeutic efforts should be focused on cancer stem cells to diminish disease recurrence and spread.

Efforts should also be focused on the persistent need for more effective therapies to overcome apoptotic resistance in metastasis. Gene and antisense therapy are clearly limited by delivery mechanisms. Small molecule, rational and *in silico* drug design to the key molecular targets that have already been identified may provide us with an improved species of chemotherapeutics for attacking metastases, or more specifically the cancer stem cells of metastatic lesions.

References

1. Entschladen F, Drell TLt, Lang K, Joseph J, Zaenker KS. Tumour-cell migration, invasion, and metastasis: navigation by neurotransmitters. Lancet Oncol 2004;5(4):254–258.
2. Weiss L. Metastatic inefficiency. Adv Cancer Res 1990;54:159–211.
3. Luzzi KJ, MacDonald IC, Schmidt EE, et al. Multistep nature of metastatic inefficiency: dormancy of solitary cells after successful extravasation and limited survival of early micrometastases. Am J Pathol 1998;153(3):865–873.
4. Malaguarnera L. Implications of apoptosis regulators in tumorigenesis. Cancer Metastasis Rev 2004;23(3-4):367–387.
5. Tenjo T, Toyoda M, Okuda J, et al. Prognostic significance of p27(kip1) protein expression and spontaneous apoptosis in patients with colorectal adenocarcinomas. Oncology 2000; 58(1):45–51.
6. Yamasaki F, Tokunaga O, Sugimori H. Apoptotic index in ovarian carcinoma: correlation with clinicopathologic factors and prognosis. Gynecol Oncol 1997;66(3):439–448.
7. Shibata H, Matsubara O. Apoptosis as an independent prognostic indicator in squamous cell carcinoma of the esophagus. Pathol Int 2001;51(7):498–503.
8. Sugamura K, Makino M, Kaibara N. Apoptosis as a prognostic factor in colorectal carcinoma. Surg Today 1998;28(2):145–150.
9. Richter EN, Oevermann K, Buentig N, Storkel S, Dallmann I, Atzpodien J. Primary apoptosis as a prognostic index for the classification of metastatic renal cell carcinoma. J Urol 2002;168(2):460–464.
10. Ohtani M, Isozaki H, Fujii K, et al. Impact of the expression of cyclin-dependent kinase inhibitor p27Kip1 and apoptosis in tumor cells on the overall survival of patients with non-early stage gastric carcinoma. Cancer 1999;85(8):1711–1718.
11. Ito Y, Matsuura N, Sakon M, et al. Both cell proliferation and apoptosis significantly predict shortened disease-free survival in hepatocellular carcinoma. Br J Cancer 1999;81(4):747–751.
12. Glinsky G, Glinsky V, Ivanova A, Hueser C. Apoptosis and metastasis: increased apoptosis resistance of metastatic cancer cells is associated with the profound deficiency of apoptosis execution mechanisms. Cancer Lett 1997;115(2):185–193.
13. Townson JL, Naumov GN, Chambers AF. The role of apoptosis in tumor progression and metastasis. Curr Mol Med 2003;3(7):631–642.
14. Holmgren L, O'Reilly M, Folkman J. Dormancy of micrometastases: balanced proliferation and apoptosis in the presence of angiogenesis suppression. Nat Med 1995;1(2):149–153.
15. Naumov GN, MacDonald IC, Chambers AF, Groom AC. Solitary cancer cells as a possible source of tumour dormancy? Semin Cancer Biol 2001;11(4):271–276.
16. Fernandez Y, Gu B, Martinez A, Torregrosa A, Sierra A. Inhibition of apoptosis in human breast cancer cells: role in tumor progression to the metastatic state. Int J Cancer 2002;101(4):317–326.
17. Khanna C, Hunter K. Modeling metastasis in vivo. Carcinogenesis 2005;26(3):513–523.
18. Eggert A, Grotzer M, Zuzak T, et al. Resistance to Tumor Necrosis Factor-related apoptosis-inducing apoptosis in neuroblastoma cells correlates with a loss of Caspase-8 expression. Cancer Res 2001;61:1314–1319.
19. Grotzer M, Eggert A, Zuzak T, et al. Resistance to TRAIL-induced apoptosis in primitive neuroectodermal brain tumor cells correlates with a loss of caspase-8 expression. Oncogene 2000;19:4604–4610.

20. Pingoud-Meier C, Lang D, Janss A, et al. Loss of Caspase-8 protein expression correlates with unfavorable survival outcome in childhood medulloblastoma. Clin Cancer Res 2003; 9:6401–6409.
21. Shin MS, Kim HS, Lee SH, et al. Alterations of Fas-pathway genes associated with nodal metastasis in non-small cell lung cancer. Oncogene 2002;21:4129–4136.
22. Koomagi R, Volm M. Relationship between the expression of caspase-3 and the clinical outcome of patients with non-small cell lung cancer. Anticancer Res 2000;20(1B):493–496.
23. Volm M, Koomagi R. Prognostic relevance of c-Myc and caspase-3 for patients with non-small cell lung cancer. Oncology Reports 2000;7(1):95–98.
24. Sun B, Zhang J, Wang B, et al. Analysis of in vivo patterns of caspase 3 gene expression in primary hepatocellular carcinoma and its relationship to p21WAF1 expression and hepatic apoptosis. World J Gastroenterol 2000;6(3):356–360.
25. Sun H, Zheng H, Yang X, et al. Expression of PTEN and Caspase-3 and their clinicopathological significance in primary gastric malignant lymphoma. Chin Med Sci J 2004;19(1):19–24.
26. Isobe N, Onodera H, Mori A, et al. Caspase-3 expression in human gastric carcinoma and its clinical significance. Oncology 2004;66(3):201–209.
27. Schimmer AD. Inhibitor of apoptosis proteins: translating basic knowledge into clinical practice. Cancer Res 2004;64(20):7183–7190.
28. Tanaka K, Iwamoto S, Gon G, Nohara T, Iwamoto M, Tanigawa N. Expression of survivin and its relationship to loss of apoptosis in breast carcinomas. Clin Cancer Res 2000;6(1): 127–134.
29. Kato J, Kuwabara Y, Mitani M, et al. Expression of survivin in esophageal cancer: correlation with the prognosis and response to chemotherapy. Int J Cancer 2001;95(2):92–95.
30. Chen W, Liu Q, Fu J, Kang S. Expression of survivin and its significance in colorectal cancer. World J Gastroenterol 2004;10(19):2886–2889.
31. Jaattela M. Escaping cell death: survival proteins in cancer. Exp Cell Res 1999;248(1): 30–43.
32. Kishi H, Igawa M, Kikuno N, Yoshino T, Urakami S, Shiina H. Expression of the survivin gene in prostate cancer: correlation with clinicopathological characteristics, proliferative activity and apoptosis. J Urol 2004;171(5):1855–1860.
33. Agui T, McConkey DJ, Tanigawa N. Comparative study of various biological parameters, including expression of survivin, between primary and metastatic human colonic adenocarcinomas. Anticancer Res 2002;22(3):1769–1776.
34. Levy-Strumpf N, Kimchi A. Death associated proteins (DAPs): from gene identification to the analysis of their apoptotic and tumor suppressive functions. Oncogene 1998;17(25): 3331–3340.
35. Raveh T, Kimchi A. DAP Kinase—a proapoptotic gene that functions as a tumor suppressor. Exp Cell Res 2001;264:185–192.
36. Inbal B, Cohen O, Polak-Charcon S, et al. DAP kinase links the control of apoptosis to metastasis. Nature 1997;390(6656):180–184.
37. Kogel D, Reimertz C, Mech P, et al. Dlk/ZIP kinase-induced apoptosis in human medullablastoma cells: requirement of the mitochondrial apoptosis pathway. Br J Cancer 2001;85(11):1801–1808.
38. Levy D, Plu-Bureau G, Decroix Y, et al. Death-associated protein kinase loss of expression is a new marker for breast cancer prognosis. Clin Cancer Res 2004;10(9):3124–3130.
39. Ng M. Death associated protein kinase: From regulation of apoptosis to tumor suppressive functions and B cell malignancies. Apoptosis 2002;7(3):261–270.
40. Kim D, Nelson H, Wiencke J, et al. Promoter methylation of DAP-kinase: association with advanced stage in non-small cell lung cancer. Oncogene 2001;20:1765–1770.

41. Gonzalez-Gomez P, Bello M, Alonso M, et al. Frequent death-associated protein-kinase promoter hypermethylation in brain metastases of solid tumors. Oncol Rep 2003;10(4): 1031–1033.
42. Baldi A, Santini D, Russo P, et al. Analysis of APAF-1 expression in human cutaneous melanoma progression. Exp Dermatol 2004;13(2):93–97.
43. Soengas M, Capodieci P, Polsky D, et al. Inactivation of the apoptosis effector Apaf-1 in malignant melanoma. Nature 2001;409(6817):201–211.
44. Umetani N, Fujimoto A, Takeuchi H, Shinozaki M, Bilchik A, Hoon D. Allelic imbalance of APAF-1 locus at 12q23 is related to progression of colorectal carcinoma. Oncogene 2004;23:8292–8300.
45. Reed J. Mechanisms of apoptosis avoidance in cancer. Cuu Opin Oncol 1999;11(1):68–75.
46. Bold RJ, Virudachalam S, McConkey DJ. BCL2 expression correlates with metastatic potential in pancreatic cancer cell lines. Cancer 2001;92(5):1122–1129.
47. Furuya Y, Krajewski S, Epstein JI, Reed JC, Isaacs JT. Expression of bcl-2 and the progression of human and rodent prostatic cancers. Clin Cancer Res 1996;2(2):389–398.
48. Takaoka A, Adachi M, Okuda H, et al. Anti-cell death activity promotes pulmonary metastasis of melanoma cells. Oncogene 1997;14(24):2971–2977.
49. Pinkas J, Martin SS, Leder P. Bcl-2-mediated cell survival promotes metastasis of EpH4 betaMEKDD mammary epithelial cells. Mol Cancer Res 2004;2(10):551–556.
50. Sierra A, Lloveras B, Castellsague X, Moreno L, Garcia-Ramirez M, Fabra A. Bcl-2 expression is associated with lymph node metastasis in human ductal breast carcinoma. Int J Cancer 1995;60(1):54–60.
51. Le MG, Mathieu MC, Douc-Rasy S, et al. c-myc, p53 and bcl-2, apoptosis-related genes in infiltrating breast carcinomas: evidence of a link between bcl-2 protein over-expression and a lower risk of metastasis and death in operable patients. Int J Cancer 1999;84(6): 562–567.
52. Zellweger T, Ninck C, Bloch M, et al. Expression patterns of potential therapeutic targets in prostate cancer. Int J Cancer 2005;113(4):619–628.
53. Martin SS, Ridgeway AG, Pinkas J, et al. A cytoskeleton-based functional genetic screen identifies Bcl-xL as an enhancer of metastasis, but not primary tumor growth. Oncogene 2004;23(26):4641–4645.
54. Fernandez Y, Espana L, Manas S, Fabra A, Sierra A. Bcl-xL promotes metastasis of breast cancer cells by induction of cytokines resistance. Cell Death Differ 2000;7(4):350–359.
55. Espana L, Fernandez Y, Rubio N, Torregrosa A, Blanco J, Sierra A. Overexpression of Bcl-x(L) in Human Breast Cancer Cells Enhances Organ-Selective Lymph Node Metastasis. Breast Cancer Res Treat 2004;87(1):33–44.
56. Gu B, Espana L, Mendez O, Torregrosa A, Sierra A. Organ-selective chemoresistance in metastasis from human breast cancer cells: inhibition of apoptosis, genetic variability and microenvironment at the metastatic focus. Carcinogenesis 2004;25(12):2293–2301.
57. Olopade OI, Adeyanju MO, Safa AR, et al. Overexpression of BCL-x protein in primary breast cancer is associated with high tumor grade and nodal metastases. Cancer J Sci Am 1997;3(4):230–237.
58. McConkey DJ, Greene G, Pettaway CA. Apoptosis resistance increases with metastatic potential in cells of the human LNCaP prostate carcinoma line. Cancer Res 1996;56(24): 5594–5599.
59. Krajewski S, Blomqvist C, Franssila K, et al. Reduced expression of proapoptotic gene BAX is associated with poor response rates to combination chemotherapy and shorter survival in women with metastatic breast adenocarcinoma. Cancer Res 1995;55(19):4471–4478.
60. Ogura E, Senzaki H, Yamamoto D, et al. Prognostic significance of Bcl-2, Bcl-xL/S, Bax and Bak expressions in colorectal carcinomas. Oncol Rep 1999;6(2):365–369.

61. Ito T, Fujieda S, Tsuzuki H, et al. Decreased expression of Bax is correlated with poor prognosis in oral and oropharyngeal carcinoma. Cancer Lett 1999;140(1-2):81–91.
62. Wu J, Shao ZM, Shen ZZ, et al. Significance of Apoptosis and Apoptotic-Related Proteins, Bcl-2, and Bax in Primary Breast Cancer. Breast J 2000;6(1):44–52.
63. Sheng S. The promise and challenge toward the clinical application of maspin in cancer. Frontiers Bioscience 2004;9:2733–2745.
64. Jiang N, Meng Y, Zhang S, Mensah-Osman E, Sheng S. Maspin sensitizes breast carcinoma cells to induce apoptosis. Oncogene 2002;21:4089–4098.
65. Liu J, Yin S, Reddy N, Spencer C, Sheng S. Bax mediates the apoptosis-sensitizing effect of maspin. Cancer Res 2004;64(5):1703–1711.
66. Li Z, Shi H, Zhang M. Targeted expression of maspin in tumor vasculatures induces endothelial cell apoptosis. Oncogene 2005;24(12):2008–2019.
67. Latha K, Zhang W, Cella N, Shi H, Zhang M. Maspin mediates increased tumor cell apoptosis upon induction of the mitochondrial permeability transition. Mol Cell Biol 2005;25(5):1737–1748.
68. Shi H, Zhang W, Liang R, et al. Modeling human breast cancer metastasis in mice: maspin as a paradigm. Histol Histopathol 2003;18:201–206.
69. Zhang M, Shi Y, Magit D, Furth PA, Sager R. Reduced mammary tumor progression in WAP-TAg/WAP-maspin bitransgenic mice. Oncogene 2000;19(52):6053–6058.
70. Shi H, Zhang W, Liang R, et al. Blocking tumor growth, invasion, and metastasis by maspin in a syngeneic breast cancer model. Cancer Res 2001;61(18):6945–6951.
71. Shi H, Liang R, Templeton N, Zhang M. Inhibition of breast tumor progression by systemic delivery of the maspin gene in a syngeneic tumor model. Mol Ther 2002;5(6):755–761.
72. Maass N, Hojo T, Rosel F, Ikeda T, Jonat W, Nagasaki K. Down regulation of the tumor suppressor gene maspin in breast carcinoma is associated with a higher risk of distant metastasis. Clin Biochem 2001;34(4):303–307.
73. Maass N, Teffner M, Rosel F, et al. Decline in the expression of the serine proteinase inhibitor maspin is associated with tumour progression in ductal carcinomas of the breast. J Pathol 2001;195(3):321–326.
74. Wang M, Yang Y, Li X, Dong F, Li Y. Maspin expression and its clinicopathological significance in tumorigenesis and progression of gastric cancer. World J Gastroenterol 2004;10(5):634–637.
75. Maass N, Biallek M, Rosel F, et al. Hypermethylation and histone deacetylation lead to silencing of the maspin gene in human breast cancer. Biochem Biophys Res Commun 2002;297(1):125–128.
76. Primeau M, Gagnon J, Momparler R. Synergistic antineoplastic action of DNA methylation inhibitor 5-AZA-2′-deoxycytidine and histone deacetylase inhibitor depsipeptide on human breast carcinoma cells. Int J Cancer 2003;103(2):177–184.
77. Akiyama Y, Maesawa C, Ogasawara S, Terashima M, Masuda T. Cell-type-specific repression of the maspin gene is disrupted frequently by demethylation at the promoter region in gastric intestinal metaplasia and cancer cells. Am J Pathol 2003;163(5):1911–1919.
78. Boltze C, Schneider-Stock R, Quednow C, et al. Silencing of the maspin gene by promoter hypermethylation in thyroid cancer. Int J Mol Med 2003;12(4):479–484.
79. Sato N, Fukushima N, Matsubayashi H, Goggins M. Identification of maspin and S100P as novel hypomethylation targets in pancreatic cancer using global gene expression profiling. Oncogene 2004;23(8):1531–1538.
80. Wang S, El-Deiry W. TRAIL and apoptosis induction by TNF-family death receptors. Oncogene 2003;22:8628–8633.
81. LeBlanc H, Lawrence D, Varfolomeev E, et al. Tumor-cell resistance to death receptor-induced apoptosis through mutational inactivation of the proapoptotic Bcl-2 homolog Bax. Nat Med 2002;8(3):274–281.

82. Deng Y, Lin Y, Wu X. TRAIL-induced apoptosis recquires Bax-dependent mitochondrial release of Smac/DIABLO. Gene Dev 2002;16:33–45.
83. Kim K, Lee Y. Amiloride augments TRAIL-induced apoptotic death by inhibiting phosphorylation of kinases and phosphatases associated with P13K-Akt pathway. Oncogene 2005; 24:355–366.
84. Harper N, Hughes M, Farrow S, Cohen G, MacFarlane M. Protein kinase C modulates tumor necrosis factor-related apoptosis-inducing ligand-induced apoptosis by targeting the apical events of death receptor signaling. J Biol Chem 2003;278(45):44,338–44,347.
85. Frese S, Pirnia F, Miescher D, et al. PG490-mediated sensitization of lung cancer cells to Apo2L/TRAIL-induced apoptosis requires activation of ERK2. Oncogene 2003;22(25): 5427–5435.
86. Thakkar H, Chen X, Tyan F, et al. Pro-survival function of Akt/protein kinase B in prostate cancer cells. Relationship with TRAIL resistance. J Biol Chem 2001;276(42): 38,361–38,369.
87. Owen-Schaub L, van Golen K, Hill L, Price J. Fas and Fas ligand interactions suppress melanoma lung metastasis. J Exp Med 1998;188(9):1717–1723.
88. Hedlund T, Duke R, Schleicher M, Miller G. Fas-mediated apoptosis in seven human prostate cancer cell lines: correlation with tumor stage. Prostate 1998;36(2):92–101.
89. Monti P, Marchesi F, Reni M, et al. A comprehensive in vitro characterization of pancreatic ductal carcinoma cell line biological behavior and its correlation with the structural and genetic profile. Virchows Arch 2004.
90. Worth L, Lafleur E, Jia S, Kleinerman E. Fas expression inversely correlates with metastatic potential in osteosarcoma cells. Oncol Rep 2002;9(4):823–827.
91. Ogawa S, Nagao M, Kanehiro H, et al. The breakdown of apoptotic mechanism in the development and progression of colorectal carcinoma. Anticancer Res 2004;24(3a): 1569–1579.
92. Wang W, Chen P, Hsiao H, Wang H, Liang W, Su Y. Overexpression of the thymosin beta-4 gene is associated with increased invasion of SW480 colon carcinoma cells and the distant metastasis of human colorectal carcinoma. Oncogene 2004;23(39):6666–6671.
93. Nagao M, Nakajima Y, Hisanaga M, et al. The alteration of Fas receptor and ligand system in hepatocellular carcinomas: How do hepatoma cells escape from the host immune surveillance in vitro? Hepatology 1999;30(2):413–421.
94. Qin L, Tang Z. The prognostic molecular markers in hepatocellular carcinoma. World J Gastroenterol 2002;8(3):385–392.
95. Munakata S, Enomoto T, Tsujimoto M, et al. Expressions of Fas ligand and other apoptosis-related genes and their prognostic significance in epithelial ovarian neoplasms. Br J Cancer 2000;82(8):1446–1452.
96. Ueno T, Toi M, Tominaga T. Circulating soluble Fas concentration in breast cancer patients. Clin Cancer Res 1999;5:3529–3533.
97. Bewick M, Conlon M, Parissenti A, et al. Soluble Fad (CD95) is a prognostic factor in patients with metastatic breast cancer undergoing high-dose chemotherapy and autologous stem cell transplantation. J Hematother Stem Cell Res 2001;10(6):759–768.
98. Mouawad R, Khayat D, Soubrane C. Plasma Fas ligand, an inducer of apoptosis, and plasma soluble Fas, an inhibitor of apoptosis, in advanced melanoma. Melanoma Res 2000;10(5):461–467.
99. Wu Y, Han B, Sheng H, et al. Clinical significance of detecting elevated serum DcR3/TR6/M68 in malignant tumor patients. Int J Cancer 2003;105:724–732.
100. Mottolese M, Buglioni S, Bracalenti C, et al. Prognostic relevance of altered Fas (CD95)-system in human breast cancer. Int J Cancer 2000;89(2):127–132.
101. Mann B, Gratchev A, Bohm C, et al. FasL is more frequently expressed in liver metastases of colorectal cancer than in matched primary carcinomas. Br J Cancer 1999;79(7-8): 1262–1269.

102. Okada K, Komuta K, Hashimoto S, Matsuzaki S, Kanematsu T, Koji T. Frequency of apoptosis of tumor-infiltrating lymphocytes induced by fas counterattack in human colorectal carcinoma and its correlation with prognosis. Clin Cancer Res 2000;6(9):3560–3564.
103. Koyama S, Koike N, Adachi S. Fas receptor counterattack against tumor-infiltrating lymphocytes in vivo as a mechanism of immune escape in gastric carcinoma. J Cancer Res Clin Oncol 2001;127(1):20–26.
104. Pan G, Ni J, Wei Y, Yu G, Gentz R, Dixit V. An antagonist decoy receptor and a death domain-containing receptor for TRAIL. Science 1997;277:815–818.
105. Takeda K, Hayakawa M, Smyth M, et al. Involvement of tumor necrosis factor-related apoptosis-inducing ligand in surveillance of tumor metastasis by liver natural killer cells. Nature Medicine 2001;7(1):94–100.
106. Cretney E, Takeda K, Yagita H, Glaccum M, Peschon J, Smyth M. Increased susceptibility to tumor initiation and metastasis in TNF-related apoptosis-inducing ligand-deficient mice. J Immunol 2002;168(3):1356–1361.
107. Abdollahi T. Potential for TRAIL as a therapeutic agent in ovarian cancer. Vitam Horm 2004;67:347–364.
108. Kaliberov S, Kaliberova L, Stockard C, Grizzle W, Buchsbaum D. Adenovirus-mediated FLT1-targeted proapoptotic gene therapy of human prostate cancer. Mol Ther 2004;10(6): 1059–1070.
109. Yamashita Y, Shimada M, Tanaka S, Okamamoto M, Miyazaki J, Sugimachi K. Electroporation-mediated tumor necrosis factor-related apoptosis-inducing ligand (TRAIL)/ Apo2L gene therapy for hepatocellular carcinoma. Hum Gene Ther 2002;13:275–286.
110. Lin T, Zhang L, Davis J, et al. Combination of TRAIL gene therapy and chemotherapy enhances antitumor and antimetastasis effects in chemosensitive and chemoresitant breast cancers. Mol Ther 2003;8(3):441–448.
111. Zhang XD, Franco A, Myers K, Gray C, Nguyen T, Hersey P. Relation of TNF-related apoptosis-inducing ligand (TRAIL) receptor and FLICE-inhibitory protein expression to TRAIL-induced apoptosis of melanoma. Cancer Res 1999;59(11):2747–2753.
112. Vigneswaran N, Wu J, Nagaraj N, Adler-Storthz K, Zacharias W. Differential susceptibility of metastatic and primary oral cancer cells to TRAIL-induced apoptosis. Int J Oncol 2005;26(1):103–112.
113. Shin M, Kim H, Lee S, et al. Mutations of tumor necrosis factor-related apoptosis-inducing ligand receptor 1 (TRAIL-R1) and receptor 2 (TRAIL-R2) genes in metastatic breast cancers. Cancer Res 2001;61(13):4942–4946.
114. Koyama S, Koike N, Adachi S. Expression of TNF-related apoptosis-inducing ligand (TRAIL) and its receptors in gastric carcinoma and tumor-infiltrating lymphocytes: a possible mechanism of immune evasion of the tumor. J Cancer Res Clin Oncol 2002;128:73–79.
115. Nyambo R, Cross N, Lippitt J, et al. Human bone marrow stromal cells protect prostate cancer cells from TRAIL-induced apoptosis. J Bone Min Res 2004;19(10):1712–1721.
116. Bharti AC, Aggarwal BB. Nuclear factor-kappa B and cancer: its role in prevention and therapy. Biochem Pharmacol 2002;64(5-6):883–888.
117. Orlowski RZ, Baldwin AS, Jr. NF-kappaB as a therapeutic target in cancer. Trends Mol Med 2002;8(8):385–389.
118. Dong G, Loukinova E, Chen Z, et al. Molecular profiling of transformed and metastatic murine squamous carcinoma cells by differential display and cDNA microarray reveals altered expression of multiple genes related to growth, apoptosis, angiogenesis, and the NF-kappaB signal pathway. Cancer Res 2001;61(12):4797–4808.
119. Huber MA, Azoitei N, Baumann B, et al. NF-kappaB is essential for epithelial-mesenchymal transition and metastasis in a model of breast cancer progression. J Clin Invest 2004; 114(4):569–581.

120. Nakshatri H, Bhat-Nakshatri P, Martin DA, Goulet RJ, Jr, Sledge GW, Jr. Constitutive activation of NF-kappaB during progression of breast cancer to hormone-independent growth. Mol Cell Biol 1997;17(7):3629–3639.
121. Biswas DK, Shi Q, Baily S, et al. NF-kappa B activation in human breast cancer specimens and its role in cell proliferation and apoptosis. Proc Natl Acad Sci USA 2004;101(27): 10,137–10,142.
122. Fujioka S, Sclabas GM, Schmidt C, et al. Function of nuclear factor kappaB in pancreatic cancer metastasis. Clin Cancer Res 2003;9(1):346–354.
123. Sunwoo JB, Chen Z, Dong G, et al. Novel proteasome inhibitor PS-341 inhibits activation of nuclear factor-kappa B, cell survival, tumor growth, and angiogenesis in squamous cell carcinoma. Clin Cancer Res 2001;7(5):1419–1428.
124. Scaife CL, Kuang J, Wills JC, et al. Nuclear factor kappaB inhibitors induce adhesion-dependent colon cancer apoptosis: implications for metastasis. Cancer Res 2002;62(23): 6870–6878.
125. Oren M. Decision making by p53: life, death and cancer. Cell Death Differ 2003;10(4): 431–442.
126. Vousden KH, Lu X. Live or let die: the cell's response to p53. Nat Rev Cancer 2002; 2(8):594–604.
127. Xu H, el-Gewely MR. P53-responsive genes and the potential for cancer diagnostics and therapeutics development. Biotechnol Annu Rev 2001;7:131–164.
128. Thompson TC, Park SH, Timme TL, et al. Loss of p53 function leads to metastasis in ras+myc-initiated mouse prostate cancer. Oncogene 1995;10(5):869–879.
129. Nikiforov MA, Hagen K, Ossovskaya VS, et al. p53 modulation of anchorage independent growth and experimental metastasis. Oncogene 1996;13(8):1709–1719.
130. Sierra A, Castellsague X, Escobedo A, et al. Bcl-2 with loss of apoptosis allows accumulation of genetic alterations: a pathway to metastatic progression in human breast cancer. Int J Cancer 2000;89(2):142–147.
131. Schuster N, Krieglstein K. Mechanisms of TGF-beta-mediated apoptosis. Cell Tissue Res 2002;307(1):1–14.
132. Wakefield LM, Roberts AB. TGF-beta signaling: positive and negative effects on tumorigenesis. Curr Opin Genet Dev 2002;12(1):22–29.
133. Siegel PM, Massague J. Cytostatic and apoptotic actions of TGF-beta in homeostasis and cancer. Nat Rev Cancer 2003;3(11):807–821.
134. Copland JA, Luxon BA, Ajani L, et al. Genomic profiling identifies alterations in TGF-beta signaling through loss of TGF-beta receptor expression in human renal cell carcinogenesis and progression. Oncogene 2003;22(39):8053–8062.
135. Gemma A, Takenaka K, Hosoya Y, et al. Altered expression of several genes in highly metastatic subpopulations of a human pulmonary adenocarcinoma cell line. Eur J Cancer 2001;37(12):1554–1561.
136. Tu WH, Thomas TZ, Masumori N, et al. The loss of TGF-beta signaling promotes prostate cancer metastasis. Neoplasia 2003;5(3):267–277.
137. Ramachandra M, Atencio I, Rahman A, et al. Restoration of transforming growth factor Beta signaling by functional expression of smad4 induces anoikis. Cancer Res 2002; 62(21):6045–6051.
138. Chen T, Carter D, Garrigue-Antar L, Reiss M. Transforming growth factor beta type I receptor kinase mutant associated with metastatic breast cancer. Cancer Res 1998; 58(21):4805–4810.
139. Wikstrom P, Stattin P, Franck-Lissbrant I, Damber JE, Bergh A. Transforming growth factor beta1 is associated with angiogenesis, metastasis, and poor clinical outcome in prostate cancer. Prostate 1998;37(1):19–29.

140. Reiss M. TGF-beta and cancer. Microbes Infect 1999;1(15):1327–1347.
141. Zhu Z, Sanchez-Sweatman O, Huang X, et al. Anoikis and metastatic potential of cloudman S91 melanoma cells. Cancer Res 2001;61(4):1707–1716.
142. Yawata A, Adachi M, Okuda H, et al. Prolonged cell survival enhances peritoneal dissemination of gastric cancer cells. Oncogene 1998;16(20):2681–2686.
143. Swan E, Jasser S, Holsinger F, Doan D, Bucana C, Myers J. Acquisition of anoikis resistance is a critical step in the progression of oral tongue cancer. Oral Oncol 2003;39: 648–655.
144. Jiang K, Sun J, Cheng J, Djeu J, Wei S, Sebti S. Akt mediates Ras downregulation of RhoB, a suppressor of transformation, invasion, and metastasis. Mol Cell Biol 2004;24(12):5565–5576.
145. Itoh N, Semba S, Ito M, Takeda H, Kawata S, Yamakawa M. Phosphorylation of Akt/PKB is required for suppression of cancer cell apoptosis and tumor progression in human colorectal carcinoma. Cancer 2002;94(12):3127–3134.
146. Walsh M, Thamilselvan V, Grotelueschen R, Farhana L, Basson M. Absence of adhesion triggers differential FAK and SAPKp38 signals in SW620 human colon cancer cells that may inhibit adhesiveness and lead to cell death. Cell Physiol Biochem 2003;13(3): 135–146.
147. Frisch S, Vuori K, Kelaita D, Sicks S. A role for Jun-N-terminal kinase in anoikis; suppression by bcl-2 and crmA. J Cell Biol 1996;135(5):1377–1382.
148. Rosen K, Shi W, Calabretta B, Filmus J. Cell detachment triggers p38 mitogen-activated protein kinase-dependent overexpression of Fas ligand J Biol Chem 2002;277(48): 46,123–46,130.
149. Marco R, Diaz-Montero C, Wygant J, Kleinerman E, McIntyre B. Alpha-integrin increases anoikis of human osteosarcoma cells. J Cell Biochem 2003;88:1038–1047.
150. Bao S, Ouyang G, Bai X, et al. Periostin potently promotes metastatic growth of colon cancer by augmenting cell survival via the Akt/PKB pathway. Cancer Cell 2004;5(4):329–339.
151. Raz A, Geiger B. Altered organization of cell-substrate contacts and membrane-associated cytoskeleton in tumor cell variants exhibiting different metastatic capabilities. Cancer Res 1982;42(12):5183–5190.
152. Rodriguez Fernandez J, Geiger B, Salomon D, Sabanay I, Zoller M, Ben-Ze'ev A. Suppression of tumorigenicity in transformed cells after transfection with vinculin cDNA. J Cell Biol 1992;119(2):427–438.
153. Subauste M, Pertz O, Adamson E, Turner C, Junger S, Hahn K. Vinculin modulation of paxillin-FAK interactions regulates ERK control survival and motility. J Cell Biol 2004; 165(3):371–381.
154. Duxbury M, Ito H, Zinner M, Ashley S, Whang E. Focal adhesion kinase gene silencing promotes anoikis and suppresses metastasis of human pancreatic adenocarcinoma cells. Surgery 2004;135(5):555–562.
155. Kurenova E, Xu L, Yang X, et al. Focal Adhesion kinase suppresses apoptosis by binding to the death domain of receptor-interacting protein. Mol Cell Biol 2004;24(10):4361–4371.
156. Windham T, Parikh N, Siwak D, et al. Src activation regulates anoikis in human colon tumor cell lines. Oncogene 2002;21:7797–7807.
157. Griffiths G, Koh M, Brunton V, et al. Expression of kinase-defective mutants of c-Src in human metastatic colon cancer cells decreases Bcl-xL and increases oxaliplatin- and Fas-induced apoptosis. J BIol Chem 2004;279(44):46,113–46,121.
158. Wei L, Yang Y, Zhang X, Yu Q. Anchorage-independent phosphorylation of p130(Cas) protects lung adenocarcinoma cells from anoikis. J Cell Biochem 2002;87(4):439–449.
159. Zeng Q, Chen S, You Z, et al. Hepatocyte growth factor inhibits anoikis in head and neck squamous cell carcinoma cells by activation of ERK and Akt signaling independent of NFkappa B. J Biol Chem 2002;277(28):25,203–25,208.

160. Duxbury M, Ito H, Zinner M, Ashley S, Whang E. EphA2: a determinant of malignant cellular behavior and a potential therapeutic target in pancreatic adenocarcinoma. Oncogene 2004;23:1448–1456.
161. Douma S, van Laar T, Zevenhoven J, Meuwissen R, van Garderen E, Peeper D. Suppression of anoikis and induction of metastasis by the neurotrophic receptor TrkB. Nature 2004; 430:1034–1040.
162. Dionne C, Camoratto A, Jani J, et al. Cell cycle-independent death of prostate adenocarcinoma is induced by the trk tyrosine kinase inhibitor CEP-751 (KT6587). Clin Cancer Res 1998;4(8):1887–1898.
163. Aoyama M, Asai K, Shishikura T, et al. Human neuroblastomas with unfavorable biologies express high levels of brain-derived neurotrophic factor mRNA and a variety of its variants. Cancer Lett 2001;164(1):51–60.
164. Eggert A, Grotzer M, Ikegaki N, et al. Expression of the neurotrophin receptor TrkB is associated with unfavorable outcome in Wilms' tumor. J Clin Oncol 2001;19(3):689–696.
165. Weeraratna A, Dalrymple S, Lamb J, et al. Pan-trk inhibition decreases metastasis and enhances host survival in experimental models as a result of its selective induction of apoptosis of prostate cancer cells. Clin Cancer Res 2001;7:2237–2245.
166. Xiong HQ, Ajani JA. Treatment of colorectal cancer metastasis: the role of chemotherapy. Cancer Metastasis Rev 2004;23(1-2):145–163.
167. Goulian M, Bleile B, Tseng BY. Methotrexate-induced misincorporation of uracil into DNA. Proc Natl Acad Sci USA 1980;77(4):1956–1960.
168. Libra M, Navolanic PM, Talamini R, et al. Thymidylate synthetase mRNA levels are increased in liver metastases of colorectal cancer patients resistant to fluoropyrimidine-based chemotherapy. BMC Cancer 2004;4(1):11.
169. Peters GJ, van Triest B, Backus HH, Kuiper CM, van der Wilt CL, Pinedo HM. Molecular downstream events and induction of thymidylate synthase in mutant and wild-type p53 colon cancer cell lines after treatment with 5-fluorouracil and the thymidylate synthase inhibitor raltitrexed. Eur J Cancer 2000;36(7):916–924.
170. de Angelis PM, Fjell B, Kravik KL, et al. Molecular characterizations of derivatives of HCT116 colorectal cancer cells that are resistant to the chemotherapeutic agent 5-fluorouracil. Int J Oncol 2004;24(5):1279–1288.
171. Cao S, Lu K, Toth K, Slocum HK, Shirasaka T, Rustum YM. Persistent induction of apoptosis and suppression of mitosis as the basis for curative therapy with S-1, an oral 5-fluorouracil prodrug in a colorectal tumor model. Clin Cancer Res 1999;5(2):267–274.
172. Mori T, Fujiwara Y, Yano M, et al. Prevention of peritoneal metastasis of human gastric cancer cells in nude mice by S-1, a novel oral derivative of 5-Fluorouracil. Oncology 2003; 64(2):176–182.
173. Hiraga T, Ueda A, Tamura D, et al. Effects of oral UFT combined with or without zoledronic acid on bone metastasis in the 4T1/luc mouse breast cancer. Int J Cancer 2003;106(6):973–979.
174. Herbst RS, Khuri FR. Mode of action of docetaxel-a basis for combination with novel anticancer agents. Cancer Treat Rev 2003;29(5):407–415.
175. Yen WC, Wientjes MG, Au JL. Differential effect of taxol in rat primary and metastatic prostate tumors: site-dependent pharmacodynamics. Pharm Res 1996;13(9):1305–1312.
176. Wang F, Cao Y, Zhao W, Liu H, Fu Z, Han R. Taxol inhibits melanoma metastases through apoptosis induction, angiogenesis inhibition, and restoration of E-cadherin and nm23 expression. J Pharmacol Sci 2003;93(2):197–203.
177. Jadeski LC, Chakraborty C, Lala PK. Role of nitric oxide in tumour progression with special reference to a murine breast cancer model. Can J Physiol Pharmacol 2002;80(2): 125–135.

178. Xie K, Huang S. Contribution of nitric oxide-mediated apoptosis to cancer metastasis inefficiency. Free Radic Biol Med 2003;34(8):969–986.
179. Dong Z, Staroselsky AH, Qi X, Xie K, Fidler IJ. Inverse correlation between expression of inducible nitric oxide synthase activity and production of metastasis in K-1735 murine melanoma cells. Cancer Res 1994;54(3):789–793.
180. Wang B, Wei D, Crum VE, et al. A novel model system for studying the double-edged roles of nitric oxide production in pancreatic cancer growth and metastasis. Oncogene 2003;22(12):1771–1782.
181. Jadeski LC, Hum KO, Chakraborty C, Lala PK. Nitric oxide promotes murine mammary tumour growth and metastasis by stimulating tumour cell migration, invasiveness and angiogenesis. Int J Cancer 2000;86(1):30–39.
182. Wei D, Richardson EL, Zhu K, et al. Direct demonstration of negative regulation of tumor growth and metastasis by host-inducible nitric oxide synthase. Cancer Res 2003;63(14): 3855–3529.
183. Shi Q, Xiong Q, Wang B, Le X, Khan NA, Xie K. Influence of nitric oxide synthase II gene disruption on tumor growth and metastasis. Cancer Res 2000;60(10):2579–2583.
184. Puhakka A, Kinnula V, Napankangas U, et al. High expression of nitric oxide synthases is a favorable prognostic sign in non-small cell lung carcinoma. Apmis 2003;111(12):1137–1146.
185. Bold R, Termuhlen P, McConkey D. Apoptosis, cancer and cancer therapy. Surg Oncol 1997;6(3):133–142.
186. Foster B, Coffey H, Morin M, Rastinejad F. Pharmacological rescue of mutant p53 conformation and function. Science 1999;286(5449):2507–2510.
187. Zhang J. Apoptosis-based anticancer drugs. Nat Rev Drug Discov 2002;1(2):101–102.
188. Sova P, Ren X, Ni S, et al. A tumor-targeted and conditionally replicating oncolytic adenovirus vactor expressing TRAIL for treatment of liver metastases. Mol Ther 2003;9(4): 496–509.
189. Reed J. Apoptosis-targeted therapies for cancer. Cancer Cell 2003;3:17–22.
190. Ashkenazi A, Pai R, Fong S, et al. Safety and antitumor activity of recombinant soluble Apo2 ligand. J Clin Invest 1999;104(2):155–162.
191. Almond J, Cohen G. The proteasome: a novel target for cancer chemotherapy. Leukemia 2002;16:433–443.
192. Singh SK, Hawkins C, Clarke ID, et al. Identification of human brain tumour initiating cells. Nature 2004;432(7015):396–401.
193. Al-Hajj M, Wicha MS, Benito-Hernandez A, Morrison SJ, Clarke MF. Prospective identification of tumorigenic breast cancer cells. Proc Natl Acad Sci USA 2003;100(7): 3983–3988.
194. Al-Hajj M, Clarke MF. Self-renewal and solid tumor stem cells. Oncogene 2004;23(43): 7274–7282.
195. Welm BE, Tepera SB, Venezia T, Graubert TA, Rosen JM, Goodell MA. Sca-1(pos) cells in the mouse mammary gland represent an enriched progenitor cell population. Dev Biol 2002;245(1):42–56.
196. Smalley M, Ashworth A. Stem cells and breast cancer: A field in transit. Nat Rev Cancer 2003;3(11):832–844.
197. Al-Hajj M, Becker MW, Wicha M, Weissman I, Clarke MF. Therapeutic implications of cancer stem cells. Curr Opin Genet Dev 2004;14(1):43–47.

4

Carcinogenesis

Balance Between Apoptosis and Survival Pathways

Dean G. Tang and James P. Kehrer

Summary

Numerous molecules and pathways have been identified that control various aspects of apoptosis and survival signaling. A dynamic balance between these opposing activities is required to fine-tune biological systems under both normal and stressed conditions. In this chapter, pathways leading to apoptosis are briefly summarized. This is followed by a discussion of various survival pathways that are activated by both prosurvival and proapoptotic stimuli. Finally, a model depicting the simultaneous engagement of both pro-survival and pro-death pathways by stress signals is presented.

Key Words: Apoptosis; survival; stress; NF-κB, reactive oxygen species (ROS); FOXO3a; lipocalin.

1. Introduction

Apoptosis, a form of cell death involving a series of well-organized events requiring active cell participation, is the basis for normal tissue remodeling as well as the end result of certain toxic insults *(1)*. Numerous molecules and pathways control various aspects of apoptosis signaling *(1,2)*, and the fine-tuning of biological functions often requires a dynamic balance between opposing activities. This is most obvious in the nervous and immune systems, but a similar balance between death and survival signals appears to exist and to be essential for normal tissue development and homeostasis. Similarly, a balance among pro- and antiapoptotic factors determines the fate of a cell being stressed by stimuli including xenobiotics (Fig. 1).

The expression or functions of genes controlling both apoptotic and survival pathways are often altered in cancer cells. These changes allow the cells to escape apoptotic signals and proliferate indefinitely. Disrupting the balance of cellular decisions regarding life and death is thus a critical factor in the progression and in the treatment of cancer *(2,3)*. Because most cancer therapies act primarily by inducing apoptosis, the resistance of cancer cells to death has serious clinical implications.

2. Extrinsic and Intrinsic Apoptotic Pathways

Apoptosis, which is the default signal for many cells, is normally suppressed by survival signals that are generated both internally and through molecules derived from neighboring cells. Apoptosis may be initiated by removing or blocking essential survival signals, or in response to the more direct activation of the extrinsic (death

From: *Apoptosis, Cell Signaling, and Human Diseases: Molecular Mechanisms, Volume 1*
Edited by R. Srivastava © Humana Press Inc., Totowa, NJ

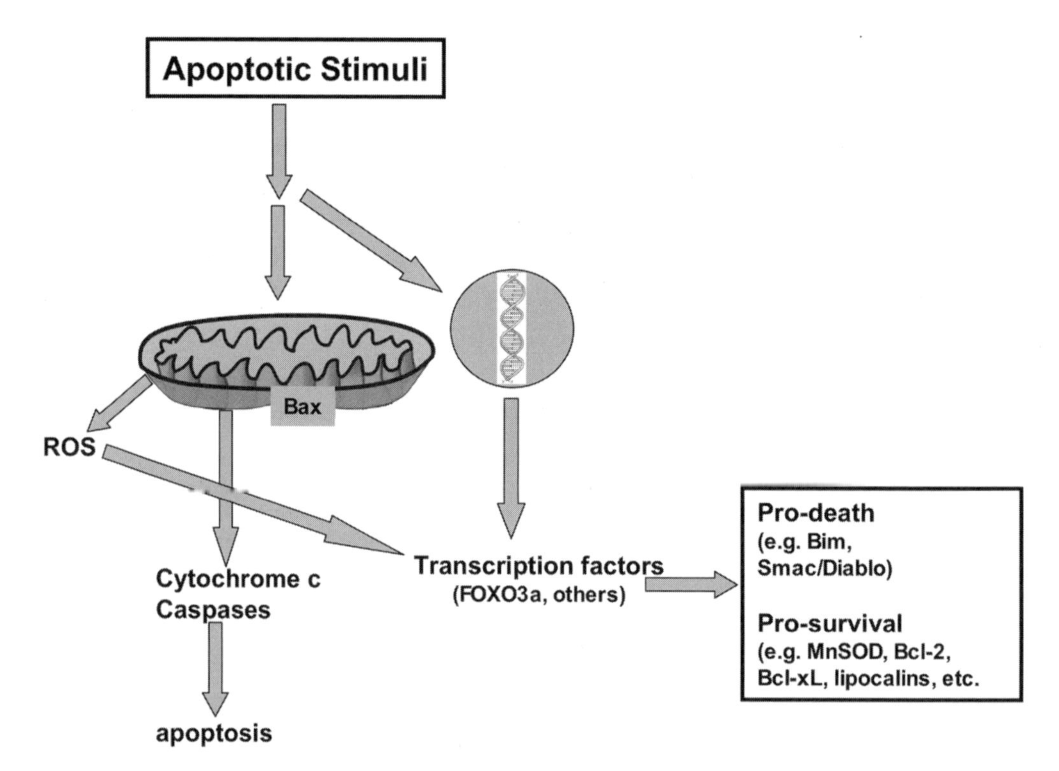

Fig. 1. Schematic depicting the balance among pro- and antiapoptotic factors which determine the fate of a cell being stressed by apoptotic stimuli.

receptor) or intrinsic (mitochondrial) death pathways *(1,4)*. Both pathways converge on the initiator and effector caspases. These enzymes carry out a proteolytic cascade that dismantles cellular structures resulting in the final demise of the cell. At the same time, however, prosurvival pathways may be activated.

All caspases are synthesized as inactive procaspases that become activated by autocatalysis or cleavage by other caspases. There are two groups within the caspase family; initiator and effector caspases *(5,6)*. Initiator caspases have long prodomains that contain a protein–protein interaction platform for the recruitment of procaspases into activating protein complexs including the death-inducing signaling complex (DISC) and the apoptosome. Initiator caspases such as caspase-8 and -10 have a death effector domain (DED) whereas caspase-1, -2, -4, -5, -9, -11, and -12 have a caspase-activating recruitment domain (CARD). Effector caspases lack the long N-terminal nonenzymatic prodomain. These caspases include caspase-3, -6, and -7. They are also known as the executioner caspases because they are responsible for the physiological manifestations of apoptosis.

In the extrinsic death pathway, activation of the death receptors by external ligands recruits initiator caspases into the DISC. Major death receptors include Fas (CD95), tumor necrosis factor (TNF) receptor, DR3 (Apo-3), DR4 (TRAIL-R1), and DR5 (TRAIL-R2) *(7)*. Once these death receptors are stimulated by their respective ligands, components of the DISC machinery are recruited through interaction of the death domains (DD). Fas-associated death domain (FADD) is a common factor that functions as a critical adaptor protein to recruit caspase-8 and -10 through its DED. FADD associates directly with Fas, DR4 and DR5 receptors, but binds to the TNF receptor through association with the TNF receptor 1-associated death domain (TRADD). The proximity of caspases-8 and -10 in the DISC results in their activation presumably by an allosteric mechanism *(8–11)*. These initiator caspases, once activated, can then cleave and activate downstream effector caspases.

The intrinsic apoptotic pathway works through mitochondria, which contribute to apoptosis in at least three ways: (1) the release of pro-apoptotic molecules, (2) increased production of reactive oxygen species (ROS), and (3) impaired ATP production. The release of apoptotic mediators from mitochondria is a critical event in the intrinsic pathway. Cytochrome *c* is one of the key factors released *(12)*. Upon its release, cytochrome *c* promotes the formation of a proapoptotic complex called the apoptosome. This complex is comprised of apoptotic protease-activating factor-1 (Apaf-1), caspase-9, and ATP. In this complex, caspase-9 is activated *(8,13)* and is then able to cleave downstream effector caspases like caspase-3.

The ability of the activated caspases to initiate apoptosis is further regulated by the inhibitor of apoptosis proteins (IAPs). IAPs (especially XIAP) inhibit both activated caspase-9 and -3 *(13)*. In addition, XIAP and IAP1/2 have a carboxyl-terminal motif found in RING-finger proteins that allows them to function as ubiquitin ligases promoting the proteasomal degradation of bound caspases *(14)*. To counter this, the mitochondria release, in addition to cytochrome *c,* other proapoptotic factors including second mitochondria-derived activator of caspases (Smac/DIABLO) and Omi/HtrA2. These factors bind to and sequester XIAP allowing activated caspase-3 to function *(15–17)*. Other proapoptotic mitochondrial intermembrane space proteins that are released during apoptosis include the apoptosis-inducing factor (AIF) and endonuclease G. These factors cause apoptosis independent of caspases. AIF translocates to the nucleus and causes chromatin condensation and large-scale DNA fragmentation while EndoG translocates into the nucleus and helps digest nuclear DNA *(18,19)*.

The mitochondrial production of ROS, and the loss of mitochondrial membrane potential, may contribute to apoptosis. When the membrane potential is lost, ATP is not synthesized and ROS accumulate *(20)*. Loss of the mitochondrial membrane potential can be caused by a permeability transition pore (PTP) which contains the inner membrane protein, adenine nucleoside translocator (ANT) and the outer membrane protein, voltage dependent anion channel (VDAC) *(21,22)*. The PTP pore has been implicated in the release of mitochondrial factors upon apoptotic stimuli. In addition, several members of the Bcl-2 family including Bax and Bak are involved in the apoptosis signaling caused by changes in the PTP (for reviews, *see* refs. *23–25*).

The role of ROS in apoptosis is still not clearly understood mechanistically. There is evidence that antioxidants can abolish the apoptotic response of many diverse stimuli. Thus, ROS may be an important mediator of apoptosis *(26)*. In addition, antiapoptotic Bcl-2 family members decrease ROS production in response to apoptotic stimuli *(27,28)*. Recent evidence highlights a specific role of ROS in the control of redox balance required for proliferation signaling. This includes the activation of the transcriptional factor activator protein-1 (AP-1) through the mitogen-activated protein kinase (MAPK) family (reviewed in ref. *29*). This ultimately leads to the transcriptional upregulation of genes involved in cellular proliferation. Thus, ROS play a dual role in mediating apoptosis as well as mitogen- or survival factor-induced cell proliferation and survival (*see* **Subheading 3.4.**).

Crosstalk between the extrinsic and intrinsic apoptosis pathways exists. This occurs mainly through cleavage of the Bcl-2 family member, Bid by active caspase-8 into a truncated form t-Bid. Once cleaved, t-Bid translocates to the mitochondria and either directly binds to and induces the oligomerization of Bax, or binds to and inactivates prosurvival Bcl-2 family proteins like Bcl-2 and Bcl-xL *(30)*. Because caspase-8 is the common target for DISC activation through the DRs, Bid therefore bridges the extrinsic and intrinsic pathway.

3. Survival Pathways

3.1. Pro- and Antiapoptotic Bcl-2 Proteins

The proapoptotic Bcl-2 proteins, Bax and Bak, function as gatekeepers to the release of apoptotic factors from the mitochondria. Opposing their activities are the gatekeeper, antiapoptotic Bcl-2 family proteins. The antiapoptotic Bcl-2 family members, including Bcl-2, Bcl-xL, Bcl-w, Mcl-1, Boo and Bcl-B, share sequence conservation in all four BH domains (i.e., BH1 to BH4). In contrast, the proapoptotic Bcl-2 proteins, Bax, Bak and Bok, share sequence similarity in only the first three BH domains. The BH3-only proteins, which include Bid, Bik, Bim, Bad, Bnip3, Blk, Bmf, Hrk, Noxa and Puma, exist in inactive forms during normal conditions and become activated or induced in response to apoptotic stimulation. The balance of these pro- and antiapoptotic Bcl-2 proteins contributes to the survival or death of a cell.

There are several models of how antiapoptotic Bcl-2 family proteins promote cell survival. One involves binding of the anti-antiapoptotic proteins to the BH3-only proteins preventing them from activating the gateway proteins Bax, Bak, or they can directly bind to Bax or Bak preventing their activation *(31,32)*. Other potential models involve Bcl-2 preventing the intracellular calcium flux, pH and ionic changes that occur early during apoptosis *(33,34)*, or binding to and inhibiting a membrane bound protein X that is an activator of Bax and Bak oligomerization *(35,36)*.

There are four mechanisms by which antiapoptotic Bcl-2 proteins can function and contribute to carcinogenesis. The first involves their subcellular localization. Bcl-2 and Bcl-xL are found on the mitochondrial outer membrane where they can inhibit the oligomerization of Bax and Bak *(37)*. Bcl-2 and Bcl-xL also localize on the ER membrane where they may affect calcium uptake or release during apoptosis *(38)*. In addition, Bcl-2 localizes in the nuclear membrane and may modulate nuclear trafficking of transcription factors such as NF-κB, AP1, CRE, and NFAT *(39)*. The second mechanism involves levels of expression. Overexpression of Bcl-2 family proteins confers resistance to many apoptotic stimuli *(40,41)*. In addition, the antiapoptotic Bcl-2 proteins are often overexpressed in cancer cells. In fact, Bcl-2 was discovered as an overexpressed gene in follicular lymphomas as a result of a translocation of the gene from chromosome 14;q32 to chromosome 18;q21 *(42)*. The third mechanism involves phosphorylation *(43)*. For example, phosphorylation of Bcl-2 at serine residues in some cell systems results in the loss of its antiapoptotic functions *(44,45)* whereas phosphorylation at serine 70 appears necessary for its antiapoptotic functions *(46)*. Other studies show that phosphorylation in the unstructured loop domain of Bcl-xL and Bcl-2 diminishes their antiapoptotic activities *(47)*. Finally, posttranslational mechanisms can affect Bcl-2 family members functions. For example, cleavage of Bcl-2 by caspase-3 results in a truncated protein with proapoptotic activity *(48)*. This occurs late during apoptosis thus playing an amplification role.

3.2. The NF–κB Pathway

The NF-κB transcription factor is a classic example of a survival pathway activated by apoptotic stimuli. NF-κB is normally sequestered in the cytoplasm by IκBs (inhibitors of NF-κB). When NF-κB is released from the IκBs, it translocates into the nucleus, binds to DNA, and transcriptionally activates various antiapoptotic genes including IAP1/2, XIAP, Bcl-xL, caspase-8-FLICE inhibitory protein (c-FLIP), and Traf1/2 *(49,50)*. These factors work together to block apoptosis at multiple steps along the apoptotic cascade. For instance, in the Fas receptor pathway, c-FLIP upregulated by NF-κB interacts with FADD and procaspase-8 to prevent the activation of procaspase-8 *(51)*. IAPs bind to and inhibit the activities of activated caspases *(52)*. TRAF1/2 are adaptor proteins involved in the TNF receptor signaling pathway *(53)*.

In order to overcome the antiapoptotic function of NF-κB, several key components of this pathway, including RelA, can be degraded by caspases *(54)*. Caspase-3 also cleaves IκBα creating a super repressor of NF-κB *(55)*. The IKK complex can be inactivated by caspases *(56)*. Lastly, receptor-interacting protein (RIP) and TRAF2, both of which are involved in NF-κB activation, are also substrates of caspases *(57,58)*. Thus, the NF-κB pathway is a complex network integrating both pro- and antiapoptotic signals upstream and downstream of this transcription factor.

3.3. The PI3K/Akt Pathway

Another complex survival network involves phosphatidylinositol 3 kinase (PI3K) and Akt. Akt is a serine/threonine kinase first identified in humans as two genes homologous to the viral oncogene *v-akt* that causes a form of mouse leukemia *(59)*. Additional studies showed that these homologues coded for a kinase similar to protein kinase C (PKC), leading to the alternate name of protein kinase B (PKB). There are now three members of this family, referred to as Akt1,2,3 (or PKBα,β,γ) *(60)*. The three genes show over 80% homology at the amino acid level and have a rather broad distribution. Recently, knockouts of each isoform have been generated (reviewed in ref. *61*). These knockouts have shown diverse phenotypes suggesting distinct regulatory roles for the isoforms. However, all individual knockouts are viable, whereas a double Akt1-Akt2 knockout is lethal shortly after birth *(61)*. This indicates functional redundancy among these proteins.

Akt is activated through a dual process involving translocation to the plasma membrane and phosphorylation at Ser473 in the C-terminus and Thr308 in the activation loop. The phosphorylation at Thr308 by PDK1 is considered essential because mutation to an alanine moiety, which cannot be phosphorylated, greatly diminishes Akt activity *(62)*. Akt phosphorylation involves the generation of phosphatidylinositol-3,4,5-trisphosphate (PIP_3) from phosphatidylinositol-4,5-bisphosphate (PIP_2) by phosphoinositide 3-kinase (PI3K). Binding of the pleckstrin homology domain of Akt to PIP_3 relocalizes the cytoplasmic Akt protein to the plasma membrane, allowing PDK1 to phosphorylate Akt at Thr308. The mechanism of Ser473 phosphorylation is unclear. Both autophosphorylation and phosphorylation by several distinct kinases (one called PDK2) have been implicated *(63,64)*. A further important regulatory feature of Akt activation is PTEN, a phosphatase whose main physiological substrate is PIP_3. PTEN thus functions as a negative regulator of PI3K signaling.

A number of kinases contain the consensus sequence for activation by PDK1 *(65)*. Thus, whereas Akt has received the most attention, other downstream kinases may also be affected when PDK1 activity is altered. One of significant interest is serum and glucocorticoid regulated protein kinase (SGK), which has 54% sequence identity in the catalytic domain with Akt *(66)*. SGK differs from Akt by lacking the pleckstrin homology domain and is thus activated in the cytosol rather than after membrane binding. In addition, the consensus sequence phosphorylated by SGK differs from that of Akt *(67)*. This, and some other differences related to the hydrophobic amino acid composition of peptide substrates, suggests that SGK and Akt have complementary rather than redundant functions *(67)*. Nevertheless, like Akt, SGK is induced by several stimuli including injury, suggesting it has a role in cell survival.

Akt activates various downstream targets that regulate apoptosis, cell cycle, DNA repair, nitric oxide production, and glycogen metabolism *(68)*. Akt protects the cell from apoptosis by phosphorylating various apoptotic regulators such as Bad, caspase-9 and FKHR1 (Forkhead receptor-1). Akt phosphorylation of Bad induces its interaction with 14-3-3 protein, which causes a conformational change in Bad leading to its phosphorylation by protein kinase A. This disrupts the ability of Bad to bind to Bcl-2 and Bcl-xL thus freeing these factors and allowing them to inhibit apoptosis *(69)*. Akt phosphorylation of FKHR1 also leads to its binding to 14-3-3 proteins in the cytosol thus preventing FKHR1 from translocating to the nucleus. This prevents FKHR1 from activating proapoptotic genes such as Bim and FasL *(70)*. Similarly, Akt phosphorylation of procaspase-9 prevents its activation by the apoptosome *(71)*. Akt is also an indirect negative regulator of p53. It phosphorylates Mdm2, increasing its ability to translocate into the nucleus where it binds to p53 and promotes its degradation *(72)*. Lastly, Akt can activate the NF-κB pathway by phosphorylating IKKα leading to the phosphorylation and degradation of IκB *(73,74)*. Therefore, Akt can promote cell survival by antagonizing both the extrinsic and intrinsic apoptotic pathways.

3.4. The MAPK/ERK Pathway

Extracellular regulated kinase (ERK) is a member of the MAPK family. Activation of ERK generally protects a cell from apoptosis through the phosphorylation and activation of downstream transcription factors that regulate antiapoptotic molecules such as IAPs, TRAF1/2 and Bcl-xL *(75)*. The Raf/MEK/ERK pathway also affects cell-cycle regulatory proteins, thereby regulating cell-cycle progression. This pathway is activated by extracellular signals that recruit the adaptor protein Grb2 (growth factor receptor bound protein 2). This protein then binds to the cytoplasmic side of cell surface receptors like EGFR and PDGFR through its Src homology 2 domain *(76)*. Grb2 is constitutively bound to son of sevenless (SOS), a guanine nucleotide exchange factor that, following activation, exchanges the GDP for GTP on RAS *(77)*. Upon Grb2 recruitment to the plasma membrane SOS gets activated *(77)*. This causes a conformational change in RAS allowing it to bind to a MAPK, RAF1. RAF1 is activated by binding to RAS and phosphorylates the MAPK, MEK1 and MEK2. These kinases in turn phosphorylate

ERK1 and ERK2. Following activation, ERKs translocate into the nucleus and phosphorylate a variety of substrates. These include the 90 kDa ribosomal S6 protein kinase ($p90^{rsk}$), the cytosolic phosphatase A2, and several transcription factors like NF-κB, c-Myc, Ets, CREB and AP-1 *(75)*.

The contributions of different MAPK family members to apoptosis have been examined following withdrawal of NGF from rat PC-12 pheochromocytoma cells *(78)*. They showed that this led to the sustained activation of JNK and p38-MAPK and the inhibition of ERKs. The combined effects of dominant negative as well as constitutively activated forms of various factors in these three pathways show that the simultaneous activation of JNK and p38MAPK and the inhibition of ERK is needed for induction of apoptosis in these cells *(78)*. This implies that ERK is a survival factor that needs to be silenced in order for apoptosis to occur.

Recently, multiple studies have revealed a more direct prosurvival mechanism of the ERK pathway. Ras/Raf/MAPK/ERK activation by, serum, endothelial growth factor (EGF), platelet-derived growth factor (PDGF), insulin, integrin-mediated adhesion, or thrombin, results in the phosphorylation of a critical BH3-only protein, Bim *(79–86)*. Phosphorylation of Bim either inhibits its interaction with Bax *(79)* or leads to proteosome-dependent degradation *(85)*.

4. Cell Survival Signaling During Apoptosis

Although great progress has been made in understanding the details of apoptosis pathways, little is known about how cells initially respond to apoptotic stimuli. For example, when a population of cycling cells is stimulated by an apoptotic signal, do they immediately enter the apoptotic mode or do they first stage a defensive response? Alternately, do the stimulated cells simultaneously activate both prosurvival and prodeath mechanisms? Is it then the balance between survival and death signals that ultimately determines when and whether the stimulated cells will die? Or is it possible that an adaptive response is activated that helps prevent excess cell death? This would explain the observation that in the presence of apoptotic stimuli, surviving cells become more resistant to apoptosis *(87)*. The work described below sheds some light on these questions.

4.1. Prosurvival Mechanisms Induced by Apoptotic Stimuli

Many stressors cause an early mitochondrial activation characterized by a rapid induction of respiration-related proteins including apocytochrome *c* and cytochrome *c* oxidase. These proteins are then rapidly imported into the mitochondria to enhance respiration leading to early membrane hyperpolarization, increased oxygen consumption, and maintenance of ATP levels *(88,89)*. These responses precede the release of holocytochrome *c* from the mitochondria and the subsequent activation of caspases *(89)*. These observations suggest that cells, upon apoptotic stimulation, rapidly mobilize defensive mechanisms in order to extend their survival.

Many drugs, chemicals, ligands, and conditions that induce apoptosis also induce an early activation of mitochondrial systems characterized by cytochrome *c* upregulation and increased mitochondrial respiration. The upregulation of the mitochondria respiratory chain proteins and increased respiratory activity most likely represent one aspect of the global mitochondrial activation aimed, perhaps, to maintain appropriate ATP levels critical for various cell activities associated with apoptosis, as well as for cell survival. In addition to maintaining ATP levels, increased apocytochrome *c* in the cytosol and upregulated holocytochrome *c* in the mitochondria also possess antiapoptotic and prosurvival functions *(90–92)*.

Another class of prosurvival molecules induced or activated by many apoptotic stimuli are antiapoptotic Bcl-2 family proteins, in particular, Bcl-2 and/or Bcl-xL *(93)*. In response to DNA damage, trophic factor deprivation, and a mitochondrial toxin, Bcl-2 and/or Bcl-xL are rapidly induced *(93)*. This induction clearly plays a prosurvival role as inhibition of the upregulation of Bcl-2 and/or Bcl-xL using specific siRNAs facilitated cell death *(93)*. Similarly, hypoxic treatment selectively upregulates Bcl-xL leading to generation of death-resistant cells *(94)*. Ultraviolet irradiation eliminates the antiapoptotic Mcl-1 but also induces increased targeting of Bcl-xL to the mitochondria *(95)*. Moreover, various apoptotic stimuli upregulate Bcl-2 protein levels via a mechanism dependent on the internal ribosomal entry site *(96)*. Finally, ultraviolet-A can upregulate the Bcl-xL protein levels by modulating the 3′-untranslated region *(97)*.

The third class of prosurvival molecules is the superoxide dismutases (SODs) that function by removing the ROS superoxide anions *(98,99)*. We found that all apoptotic stimuli tested upregulated the levels of both MnSOD and Cu/ZnSOD at about the same time when proapoptotic Bim is induced *(93)*. MnSOD siRNA blocked the induction of MnSOD and also enhanced apoptosis *(93)*, suggesting that the upregulated MnSOD is serving a prosurvival function.

A fourth class of prosurvival molecules induced by apoptotic stimuli (and stressful stimuli in general) includes various chaperone and cochaperone proteins. Multiple heat-shock proteins (Hsps) are cytoprotective *(100–103)*. Indeed, we have observed that Hsp60 is rapidly upregulated and/or released from the mitochondria to the cytosol in response to apoptotic stimulation *(89)*, presumably to extend cell survival. The mitochondrial Hsp70 interacts with $p66^{Shc}$, a molecule implicated in determining the cell's lifespan, to extend cell survival in the presence of stress signals *(104)*. Similarly, the bacterial Hsp60 (GroEL) protects epithelial cells from apoptosis induction via activation of the ERK pathway *(105)*.

Cell cycle-arrested cells generally survive better than proliferating cells *(106)*. Thus, although the lower rates of metabolism in resting cells is likely important in their improved survival, cell-cycle inhibitors such as $p27^{KIP1}$ and $p21^{WAF-1}$, may also represent prosurvival molecules. For example, $p21^{WAF-1}$ is a critical prosurvival factor transactivated by p53 *(107,108)*; overexpression of p21 confers on colon cancer cells resistance to apoptosis induction by chemicals in both p53-depedent and p53-independent manners *(109)*. In contrast, decreased p21 expression sensitizes cells to apoptosis *(110,111)*. Indeed, we have observed an inverse correlation between p21 expression and apoptosis. When LNCaP cells, a p53-wt prostate cancer cell line, are stimulated with γ-irradiation, etoposide, doxorubicin, or taxol, γ-irradiation and etoposide significantly induced p21 protein expression with little cell death whereas doxorubicin and taxol did not upregulate p21 but caused obvious cell death (unpublished observations).

4.2. Secreted Proteins

A role for secreted proteins in how cells respond to apoptotic stimuli has been little studied. However, recent data demonstrate that the lipocalins 24p3/NGAL (neutrophil gelatinase associated lipocalin) are profoundly induced in response to several proapoptotic xenobiotics *(112)*. As a secreted protein, 24p3/NGAL may facilitate the survival of surrounding cells.

Lipocalins are a diverse family of more than 20 small soluble, and often secreted, proteins. These proteins have a large degree of diversity at the sequence level (only 20% identity), although most share three conserved motifs. These proteins are defined as lipocalins largely on the basis of their three-dimensional structure, which is comprised of a single eight-stranded, continuously hydrogen-bonded anti-parallel β-barrel. The functions of the lipocalins are not well understood. The cavity in the β-barrel is thought to bind and transport a variety of lipophilic substances *(113)*. In addition, they can bind to specific cell-surface receptors and may deliver ligands, including iron, growth, or survival factors to the cell by receptor-mediated endocytosis *(114,115)*. Importantly, lipocalins appear to have roles in cell regulation, both in terms of proliferation and differentiation *(116)*. This has suggested a number of functions related to cancer where elevated expression of lipocalins is often observed *(117)*. Several lipocalins such as α_1-microglobulin, $\alpha_2\mu$-globulin, and orsomucoid are expressed in different forms of cancer and it has been proposed that lipocalins mediate cell signaling either by delivering lipophilic ligands such as retinoids and fatty acids, or by inhibiting proteases *(117)*.

24p3 was first identified as an overexpressed gene in SV40 infected primary kidney cells from mice *(118)*. NGAL, the human analog of 24p3, (NGAL and 24p3 are 62% identical in amino acid sequence *[119]*, but only about 20% identical in nucleotide sequence), was first purified from human neutrophils because of its association with gelatinase *(120)*. Although the association with neutrophil gelatinase suggests the potential to modulate this enzyme, various studies have shown this is not true *(120)*.

Several functions of 24p3/NGAL have been identified but precisely which, if any, are important under normal circumstances is unknown. Functions related to cancer have been suggested *(117)*, but overall little is known about the role of 24p3/NGAL in cell signaling, proliferation, and apoptosis. Some data suggested a role in inflammation, whereas other data indicate NGAL has an important role in iron metabolism *(121)*. A number of inducers of this gene have been found including serum, lipopolysaccharide, various growth factors, retinoic acid, glucocorticoids, and phorbol esters *(117,119)*. In addition, we have demonstrated that several xenobiotics that induce apoptosis, are potent inducers of 24p3/NGAL *(112, unpublished data)*.

24p3 is expressed at high levels in the mammary gland and uterus where it is proposed to induce neutrophil apoptosis thereby allowing selected cells to survive involution *(122)*. The work of Divireddy et al. *(123)* suggested that extracellular 24p3 had biologic activity in terms of stimulating apoptosis in mouse pro-B FL5.12 cells. On the other hand, as a lipocalin, this protein can bind small lipophilic substances and may thus play a role in transport of signaling species, particularly survival/growth factors. Although the ability of secreted 24p3 to induce apoptosis is uncertain, NGAL appears to have a survival function because recombinant NGAL protein was nontoxic to cells, an antibody to NGAL was a highly effective apoptosis inducer, overexpression of NGAL was protective, and underexpression of NGAL enhanced drug-induced apoptosis (unpublished data). The expression of 24p3/NGAL in all tissues examined *(119)*, and the high levels of expression of 24p3/NGAL by various cancers *(117)* as well as in normal mammary gland, uterus *(122)*, and testes, is also consistent with a survival function. Finally, the down regulation of *ExFABP,* a secreted lipocalin that binds fatty acids, induces apoptosis *(124)*, and is supportive of a survival function.

4.3. Transcription Factors Involved in Apoptotic Stimuli-Activated Prosurvival Genes

Other prosurvival signaling mechanisms may exist in stressed cells. For example, apoptotic stimulation may result in a rapid phosphorylation of the translation initiation factor-2, leading to the cessation of *de novo* protein synthesis and providing cytoprotection *(125)*. Overall, induction of prosurvival mechanisms by apoptotic stimuli seems to represent a general phenomenon. Indeed, even apoptosis induced by TNF-α and Fas (discussed below) and death kinase PKR *(126)* is preceded by an early phase of NF-κB-mediated prosurvival activities. The induction of prosurvival molecules apparently plays a critical role in extending cell survival as prevention of the induction of Bcl-2, Bcl-xL, or MnSOD by apoptotic stimuli accelerates cell death *(93)*.

In the above examples, the pro-survival molecules are induced either simultaneously with or slightly prior to the induction of various prodeath molecules. Our recent work demonstrates that this occurs at the transcriptional level *(93)*, particularly involving master transcription factors such as NF-κB, FOXO3a, p53, Rb, E2F1, and c-Myc.

NF-κB is clearly an important transcription factor in mediating the pro-survival signaling in response to inflammatory, apoptotic, and stress stimuli. Although NF-κB activation on some genetic backgrounds can lead to apoptosis by, for example, stabilizing p53 *(127)*, repressing the induction of antiapoptotic genes *(128)*, or collaborating with the adenine nucleotide translocator *(129)*, most experimental data suggest that increased NF-κB activity is associated with resistance to apoptosis *(130–137)*. The prosurvival function of NF-κB is mainly associated with its induction of prosurvival genes such as IAP, Bcl-xL, and FLIP, as well as SNF1/AMP kinase-related kinase *(138)* and the ferritin heavy chain *(139)*. The prosurvival function of NF-κB is well illustrated by the response of cells to the TNF-α family proteins *(140–146)*. Two sequential signaling complexes are formed upon TNF-α binding to the TNF receptor 1 *(141)*. The plasma membrane-bound complex I is rapidly formed upon receptor activation and contains TNF receptor 1, adaptor protein TRADD, DD-containing kinase RIP1, and TRAF-2, leading to NF-κB activation. Then, complex I exits the receptor and forms a different, long-lived complex II. This complex localizes mainly in the cytosol and contains apoptotic proteins FADD, caspase-8, and caspase-10 in addition to TRADD, RIP1, and TRAF-2. The activation of complex II results in cell death *(141)*. Thus, TNF-α induces the complex II-mediated apoptosis only when the complex I-initiated pro-survival signal (i.e., NF-κB) is not activated.

Recent evidence indicates that CD95 (Fas ligand) and TRAIL, TNF-α family members conventionally thought to be solely proapoptotic, also activate NF-κB prior to activating the DISC and caspase-8 *(140,142–146)*. This explains why these death ligands do not induce apoptosis in many cancer cells. In fact, most epithelial cancer cells appear to be so-called type II cells *(140)* and CD95 stimulation of these cells not only fails to kills them but actually promotes cell migration and invasion *(146)*, possibly through NF-κB activated urokinase-type plasminogen activator and the SNF1/AMP kinase-related kinase *(138)*. Intriguingly, different from the activation of NF-κB by TNF-α *(141)*, NF-κB activation by Fas is mediated through FADD, caspase-8, and RIP and is inhibited by FLIP *(144,145)*. The concomitant prosurvival and prodeath functions of death ligands such as CD95 and TRAIL are not unique. An increasing number of molecules once thought of as proapoptotic only, including the BH3-only proteins Bad *(147,148)* and BNIP *(149)*, the mitochondrial protease Omi *(150)*, multi-BH protein Bak *(151,152)*, and activated caspases *(153,154)* have functions unrelated to apoptosis; even pro-survival functions.

FOXO3a has emerged as a critical regulator of cell death and survival. FOXO3a is a mammalian homologue of *Caenorhabditis elegans* DAF-16 and one of the FOXO (Forkhead box, class O) subclass of Forkhead transcription factor family. FOXO3a plays a critical role in coordinating cell survival and death and regulating stress response and longevity *(155)*. The activities of the FOXO proteins are regulated by phosphorylation. ISGK *(67)* as well as Akt can phosphorylate (and thus inactivate) *(18,25)* Forkhead in rhabdomyosarcoma (FOXO1), FKHR-Like 1 (FOXO3a), and acute-lymphocytic-leukemia-1 fused gene from chromosome X (FOXO4) *(156)*. Phosphorylation appears to take place in the nucleus causing FOXOs to bind 14-3-3 proteins. The FOXO/14-3-3 complex is exported to the cytoplasm where the inactive FOXO proteins remain bound to 14-3-3 proteins. In the absence of survival signals, cytoplasmic FOXO is dephosphorylated by PTEN *(156)*, causing dissociation from 14-3-3 proteins and allowing nuclear import, which activates gene expression presumably through a highly conserved monomeric DNA-binding domain *(157)*.

There is a general consensus that phosphorylation by Akt is critical in regulating the function of FOXO factors, and as Akt is inactivated, these transcription factors can become dominant *(158)*. Overexpression of FOXO members can activate some proapoptotic genes and induce apoptosis, although these effects seem to be largely restricted to nontransformed hematopoietic cells *(156)*. FOXO transcription factors also have a role in the regulation of some antiapoptotic genes and can control genes that induce cells to enter a more quiescent cell state where protective mechanisms can act *(159)*.

FOXO3a can transcriptionally activate proapoptotic Bim, TRAIL and TRADD and antiapoptotic MnSOD and cyclin-dependent kinase inhibitor $p27^{KIP1}$ *(93,155)*. FOXO3a also inhibits cell-cycle progression by downregulating cyclin D1 *(93,155)*. We found that FOXO3a is involved in directly regulating the apoptotic stimuli activated Bim and MnSOD *(93)*, two molecules that contain the FOXO3a sites in their promoter regions. Interestingly, several other prosurvival molecules including Cu/ZnSOD, Bcl-2, Bcl-xL, and cytochrome *c* also appear to be partially regulated by FOXO3a as their induction is inhibited by downregulating this transcription factor (dominant negative or siRNA) or in $FOXO3a^{-/-}$ fibroblasts *(93)*. Whether FOXO3a directly or indirectly regulates these molecules remains to be determined.

Other transcription factors, either individually or in combination, may also be involved in the transcriptional activation of both pro- and antiapoptotic molecules in response to apoptotic stimuli. For example, p53 transactivates both prosurvival $p21^{WAF\text{-}1}$ *(107–111)* and multiple proapoptotic molecules such as BH3-only proteins (Bid, PUMA, and Noxa), Bax, and procaspases *(160,161)*. p53 may also protect cells from apoptosis by inhibiting JNK activation *(162)*. Rb similarly regulates the transcription of several life-and-death genes *(163)*. E2F1 not only transcriptionally regulates cell cycle-related genes but also cell death genes including Apaf-1 and caspases *(164)*. c-Myc transcriptionally regulates molecules involved in cell cycle progression, survival, and death *(106)*. Finally, the transcription factor Nrf2 transcriptionally activates pro-survival genes during apoptotic stimulation, in particular, during endoplasmic reticular stress *(165,166)*.

Recent data point to crosstalk between these master transcription factors. For example, activation of NF-κB can lead to decreased stabilization of p53 and further enhance cell survival *(130)*. On the other hand, p53 activation can lead to phosphorylation of FOXO3a and changes in its subcellular localization resulting in inhibition of FOXO3a transcriptional activity *(167)*. Newly emerging evidence also makes a connection between FOXO3a and NF-κB. One study suggests that IκB kinase inhibits FOXO3a through physical interaction and phosphorylation independent of Akt, which promotes FOXO3a proteolysis via the Ub-dependent proteasome pathway *(168)*. Another study suggests that FOXO3a negatively regulates NF-κB and FOXO3a deficiency results in NF-κB hyperactivation and T-cell hyperactivity *(169)*. Overall, several transcription factors activated in response to apoptotic or stressful stimuli clearly mediate the activities of both survival and death pathways.

4.4. Activators Upstream of the Transcription Factors

Master transcription factors such as FOXO3a are themselves induced at the transcriptional level by apoptotic stimuli with distinct mechanisms of action in several genetically distinct cell types *(93)*. This suggests that a common mechanism may be operating to mediate the transcriptional activation of these factors. One possible mechanism involves ROS. These reactive species appear to function as critical apical signaling molecules to activate FOXO3a and perhaps other multifunctional transcription factors *(93)*. This concept is supported by several lines of evidence. First, increased ROS are an early event following many different apoptotic stimuli *(88,89,93)*. Second, many of the antiapoptotic molecules induced by proapoptotic stimuli, including cytochrome *c*, Bcl-2, Bcl-xL, and the SODs, are related to, or induced by, oxidative stress. This raises the possibility that these molecules are induced during apoptotic stimulation to protect against further increases in ROS. Third, suppression of ROS by inhibitors or scavengers inhibits apoptotic signal-induced upregulation of FOXO3a as well as its prodeath and prolife targets. Conversely, artificially generated oxidative stress upregulates FOXO3a and its targets (93). Fourth, the function of ROS as signaling molecules that activate multifunctional transcription factors and ultimately determine the life and death of a cell is consistent with the well established dual survival and death functions of ROS *(99,170–174)*. Finally, FOXO3a *(175–180)*, NF-κB *(99,181)*, and p53 *(99,182)* are well-known to be regulated by and also respond to oxidative stress.

5. A Model and Its Implications

The findings discussed above support our proposal that proapoptotic stimuli cause an early mitochondrial activation leading to rapid generation of ROS, which activate master transcription factors such as FOXO3a and NF-κB. These transcription factors, in turn, activate multiple molecular targets with both proapoptotic and prosurvival functions. These targets appear to be activated as soon as the cell senses stress, and the activation is independent of the extent of the stress and whether or not the final outcome is apoptosis *(93)*. The determination of the ultimate outcome for a cell (survival or death) seems to be a function of the strengths and timings of the different prosurvival and prodeath signals *(93)*. Presumably, the cell can integrate a myriad of signals leading to the final decision of life or death.

This model (*see* ref. *93* and Fig. 8 therein) has several important implications. First, because both prodeath and prosurvival molecules are activated, the sensitivity of any target cells to the induction of apoptosis will be determined by the balance of these two opposing signals. Second, the induction of prosurvival molecules may occur prior to induction of prodeath molecules. Significant cell killing will occur *only when* proapoptotic signals overwhelm the prosurvival signals or when the latter are eliminated. This prediction is consistent with the recent demonstration that apoptosis elicited by TNF-α proceeds in two steps: an early step, where prosurvival signaling mediated by NF-κB dominates, and a later step, where prodeath signaling mediated by caspase-8/10 dominate. Cell death occurs only when step two is activated or when step one is inactivated *(141)*. Third, because cells respond to apoptotic stimulation asynchronously and differently, some cells may preferentially upregulate prosurvival molecules, rendering them relatively resistant to further apoptotic stimulation, as often observed in therapy-resistant cancer cells. Similarly, such resistant cells may secrete molecules that protect surrounding cells. Finally, these observations suggest that simultaneously promoting apoptosis and suppressing prosurvival mechanisms will yield the most effective anticancer therapies *(183,184)*.

Acknowledgments

This work was supported in part by NIH grants CA90297 and AG023374, ACS grant RSG MGO-105961, DOD grant DAMD17-03-1-0137 (DGT), and CA83701 (JPK). Support from NIEHS Center Grant ES07784 is also acknowledged.

References

1. Adams JM. Ways of dying: multiple pathways to apoptosis. Gene Dev 2003;17:2481–2495.
2. Danial NN, Korsmeyer SJ. Cell death: critical control points. Cell 2004;116:205–219.
3. Waxman DJ, Schwartz PS. Harnessing apoptosis for improved anticancer gene therapy. Cancer Res 2003;63:8563–8572.
4. Sprick MR, Walczak H. The interplay between the Bcl-2 family and death receptor-mediated apoptosis. Biochim Biophys Acta 2004;1644:125–132.
5. Wang J, Lenardo MJ. Role of caspases in apoptosis, development, and cytokine maturation revealed by homozygous gene deficiencies. J Cell Sci 2000;113:753–757.
6. Salveson GS. Caspases and apoptosis. Essays Biochem 2002;38:9–19.
7. Fulda S, Debatin KM. Signaling through death receptors in cancer therapy. Curr Opin Pharmacol 2004;4:327–332.
8. Chang DW, Ditsworth D, Liu H, Srinivasula SM, Alnemri ES, Yang X. Oligomerization is a general mechanism for the activation of initiator and inflammatory procaspases. J Biol Chem 2003;278:16,466–16,469.
9. Chen M, Oroszo A, Spencer DM, Wang J. Activation of initiator caspases through a stable dimeric intermediate. J Biol Chem 2002;277:50,761–50,767.
10. Donepudi M, Sweeney AM, Briand C, Grutter MG. Insights into the regulatory mechanism for caspase-8 activation. Mol Cell 2003;11:543–549.
11. Boatright KM, Renatus M, Scott FL, et al. A unified model for apical caspase activation. Mol Cell 2003;11:529–541.

12. Wang X. The expanding role of mitochondria in apoptosis. Genes Dev 2001;15:2922–2933.
13. Shiozaki EN, Chai J, Rigotti DJ, et al. Mechanism of XIAP-mediated inhibition of caspase-9. Mol Cell 2003;11:519–527.
14. Joazeiro CA, Weissman AM. RING finger proteins: Mediators of ubiquitin ligase activity. Cell 2000;102:549–552.
15. Verhagen AM, Ekert PG, Pakusch M, et al. Identification of DIABLO, a mammalian protein that promotes apoptosis by binding to an antagonizing IAP proteins. Cell 200;102:43–53.
16. Du C, Fang M, Li Y, Li L, Wang X. Smac, a mitochondrial protein that promotes cytochrome c-dependent caspase activation by eliminating IAP inhibition. Cell 2000;102: 33–42.
17. Verhagen AM, Silke J, Ekert PG, et al. HtrA2 promotes cell death through its serine protease activity and its ability to antagonize inhibitor of apoptosis proteins. J Biol Chem 2002;277:445–454.
18. Susin SA, Lorenzo HK, Zamzani N, et al. Molecular characterization of mitochondrial apoptosis-inducing factor. Nature 1999;397:441–446.
19. Li LY, Luo X, Wang X Endonuclease G is an apoptotic DNase when released from mitochondria. Nature 2001;412:95–99.
20. Zamzami N, Marchetti P, Castedo M, et al. Sequential reduction of mitochondrial transmembrane potential and generation of reactive oxygen species in early programmed cell death. J Exp Med 1995;182:367–377.
21. Gross A, McDonnell JM, Korsmeyer SJ. BCL-2 family members and the mitochondria in apoptosis. Genes Dev 1999;13:1899–1911.
22. Degterev A, Boyce M, Yuan J. The channel of death. J Cell Biol 2001;155:695–697.
23. Scorrano L, Korsmeyer SJ. Mechanisms of cytochrome c release by proapoptotic BCL-2 family members. Biochim Biophys Res Comm 2003;304:437–444.
24. Eposti MD, Dive C. Mitochondrial membrane permeabilization by Bax/Bak. Biochim Biophys Res Comm 2003;304:455–461.
25. Curtin JF, Donovan M, Cotter TG. Regulation and measurement of oxidative stress in apoptosis. J Immunol Methods 2002;265:49–72.
26. Li PF, Dietz R, von Harsdorf R. p53 regulates mitochondrial membrane potential through reactive oxygen species and induces cytochrome c-independent apoptosis blocked by Bcl-2. EMBO J 1999;18:6027–6036.
27. Gottlieb E, Vander Heiden MG, Thompson CB. Bcl-x(L) prevents initial decrease in mitochondrial membrane potential and subsequent reactive oxygen species production during tumor necrosis factor alpha-induced apoptosis. Mol Cell Biol 2000;20:5680–5689.
28. Behrend L, Henderson G, Zwacka RM. Reactive oxygen species in oncogenic transformation. Biochem Soc Trans 2003;31:1441–1444.
29. Gottlieb RA. Mitochondria: Execution central. FEBS Lett 2000;482:6–12.
30. Li H, Zhu H, Xu CJ, Yuan J. Cleavage of BID by caspase 8 mediates the mitochondrial damage in the Fas pathway of apoptosis. Cell 1998;94:491–501.
31. Sattler M, Liang H, Nettesheim D, et al. Structure of Bcl-xL-Bak peptide complex: Recognition between regulators of apoptosis. Science 1997;275:983–986.
32. Oltvai ZN, Milliman CL, Korsmeyer SJ. Bcl-2 heterodimerizes in vivo with a conserved homolog, Bax, that accelerates programmed cell death. Cell 1993;74:609–619.
33. Matsuyama S, Reed JC. Mitochondria dependent apoptosis and cellular pH damage. Cell Death Diff 2000;7:1155–1165.
34. Yu SP, Canzoniero LM, Choi DW. Ion homeostasis and apoptosis. Curr Opin Cell Biol 2001;13:405–411.
35. Hsu YT, Youle RJ. Nonionic detergents induce dimerization among members of the Bcl-2 family. J Biol Chem 1997;272:13,829–13,834.

36. Wilson-Annan J, O'Reilly LA, Crawford SA, et al. Proapoptotic BH3-only proteins trigger membrane integration of prosurvival Bcl-w and neutralize its activity. J Cell Biol 2003; 162:877–888.
37. Nakai M, Takeda A, Cleary ML, Endo T. The Bcl-2 protein is inserted into the outer membrane but not into the inner membrane of rat liver mitochondria in vitro. Biochem Biophys Res Comm 1993;196:233–239.
38. Distelhorst CW, Shore GC. Bcl-2 and calcium: controversy beneath the surface. Oncogene 2004;23:2875–2880.
39. Massaad CA, Portier BP, Taglialatela G. Inhibition of transcription factor activity by nuclear compartment-associated Bcl-2. J Biol Chem 2004;279:54,470–54,478.
40. Kitada S, Andersen J, Akar S, et al. Expression of apoptosis-regulating proteins in chronic lymphocytic leukemia: correlations with in vitro and in vivo chemoresponses. Blood 1998;91:3379–3389.
41. McDonnell TJ, Beham A, Sarkiss M, Andersen MM, Lo P. Importance of Bcl-2 family in cell death regulation. Experentia 1996;52:1008–1017.
42. Robertson LE, Plunkett W, McConnell K, Keating MJ, McDonnell TJ. Expression in chronic lymphocytic leukemia and its correlation with the induction of apoptosis and clinical outcome. Leukemia 1996;10:456–459.
43. Pratesi G, Perego P, Zunino F. Role of Bcl-2 and its post-transcriptional modification in response to antitumor therapy. Biochem Pharmacol 2001;61:381–386.
44. Haldar S, Jena N, Croce CM. Inactivation of Bcl-2 by phosphorylation. Proc Natl Acad Sci USA 1995;92:4507–4511.
45. Haldar S, Chintapalli J, Croce CM. Taxol induces Bcl-2 phosphorylation and death of prostate cancer cells. Cancer Res 1996;56:1253–1255.
46. Ito T, Deng X, Carr B, May WS. Bcl-2 phosphorylation required for its anti-apoptotic function. J Biol Chem 1997;272:11,671–11,673.
47. Chang BS, Minn AJ, Muchmore SW, Fesik SW, Thompson CB. Identification of a novel regulatory domain in Bcl-X(L) and Bcl-2. EMBO J 1997;16:968–977.
48. Kirsch DG, Doseff A, Chau BN, et al. Caspase-3-dependent cleavage of Bcl-2 promotes release of cytochrome c. J Biol Chem 1999;274:21,155–21,161.
49. Karin M, Lin A. NF-κB at the crossroads of life and death. Nat Immunol 2002;3:221–227.
50. Lin A, Karin M. NF-κB in cancer: a marked target. Sem Cancer Biol 2003;13:107–114.
51. Micheau O, Lens S, Gaide O, Alevizopoulos K, Tschopp J. NF-κB signals induce the expression of c-FLIP. Mol Cell Biol 2001;21:5299–5305.
52. Deveraux QL, Roy N, Stennicke HR, et al. IAPs block apoptotic events induced by caspase-8 and cytochrome c by direct inhibition of distinct caspases. EMBO J 1998;17:2215–2223.
53. Baldwin AS, Jr. The NF-kappa B and I kappa B proteins: new discoveries and insights. Annu Rev Immunol 1996;14:649–683.
54. Levkau B, Scatena M, Giachelli CM, Ross R, Raines EW. Apoptosis overrides survival signals through a caspase-mediated dominant-negative NF-kappa B loop. Nat Cell Biol 1999;1:227–233.
55. Reuther JY, Baldwin AS, Jr. Apoptosis promotes a caspase-induced amino-terminal truncation of IkappaBalpha that functions as a stable inhibitor of NF-kappaB. J Biol Chem 1999;274:20,664–20,670.
56. Tang G, Yang J, Minemoto Y, Lin A. Blocking caspase-3-mediated proteolysis of IKKbeta suppresses TNF-alpha-induced apoptosis. Mol Cell 2001;8:1005–1016.
57. Hong SY, Yoon WH, Park JH, Kang SG, Ahn JH, Lee TH. Involvement of two NF-kappa B binding elements in tumor necrosis factor alpha-, CD40-, and epstein-barr virus latent membrane protein 1-mediated induction of the cellular inhibitor of apoptosis protein 2 gene. J Biol Chem 2000;275:18,022–18,028.

58. Arch RH, Gedrich RW, Thompson CB. Translocation of TRAF proteins regulates apoptotic threshold of cells. Biochem Biophys Res Commun 2000;272:936–945.
59. Staal SP. Molecular cloning of the akt oncogene and its human homologues AKT1 and AKT2: amplification of AKT1 in a primary human gastric adenocarcinoma. Proc Natl Acad Sci USA 1987;84:5034–5037.
60. Nicholson KM, Anderson NG. The protein kinase B/Akt signaling pathway in human malignancy. Cell Signal 2002;14:381–395.
61. Brazil DP, Yang ZZ, Hemmings BA. Advances in protein kinase B signaling: AKTion on multiple fronts. Trends Biochem Sci 2004;29:233–242.
62. Alessi DR, Andjelkovic M, Caudwell B, Cron P, Morrice N, Cohen P, Hemmings BA. Mechanism of activation of protein kinase B by insulin and IGF-1. EMBO J 1996;15: 6541–6551.
63. Scheid MP, Woodgett JR. Unravelling the activation mechanisms of protein kinase B/Akt. FEBS Lett 2003;546;108–112.
64. Franke TF, Hornik CP, Segev L, Shostak GA, Sugimoto C. PI3K/Akt and apoptosis: size matters. Oncogene 2003;22:8983–8998.
65. Mora A, Komander D, van Aalten DMF, Alessi DR. PDK1, the master regulator of AGC kinase signal transduction. Sem Cell & Develop Biol 2004;15:161–170.
66. Kobayashi T, Cohen P. Activation of serum- and glucocorticoids-regulated protein kinase by agonists that activate phosphatidyl 3-kinase is mediated by 3-phosphoinositide-dependent protein kinase-1 (PDK1) and PDK2. Biochem J 1999:339;319–328.
67. Brunet A, Park J, Tran H, Hu LS, Hemmings BA, Greenberg ME. Protein kinase SGK mediates survival signals by phosphorylating the Forkhead transcription factor FKHRL1 (FOXO3a). Mol Cell Biol 2001;21:952–965.
68. Vivanco I, Sawyers CL. The phosphatidylinositol 3-Kinase AKT pathway in human cancer. Nat Rev Cancer 2002;2:489–501.
69. Hermeking H. The 14-3-3 cancer connection. Nat Rev Cancer 2003;3:942–943.
70. Brunet A, Bonni A, Zigmond MJ, et al. Akt promotes cell survival by phosphorylating and inhibiting a Forkhead transcription factor. Cell 1999;96:857–868.
71. Cardone MH, Roy N, Stennicke HR, et al. Regulation of cell death protease caspase-9 by phosphorylation. Science 1998;282:1318–1321.
72. Sordet O, Khan Q, Kohn KW, Pommier Y. Apoptosis induced by topoisomerase inhibitors. Curr Med Chem Anticancer Agents 2003;3:271–290.
73. Ozes ON, Mayo LD, Gustin JA, Pfeffer SR, Pfeffer LM, Donner DB. NF-kappaB activation by tumour necrosis factor requires the Akt serine-threonine kinase. Nature 1999;401:82–85.
74. Romashkova JA, Makarov SS. NF-kappaB is a target of AKT in anti-apoptotic PDGF signalling. Nature 1999;401:86–90.
75. Chang F, Steelman LS, Shelton JG, et al. Regulation of cell cycle progression and apoptosis by the Ras/Raf/ERK pathway. Int J Oncol 2003;22:469–480.
76. Lowenstein EJ, Daly RJ, Batzer AG, et al. The SH2 and SH3 domain-containing protein GRB2 links receptor tyrosine kinases to ras signaling. Cell 1992;70:431–442.
77. Chardin P, Camonis JH, Gale NW, et al. Human SOS1: a guanine nucleotide exchange factor for Ras that binds to GRB2. Science 1993;260:1338–1343.
78. Xia Z, Dickens M, Raingeaurd J, Davis RJ, Greenberg ME. Opposing effects of ERK and JNK-p38 MAP kinases on apoptosis. Science 1995;270:1326–1331.
79. Harada H, Quearry B, Ruiz-Vela A, Korsmeyer SJ. Survival factor-induced extracellular signal-regulated kinase phosphorylates BIM, inhibiting its association with BAX and proapoptotic activity. Proc Natl Acad Sci USA 2004;101:15,313–15,317.
80. Marani M, Hancock D, Lopes R, Tenev T, Downward J, Lemoine NR. Role of Bim in the survival pathway induced by Raf in epithelial cells. Oncogene 2004;23:2431–2441.

81. Wang P, Gilmore AP, Streuli CH. Bim is an apoptosis sensor that responds to loss of survival signals delivered by epidermal growth factor but not those provided by integrins. J Biol Chem 2004;279:41,280–41,285.
82. Chalmers CJ, Balmanno K, Hadfield K, Ley R, Cook SJ. Thrombin inhibits Bim (Bcl-2-interacting mediator of cell death) expression and prevents serum-withdrawal-induced apoptosis via protease-activated receptor 1. Biochem J 2003;375:99–109.
83. Reginato MJ, Mills KR, Paulus JK, et al. Integrins and EGFR coordinately regulate the pro-apoptotic protein Bim to prevent anoikis. Nat Cell Biol 2003;5:733–740.
84. Molton SA, Todd DE, Cook SJ. Selective activation of the c-Jun N-terminal kinase (JNK) pathway fails to elicit Bax activation or apoptosis unless the phosphoinositide 3′-kinase (PI3K) pathway is inhibited. Oncogene 2003;22:4690–4701.
85. Ley R, Balmanno K, Hadfield K, Weston C, Cook SJ. Activation of the ERK1/2 signaling pathway promotes phosphorylation and proteasome-dependent degradation of the BH3-only protein, Bim. J Biol Chem 2003;278:18,811–18,816.
86. Weston CR, Balmanno K, Chalmers C, et al. Activation of ERK1/2 by deltaRaf-1:ER* represses Bim expression independently of the JNK or PI3K pathways. Oncogene 2003;22: 1281–1293.
87. LeGrand, EK. An adaptationist view of apoposis. Quart Rev Biol 1997;72:135–147.
88. Joshi B, Li L, Taffe BG, et al. Apoptosis induction by a novel anti-prostate cancer compound, BMD188 (a fatty acid-containing hydroxamic acid), requires the mitochondrial respiratory chain. Cancer Res 1999;59:4343–4355.
89. Chandra D, Liu JW, Tang DG. Early mitochondrial activation and cytochrome c up-regulation during apoptosis. J Biol Chem 2002;277:50,842–50,854.
90. Martin AG, Fearnhead HO. Apocytochrome c blocks caspase-9 activation and Bax-induced apoptosis. J Biol Chem 2002;277:50,834–50,841.
91. Zhao Y, Wang Z-B, Xu J-X. Effect of cytochrome *c* on the generation and elimination of O^{-}_{2} and H_2O_2 in mitochondria. J Biol Chem 2003;278:2356–2360.
92. Martin AG, Nguyen J, Wells JA, Fearnhead HO. Apo cytochrome c inhibits caspases by preventing apoptosome formation. Biochim Biophys Res Comm 2004;319:944–950.
93. Liu J-W, Chandra D, Rudd MD, et al. Induction of pro-survival molecules by apoptotic stimuli: Involvement of FOXO3a and ROS. Oncogene 2005;24:2020–2031.
94. Dong Z, Wang J. Hypoxia selection of death-resistant cells: A role for Bcl-xL. J Biol Chem 2004;279:9215–9221.
95. Nijhawan D, Fang M, Traer E, et al. Elimination of Mcl-1 is required for the initiation of apoptosis following ultraviolet irradiation. Genes Dev 2003;17:1475–1486.
96. Sherrill KW, Byrd MP, Van Eden ME, Lloyd RE. BCL-2 translation is mediated via internal ribosome entry during cell stress. J Biol Chem 2004;279:29,066–29,074.
97. Bachelor MA, Bowden TG. Ultraviolet A-induced Modulation of Bcl-XL by p38 MAPK in Human Keratinocytes. Post-transcriptional regulation through the 3′-untranslated region. J Biol Chem 2004;279:42,658–42,668.
98. Kinnula VL, Crapo JD. Superoxide dismutases in the lung and human lung diseases. Am J Respir Crit Care Med 2003;167:1600–1619.
99. Mikkelsen RB, Wardman P. Biological chemistry of reactive oxygen and nitrogen and radiation-induced signal transduction mechanisms. Oncogene 2003;22:5734–5754.
100. Chen J-G, Yang C-PH, Cammer M, Horwitz SB. Gene expression and mitotic exit induced by microtubule-stabilizing drugs. Cancer Res 2003;63:7891–7899.
101. Takayama S, Reed JC, Homma S. Heat-shock proteins as regulators of apoptosis. Oncogene 2003;22:9041–9047.
102. Parcellier A, Schmitt E, Gurbuxani S, et al. HSP27 is a ubiquitin-binding protein involved in I-kappaBalpha proteasomal degradation. Mol Cell Biol 2003;23:5790–5802.

103. Garrido C, Solary E. A role of HSPs in apoptosis through "protein triage"? Cell Death Differ 2003;10:619–620.
104. Parcellier A, Gurbuxani S, Schmitt E, et al. The life span determinant $p66^{Shc}$ localizes to mitochondria where it associates with mitochondrial heat shock protein 70 and regulates transmembrane potential. J Biol Chem 2004;279:25,689–25,695.
105. Zhang L, Pelech S, Uitto V-J. Bacterial Gro-EL heat shock protein 60 protects epithelial cells from stress-induced death through activation of ERK and inhibition of caspase 3. Exp Cell Res 2004;292:231–240.
106. Green DR, Evan GI. A matter of life and death. Cancer Cell 2001;1:19–30.
107. Shibue T, Takeda K, Oda E, et al. Integral role of Noxa in p53-mediated apoptotic response. Genes & Dev 2003;17:2233–2238.
108. Yu J, Wang Z, Kinzler KW, Vogelstein B, Zhang L. PUMA mediates the apoptotic response to p53 in colorectal cancer cells. Proc Natl Acad Sci USA 2003;100:1931–1936.
109. Mahyar-Roemer M, Roemer K. p21 Waf/Cip1 can protect human colon carcinoma cells against p53-dependent and p53-independent apoptosis induced by natural chemopreventive and therapeutic agents. Oncogene 2001;20:3387–3398.
110. Javelaud D, Besancon F. Inactivation of $p21^{WAF-1}$ sensitizes cells to apoptosis via an increase of both $p14^{ARF}$ and p53 levels and alteration of the Bax/Bcl-2 ratio. J Biol Chem 2002;277:37,949–37,954.
111. Spierings GE, de Vries E, Stel AJ, Riestap NT, Vellenga E, de Jong S. Low $p21^{Waf1/Cip1}$ protein level sensitizes testicular germ cell tumor cells to Fas-mediated apoptosis. Oncogene 2004;23:4862–4872.
112. Tong Z, Wu X, Kehrer JP. Increased expression of the lipocalin 24p3 as an apoptotic mechanism for MK886. Biochem J 2003;372:203–210.
113. Flower DR. The lipocalin protein family: structure and function. Biochem J 1996;318:1–14.
114. Flower DR. Beyond the superfamily: the lipocalin receptors. Biochim Biophys Acta 2000;1482:327–336.
115. Elangovan N, Lee Y-C, Tzeng W-F, Chu S-T. Delivery of ferric ion to mouse spermatozoa is mediated by lipocalin internalization. Biochem Biophys Res Commun 2004;319: 1096–1104.
116. Flower DR. The lipocalin protein family: a role in cell regulation. FEBS Lett 1994;354: 7–11.
117. Bratt T. Lipocalins and cancer. Biochim Biophys Acta 2000;1482:318–326.
118. Hraba-Renevey S, Türler H, Kress M, Salomon C, Weil R. SV40-induced expression of mouse gene 24p3 involves a post-transcriptional mechanism. Oncogene 1989;4:601–608.
119. Kjeldsen L, Cowland JB. Borregaard N. Human neutrophil gelatinase-associated lipocalin and homologous proteins in rat and mouse. Biochim Biophys Acta 2000;1482:272–283.
120. Kjeldsen L, Johnsen AH, Sengeløv H, Borregaard N. Isolation and primary structure of NGAL, a novel protein associated with human neutrophil gelatinase. J Biol Chem 1993; 268:10,425–10,432.
121. Yang J, Goetz D, Li J-Y, et al. An iron delivery pathway mediated by a lipocalin. Mol Cell 2002;10:1045–1056.
122. Ryon J, Bendickson L, Nielsen-Hamilton M. High expression in involuting reproductive tissues of uterocalin/24p3, a lipocalin and acute phase protein. Biochem J 2002;367:271–277.
123. Devireddy LR, Teodoro JG, Richard FA, Green MR. Induction of apoptosis by a secreted lipocalin that is transcriptionally regulated by IL-3 deprivation. Science 2001;293:829–834.
124. Di Marco E, Sessarego N, Zerega B, Cancedda R, Cancedda FD. Inhibition of cell proliferation and induction of apoptosis by *ExFABP* targeting. J Cell Physiol 2003;196:464–473.
125. Lu PD, Jousse C, Marciniak SJ, et al. Cytoprotection by pre-emptive conditional phosphorylation of translation initiation factor 2. EMBO J 2004;23:169–179.

126. Donze O, Deng J, Curran J, Sladek R, Picard D, Sonerberg N. The protein kinase PKR: a molecular clock that sequentially activates survival and death programs. EMBO J 2004;23: 564–571.
127. Fujioka S, Schmidt C, Sclabas GM, et al. Stabilization of p53 is a novel mechanism for proapoptotic function of NF-κB. J Biol Chem 2004;279:27,549–27,559.
128. Campbell KJ, Rocha S, Perkins ND. Active repression of antiapoptotic gene expression by RelA(p65) NF-κB. Mol Cell 2004;13:853–865.
129. Zamora M, Merono C, Vinas O, Mampel T. Recruitment of NF-κB into mitochondria is involved in adenine nucleotide translocase 1 (ANT1)-induced apoptosis. J Biol Chem 2004;279:38,415–38,423.
130. Tergaonkar V, Pando M, Vafa O, Wahl G, Verma I. p53 stabilization is decreased upon NF-κB activation: a role for NFκB in acquisition of resistance to chemotherapy. Cancer Cell 2002; 1:493–503.
131. Jang J-H, Surh Y-J. Bcl-2 attenuation of oxidative cell death is associated with up-regulation of g-glutamylcysteine ligase via constitutive NF-κB activation. J Biol Chem 2004;279: 38,779–38,786.
132. Mabuchi S, Ohmichi M, Nishio Y, et al. Inhibition of NFκB increases the efficacy of cisplatin in *in vitro* and *in vivo* ovarian cancer models. J Biol Chem 2004;279:23,477–23,485.
133. Balkwill F, Coussens LM. Cancer: an inflammatory link. Nature 2004;431:405–406.
134. Pikarsky E, Porat RM, Stein I, et al. NF-kappaB functions as a tumour promoter in inflammation-associated cancer. Nature 2004;431:461–466.
135. Clevers H. At the crossroads of inflammation and cancer. Cell 2004;118:671–674.
136. Luo JL, Maeda S, Hsu LC, Yagita H, Karin M. Inhibition of NF-kappaB in cancer cells converts inflammation-induced tumor growth mediated by TNFalpha to TRAIL-mediated tumor regression. Cancer Cell 2004;6:297–305.
137. Aggarwal BB. Nuclear factor-kappaB: the enemy within. Cancer Cell 2004;6:203–208.
138. Legembre P, Schickel R, Barnhart BC, Peter ME. Identification of SNF1/AMP kinase-related kinase as an NF-B-regulated anti-apoptotic kinase involved in CD95-induced motility and invasiveness. J Biol Chem 2004;279:46,742–46,747.
139. Pham CG, Bubici C, Zazzeroni1 F, et al. Ferritin heavy chain upregulation by NF-κB inhibits TNF-induced apoptosis by suppressing reactive oxygen species. Cell 2004;119:529–542.
140. Algeciras-Schmnich A, Pietras EM, Barnhart BC, et al. Two CD95 tumor classes with different sensitivities to antitumor drugs. Proc Natl Acad Sci USA 2003;100:11,445–11,450.
141. Micheau O, Tschopp J. Induction of TNF receptor I-mediated apoptosis via two sequential signaling complexes. Cell 2003;114:181–190.
142. Ehrhardt H, Fulda S, Schmid I, Hiscott J, Debatin K-M, Jeremias I. TRAIL induced survival and proliferation in cancer cells resistant towards TRAIL-induced apoptosis mediated by NF-κB. Oncogene 2003;22:3842–3852.
143. Huerta-Yepez S, Vega M, Jazirehi A, et al. Nitric oxide sensitizes prostate carcinoma cell lines to TRAIL-mediated apoptosis via inactivation of NF-κB and inhibition of Bcl-xL expression. Oncogene 2004;23:4993–5003.
144. Kreuz S, Siegmund D, Rumpf J-J, et al. NF-κB activation by Fas is mediated through FADD, caspase-8, and RIP and is inhibited by FLIP. J Cell Biol 2004;166:369–380.
145. Imamura R, Konaka K, Matsumoto N, et al. Fas ligand induces cell-autonomous NF-κB activation and interleukin-8 production by a mechanism distinct from that of tumor necrosis factor-a. J Biol Chem 2004;279:46,415–46,423.
146. Barnhart BC, Legenbre P, Pietras E, Bubici C, Franzoso G, Peter ME. CD95 ligand induces motility and invasiveness of apoptosis-resistant tumor cells. EMBO J 2004;23:3175–3185.
147. Danial NN, Gramm CF, Scorrano L, et al. BAD and glucokinase reside in a mitochondrial complex that integrates glycolysis and apoptosis. Nature 2003;424:952–956.

148. Seo SY, Chen Y, Ivanovska I, et al. BAD is a pro-survival factor prior to activation of its proapoptotic function. J Biol Chem 2004;279:42,240–42,249.
149. Nakajima K, Hirose H, Taniguchi M, et al. Involvement of BNIP1 in apoptosis and endoplasmic reticulum membrane fusion. EMBO J 2004;23:3216–3226.
150. Jones JM, Datta P, Srinivasula SM, et al. Loss of Omi mitochondrial protease activity causes the neuromuscular disorder of mnd2 mutant mice. Nature 2003;425:721–727.
151. Cheng EH-Y, Sheiko TV, Fisher JK, Craigen WJ, Korsmeyer SJ. VDAC2 inhibits BAK activation and mitochondrial apoptosis. Science 2003;301:513–517.
152. Fannjiang Y, Kim CH, Huganir RL, et al. BAK alters neuronal excitability and can switch from anti- to pro-death function during postnatal development. Dev Cell 2003;4:575–585.
153. Yang L, Cao Z, Yan H, Wood WC. Coexistence of high levels of apoptotic signaling and inhibitor of apoptosis proteins in human tumor cells: implication for cancer specific therapy. Cancer Res 2003;63:6815–6824.
154. Newton K, Strasser A. Caspases signal not only apoptosis but also antigen-induced activation in cells of the immune system. Genes Dev 2003;17:819–825.
155. Birkenkamp KU, Coffer PJ. FOXO transcription factors as regulators of immune homeostasis: Molecules to die for? J Immunol 2003;171:1623–1629.
156. Accili D, Arden KC. FoxOs at the crossroads of cellular metabolism, differentiation, and transformation. Cell 2004;117:421–426.
157. Burgering BMT, Kops GJPL. Cell cycle and death control: long live Forkheads. Trends Biochem Sci 2002;27:352–360.
158. Rokudal S, Fujita N, Kitahara O, Nakamura Y, Tsuruo T. Involvement of FKHR-dependent TRADD expression in chemotherapy drug-induced apoptosis. Mol Cell Biol 2002;22: 8695–8708.
159. Burgering BMT, Medema RH. Decisions on life and death: FOXO Forkhead transcription factors are in command when PKB/Akt is off duty. J Leukocyte Biol 2003;73:689–701.
160. El-Deiry W. The role of p53 in chemosensitivity and radisensitivity. Oncogene 2003;22: 7486–7495.
161. Fridman JS, Lowe SW. Control of apoptosis by p53. Oncogene 2003;22:9030–9040.
162. Lo PK, Huang SZ, Chen HC, Wang FF. The prosurvival activity of p53 protects cells from UV-induced apoptosis by inhibiting c-Jun NH2-terminal kinase activity and mitochondrial death signaling. Cancer Res 2004;64:8736–8745.
163. Chau BN, Wang YJ. Coordinated regulation of life and death by RB. Nat Rev Cancer 2003;3:130–138.
164. Bell LA, Ryan KM. Life and death decisions by E2F-1. Cell Death and Differ 2003;10: 1–6.
165. Dhakshinamoorthy S, Porter AG. Nitric oxide-induced transcriptional up-regulation of protective genes by Nrf2 via the antioxidant response element counteracts apoptosis of neuroblastoma cells. J Biol Chem 2004;279:20,096–20,107.
166. Cullinan SB, Diehl JA. ERK-dependent activation of Nrf2 contributes to redox homeostasis and cell survival following endoplasmic reticulum stress. J Biol Chem 2004;279:20,108–20,117.
167. You H, Jang Y, You-Ten AI, et al. p53-dependent inhibition of FKHRL1 in response to DNA damage through protein kinase SGK1. Proc Natl Acad Sci USA 2004;101:14,057–14,062.
168. Hu MC, Lee DF, Xia W, et al. IkappaB kinase promotes tumorigenesis through inhibition of forkhead FOXO3a. Cell 2004;117:225–237.
169. Lin L, Hron JD, Peng SL. Regulation of NF-kappaB, Th activation, and autoinflammation by the forkhead transcription factor Foxo3a. Immunity 2004;21:203–213.
170. Huang H-L, Fang L-W, Lu S-P, Chou C-K, Luh T-Y, Lai M-Z. DNA-damaging reagents induce apoptosis through reactive oxygen species-dependent Fas aggregation. Oncogene 2003;22:8168–8177.

171. Katoh I, Tomimori Y, Ikawa Y, Kurata S. Dimerization and processing of procaspase-9 by redox stress in mitochondria. J Biol Chem 2004;279:15,515–15,523.
172. Sattler M, Winkler T, Verma S, et al. Hematopoietic growth factors signal through the formation of reactive oxygen species. Blood 1999;93:2928–2935.
173. Sundaresan M, Yu ZX, Ferrons VJ, Irani K, Finkel T. Requirement for generation of H_2O_2 for platelet-derived growth factor signal transduction. Science 1995;270:296–299.
174. Vaquero EC, Edderkaoui M, Pandol SJ, Gukovsky I, Gukovskaya AS. Reactive oxygen species produced by NAD(P)H oxidase inhibit apoptosis in pancreatic cancer cells. J Biol Chem 2004;279:34,643–34,654.
175. Furukawa-Hibi Y, Yoshida-Araki K, Ohta T, Ikeda K, Motoyama N. FOXO Forkhead transcription factors induce G_2-M checkpoint in response to oxidative stress. J Biol Chem 2002;277:26,729–26,732.
176. Kops GJPL, Dansen TB, Polderman PE, et al. Forkhead transcription factor FOXO3a protects quiescent cells from oxidative stress. Nature 2002;419:316–321.
177. Nemoto S, Finkel T. Redox regulation of forkhead proteins through a *p66shc*-dependent signaling pathway. Science 2002;295:2450–2452.
178. Brunet A, Sweeney LB, Sturgill JF, et al. Stress-dependent regulation of FOXO transcription factors by the SIRT1 deacetylase. Science 2004;303:2011–2015.
179. Essers MA, Weijzen S, de Vries-Smits AM, et al. FOXO transcription factor activation by oxidative stress mediated by the small GTPase Ral and JNK. EMBO J. 2004;23:4802–4812.
180. Yin L, Huang L, Kufe D. MUC1 oncoprotein activates the FOXO3a transcription factor in a survival response to oxidative stress. J Biol Chem 2004;279:45,721–45,727.
181. Nakata S, Matsumura I, Tanaka H, et al. NFκB family proteins participate in multiple steps of hematopoiesis through elimination of reactive oxygen species. J Biol Chem. 2004; 279:55,578–55,586.
182. Macip S, Igarashi M, Berggren P, Yu J, Lee SW, Aaronson SA. Influence of induced reactive oxygen species in p53-mediated cell fate decisions. Mol Cell Biol 2003;23:8576–8585.
183. McCormick F. Survival pathways meet their ends. Nature 2004;428:267–269.
184. Wendel H-G, de Stanchina E, Fridman JS, et al. Survival signaling by Akt and eIF4E in oncogenesis and cancer therapy. Nature 2004;428:332–337.

5

Aberrations of DNA Damage in Checkpoints in Cancer

Marikki Laiho

Summary

Mutations in gene products controlling DNA damage checkpoints and repair pathways cause predisposition to a large number of sporadic cancers, hereditary cancer syndromes, and developmental defects. This underscores the vital need for the fidelity of checkpoint control and efficiency for the repair machineries. The checkpoint functions are ensured by multiple, often parallel, pathways and show specificity regarding the nature of the damage, cell-cycle phase, and the subsequent cellular response. The checkpoint control mechanisms also link to other cellular responses such as apoptosis to initiate a death program in the event of unsuccesful repair. It is striking that several checkpoint mutations are associated with developmental abnormalities and cancer syndromes, such as the Nijmegen breakage syndrome and Fanconi anemia, indicating that the maintenance of the genome integrity is essential throughout development. Though several critical DNA maintenance proteins have been identified and their links to tumor progression have been established, alterations of several known checkpoint-associated proteins (e.g., 53BP1, Mdc1, SMC1) in cancer are still undiscovered. Knowledge of the DNA damage checkpoint pathways and pathways sensing the damage and instigating repair will pave the way to improved diagnostics, identification of genetic susceptibility, and, in future, rational therapy of cancer.

Key Words: Checkpoint; cell cycle; DNA damage; p53; H2AX; hereditary cancer syndrome; repair.

1. Introduction

DNA lesions occur as inherent mistakes in the metabolism of DNA, typically during replication, or by exogenous damage caused by either natural environmental causes (ultraviolet [UV]-radiation, ionizing radiation, reactive oxygen species [ROS]) or synthetic genotoxic chemicals such as those present in tobacco smoke. Depending on the chemical nature of the damage on DNA, different types of repair programes are evoked. UV radiation causes the formation of bulky DNA adducts, which stall replication and transcription machineries. This activates nucleotide excision repair (NER), a process involving a complex machinery of enzymes in damage recognition and repair *(1,2)*. Damage causing by double-strand breaks (DSB), typically ionizing radiation (IR), is repaired by homologous recombination (HR) and by nonhomologous end-joining (NHEJ) *(3,4)*.

DNA lesions are mainly recognized by damage-type specific sensor molecules. For example, the primary sensor activated by IR is ataxia teleangiectasia mutated protein (ATM), which phosphorylates several key checkpoint proteins and initiates a protein kinase cascade leading to the phosphorylation of a large number of its downstream targets. UV damage and stalled replication forks activate an analogous, but different

From: *Apoptosis, Cell Signaling, and Human Diseases: Molecular Mechanisms, Volume 1*
Edited by R. Srivastava © Humana Press Inc., Totowa, NJ

pathway involving kinases ataxia teleangiectasia and Rad3-related (ATR) and Chk1. In both cases, the respective protein kinase pathways dictate the subcellular responses and initiate damage repair *(5–10)*.

Exposure of the cells to the various DNA lesions also causes other rapid cellular responses. On one hand, intracellular signaling cascades, such as JNK, p38, and growth factor receptor induced pathways are activated, and, on the other hand, cell-cycle checkpoints are enforced causing the arrest of the cells in specific phases of the cell cycle. While γ-radiation provokes both G_1, intra-S, and G_2/M arrest, UV damage causes mainly a replicative arrest without distinct accumulation of the cells at any specific cell-cycle phase *(11,12)*. Though the upstream activating events vary, the γ-radiation induced cell-cycle responses are largely mediated by a similar set of proteins controlling the normal cell cycle. The arrest of the cells in the cycle provides time for the damage repair, followed by activation of mechanisms mediated by polo-like kinase to reactivate the cycle *(13)*. Thus, not only are the machineries ensuring adequate damage repair essential, but also the cell-cycle exit pathways are controlled.

Several of these features and molecules involved in the processes are currently being exploited for diagnostics and new strategic therapies, which involve attempts to both recover the repair deficiences and to enhance tumor cell killing by provoking unscheduled cell death by preventing succesful transcription-coupled repair. These new treatment modalities may offer effective therapeutic opportunities even in cases of severe mutational loads, such as in hereditary cancer repair syndromes. This chapter discusses the pathways involved in sensing the DNA damage and damage repair, the overlap between the cell cycle and DNA damage checkpoints, their defects in human syndromes, and how this knowledge is currently being exploited in therapeutic approaches.

2. Sensing the Damage

Cellular responses detecting DNA damage couple directly with DNA damage checkpoints and form a fast acting complex network with partially overlapping interactions. One of the key sensors is the ATM protein, which is a large phosphoinositide-3 kinase like kinase normally activated upon DSBs, and absent in AT patients because of its mutations *(5,8,10)*. Activation of ATM leads to its dimerization and rapid autophosphorylation within minutes of the damage *(14)*. ATM has a number of target proteins which function in the ensuing damage response and repair pathways *(8,10)*. Interestingly, the fast activation kinetics of ATM and its absence from the DNA damage induced foci are indicative that ATM does not associate with the DNA lesions itself, but is rather activated by changes in chromatin conformation and orchestrates the subsequent events. ATM targets include proteins responsible for G_1 (p53, Mdm2, Chk2), intra-S (Nbs1, FancD2, SMC1), as well as G_2/M checkpoints (Brca1, Rad17) (*see* Chapter 4). The phosphorylation of histone γ-H2AX (H2AX) by ATM *(15)* leads to H2AX accumulation at the sites of DSB and formation of nuclear foci. Similarly, H2AX is phosphorylated by ATR in response to replicational stress *(16)*. Whereas the lack of H2AX does not affect the DNA damage checkpoints per se, H2AX is facilitating the accrual of Nijmegen breakage syndrome 1 (Nbs1), p53 binding protein 1 (53BP1), and breast cancer-1 (Brca1) to the sites of the damage *(17–19)*. Hierarchially, the recruitment of 53BP1, originally identified as a p53 interacting protein *(20)*, to these foci *(21,22)*, is essential for the subsequent accumulation and phosphorylation of other ATM targets Nbs1, cohesin protein structural maintenance of chromosomes

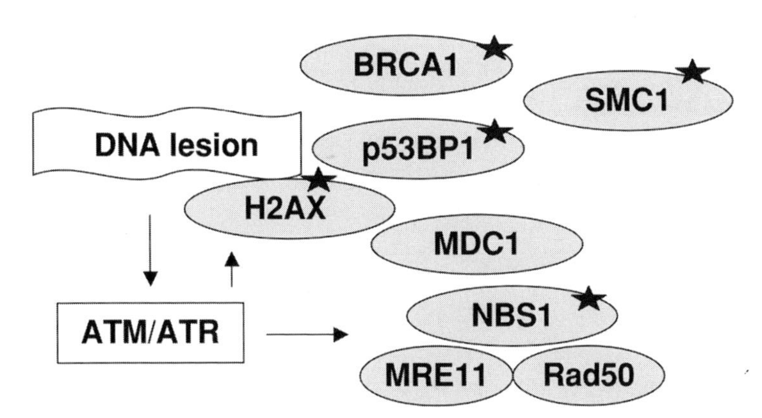

Fig. 1. DNA lesion induced recruitation of DNA damage response proteins depend on ATM/ATR initiated phosphorylation cascades. Proteins marked with star are targeted by ATM/ATR.

protein 1 (SMC1) and Brca1 and, at least partly, for phosphorylation of Chk2 *(23–28;* for review, *see* ref. *29*) (Fig. 1). However, whereas H2AX in not required for the initial 53BP1 foci formation, it is essential for the maintenance of 53BP1 at or in the vicinity of the damage *(21)*. Support for models placing ATM both upstream and downstream of 53BP1 exist *(22)*. Foci containing H2AX, 53BP1, Nbs1, and mediator of DNA damage checkpoint 1/NFBD1 (Mdc1) are also detected in senescent cells and apparently represent activation of the DNA damage responses by increased number of telomere breakage *(30)*. Mdc1 may further determine the sustained interactions of Nbs1 with DSBs and appears therefore essential for the dynamic interactions at the site of the damage *(31)*.

A close homolog of ATM, ATR is activated by DNA helix distorting events, such as UV damage and DNA topoisomerase inhibitors, and shares an apparently similar substrate spectrum as ATM *(5,10)*. However, ATR, but not ATM, is essential during embryonal development indicating that they also have diverse functions *(32,33)*. ATR couples constitutively with its regulatory partner ATRIP *(34)* and the complex accumulates to the arrested replication forks *(35)*. The ATR-dependent checkpoint activation is dependent on replication protein A (RPA)-mediated loading of ATR-ATRIP, Rad9-Rad1-Hus1 complex, Claspin and DNA polymerase α to the site of stalled replication causing the activation of the ATR dependent checkpoint *(10,36)*.

3. Repairing the Damage

3.1. Repair of DNA Doublestrand Breaks

DSBs occur through exogenous DNA damage, typically by ionizing radiation or chemicals, or by replication of single-strand breaks (SSBs). Unrepaired DSBs sensitize the cells to chromosomal aberrations, aneuploidy, and deletions. DSBs are repaired by both HR and NHEJ. Whereas HR is essentially error free, NHEJ allows loss or gain of few nucleotides *(2–4)*. In HR, an intact template of the sister chromatid is used to repair the lesion, whereas NHEJ operates by joining the DNA ends without sequence homology with the aid of Ku70/80 complex, DNA-PK catalytic subunit and XRCC4 ligase 4 *(2,3)*. HR appears significant for the re-establishment of stalled replication forks due to DSBs, therefore allowing cell survival. Although NHEJ is constitutive, HR is more cell cycle-restricted and dependent on exogenous stress signaling *(37)*. HR is activated by the ATM/ATR pathway and requires a large number of molecules including Rad51 and its paralogs, Rad52, Rad54, Brca1, Brca2, XRCC2, XRCC3, and the MRE11, Rad50, Nbs1 (MNR) complex *(2,3)*. The phosphorylation of Rad51 by Chk1 and their interaction is essential for succesful HR again establishing a connection between checkpoint control and repair *(38)*.

3.2. Repair of UV-Induced DNA Lesions

Nucleotide excision repair (NER) is activated in mammalian cells by bulky DNA helix-distorting lesions, such as pyrimidine dimers and 6-4 photoproducts caused by UV damage, or other adducts caused by certain chemicals and oxidative damage *(39)*. NER is accomplished by the coordinated action of 20 to 30 proteins each with specific functions in damage recognition, unwinding of the DNA helix, excision of the damage site, ligation, and rewinding of the DNA helix *(2,39)*. Most proteins in the recognition and incision steps have been identified and named according to the seven complementation groups of *Xeroderma pigmentosum* (XP, XPA to XPG), a photosensitivity disorder resulting from deficient NER *(2)*. Importantly, the TFIIH transcription factor is involved not only in general cellular transcription, but also in the local unwinding around the injury through its DNA helicase activity *(2)*. NER is accomplished via two mechanically distinct processes—transcription-coupled repair (TCR) and global genomic repair (GGR) *(40,41)*. TCR is activated through stalling of the RNA polymerase II as a result of the bulky DNA lesions *(42)*, whereas GGR is launched by XPC-hHR23B protein complex recognizing the UV type of DNA damage *(43)*. TCR occurs rapidly on the transcriptionally active DNA template strand, and may show selectivity towards faster repair of shorter genes *(44)*. GGR is slower and responsible for the repair of both nontemplate strand and the nontranscribed areas. In the event of unsuccesful TCR, the cell is destined toward apoptosis, thus effectively removing cells with irrepairable damage *(40,41)*. However, the genomic fidelity is influenced more by GGR, because the failure of GGR fidelity leads to passing of the lesions to the daughter cells and thus accumulation of the genetic defects.

During TCR, when repair enzymes are recruited, they must remove the RNA polymerase II to access the lesion. This occurs either by ubiquitination and the subsequent degradation of RNA polymerase II *(45)*, its phosphorylation and unstable association with DNA *(46)*, or by some other, still unresolved mechanism. In addition, TCR requires Cockayne syndrome proteins CSA and CSB. Whereas these are unessential for GGR, their deficiency results in the UV radiation hypersensitivity disorder Cockayne syndrome *(47)*. Furthermore, UV damage causes local inhibition of transcription around the damaged areas in the nucleus through deprivation of TFIIH, which is required for both NER and transcription initiation *(48)*.

4. Coupling of DNA Damage and Cell-Cycle Checkpoints

DNA damage checkpoints have evolved to ensure the genetic fidelity. In essence they are positions within the cell cycle when the surveillance mechanisms are active and scan for the presence of damage. If damage is detected, the checkpoint is activated *(49–51)*. The activation of mammalian checkpoints provoke distinct cellular responses, including cell-cycle arrest at various points in the cycle (G_1, intra-S, G_2/M) and/or apoptosis. In principle, the enforcement of these checkpoints is believed to yield time for the cell for efficient repair and to prevent accumulation of the genetic lesions by forcing the cells into apoptosis *(52)*.

4.1. G_1 and G_1/S Transition Checkpoints—The ATM, Chk1/2, p53 Cascade

TP53 gene is the most frequently mutated gene in all human cancers. p53 protein is activated by different forms of cellular stress like hypoxia, nucleotide imbalance, replicative stress, pH-change, and perhaps most importantly, by DNA damage *(53)*. Activation of p53 leads to cell-cycle arrest, rendering time for damage repair, or induction of apoptosis to prevent division of the damaged cells. Therefore its key function is considered as a sensor for the integrity of the genome *(54)*. p53 is negatively regulated by three E3 ubiquitin ligases, murine double minute (MDM)2, Pirh2, and COP1 initiating p53 proteasomal degradation *(55–57)*. Of these, Mdm2 and COP1 are direct p53 transcriptional targets and form important regulatory loops controlling p53 function. p53 and MDM2 are directly phosphorylated by ATM following IR and p53 by ATR (*58–60*; reviewed in ref. *61*). Chk2 further phosphorylates p53 on its aminoterminus *(62,63)*. The phosphorylation of p53 to multiple sites may negatively regulate its association with MDM2 and affect the subsequent p53 transactivation and its interaction with other modulators like p300 histone acetyltransferase *(61,64)*. Though the lack of Chk2 does not affect p53 stabilization following damage, Chk2 null cells and animals display a defect in p53-dependent transcription and apoptotic response to IR *(65)*.

p53 excerts its functions by acting as a transcription factor regulating multiple target genes involved in cell-cycle regulation and apoptosis, but also in cellular signaling and regulation of extracellular matrix and cellular structures *(66)*. The key p53 target leading to a cell-cycle arrest is p21Cip1/Waf1, which inhibits the activities of cyclin-dependent kinase (Cdk)2-cyclin complexes required for the S-phase entry of the cells *(67,68)*. The major function of p21 is to provide a sustained downregulation of the Cdk2/cyclin E complex activity both in G_1/S transition and in S-phase (Fig. 2).

p53 apoptotic targets include several mitochondrial proapoptotic genes, NOXA, PUMA, BAX, DR5 (reviewed in ref. *69*). Moreover, p53 itself is directly localized to the mitochondrial inner membrane, binds with BclXL and Bcl2 *(70)* and releases proapoptotic proteins Bak and Bax *(71,72)*. Therefore, in mammals the activation of the p53 function may lead to multiple cellular responses that affect the decision between cell-cycle arrest and apoptosis and have therefore therapeutic implications *(73–75)*.

In *Drosophila*, the cellular DNA damage responses are dependent on the developmental stage, and presented with either a delayed mitosis or a premitotic block. The ATM provoked signaling cascade is fully conserved in *Drosophila*, and signaling through *Mei-41* (ATM/ATR), *Mnk* (Chk2), and *Dmp53* (p53) is required for IR induced apoptosis *(76)*. Furthermore, the evaluation of DNA damage responses in *Drosophila* has been very informative of the functions of *Dmp53* in apoptosis. In response to γ-radiation, *Dmp53* is required for the apoptotic response, but not G_2 arrest. *Dmp53* transcriptional targets, induced by γ-radiation, include proapoptotic genes *reaper*, *sickle, hid,* and *Eiger*, but classical p53 cell cycle targets, like *Dacapo* (p21/p27) are missing *(76)*. Furthermore, the *Drosophila* apoptotic response is dependent on *mnk*, the Chk2 homolog, phosphorylating p53. Therefore, the *Drosophila* fully recapitulates the upstream p53

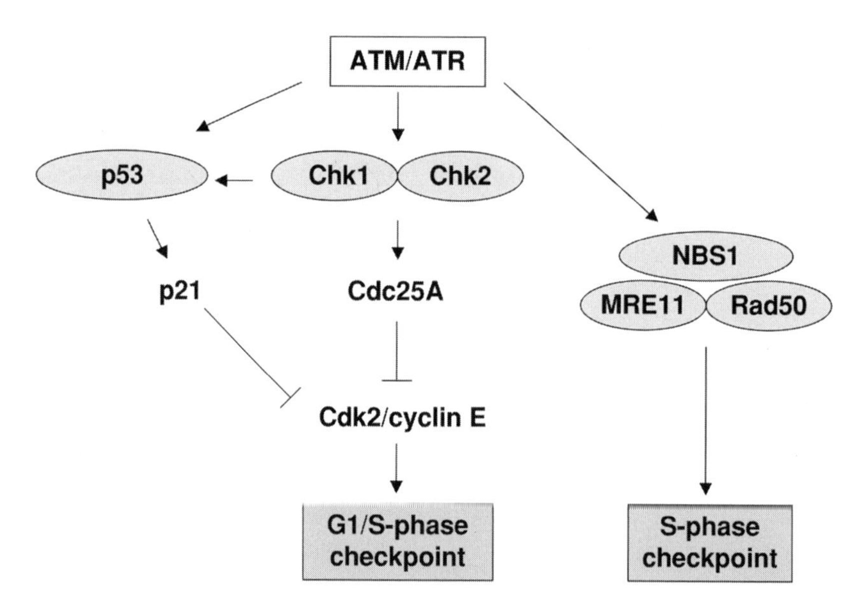

Fig. 2. G_1/S transition and intra-S phase checkpoints are enforced by parallel pathways.

DNA damage activation pathway and provides a model for further studies in damage checkpoint functions.

Progression of the mammalian cell cycle requires the activity of Cdc25 family of phosphatases, which remove the inhibitory phosphorylations of Cdks and thus enable Cdk-activities required for both S-phase (Cdk2) and mitosis (Cdk1) together with their cognate cyclins *(77)*. During normal cell-cycle Cdc25A is phosphorylated by Chk1 and targeted for degradation, thus controlling the entry of the cells into S-phase *(78,79)*. In response to both IR and replication blocks, Cdc25A is phosphorylated and degraded by the proteasome *(80–82)*. IR-induced degradation of Cdc25A is provoked by ATM-Chk2 kinases, whereas the UVC-induced degradation occurs through the ATR-Chk1 kinase cascade *(80,83,84)*. In consequence, the replicative cyclin E-Cdk2 complex is inactivated, S-phase arrest ensues *(80)* and prevents the loading of Cdc45 to replication origins *(85)* (Fig. 2). Both Chk2 and Chk1 are required for Cdc25A degradation in response to IR *(84,86)*, whereas the activity of Chk1 is sufficient following UV or stalled replication forks *(80)*. On the other hand, activation of both Chk1 and Chk2 occurs in senescent cells as a result of the presence of uncapped telomeres provoking a DNA damage response *(30)*. These findings therefore couple cell-cycle control and damage control.

Interestingly, embryonic stem cells do not display a proper G_1 arrest following IR apparently because of centrosomal sequestration of Chk2, causing increased cell apoptosis *(87)*. Also Chk1 is observed on centrosomes and controls cdk1–cyclin B complex activity *(88)*. Chk2 targets also transcription factors E2F and promyelocytic leukemia protein PML, which through their activities may further modulate the cellular reponses by promoting, for example, apoptosis *(89,90)*. Indeed, by phosphorylating and activating E2F1 following DNA damage both Chk1 and Chk2 upregulate p73, a close family member of p53 *(91)*.

4.2. Intra-S-Phase Checkpoints

During S-phase, the cell monitors the progression of DNA replication. Stalled replication forks, either caused by damage to DNA or depletion of ribonucleotides, generate a damage response and arrest of the cells before entry to G_2. UV-damage causes a reduction at the rate of DNA synthesis manifested by a decrease in chain elongation and replication initiation *(11)*. However, in the presence of DSB, replication is not completely halted, but the S-phase progression caused by inhibition of origin firing is delayed *(92)*. Thus, mechanistically, IR and, e.g., UV radiation impose distinct cellular stresses and damage responses.

Several ATM initiated pathways or targets are activated during S-phase following DNA damage either to ensure an intra-S-phase delay or to initiate repair processes. These include the Chk1 and Chk2 initiated phosphorylation cascades leading to degradation of Cdc25A *(83,84)* and a S-phase delay by maintaining the Wee1 mediated inhibitory Thr14/Tyr15 phosphorylation on Cdk2 thus inhibiting the Cdk2/cyclin E kinase activities *(79)* (Fig. 2). In consequence, the maintenance of the retinoblastoma protein in an E2F suppressive mode ensures further that E2F targets, several of which are required for the orderly progression through S-phase, are not upregulated. This is however contrasted by the findings that DNA damage also activates E2F through Chk2 mediated phosphorylation and promotes apoptosis *(90,93)*. During normal cell cycle, the selection between apoptotic and cell cycle-promoting activities can be mediated by TopBP1 which, by recruiting Brg1/Brm, a SWI/SNF chromatin-remodeling complex factor, to E2F1-responsive promoters, represses E2F1 activity *(94)*.

The MRN complex, composed of Mre11, RAD50, and Nbs1 subunits, ensures the fidelity of the replication complex *(95)*. The trimeric MRN complex is formed upon Nbs1 phosphorylation in an ATM/H2AX-dependent manner. Another ATM substrate, Mdc1 is localized to DSB foci and functions to mediate the interactions of H2AX and the MRN complex *(96–98)* (Fig. 2). Nbs1 is rapidly recruited to the site of the damage, whereas Chk2, though contacting the DSBs, appears to be a mobile transmitter of the damage signaling cascade *(99)*. The MRN complex enforcing an intra-S-phase dependent checkpoint acts thus in parallel with Chk2-Cdc25A-Cdk2 to mediate S-phase arrest *(85)* (Fig. 2).

The capacity of cells to pass unrestricted through S-phase in the presence of DNA damage is called radioresistant DNA synthesis (RDS). This is a property shared by ATM-defective cells and cells null or carrying mutations of 53BP1, Brca1, Mre11, and FancD2 *(22)*, whereas Nbs1 and ATLD cells show partial RDS *(85)*. Interestingly, radioresistant DNA synthesis couples often with radiosensitivity of the cells apparently reflecting their increased genomic instability. Thus ATM, Chk1, 53BP1, H2AX, Mdc1, Nbs1, SMC1, Brca1, and FancD2 defective cells display increased radiosensitivity *(10,22,79,100)*. FancD2, mutated in Fanconi anemia (FA), a chromosomal instability disorder with high frequency of cancers, is also a target of ATM *(101)*, and interacts with Nbs1 *(102)*. Thus, AT, NBS, and FA patients and 53BP1 null mice are cancer prone, suggesting that they have a critical role in damage provoked checkpoints *(101,103)*. The increased radiosensitivity in the presence of a checkpoint defect could reflect unsuccesful repair, increased transcriptional stress and/or mitotic catastrophe. Yet, at the same time these features also predispose to increased genomic instability and malignant conversion.

ATR, on the other hand, appears to act as a surveillance mechanism of both DNA replication and for stalled replication forks. Similar to the ATR-ATRIP complex binding to ssDNA during replicative stress, ATR, Claspin, Rad9-Hus1 complex are required during the normal replicative cycle *(36)*. In *Xenopus* extracts, DNA replication is a prerequisite for recruitment of ATR and Rad1 to chromatin and for damage-induced phosphorylation of Chk1 *(104)*. Although the UV damage induced S-phase checkpoint response is mediated through ATR signaling through Chk1, it is independent of ATM, Nbs1, and Mre11 *(105)*. Whether replicative stress causes further activation of ATR or changes its substrate spetrum remains to be determined.

4.3. Mitotic Checkpoint Control

The error-free segregation of the sister chromatids during mitosis is essential for the division of equal copy numbers to the daugther cells. Frequent aneuploidy in tumors is a characteristic feature of dysfunction of this critical checkpoint. In response to unreplicated DNA or DNA lesions, cells arrest before entry into mitosis and accumulate to G_2 *(106)*. Later in mitosis, the spindle assembly checkpoint is triggered if unattached kinetochores are present and to inhibit the anaphase entry. Kinases (Chk1, Chk2) and phosphatases (Wee1) acting on the absolutely essential mitotic kinase, Cdk1, are the core machinery of this checkpoint *(106)*. Following DSB or replication stress, Chk1/Chk2 phosphorylate dual specificity phosphatases Cdc25C and Cdc25A, and provoke their subsequent proteasomal degradation (*80*; reviewed in ref. *107*). This leads to inactivation of the mitotic kinase Cdk1 by maintaining its inactivating Thr14/Tyr15 phosphorylations and G_2/M-phase arrest *(86)*. Furthermore, 14-3-3 proteins associate with and may mediate the relocalization of both Cdc25B and Cdc25C to the cytoplasmic compartment *(107,108)*. p53 may contribute to this event by inducing the expression of 14-3-3sigma *(109)*. Conversely, during mitosis, cyclin B-Cdk1 phosphorylates Cdc25A, leading to its stabilization and allowing its mitotic activity *(110)*. Polo-like kinase, a mitotic kinase which activates Cdk1-cyclin B complex, is inhibited following DNA damage *(111)* further ensuring the mitotic arrest.

BRCA1 deficiency leads to gross chromosomal changes and mitotic abnormalities, which correlate with a deficiency in homologous recombination and DSB repair *(100)*. Both Brca1 and Brca2 are expressed during S- and G_2-phases of the cell cycle. Following DSB, Brca1 translocates to the lesions and is phosphorylated by ATM, ATR, and Chk2 *(17,23,60)*. Furthermore, it colocalizes with FancD2 and Rad51 and interacts with the MRN complex *(100)*. The multiple Brca1 interactions may be organized through recognition and binding of phosphoprotein domains via its BRCT-domain *(112,113)*, present also in 53BP1 and Mdc1. Interestingly, Brca1 is required for the activation of Chk1, down-regulation of Cdc25C and the mitotic checkpoint *(114)*, implicating extensive crosstalk in the damage response pathways. Brca2 can also be found in the DNA damage foci containing Brca1, FancD2, and Rad51, which occurs through FancD2 mediated monoubiquitination of Brca2 *(115)*. Brca2 and FancD2 null cells show chromosomal instability indicating that they participate in transcriptional repair *(115)*.

The spindle assembly checkpoint is governed by proteins that sense and bind unattached kinetochores such as the Mad and Bub proteins *(116)*. Their function is required for the mitotic checkpoint in both yeast and mammalian cells. Bub1 is mutated in a subset of colorectal cancers *(117)* suggesting that its dysfunction in mitotic control predisposes to tumorigenesis. Furthermore, 53BP1 localizes to the kinetochores of mitotic cells suggesting that it may function also in the mitotic checkpoint. Indeed, cells deficient for 53BP1 manifest a G_2/M checkpoint defect following low, but not high doses of IR *(25,26)*. This suggests that high doses of DNA damage invoke additional surveillance mechanisms independent of 53BP1.

5. Hereditary Syndromes With Increased Predisposition to Cancer Couple With Defects in Damage Sensing and Repair

Several rare autosomal recessive disorders are characterized by spontaneous chromosome breakage. AT patients have an increased risk of cancers, but also progressive cerebellar ataxia, immune deficiencies, and other malformations *(5,8)* (Table 1). Nijmegen breakage syndrome shares several clinical features with AT, but the NBS patients have also microcephaly and distinquished facial features *(118)*. AT-like disorder (ATLD) patients with mutations in the *MRE11* subunit of the MRN complex, resemble those of AT patients and have also chromosome instability (Table 1). Whereas AT, ATLD, and NBS have partially overlapping phenotypic features, a mutation of ATR in Seckel syndrome produces mostly congenital abnormalities with severe growth and mental retardation and microcephaly. Fanconi anemia (FA) is a multigenic syndrome with chromosome instability and overlapping clinical features including skeletal abnormalities and bone marrow failure *(119)*. Hereditary breast cancer associated genes, *BRCA1*, *BRCA2/FANCD2*, and *CHK2* are intimately coupled with the damage signaling networks through their direct associations and by coupling with the recombination and repair machineries. Lastly, Li-Fraumeni syndrome patients, with multiple types of early-onset cancers, have germline mutations of *TP53* and *CHK2 (120)*. It is striking that of the genes involved in the DNA damage response and damage checkpoint pathways, only *MDC1*, *SMC1*, *53BP1*, and *H2AX* are yet to be coupled to either human cancer syndromes or other developmental abnormalities.

Mutations of DNA repair-associated genes in hereditary syndromes have been reviewed recently and are not discussed here *(2,115)*.

6. New Therapeutic Approaches Through Interference of the DNA Damage Response Pathways

The sensitivity of the damage checkpoint pathways have presented the possibility that revoking their activities connecting with the apoptosis pathways could be utilized in strategies for new treatments. One such example is β-lapachone, an S-phase checkpoint activator and topoisomerase I inhibitor, which activates E2F1 and selectively kills tumor cells by inducing apoptosis *(121)*. Attempts have also been based on specific abrogation of the damage sensory processes or damage checkpoints. There are numerous examples of radiosensitivity in cells with defective damage responses and repair mechanisms. Therefore, several approaches have been undertaken to search for compounds abrogating the proper checkpoints thus causing chemosensitivity. For example, UCN-01, which selectively inhibits Chk1, is being tested as a chemosensitizing agent in clinical phase I/II trials in combination with several topoisomerase inhibitors and antimetabolites *(122)*. Several lines of experimental evidence have provided support that downregulation or inhibition of the pathway mediators, like silencing of ATM, ATR, or DNA-PKcs by RNAi based methods *(123)*, provides enhanced killing of the

Table 1
Hereditary Cancer Syndromes With Mutations in DNA Damage Sensory and Checkpoint Proteins

Syndrome	Gene	Locus	Protein	Function	Clinical syndrome
Ataxia teleangiectasia	*ATM*	11q22	ATM	Damage sensor	Cerebellar ataxia, teleangiectasia immune defects, lymphoma, leukemia
ATR-Seckel	*ATR*	3q22-24	ATR	Damage sensor	Growth retardation, microcephaly, mental retardation
AT-like disorder (ATLD)	*MRE11*	11q21	Mre11	Damage signaling	Cerebellar ataxia, chromosomal instability
Nijmegen breakage	*NBS1*	8q21	Nbs1	Damage signaling	Growth retardation, immunodeficiency, microcephaly, chromosomal instability, lymphoma
Fanconi anemia*	*FANC*	3p25.3	FancD2	Checkpoint control and homologous recombination	Congenital abnormalities and malformations, bone-marrowfailure, leukaemia, skin changes
Hereditary breast cancer	*BRCA1* *CHK2* *BRCA2* (*FANCD1*)	17q21 22q12 13q12-13	Brca1 Chk2 Brca2	Signal processor/ repair Damage signaling Repair	Breast and ovarian cancer
Li-Fraumeni	*TP53* *CHK2*	17p13 22q12	p53 Chk2	Transcription factor Damage signaling	Breast carcinoma, soft tissue sarcoma, brain tumors, osteosarcoma, leukemia

*Major known mutated genes are shown.

tumor cells. However, examples also exist of the presence of multiple pathway defects in tumor cells, affecting G_1, S, G_2/M checkpoints (p53 and Brca1 mutations) which do not sensitize the cells to enhanced death *(124)*.

Miscuing of the cell-cycle checkpoints can also enhance cell killing. For example, the premature activation of Cdk1 has been shown to lead to apoptosis *(125)* and inhibition of Wee1 activity by PD166285 radiosensitizes p53 mutant cells lacking a functional G_1 checkpoint *(126)*. The PD (pyrido [2,3-d] pyrimidine) compounds were initially synthesized as ATP competitive tyrosine kinase inhibitors and found to possess activities against Src, PDGF receptor and Abl-family tyrosine kinases *(127)*. These compounds inhibit either S-phase (PD179483, PD166326) or mitosis (PD173955) *(128)*. The S-phase inhibitors act in vitro as Wee1 inhibitors, have demonstrated activity in murine models of chronic myeloic leukemia *(129)* and could therefore be useful in combination therapies.

Several transcriptional inhibitors may in fact promote cell death by provoking transcriptional stress and thereby activation of the ATR pathway. Cdk-inhibitors flavopiridol and roscovitine block the catalytic activity of Cdks and also inhibit RNA polymerase II C-terminal phosphorylation thereby causing global inhibition of transcription. Therefore they not only arrest cell-cycle progession but also inflict transcriptional stress *(130,131)*. Several cytotoxic drugs currently in clinical use act by causing DNA damage and by inhibiting transcription (alkylating agents, antimetabolites, topoisomerase inhibitors) *(122,131)*. However, these have severe side effects, which calls for the need of more specific drugs or combination therapies.

There is substantial evidence that *TP53* mutations associate with increased radioresistance and chemoresistance *(73–75)*. On the other hand, if p53 function is excessively activated during therapy, it can inappropriately promote death of radiosensitive normal cells. Therefore the p53 pathway has been an attractive target either in attempts to activate the mutant forms of p53 or to prevent the activity wt p53 *(132)*. Examples exist for compounds activating wild type (Nutlins) *(133)* and mutant forms of p53 (Prima) *(134)* or inhibiting wild type p53 (Pifithrin) *(135)*.

Lastly, because several hereditary repair syndromes have specific defects in one or more of the repair proteins, their replacement or increased repair by other means could provide a basis for effective therapy. In mouse models, the expression of photolyase, a cyclobutane pyrimidine dimer correcting enzyme absent in mammals, reduces skin damage (erythema, hyperplasia, apoptosis) *(136,137)*. On the other hand, topical administration of pTT thymidine dimers mimick the UV-induced pyrimidine dimers and provoke an extensive damage response with upregulation of p53 and apoptosis *(138)*. This treatment effectively enhances the rate of repair and diminishes the amount of UV induced skin cancer in mouse *(139)*.

7. Conclusion

A substantial amount of information has been acquired on the players and pathways reacting to DNA damage. Detailed information has been accumulated on how dysfunction of the various components affect the cellular damage responses and predisposition to cancer. New strategies for treatment have been and are being designed to utilize this knowledge in various ways, including both damage reparative measures and measures imposing excessive damage or stress to provoke cell death. It is anticipated that these approaches will provide more targeted and effective treatments with less drug-related adverse reactions.

References

1. Friedberg EC. How nucleotide excision repair protects against cancer. Nat Rev Cancer 2001;1:22–33.
2. Hoeijmakers JHJ. Genome maintenance mechanisms for preventing cancer. Nature 2001;411:366–374.
3. West SC. Molecular views of recombination proteins and their control. Nat Rev Mol Cell Biol 2003;4:435–445.
4. Valerie K, Povirk LF. Regulation and mechanisms of mammalian double-strand break repair. Oncogene 2003;22:5792–5812.
5. Kastan MB, Lim DS. The many substrates and functions of ATM. Nat Rev Mol Cell Biol 2000;1:179–186.
6. Kastan MB, Bartek J. Cell cycle checkpoints and cancer. Nature 2004;432:316–323.
7. Bartek J, Lukas J. Chk1 and Chk2 kinases in checkpoint control and cancer. Cancer Cell 2003;3:421–429.
8. Shiloh Y. ATM and related protein kinases: safeguarding genome integrity. Nat Rev Cancer 2003;3:155–168.
9. Laiho M, Latonen L. Cell cycle control, DNA damage checkpoints and cancer. Ann Med 2003;35:391–397.
10. Bakkenist CJ, Kastan MB. Initiating cellular stress responses. Cell 2004;118:9–17.
11. Kaufmann WK, Cleaver JE, Painter RB. Ultraviolet radiation inhibits replicon initiation in S phase human cells. Biochim Biophys Acta 1980;608:191–195.
12. Latonen L, Laiho M. Cellular UV damage responses—functions of tumor suppressor p53. *BBA Reviews in Cancer* 2005;1755:71–89.
13. van Vugt MA, Bras A, Medema RH. Polo-like kinase-1 controls recovery from a G2 DNA damage-induced arrest in mammalian cells. Mol Cell 2004;15:799–811.
14. Bakkenist CJ, Kastan MB. DNA damage activates ATM through intermolecular autophosphorylation and dimer dissociation. Nature 2003;421:499–506.
15. Burma S, Chen BP, Murphy M, Kurimasa A, Chen DJ. ATM phosphorylates histone H2AX in response to DNA double-strand breaks. J Biol Chem 2001;276:42,462–42,467.
16. Ward IM, Chen J. Histone H2AX is phosphorylated in an ATR-dependent manner in response to replicational stress. J Biol Chem 2001;276:47,759–47,762.
17. Cortez D, Wang Y, Qin J, Elledge SJ. Requirement of ATM-dependent phosphorylation of BRCA1 in the DNA damage response to double-strand breaks. Science 1999;286: 1162–1166.
18. Celeste A, Petersen S, Romanienko PJ, et al. Genomic instability in mice lacking histone H2AX. Science 2002;296:922–927.
19. Celeste A, Fernandez-Capetillo O, Kruhlak MJ, et al. Histone H2AX phosphorylation is dispensable for the initial recognition of DNA breaks. Nat Cell Biol 2003;5:675–679.
20. Iwabuchi K, Bartel PL, Li B, Marraccino R, Fields S. Two cellular proteins that bind to wild-type but not mutant p53. Proc Natl Acad Sci USA 1994;91:6098–6102.
21. Schultz LB, Chehab NH, Malikzay A, Halazonetis TD. p53 binding protein 1 (53BP1) is an early participant in the cellular response to DNA double-strand breaks. J Cell Biol 2000;151:1381–1390.
22. Mochan TA, Venere M, DiTullio RA, Jr, Halazonetis TD. 53BP1, an activator of ATM in response to DNA damage. DNA Repair 2004;3:945–952.
23. Lee JS, Collins KM, Brown AL, Lee CH, Chung JH. hCds1-mediated phosphorylation of BRCA1 regulates the DNA damage response. Nature 2000;404:201–204.

24. Matsuoka S, Rotman G, Ogawa A, Shiloh Y, Tamai K, Elledge SJ. Ataxia telangiectasia-mutated phosphorylates Chk2 in vivo and in vitro. Proc Natl Acad Sci USA 2000;97: 10,389–10,394.
25. Wang B, Matsuoka S, Carpenter PB, Elledge SJ. 53BP1, a mediator of the DNA damage checkpoint. Science 2002;298:1435–1438.
26. Fernandez-Capetillo O, Chen H-T, Celeste A, et al. DNA damage-induced G2-M checkpoint activation by histone H2AX and 53BP1. Nat Cell Biol 2002;4:993–997.
27. DiTullio RA, Jr, Mochan TA, Verene M, Bartkova J, Sehested M, Bartek J, Halazonetis TD. 53BP1 functions in an ATM-dependent checkpoint pathway that is constitutively activated in human cancer. Nat Cell Biol 2002;4:998–1002.
28. Kitagawa R, Bakkenist CJ, McKinnon PJ, Kastan MB. Phosphorylation of SMC1 is a critical downstream event in the ATM-NBS1-BRCA1 pathway. Genes Dev 2004;18: 1423–1438.
29. Cline SD, Hanawalt PC. Who's on first in the cellular response to DNA damage? Nat Rev Mol Cell Biol 2003;4:361–372.
30. d'Adda di Fagagna F, Reaper PM, Clay-Farrace L, et al. A DNA damage checkpoint response in telomere-initiated senescence. Nature 2003;426:194–198.
31. Lukas C, Melander F, Stucki M, et al. Mdc1 couples DNA double-strand break recognition by Nbs1 with its H2AX-dependent chromatin retention. EMBO J 2004;23:2674–2683.
32. Liu Q, Guntuku S, Cui XS, et al. Chk1 is an essential kinase that is regulated by Atr and required for the G(2)/M DNA damage checkpoint. Genes Dev 2000;14:1448–1459.
33. Brown EJ, Baltimore D. ATR disruption leads to chromosomal fragmentation and early embryonic lethality. Genes Dev 2000;14:397–402.
34. Cortez D, Guntuku S, Qin J, Elledge SJ. ATR and ATRIP: partners in checkpoint signaling. Science 2001;294:1713–1716.
35. Zou L, Elledge SJ. Sensing DNA damage through ATRIP recognition of RPA-ssDNA complexes. Science 2003;300:1542–1548.
36. Sørensen CS, Syljuåsen RG, Lukas J, Bartek J. ATR, Claspin and the Rad9-Rad1-Hus1 complex regulate Chk1 and Cdc25A in the absence of DNA damage. Cell Cycle 2004;3: 941–945.
37. Takata M, Sasaki MS, Sonoda E, et al. Homologous recombination and non-homologous end-joining pathways of DNA double-strand break repair have overlapping roles in the maintenance of chromosomal integrity in vertebrate cells. EMBO J 1998;17:5497–5508.
38. Sørensen CS, Hansen LT, Dziegielewski J, et al. The cell-cycle checkpoint kinase Chk1 is required for mammalian homologous recombination repair. Nat Cell Biol 2005;7:195–201.
39. de Laat WL, Jaspers NG, Hoeijmakers JH. Molecular mechanism of nucleotide excision repair. Genes Dev 1999;13:768–785.
40. Hanawalt PC. Subpathways of nucleotide excision repair and their regulation. Oncogene 2002;21:8949–8956.
41. Svejstrup JQ. Mechanisms of transcription-coupled DNA repair. Nat Rev Mol Cell Biol 2002;3:21–29.
42. Tornaletti S, Hanawalt PC. Effect of DNA lesions on transcription elongation. Biochimie 1999;81:139–146.
43. Sugasawa K, Ng JM, Masutani C, et al. Xeroderma pigmentosum group C protein complex is the initiator of global genome nucleotide excision repair. Mol Cell 1998;2:223–232.
44. McKay BC, Stubbert LJ, Fowler CC, Smith JM, Cardamore RA, Spronck JC. Regulation of ultraviolet light-induced gene expression by gene size. Proc Natl Acad Sci USA 2004;101:6582–6586.
45. Ratner JN, Balasubramanian B, Corden J, Warren SL, Bregman DB. Ultraviolet radiation-induced ubiquitination and proteasomal degradation of the large subunit of RNA

polymerase II. Implications for transcription-coupled DNA repair. J Biol Chem 1998; 273:5184–5189.
46. Rockx DA, Mason R, van Hoffen A, et al. UV-induced inhibition of transcription involves repression of transcription initiation and phosphorylation of RNA polymerase II. Proc Natl Acad Sci USA 2000;97:10,503–10,508.
47. Venema J, Mullenders LH, Natarajan AT, van Zeeland AA, Mayne LV. The genetic defect in Cockayne syndrome is associated with a defect in repair of UV-induced DNA damage in transcriptionally active DNA. Proc Natl Acad Sci USA 1990;87:4707–4711.
48. Mone MJ, Bernas T, Dinant C, et al. In vivo dynamics of chromatin-associated complex formation in mammalian nucleotide excision repair. Proc Natl Acad Sci USA 2004;101: 15,933–15,937.
49. Hartwell LH, Weinert TA. Checkpoints: controls that ensure the order of cell cycle events. Science 1989;246:629–634.
50. Hartwell LH, Kastan MB. Cell cycle control and cancer. Science 1994;266:1821–1828.
51. Zhou BB, Elledge SJ. The DNA damage response: putting checkpoints in perspective. Nature 2000;408:433–439.
52. Paulovich AG, Toczyski DP, Hartwell LH. When checkpoints fail. Cell 1997;88:315–321.
53. Wahl GM, Carr AM. The evolution of diverse biological responses to DNA damage: insights from yeast and p53. Nat Cell Biol 2001;3:277–286.
54. Levine AJ. p53, the cellular gatekeeper for growth and division. Cell 1997;88:323–331.
55. Honda R, Tanaka H, Yasuda H. Oncoprotein MDM2 is a ubiquitin ligase E3 for tumor suppressor p53. FEBS Lett 1997;420:25–27.
56. Leng RP, Lin Y, Ma W, et al. Pirh2, a p53-induced ubiquitin-protein ligase, promotes p53 degradation. Cell 2003;112:779–791.
57. Dornan D, Wertz I, Shimizu H, Arnott D, Frantz GD, Dowd P, O'Rourke K, Koeppen H, Dixit VM. The ubiquitin ligase COP1 is a critical negative regulator of p53. Nature 2004; 429:86–92.
58. Banin S, Moyal L, Shieh S, et al. Enhanced phosphorylation of p53 by ATM in response to DNA damage. Science 1998;281:1674–1677.
59. Canman CE, Lim DS, Cimprich KA, et al. Activation of the ATM kinase by ionizing radiation and phosphorylation of p53. Science 1998;281:1677–1679.
60. Tibbetts RS, Brumbaugh KM, Williams JM, et al. A role for ATR in the DNA damage-induced phosphorylation of p53. Genes Dev 1999;13:152–157.
61. Meek DW. The p53 response to DNA damage. DNA Repair 2004;3:1049–1056.
62. Chehab NH, Malikzay A, Appel M, Halazonetis TD. Chk2/hCds1 functions as a DNA damage checkpoint in G(1) by stabilizing p53. Genes Dev 2000;14:278–288.
63. Shieh SY, Ahn J, Tamai K, Taya Y, Prives C. The human homologs of checkpoint kinases Chk1 and Cds1 (Chk2) phosphorylate p53 at multiple DNA damage-inducible sites. Genes Dev 2000;14:289–300.
64. Lakin N, Jackson SP. Regulation of p53 in response to DNA damage. Oncogene 1999;18: 7644–7655.
65. Hirao A, Kong YY, Matsuoka S, et al. DNA damage-induced activation of p53 by the checkpoint kinase Chk2. Science 2000;287:1824–1827.
66. Zhao R, Gish K, Murphy M, et al. Analysis of p53-regulated gene expression patterns using oligonucleotide arrays. Genes Dev 2000;14:981–993.
67. el-Deiry WS, Tokino T, Velculescu VE, et al. WAF1, a potential mediator of p53 tumor suppression. Cell 1993;75:817–825.
68. Dulic V, Kaufmann WK, Wilson SJ, et al. p53-dependent inhibition of cyclin-dependent kinase activities in human fibroblasts during radiation-induced G1 arrest. Cell 1994;76: 1013–1023.

69. Vousden KH, Lu X. Live or let die: the cell's response to p53. Nat Rev Cancer 2002;2:594–604.
70. Mihara M, Erster S, Zaika A, Petrenko O, Chittenden T, Pancoska P, Moll UM. p53 has a direct apoptogenic role at the mitochondria. Mol Cell 2003;11:577–590.
71. Leu JI, Dumont P, Hafey M, Murphy ME, George DL. Mitochondrial p53 activates Bak and causes disruption of a Bak-Mcl1 complex. Nat Cell Biol 2004;6:443–450.
72. Chipuk JE, Kuwana T, Bouchier-Hayes L, et al. Direct activation of Bax by p53 mediates mitochondrial membrane permeabilization and apoptosis. Science 2004;303:1010–1014.
73. Gudkov AV, Komarova EA. The role of p53 in determining sensitivity to radiotherapy. Nat Rev Cancer 2003;3:117–129.
74. Fei P, El-Deiry WS. P53 and radiation responses. Oncogene 2003;22:5774–83.
75. El-Deiry WS. The role of p53 in chemosensitivity and radiosensitivity. Oncogene 2003;22:7486–7495.
76. Brodsky MH, Weinert BT, Tsang G, et al. Drosophila melanogaster MNK/Chk2 and p53 regulate multiple DNA repair and apoptotic pathways following DNA damage. Mol Cell Biol 2004;24:1219–1231.
77. Nilsson I, Hoffmann I. Cell cycle regulation by the Cdc25 phosphatase family. Prog Cell Cycle Res 2000;4:107–114.
78. Shimuta K, Nakajo N, Uto K, Hayano Y, Okazaki K, Sagata N. Chk1 is activated transiently and targets Cdc25A for degradation at the Xenopus midblastula transition. EMBO J 2002;21:3694–3703.
79. Bartek J, Lukas C, Lukas J. Checking on DNA damage in S phase. Nat Rev Mol Cell Biol 2004;5:792–804.
80. Mailand N, Falck J, Lukas C, et al. Rapid destruction of human Cdc25A in response to DNA damage. Science 2000;288:1425–1429.
81. Molinari M, Mercurio C, Dominguez J, Goubin F, Draetta GF. Human Cdc25 A inactivation in response to S phase inhibition and its role in preventing premature mitosis. EMBO Rep 2000;1:71–79.
82. Hassepass I, Voit R, Hoffmann I. Phosphorylation at serine 75 is required for UV-mediated degradation of human Cdc25A phosphatase at the S-phase checkpoint. J Biol Chem 2003;278:29,824–29,829.
83. Falck J, Mailand N, Syljuasen RG, Bartek J, Lukas J. The ATM-Chk2-Cdc25A checkpoint pathway guards against radioresistant DNA synthesis. Nature 2001;410:842–847.
84. Sørensen CS, Syljuåsen RG, Falck J, et al. Chk1 regulates the S phase checkpoint by coupling the physiological turnover and ionizing radiation-induced accelerated proteolysis of Cdc25A. Cancer Cell 2003;3:247–258.
85. Falck J, Petrini JH, Williams BR, Lukas J, Bartek J. The DNA damage-dependent intra-S phase checkpoint is regulated by parallel pathways. Nat Genet 2002;30:290–294.
86. Zhao H, Watkins JL, Piwnica-Worms H. Disruption of the checkpoint kinase 1/cell division cycle 25A pathway abrogates ionizing radiation-induced S and G2 checkpoints. Proc Natl Acad Sci USA 2002;99:14,795–14,800.
87. Hong Y, Stambrook PJ. Restoration of an absent G1 arrest and protection from apoptosis in embryonic stem cells after ionizing radiation. Proc Natl Acad Sci USA 2004;101: 14,443–14,448.
88. Krämer A, Mailand N, Lukas C, et al. Centrosome-associated Chk1 prevents premature activation of cyclin-B-Cdk1 kinase. Nat Cell Biol 2004;6:884–891.
89. Yang S, Kuo C, Bisi JE, Kim MK. PML-dependent apoptosis after DNA damage is regulated by the checkpoint kinase hCds1/Chk2. Nat Cell Biol 2002;4:865–870.
90. Stevens C, Smith L, La Thangue NB. Chk2 activates E2F-1 in response to DNA damage. Nat Cell Biol 2003;5:401–409.

91. Urist M, Tanaka T, Poyurovsky MV, Prives C. p73 induction after DNA damage is regulated by checkpoint kinases Chk1 and Chk2. Genes Dev 2004;18:3041–3054.
92. Painter RB, Young BR. Radiosensitivity in ataxia-telangiectasia: a new explanation. Proc Natl Acad Sci USA 1980;77:7315–7317.
93. Stevens C, La Thangue NB. The emerging role of E2F-1 in the DNA damage response and checkpoint control. DNA Repair 2004;3:1071–1079.
94. Liu K, Luo Y, Lin FT, Lin WC. TopBP1 recruits Brg1/Brm to repress E2F1-induced apoptosis, a novel pRb-independent and E2F1-specific control for cell survival. Genes Dev 2004;18:673–686.
95. D'Amours D, Jackson SP. The Mre11 complex: at the crossroads of DNA repair and checkpoint signalling. Nat Rev Mol Cell Biol 2002;3:317–327.
96. Goldberg M, Stucki M, Falck J, et al. MDC1 is required for the intra-S-phase DNA damage checkpoint. Nature 2003;421:952–956.
97. Lou Z, Minter-Dykhouse K, Wu X, Chen J. MDC1 is coupled to activated CHK2 in mammalian DNA damage response pathways. Nature 2003;421:957–961.
98. Stewart GS, Wang B, Bignell CR, Taylor AM, Elledge SJ. MDC1 is a mediator of the mammalian DNA damage checkpoint. Nature 2003;421:961–966.
99. Lukas C, Falck J, Bartkova J, Bartek J, Lukas J. Distinct spatiotemporal dynamics of mammalian checkpoint regulators induced by DNA damage. Nat Cell Biol 2003;5: 255–260.
100. Venkitaraman AR. Cancer susceptibility and the functions of BRCA1 and BRCA2. Cell 2002;108:171–182.
101. Taniguchi T, Garcia-Higuera I, Xu B, et al. Convergence of the Fanconi anemia and Ataxia telangiectasia signaling pathways. Cell 2002;109:459–472.
102. Nakanishi K, Taniguchi T, Ranganathan V, et al. Interaction of FANCD2 and NBS1 in the DNA damage response. Nat Cell Biol 2002;4:913–920.
103. Ward IM, Minn K, van Deursen J, Chen J. p53 Binding protein 53BP1 is required for DNA damage responses and tumor suppression in mice. Mol Cell Biol 2003;23:2556–2563.
104. Lupardus PJ, Byun T, Yee MC, Hekmat-Nejad M, Cimprich KA. A requirement for replication in activation of the ATR-dependent DNA damage checkpoint. Genes Dev 2002;16: 2327–2332.
105. Heffernan TP, Simpson DA, Frank AR, et al. An ATR- and Chk1-dependent S checkpoint inhibits replicon initiation following UVC-induced DNA damage. Mol Cell Biol 2002;22: 8552–8561.
106. Nigg EA. Mitotic kinases as regulators of cell division and its checkpoints. Nat Rev Mol Cell Biol 2001;2:21–32.
107. Donzelli M, Draetta GF. Regulating mammalian checkpoints through Cdc25 inactivation. EMBO Rep. 2003;4:671–677.
108. Conklin DS, Galaktionov K, Beach D. 14-3-3 proteins associate with cdc25 phosphatases. Proc Natl Acad Sci USA 1995;92:7892–7896.
109. Hermeking H, Lengauer C, Polyak K, He TC, Zhang L, Thiagalingam S, Kinzler KW, Vogelstein B. 14-3-3 sigma is a p53-regulated inhibitor of G2/M progression. Mol Cell 1997;1:3–11.
110. Mailand N, Podtelejnikov AV, Groth A, Mann M, Bartek J, Lukas J. Regulation of G(2)/M events by Cdc25A through phosphorylation-dependent modulation of its stability. EMBO J 2002;21:5911–5920.
111. Smits VA, Klompmaker R, Arnaud L, Rijksen G, Nigg EA, Medema RH. Polo-like kinase-1 is a target of the DNA damage checkpoint. Nat Cell Biol 2000;2:672–676.
112. Manke IA, Lowery DM, Nguyen A, Yaffe MB. BRCT repeats as phosphopeptide-binding modules involved in protein targeting. Science 2003;302:636–639.

113. Yu X, Chini CC, He M, Mer G, Chen J. The BRCT domain is a phospho-protein binding domain. Science 2003;302:639–642.
114. Yarden RI, Pardo-Reoyo S, Sgagias M, Cowan KH, Brody LC. BRCA1 regulates the G2/M checkpoint by activating Chk1 kinase upon DNA damage. Nat Genet 2002;30:285–289.
115. Risinger MA, Groden J. Crosslinks and crosstalk: human cancer syndromes and DNA repair defects. Cancer Cell 2004;6:539–545.
116. Musacchio A, Hardwick KG. The spindle checkpoint: structural insights into dynamic signalling. Nat Rev Mol Cell Biol 2002;3:731–741.
117. Cahill DP, Lengauer C, Yu J, Riggins GJ, Willson JK, Markowitz SD, Kinzler KW, Vogelstein B. Mutations of mitotic checkpoint genes in human cancers. Nature 1998;392: 300–303.
118. Tauchi H, Matsuura S, Kobayashi J, Sakamoto S, Komatsu K. Nijmegen breakage syndrome gene, NBS1, and molecular links to factors for genome stability. Oncogene 2002;21: 8967–8980.
119. D'Andrea AD. The Fanconi road to cancer. Genes Dev 2003;17:1933–1936.
120. Varley J. TP53, hChk2, and the Li-Fraumeni syndrome. Methods Mol Biol 2003;222: 117–129.
121. Li Y, Sun X, LaMont JT, Pardee AB, Li CJ. Selective killing of cancer cells by beta -lapachone: direct checkpoint activation as a strategy against cancer. Proc Natl Acad Sci USA 2003; 100:2674–2678.
122. Zhou BB, Bartek J. Targeting the checkpoint kinases: chemosensitization versus chemoprotection. Nat Rev Cancer 2004;4:216–225.
123. Collis SJ, Swartz MJ, Nelson WG, DeWeese TL. Enhanced radiation and chemotherapy-mediated cell killing of human cancer cells by small inhibitory RNA silencing of DNA repair factors. Cancer Res 2003;63:1550–1554.
124. Xu B, O'Donnell AH, Kim ST, Kastan MB. Phosphorylation of serine 1387 in Brca1 is specifically required for the Atm-mediated S-phase checkpoint after ionizing irradiation. Cancer Res 2002;62:4588–4591.
125. Shi L, Nishioka WK, Th'ng J, Bradbury EM, Litchfield DW, Greenberg AH. Premature p34cdc2 activation required for apoptosis. Science 1994;263:1143–1145.
126. Wang Y, Li J, Booher RN, et al. Radiosensitization of p53 mutant cells by PD0166285, a novel G(2) checkpoint abrogator. Cancer Res 2001;61:8211–8217.
127. Bridges AJ. The rationale and strategy used to develop a series of highly potent, irreversible, inhibitors of the epidermal growth factor receptor family of tyrosine kinases. Curr Med Chem 1999;6:825–843.
128. Mizenina OA, Moasser MM. S-phase inhibition of cell cycle progression by a novel class of pyridopyrimidine tyrosine kinase inhibitors. Cell Cycle 2004;3:796–803.
129. Wolff NC, Veach DR, Tong WP, Bornmann WG, Clarkson B, Ilaria RL. PD166326, a novel tyrosine kinase inhibitor, has greater anti-leukemic activity than imatinib in a murine model of chronic myeloid leukemia. Blood 2005;105:3995–4003.
130. Ljungman M, Paulsen MT. The cyclin-dependent kinase inhibitor roscovitine inhibits RNA synthesis and triggers nuclear accumulation of p53 that is unmodified at Ser15 and Lys382. Mol Pharmacol 2001;60:785–789.
131. Ljungman M, Lane DP. Transcription - guarding the genome by sensing DNA damage. Nat Rev Cancer 2004;4:727–737.
132. Lane DP, Hupp TR. Drug discovery and p53. Drug Discov Today 2003;8:347–355.
133. Vassilev LT, Vu BT, Graves B, et al. In vivo activation of the p53 pathway by small-molecule antagonists of MDM2. Science 2004;303:844–848.
134. Bykov VJ, Issaeva N, Shilov A, et al. Restoration of the tumor suppressor function to mutant p53 by a low-molecular-weight compound. Nat Med 2002;8:282–288.

135. Komarov PG, Komarova EA, Kondratov RV, et al. A chemical inhibitor of p53 that protects mice from the side effects of cancer therapy. Science 1999;285:1733–1737.
136. Schul W, Jans J, Rijksen YM, et al. Enhanced repair of cyclobutane pyrimidine dimers and improved UV resistance in photolyase transgenic mice. EMBO J 2002;21:4719–4729.
137. Jans J, Schul W, Sert YG, et al. Powerful skin cancer protection by a CPD-photolyase transgene. Curr Biol 2005;15:105–115.
138. Eller MS, Maeda T, Magnoni C, Atwal D, Gilchrest BA. Enhancement of DNA repair in human skin cells by thymidine dinucleotides: evidence for a p53-mediated mammalian SOS response. Proc Natl Acad Sci USA 1997;94:12,627–12,632.
139. Goukassian DA, Helms E, van Steeg H, van Oostrom C, Bhawan J, Gilchrest BA. Topical DNA oligonucleotide therapy reduces UV-induced mutations and photocarcinogenesis in hairless mice. Proc Natl Acad Sci USA 2004;101:3933–3938.

6

c-Myc, Apoptosis, and Disordered Tissue Growth

Michael Khan and Stella Pelengaris

Summary

Deregulated expression of *c-Myc* is present in most, if not all, human cancers and is associated with a poor prognosis. The *c-Myc* proto-oncogene is essential for both cellular growth and proliferation, but paradoxically may also promote cell death. The study of this "dual potential" of c-Myc over the past two decades has provided a paradigm for exploring the role of other mitogenic proteins, such as E2F, many of which have now also been shown to have such intrinsic "tumor suppressor" properties. In fact, it may be a general feature of proteins that promote cell cycle that the oncogenic potential of deregulated expression is restrained by concurrent activation of processes, such as apoptosis or senescence, which effectively prevent propagation of the "damaged" cell. By implication, these in-built "failsafe" mechanisms must be overcome during tumorigenesis. However, once prevented, for instance by inactivation of the p53 or Rb pathways or upregulation of antiapoptotic proteins, the potentially devastating oncogenic potential of proteins such as c-Myc is unmasked.

It is also likely that highly conserved proteins such as c-Myc, situated upstream of signaling pathways regulating both cellular replication/growth on the one hand and apoptosis/growth arrest, act as key integrators of processes determining cell numbers and tissue size during normal development and adult tissue homeostasis. Therefore, maybe not surprisingly, c-Myc may also contribute to disease by way of stimulating apoptotic pathways, the very process that also restrains their oncogenic potential; an expanding body of recent evidence implicates c-Myc as a contributor to the death of insulin-secreting β-cells in diabetes.

In trying to unravel the diverse roles of these enigmatic proteins and to explore their potential as therapeutic targets, the application of regulatable transgenic mouse models has proved invaluable. Such models have led us to reassess prevailing views of oncogenesis—particularly as they suggest that cancer formation may occur on a "minimal platform" of as few as two strategically placed oncogenic lesions. This has become the source of much debate, particularly following elegant studies showing that transformation of human cells in vitro requires a greater number of genetic lesions than equivalent rodent cells, a view that would seem to challenge the "minimal platform" model. It now seems likely that a greater number of individual signaling pathways require targeting for oncogenesis in man (including p53, Rb, telomere length, and others—at least in cultured cells) compared with mouse, though this does not seem to be the case if c-Myc expression is deregulated. So how do we reconcile this? c-Myc appears to be a special case (though there are likely others) as deregulation clearly interferes with multiple cancer-relevant signaling pathways and thereby induce multiple "hallmark" features of cancer. For this reason, c-Myc is also a potentially key therapeutic target, a view that is encouraged by the regression of even advanced and genetically unstable tumors, seen following deactivation of c-Myc in various models. Increasingly, and based on the logical assumption that the range of c-Myc induced phenotypes are driven primarily by the transcriptional activation of various target genes, many researchers are exploiting post-genome era technologies, for global gene expression and proteomic analyses, in order to define the activities of c-Myc in physiology and disease at a molecular level.

However, before we get too carried away, it must be emphasized that conditional mouse cancer models are increasingly reproducing the more unfortunate aspects of their human counterparts;

From: *Apoptosis, Cell Signaling, and Human Diseases: Molecular Mechanisms, Volume 1*
Edited by R. Srivastava © Humana Press Inc., Totowa, NJ

including the development of "escapers," cancer cells that have become independent of the initiating mutations and thereby fail to regress, and the presence of so-called dormant cancer cells, which remain in apparently fully regressed tumors and eventually contribute to tumor recurrence. There is still a long way to go. This chapter will discuss these issues with a particular focus on *c-Myc*.

Key Words: *c-Myc*; apoptosis; growth; cell cycle; differentiation; cancer; diabetes; regression; gene expression.

1. Introduction

The *c-Myc* proto-oncogene encodes a transcription factor, c-Myc, the cellular homolog to the viral oncogene (v-*Myc*) of the avian myelocytomatosis retrovirus *(1)*. It was shown some time ago that activated oncogenic *c-Myc* is instrumental in the progression of a human cancer, Burkitt's lymphoma *(2)*. More recently, elevated or deregulated expression of c-Myc has been detected in a wide range of human cancers, and is often associated with aggressive, poorly differentiated tumors *(3–6)*.

The normal key biological function of c-Myc is to promote cell-cycle progression. However, as will be discussed, this enigmatic protein appears to be a key player in various other biological processes, such as differentiation, cell death, and angiogenesis. c-Myc is a transcription factor that requires dimerization with another protein, Max, in order to become transcriptionally active. Thus, most of its effects on cell behavior are likely to be the result of expression (and in some cases repression) of various target genes.

In the intact organism, c-Myc is expressed ubiquitously during development in growing tissues *(7,8)*. Likewise, in the adult body, c-Myc is expressed in tissue compartments possessing high proliferative capacity (e.g., skin epidermis and gut), whereas it is undetected in cells that have exited the cell cycle. There also exist other homologs of c-Myc (i.e., L-Myc and N-Myc), and the expression of each is normally restricted to certain tissues. Aside from c-Myc expression correlating well with proliferating tissues in vivo, this protein is crucial for normal development, as deletion of both c-Myc alleles in the mouse leads to embryonic lethality *(9)*.

With respect to human cancer, oncogenic *c-Myc* is instrumental in the progression of Burkitt's lymphoma resulting from a translocational event. In addition, elevated or deregulated expression of c-Myc has been detected in numerous human cancers, such as breast, colon, cervical, small cell lung carcinomas, osteosarcomas, glioblastomas, melanoma, and myeloid leukaemias. In most cases the causes of c-Myc overexpression are not known, but include gene amplification, whereby multiple copies of the *c-Myc* gene arise under the influence of genome instability.

Not surpringly, given the well described dangers of deregulated c-Myc expression there are a myriad of cellular processes able to restrain c-Myc. Thus, transcription of the *c-Myc* gene is regulated by positive and negative growth signals and subsequently various factors influence the ability of c-Myc to regulate gene expression. The c-Myc protein is subject to phosphorylation, acetylation ubiquitinylation, and rapid degradation or may be expelled from the nucleus or even placed under "house arrest" in the nucleolus. Thus, many factors able to interfere with this system can also affect c-Myc levels, but their role in cancer remains to be fully described.

Not surprisingly, considerable efforts are being devoted to developing therapeutics to target aspects of c-Myc activity. To this end, proof of hypothesis and much important information has derived from regulatable mouse tumor models (discussed later)—these models have exposed various oncogenic properties of c-Myc during tumor progression as well as determining whether inactivating c-Myc in tumors can lead to their regression. Not surprisingly, a major challenge is achieving specific delivery of the c-Myc inhibitor only to the nucleus of tumor cells in vivo, an issue that is relevant for many therapeutic agents that target specific tumorigenic proteins.

Even, where c-Myc itself may not prove a viable therapeutic target, some of those genes/proteins whose expression is regulated by c-Myc might do. Over the past few years a large number of studies have employed gene chip microarrays, serial analysis of gene expression (SAGE), proteomics, and other high-throughput techniques to identify potential c-Myc-targets in cultured cells in vitro. In general, as recently proposed following elegant studies from the laboratories of Sedivy and colleagues; c-Myc-activated genes are indicative of a physiological state geared toward the rapid utilization of carbon sources, the biosynthesis of precursors for macromolecular synthesis, and the accumulation of cellular mass. In contrast, the majority of c-Myc-repressed genes are involved in the interaction and communication of cells with their external environment, and several are known to possess antiproliferative or antimetastatic properties *(10)*. Gene expression has been examined in c-Myc activated basal keratinocytes in vivo, but apoptosis is not a feature of this model *(11)*; results are in line with predictions about growth regulatory properties of c-Myc. In fact, based on studies where c-Myc is activated in cells previously rendered c-Myc null, it has been estimated that as many as 10% of all genes may be regulated by c-Myc activation in some way *(12,13)*. However, all these studies, despite their undoubted utility, do not give us any direct indication of which of the many genes potentially regulated by c-Myc are actually responsible for mediating the different and even opposing outcomes of c-Myc activation. That this must be undertaken by examining the full repertoire of c-Myc phenotyopes in vivo is essential; it is clear that c-Myc-induced cellular phenotype is absolutely dependent on tissue location in vivo *(14,15)*; as will be discussed later, even the decision of whether c-Myc activation will result in growth and replication or death of the cell depends on environmental cues and epigenetic context. Thus, at present, despite important recent advances in describing putative c-Myc target genes/proteins many key questions remain unanswered.

In order to identify new therapeutic targets in c-Myc-induced responses we must first answer the following key questions.

1. How does c-Myc exert these distinct cellular outcomes—namely proliferation, growth, loss of differentiation, and apoptosis? Are they dependent on the transcription or repression of specific distinct or overlapping sets of c-Myc target genes?
2. To what extent do cellular responses to c-Myc depend on cell type or, more importantly, on tissue location, which will dictate the signals a given cell type receives from cellular interactions between extracellular matrix and neighboring cells?
3. How might circulating or paracrine/autocrine signals, able to mitigate particular c-Myc functions, be regulated?

The recent genetic construction of mice in which the expression or activation of a given gene/protein can be switched on or off in vivo, has provided a means to assess the physiological roles of individual (or combinations of) genes and their protein products in tumorigenesis in the all-important context of the intact adult organism (reviewed in ref. *16*). Thus, we may now address how host and tumor cells communicate in induced tumors, as well as validate a given gene/protein as a therapeutic target by simply switching off expression of the oncogene and looking for phenotype reversal (such as tumor regression). Such studies will have important implications for designing future therapies targeted to cancer cells.

2. Regulation of c-Myc Activity

Transcription of the c-*myc* gene is increased markedly at the G_0–G_1 transition, when resting cells are induced to proliferate, and is normally tightly controlled by external signals including growth factors, mitogens, and β-catenin, which promote and factors such as transforming growth factor β (TGF-β), which inhibit.

When a normal cell in vitro receives a signal to proliferate, such as that following binding of a growth factor to its transmembrane receptor, the result is activation of signaling pathways such as the Raf-mitogen activated protein kinase (MAPK) cascade inside the cell that will ultimately lead to changes in gene expression. Consequently, the induction of specific genes will activate the cell-cycle engine. The *c-Myc* proto-oncogene is one of these induced genes, known as an early-response gene (induced within 15 min of growth factor treatment in vitro) and plays a crucial role in allowing cells to exit G_0 and to make the transition from G_1 to S.

Various other factors may influence activation of c-Myc. β-catenin plays a signal-integrating role in Wnt- and growth factor-dependent proliferation events in mammalian development by both derepressing several classes of repressors and by activating Pitx2, which in turn regulates the activity of several growth control genes, including c-Myc *(17)*. The activation of lymphoid enhancer factor (LEF)/T-cell factor (TCF)-mediated transcription by sustained expression of β-catenin and the loss of TGF-β signaling are essential steps in carcinogenesis, particularly for cancers of the colon, breast, and liver. Both of these signaling pathways can target c-Myc. LEF/TCF-responsive elements are found in the promoter of the human *c-Myc* gene and β-catenin activates the transcriptional activity of the c-Myc promoter by binding to this element in various cell lines. When TCF, but interestingly not LEF, is bound to this element, TGF-β dissociates β-catenin and represses the transcriptional activity of the *c-Myc* promoter *(18)*, suggesting that under situations where LEF predominates the ability to downregulate c-Myc by TGF could be impaired. c-Myc is also a target gene of the c-Myb transcription factor, which is often over expressed in leukaemia and epithelial cancers *(19)*.

In most cases the causes of c-Myc overexpression remain to be specifically described, but include gene amplification, whereby multiple copies of the *c-Myc* gene arise under the influence of genome instability, which accompanies advancing neoplasia. Recent data suggest that the stability of the c-Myc protein may itself be regulated. Cells have evolved a number of mechanisms to limit the activity and accumulation of c-Myc. One of the most striking of these mechanisms is through the ubiquitin-proteasome pathway, which typically destroys c-Myc within minutes of its synthesis, thus many factors able to interfere with this system can also affect c-Myc degradation *(20–22)* (*see* Fig. 1).

2.1. c-Myc Turnover

The turnover of c-Myc is at least in part dependent on phosphorylation of two highly conserved residues in MB1 that are mutated in v-Myc (the viral oncogenic Myc) and in various cancers; phosphorylation at Ser 62 stabilizes c-Myc, whereas subsequent

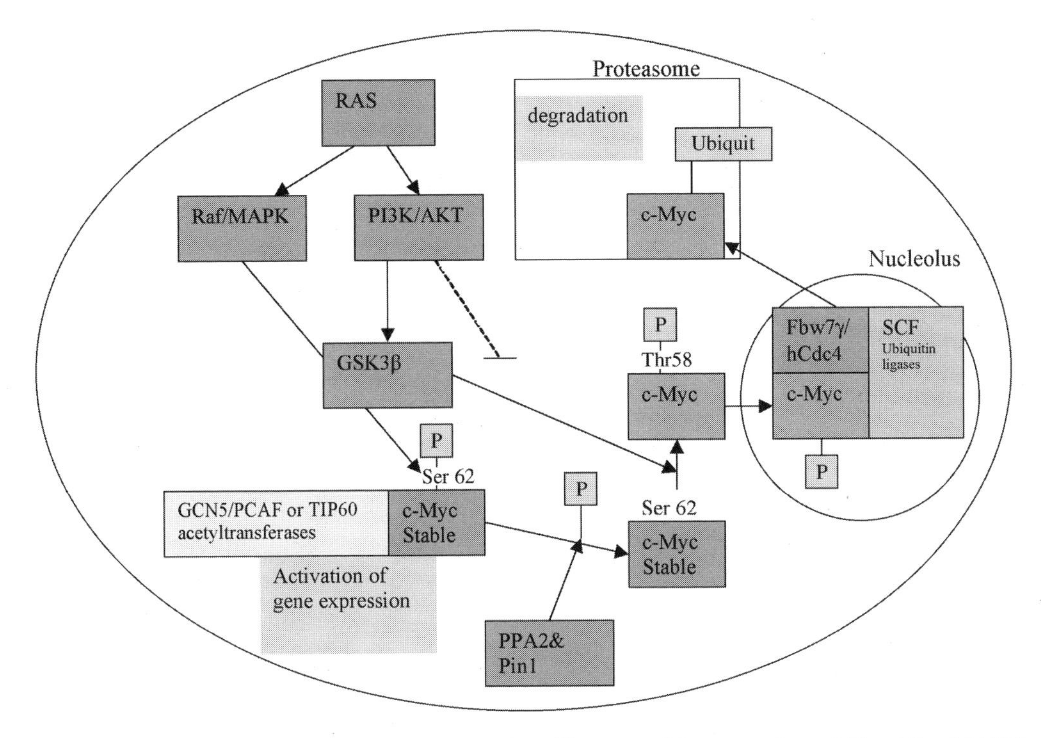

Fig. 1. Effects of acetylation and of Ras signalling on c-Myc protein stability. The MAPK pathway can stabilize c-Myc by phosphorylation of Ser62. Acetylation by acetyltransferases can also stabilize c-Myc. Conversely, c-Myc is destabilized by removal of the phosphate on Ser 62, by the phosphatases PPA2 and Pin1, and thence by phosphorylation of Thr 58 by the enzyme GSK3β. Thr 58 phosphorylation targets c-Myc for ubiquitinylation by Fbw7 and SCF ligases. This ubiquitinylation may in part take place in the nucleolus. PI3K may contribute to stabilisation of c-Myc by preventing phosphorylation on Thr58.

phosphorylation at Thr 58 and dephosphorylation at Ser 62 are needed prior to ubiquitinylation of c-Myc and degradation in the proteasome. Several pathways have now been identified that play key roles in determining c-Myc phosphorylation, particularly those regulated by Ras. Thus, via the Raf–MAPK cascade, c-Myc can be phosphorylated on Ser 62, which stabilizes c-Myc, but may also be a necessary step prior to further phosphorylation at Thr 58 by glycogen synthase kinase 3β (GSK-3β). Removal of the phosphate at Ser 62 is mediated bythe protein phosphatase PPA2 and the isomerase Pin1, two other Ras-regulated enzymes *(23)*. Conversely, via the PI3K/Akt pathway, Ras can inhibit GSK-3β, during early G_1, preventing phosphorylation of Thr 58 and degradation of c-Myc. However, the subsequent decline in Akt later in G_1, allows Thr 58 phosphorylation and c-Myc degradation.

The ability of c-Myc to activate transcription depends on the recruitment of several cofactor complexes including histone acetyltransferases (discussed later). In fact, acetylation of nucleosomal histones has long been recognized as a major regulator of gene transcription, but only recently has it been discovered that acetylation may also regulate subcellular localization and protein turnover. Thus, in a very recent study it has been demonstrated that degradation of c-Myc may also be regulated by acetylation by the acetytransferases mGCN5/PCAF and TIP60, which enhance protein stability *(24)*. As these same enzymes, alongside others, seem important in mediating transcriptional activation by c-Myc this raises the intriguing possibility that at least in part this may involve stabilizing of protein levels.

The final step in c-Myc degradation is ubiquitinylation, which targets the protein to the proteasome for degradation. The human tumor suppressor Fbw7/hCdc4 acts in a complex with SCF ubiquitin ligases to catalyze the ubiquitinylation of c-Myc. One particular isoform the Fbw7 γ may regulate nucleolar c-Myc accumulation and ubiquitinylation by binding the c-Myc box1 (MB1) domain *(22)*. Recent studies have indicated how complex the regulation of c-Myc may prove to be, as not all c-Myc is necessarily subjected to the same rates of turnover; a pool of c-Myc that is metabolically stable has been identified *(25)*.

2.2. Transcriptional Activity of c-Myc

c-Myc activates a diverse group of genes as part of a heterodimeric complex with its partner protein, Max. The C-terminal basic helix-loop-helix leucine zipper (bHLH-LZ) domain of c-Myc binds Max, also a bHLH-LZ protein, to form c-Myc-Max heterodimers that are capable of binding specific DNA sequences, such as the E-box sequence CACGTG *(26)*. The 150 amino acid N-terminal region of c-Myc contains two highly conserved elements, c-MycBoxI (MBI) and MBII, and is required for transactivation of target genes. Importantly, mutations in the c-Myc transactivation region, or in the bHLH-LZ domain required for dimerization with Max and DNA binding abolish c-Myc's effects on cell proliferation, transformation and apoptosis *(27)*. c-Myc's functional domains as well as various c-Myc-interacting proteins involved in transcriptional regulation *(28)*, are shown in Fig. 2. One way in which c-Myc-Max heterodimers can activate gene expression is by displacement of the putative tumor suppressor Mnt from target genes *(29)*.

Exciting data provide further understanding of how c-Myc-Max complexes regulate gene activation—through chromatin remodeling *(30–32)*. c-Myc was found to associate with the coactivator TRRAP, a component of a large complex that contains histone acetyltransferase (HAT) activity. Bouchard et al, *(33)* showed that upon binding the cyclin D2 promoter, c-Myc recruits TRRAP and induces the preferential acetylation of histone H4 at a single nucleosome. This would alter chromatin structure, thus allowing accessibility and binding of c-Myc-Max transcriptional activator complexes to target DNA sequences. Furthermore, c-MycBoxII in the N-terminal region of c-Myc is required for TRRAP recruitment, histone acetylation, and transcriptional activation at the cyclin D2 locus. Correlation of H4 acetylation with induction of gene expression has been extended to several other c-Myc target genes *(34)*. However, work from the laboratory of Farnham and coworkers has shown that acetylated H3 and H4 may not be augmented by c-Myc, at least with respect to the *cad* and *tert* genes *(35)*.

Very recent studies provide further information on how c-Myc may silence gene expression by interfering with the function of other transcriptional activators by the active recruitment of corepressor proteins *(36)*. c-Myc can bind the corepressor Dnmt3a and influence DNA methyltransferase activity in vivo. In cells with reduced Dnmt3a levels, c-Myc-repressed genes such as *p21Cip1* are reactivated, whereas there is no apparent effect on expression of c-Myc-activated genes. Importantly, the authors noted that c-Myc, Dnmt3a and Miz-1 form a ternary complex that may be responsible for corepression of the *p21Cip1* promoter.

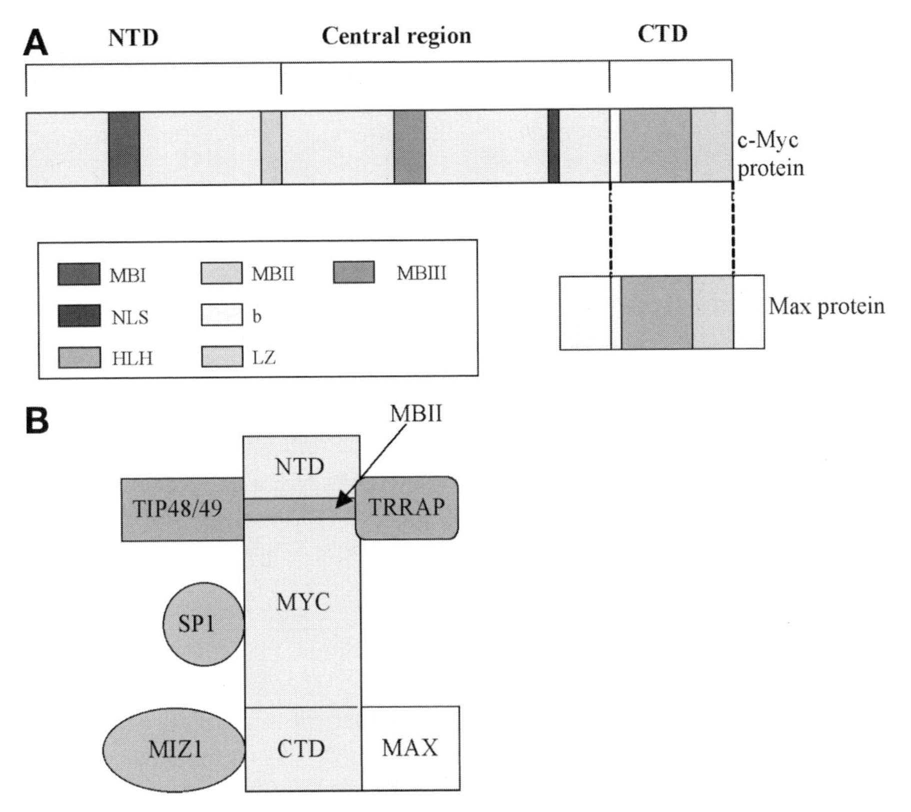

Fig. 2. Functional domains of human c-Myc protein. **(A)** The C-terminal domain (CTD) of human c-Myc protein harbors the basic **(b)** helix-loop-helix leucine zipper (bHLH-LZ) motif for dimerization with its partner, Max, and subsequent DNA binding of c-Myc-Max heterodimers. The N-terminal domain (NTD) harbours conserved "c-Myc boxes" I and II (MBI and MBII), essential for transactivation of *c-Myc*-target genes. Recently the MBIII situated in the central region has been found to be important for negatively regulating the apoptotic response. **(B)** Some major c-Myc-interacting proteins that may or may not bind simultaneously to c-Myc. These include coactivator TRRAP, part of a complex possessing histone acetylase activity (HAT), which interacts with MBII region and mediates chromatin remodelling. TIP48 and TIP49 proteins interact with the NTD of c-Myc and are implicated in chromatin remodelling as a result of their ATP-hydrolyzing and helicase activities. Proteins involved in transcriptional regulation, such as Miz-1 interact with the CTD of c-Myc, whereas SP1 interacts with the central region of c-Myc. NLS; nuclear localization signal.

Recent studies have begun to reveal an important role for the highly conserved c-Myc box III (MbIII) region, situated in the central region of the protein. MbIII is important for transcriptional repression by c-Myc, and for transformation in vitro and lymphoma development in vivo. Conversely, disruption of MbIII prevents transformation by increasing the apoptotic activity of c-Myc, suggesting this region may be an important negative regulator of apoptosis activity of c-Myc *(37)*.

2.3. c-Myc; Cell Growth and Proliferation

One of the key biological functions of c-Myc is its ability to promote cell-cycle progression *(3,38–40)*. In quiescent cells in vitro, c-Myc expression is virtually undetectable. However, upon mitogenic or serum stimulation, c-Myc mRNA and protein are rapidly induced and cells enter the G_1 phase of the cell cycle. Thereafter, *c-Myc* mRNA and protein decline to low but detectable steady-state levels in proliferating cells. If serum or growth factors are removed, c-Myc levels decline to undetectable levels and cells arrest *(41)*. During development, enhanced c-Myc expression correlates well with active proliferation and its down-regulation accompanies mitotic arrest and onset of differentiation; targeted gene disruption of both c-Myc alleles in embryonic stem cells, leads to embryonic lethality at days 9.5 to 10.5 *(7,8)*.

More recently, the cell-cycle effects of ablating c-Myc have been investigated. A rat fibroblast cell line in which both alleles of c-Myc were ablated shows greatly reduced rates of cell proliferation *(42,43)*. A further prominent effect was the marked deficiency in cell growth, defined by decreased global mRNA and protein synthesis. Cell-cycle elongation is caused by a major lengthening of G_1 phase (four- to fivefold) and associated with significantly delayed phosphorylation of the retinoblastoma protein (Rb), and a more limited lengthening of G_2 phase (twofold), whereas S phase duration is largely unaffected. Progression from mitosis to the G_1 restriction point and the subsequent progression from the restriction point into S phase are both drastically delayed, suggesting that c-Myc directly affects cell growth (accumulation of mass) and cell proliferation (the cell-cycle machinery) by independent pathways.

The first notion that c-Myc influenced cell growth came from the correlation between c-Myc and the expression of the rate-limiting translation initiation factors eIF4E and eIF2α *(44)* now known to be direct c-Myc targets *(45,46)*. More recently, c-Myc has been shown to have a direct role in the growth of invertebrate *(47)* and mammalian cells *(48–50)*. Similarly, overexpression of c-Myc in B-lymphocytes of Eμ-*Myc* transgenic mice resulted in cell growth in the absence of cell-cycle progression. Important information is accumulating as to how c-Myc may be mediating effects on cell growth. RNA polymerase III (pol III) is involved in the generation of transfer RNA and 5S ribosomal RNA required for protein synthesis in growing cells and is activated by c-Myc via binding to TFIIIB, a pol III-specific general transcription factor *(51)*. In fact, recent studies suggest that c-Myc may regulate the activity of all three known nuclear RNA polymerases *(52)*. It is indeed plausible that c-Myc's role in regulating cell proliferation could at least in part be mediated through its effects on ribosomal biogenesis and cell growth (reviewed in ref. *53*).

Insights into how c-Myc might promote cell proliferation have resulted from a number of important studies revealing c-Myc's ability to activate or repress target genes involved in cell-cycle progression (*see* Fig. 3A).

G_1–S progression of eukaryotic cells is controlled by the activities of the cyclin-dependent kinase (CDK) complexes, cyclin D–CDK4 and cyclin E–CDK2. c-Myc induces cyclin E–CDK2 activity early in the G_1 phase of the cell cycle, which is regarded as an essential event in c-Myc-induced G_1-S progression *(54–55)*. But how does c-Myc activate cyclin E–CDK2? It was recently shown that *CCND2* (which encodes cyclin D2) and *CDK4* are direct target genes of c-Myc *(56,57)*. Expression of *CCND2* and *CDK4* leads to sequestration of the CDK inhibitor KIP1 (also known as p27) in cyclin D2–CDK4 complexes. The subsequent degradation of KIP1 has recently been shown to involve two other c-Myc target genes, *CUL-1* and *CKS (58)*. By preventing the binding of KIP1 to cyclin E–CDK2 complexes, c-Myc allows inhibitor-free cyclin E–CDK2 complexes to become accessible to phosphorylation by cyclin activating kinase (CAK) *(59,60)*. Increased CDK2 and CDK4 activities would result in

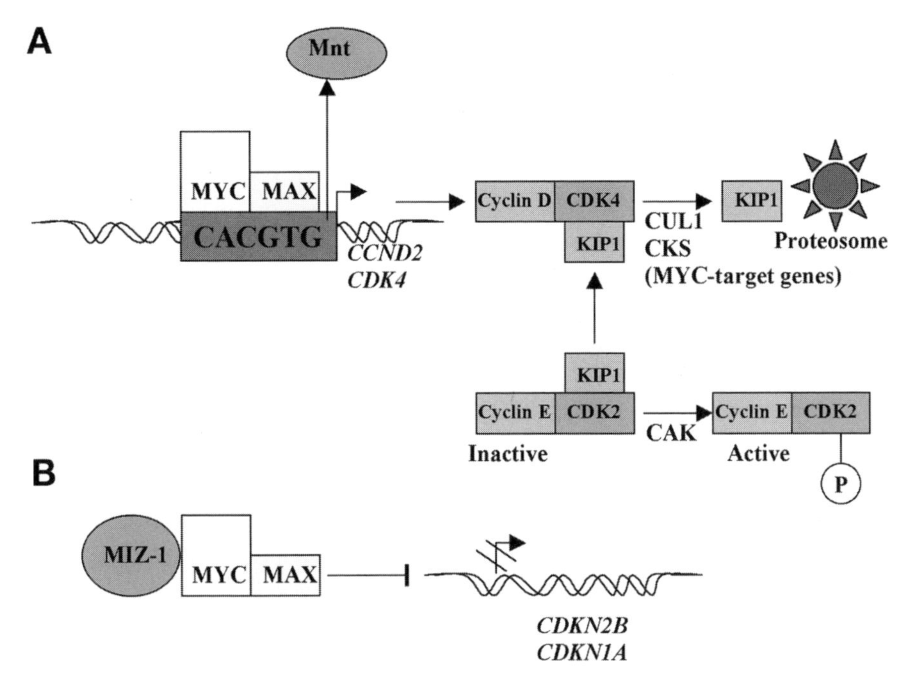

Fig. 3. c-Myc promotes G1–S progression through gene activation and repression. **(A)** c-Myc-Max heterodimers activate target genes, cyclin D2 and cyclin-dependent kinase 4 (Cdk4), which leads to sequestration of Cdk inhibitor KIP1 (p27) in cyclin D2-Cdk4 complexes. Subsequent degradation of KIP1 involves two further c-Myc-target genes, Cul-1 and Cks. In so doing, KIP1 is not available to bind to and inhibit cyclin E-Cdk2 complexes thereby allowing cyclin E-Cdk2 to be phosphorylated by cyclin activating kinase (CAK). **(B)** c-Myc-Max heterodimers repress Cdk inhibitors, INK4B (p15) and WAF1 (p21), which are involved in cell cycle arrest. By interacting with transcription factors, Miz-1 (and/or Sp1), c-Myc-Max prevents transactivation of INK4B (CDKN2B) and WAF1 (CDKN1A), the latter and Inr pathway being predominantly regulated by Miz-1.

Rb hyperphosphorylation and subsequent release of E2F from Rb. Although cyclin D2 is known to be an essential downstream effector of c-Myc in promoting cell proliferation, D cyclins do not appear to be required for c-Myc-induced apoptosis, at least in vitro *(59)*, indicating that these two major functions of c-Myc—proliferation and apoptosis—may involve, at least in part, distinct sets of downstream mediators. It will be interesting to see if this is the case in vivo.

Recent studies support the idea that c-Myc may also exert important influences on the cell cycle by repressing genes, such as the CDK inhibitors INK4B and WAF1 (also known as P15 and P21, respectively)—that are involved in cell-cycle arrest, through the c-Myc-Max heterodimer interacting with positively acting transcription factors such as Miz-1 and SP1 (*see* Fig. 3B). The interaction of c-Myc-Max with Miz-1 blocks the association of Miz-1 with its own coactivator (P300 protein), with the subsequent down-regulation of INK4B *(60)* and WAF1 *(61)*.

Much is still being learnt about repression of various genes by c-Myc and this area will undoubdtedly be an increasingly interesting one for future studies and therapeutic developments *(62–68)*.

2.4. c-Myc and Cell Differentiation—Role of the Mad Protein Family

Activation of c-Myc and subsequent cell-cycle entry is generally incompatible with terminal differentiation. However, the ability of c-Myc to block cell differentiation can be uncoupled from its ability to drive cell proliferation *(69,70)*. Numerous studies have highlighted the importance of the c-Myc/Max/Mad network in regulating cell proliferation and differentiation *(71,72)*. Several lines of evidence support the view that the Mad/Mxi1 bHLH-LZ transcription factor proteins antagonize c-Myc function, for example, ectopic expression of Mad/Mxi1 proteins leads to growth arrest, and suppresses c-Myc-dependent transformation, and growth of tumors *(72)*.

In general, expression of different members of the Mad/Mxi1 protein family coincides with downregulation of c-Myc expression and cells begin to exit the cell cycle and acquire a terminally differentiated phenotype, although there are some exceptions *(73–75)*. In general, c-Myc protein is readily detected in immature proliferating cells, whereas Mad proteins are restricted to post mitotic differentiating cells *(76–78)*. Whether down-regulation of c-Myc, however, is the trigger for differentiation or a consequence of this cell fate is still unclear. Intriguingly, some studies indicate that onset of differentiation does not always involve cell-cycle arrest; c-Myc may play a role in advancing cells along pathways of epidermal and hematopoietic differentiation *(79–81)*. Whether an increase in cell growth and metabolism induced by c-Myc is important for lineage commitment awaits further investigation.

Mad/Mxi proteins heterodimerize with Max and subsequently repress transcription by recruiting a chromatin-modifying corepressor complex to E-box sites on the same target genes as c-Myc/Max, such that c-Myc-Max complexes can no longer activate its target genes. Mad proteins contain an N-terminal repression domain that associates directly with the corepressor SIN3 *(82,83)*, which in turn binds the class I histone deactylases HDAC1 and HDAC2. This association suggests that Mad–Max recruits HDACs to its specific target DNA to result in local histone deacetylation within nucleosomes, thereby decreasing the accessibility of DNA to transactivation factors. Recent results from the laboratory of Bob Eisenmann have suggested that the situation may be more complex than previously appreciated; c-Myc and Mad, although possessing identical in vitro DNA-binding specificities, do not have an identical set of target genes in vivo. In particular, apoptosis is one biological outcome in which the transcriptional effects of c-Myc are not directly antagonized by those of Mad *(84)*. The importance of tight control over c-Myc activity by members of the Mad family, is emphasized by the phenotypes of *Mad1* and *Mxi1* knockout mice *(72)*.

Recent studies have suggested that c-Myc activity may play an important part in maintaining stem cell pluripotency and self-renewal *(85)*. Transcriptional control by LIF and suppression of T58 phosphorylation are crucial for regulation of c-Myc activity in ES cells and therefore in promoting self-renewal. Conversely the c-Myc antagonist Mad and the CDKI, p27 can antagonise this action in hematopoietic stem cells *(86,87)*.

2.5. c-Myc and Apoptosis

Putative cancer cells must avoid apoptosis in order for tumors to arise—the net expansion of a clone of transformed cells is achieved by an increased proliferative index and by a decreased apoptotic rate. In the early 1990s, several laboratories made an intriguing discovery: oncoproteins such as c-Myc and the adenovirus E1A—both potent inducers of cell proliferation—were shown to possess apoptotic activity *(88–90)*. The most widely held view of oncoprotein-induced apoptosis is that the induction of cell-cycle entry "sensitizes" the cell to apoptosis: cell proliferative and apoptotic pathways are coupled. However, the apoptotic pathway is suppressed so long as appropriate survival factors deliver antiapoptotic signals. Given this, the predominant outcome of these contradictory processes will depend on the availability of survival factors.

Since these early experiments, other promoters of cell proliferation (e.g., E2F) have been found to possess proapoptotic activity *(91–93)*. The notion that cells acquiring growth-deregulating mutations in vivo possess an "in-built" tumor suppressor function, which hinders expansion of potentially malignant cells, is a fascinating one. Such a cell population would be unable to outgrow its environment unless apoptosis was inhibited. Indirect evidence supports this idea, as shown by the dramatic synergy between oncoproteins such as c-Myc and mechanisms that suppress apoptosis, for example overexpression of antiapoptotic proteins such as Bcl-2 or Bcl-X_L, or loss of ARF or P53 tumor suppressors *(94–99)*. Interestingly, stimulation of apoptosis by c-Myc may not invariably be a direct effect linked to cell cycling but can also arise through indirect actions culminating in DNA damage. Thus, recent data in vitro links c-Myc to the accumulation of reactive oxygen species (ROS) via E2F1-mediated inhibition of NFκB, though the consequences, either apoptosis or growth arrest, may be critically dependent on cell type *(100,101)*.

At least two separate pathways are involved in the induction of apoptosis: an intrinsic pathway regulated by various factors such as DNA damage, stress, and imbalances between growth promoting and survival promoting factors that act through mitochondrial permeability, and an extrinsic pathway, utilized, for example, by cytotoxic T cells involving ligation of cell surface death receptors (e.g., Fas and TNFR) and activation of the death inducing signaling complex (DISC). The intrinsic pathway involves activation of the apical procaspase-9, whereas the extrinsic pathway involves activation of procaspase-8. This separation of pathways is complicated, however, by the requirement in some cells for co-opting of the intrinsic pathway by caspase-8 mediated activation of the BH3-only protein Bid. Various proteins regulate these apoptotic pathways, prominent amongst which are the anti- and proapoptotic members of the Bcl-2 family, and the inhibitors of apoptosis (IAPs), such as XIAP. Both pathways have been implicated in mediating the action of c-Myc on apoptosis in various studies. A summary of the various mechanisms implicated in c-Myc-induced apoptosis is shown (*see* Fig. 4). However, the relative contributions of these mechanisms for a given tissue or cell-type or critically within an individual developing tumor are generally not well known.

How does the c-Myc oncoprotein induce or "sensitize" cells to apoptosis? Most of the key experiments used to answer this question have come from cell culture. Although much of the apoptotic machinery has been identified, we are only beginning to define the mechanisms by which c-Myc engages such machinery. As is often the case for cell signalling events, there is not just one mechanism by which c-Myc engages or activates the apoptotic machinery. The mechanism chosen most likely depends on factors, such as, cell type, signals received from the cells environment, or whether the cell has acquired DNA damage or not. Expression of c-Myc sensitizes cells to a wide range of pro-apoptotic stimuli—such as hypoxia, DNA damage and depleted survival factors *(8–91)*—as well as enhancing sensitivity to signalling through the CD95 *(102)*, TNF

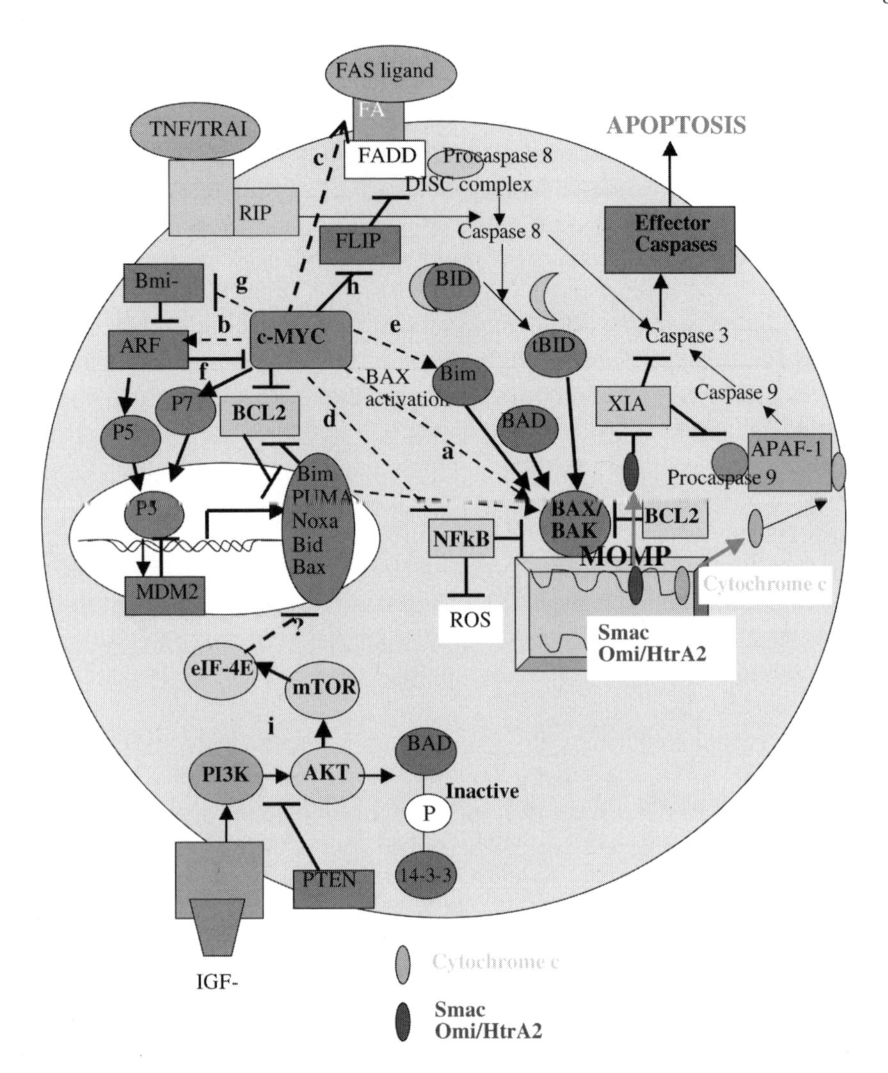

Fig. 4. Pathways involving c-Myc and apoptosis. c-Myc protein sensitises cells to a wide range of proapoptotic stimuli (e.g., hypoxia, DNA damage, depleted survival factors, signaling through CD95, TNF, and TRAIL receptors). Moreover, as c-Myc can also transcriptionally activate other "dual function" mitogenic proteins, such as some E2F family members, the potential repertoire of mediators of Myc's role in apoptosis may be even more complex than suggested in this schematic. During apoptosis c-Myc induces release of cytochrome *c* from the mitochondria into the cytosol possibly through activation of proapoptotic molecule, Bax (**a**). Activated Bax within the mitochondrial membrane leads to creation or alteration of membrane pores, resulting in mitochondrial outer membrane permeabilization (MOMP). Once released into the cytosol, cytochrome *c* associates with Apaf-1 protein and pro-caspase 9 to form the apoptosome ("wheel of death"). In the presence of ATP, caspase 9 is activated leading to activation of downstream effector caspases including caspase 3, which ultimately leads to degradation of cell components and demise of the cell. Also released by MOMP, Smac and Omi can inhibit the action of IAPs, such as XIAP, that otherwise normally prevent activation of the apoptosome and also of effector caspases. Other pathways involving c-Myc-induced cytochrome *c* release and apoptosis include indirect activation of p53 tumor suppressor via p19ARF leading to transcription of Bax. ARF can also inhibit c-Myc by blocking transcriptional activation of genes involved in growth whilst not affecting c-Myc

(103), and TRAIL *(104)* death receptors. In fact, it has recently been shown that c-Myc can lead to down-regulation of the inhibitor of caspase activation, FLICE inhibitory protein (FLIP). FLIP is an inhibitor of the extrinsic pathway, which normally competes with caspase-8 for binding to the DISC- releasing FLIP inhibition may at least in part explain the ability of c-Myc to sensitize cells to death receptor stimuli *(105)*.

Apoptosis can be activated through ligation of cell-surface death-receptors such as Fas/CD95. In some cell types, this alone is sufficient to trigger the downstream caspases and eventual cell destruction. However, in other cells, the death-receptor-induced apoptosis also requires the mitochondrial pathway. c-Myc sensitizes cells to death-receptor signaling by recruiting the mitochondrial pathway.

First insight into c-Myc's role in apoptosis came from studies showing that c-Myc could induce the release of cytochrome *c* from the mitochondria during apoptosis, and that ectopic addition of cytochrome *c* sensitized cells to undergoing apoptosis *(106)*. Once released into the cytoplasm, cytochrome *c* associates with another protein called apoptotic protease-activating factor 1 (APAF-1) to create the apoptosome, which acts as a scaffold for activating procaspase-9 *(107)*. In the presence of cytochrome *c* and either ATP or ADP, caspase-9 is autocatalytically activated, which then activates the downstream caspase effector cascade involving caspases-2,-3,-6,-7,-8, and -10, which break up the cell's components leading to demise of the cell.

---

mediated gene repression, resulting in increased apoptosis. This feedback mechanism operates independently of p53 **(b)**. In some tumor models, deregulated expression of the inhibitor of ARF, Bmi-1 can co-operate with Myc in tumorigenesis. Ligation of death receptor, CD95/Fas, triggers association of the intracellular adaptor protein FADD with the CD95 receptor and form the DISC **(c).** FADD then recruits pro-caspase 8 resulting in autoactivation of the pro-caspase, which cleaves and activates executioner caspases. Activation of caspase-8 may be negatively regulated by FLIP, which competes for binding to the DISC. Caspase-8 may also activate the proapoptotic protein, Bid, which may promote MOMP, thereby linking the extrinsic and intrinsic pathways of apoptosis. Recently c-Myc has been reported to mediate apoptosis in some cells by a mechanism involving generation of ROS and suppression of the survival promoting activities of NF-κB **(d)**. In some cell types Myc can induce expression of the BH3-only pro-apoptotic protein Bim and suppress expression of the antiapoptotic proteins Bcl2 and BclxL **(e)**. c-Myc may increase expression of the p53 family member p73, which might act to direct p53 itself towards proapoptotic genes **(f)**. One question that remains unanswered is how activation of oncogenes such as c-Myc results in expression of ARF, which is not usually activated during normal cell replication. One possibility is through inhibition of proteins like Bmi-1 **(g)**, though this is speculative. Lastly, the inhibitor of caspase activation, FLICE inhibitory protein (FLIP) is transcriptionally repressed by c-Myc, providing a mechanism whereby c-myc could sensitise cells to a wide-range of death receptor stimuli **(h)**. Survival signals that serve to block c-Myc-induced apoptosis **(i)**, include signalling via the IGF1 receptor and PI3 kinase (PI3K) or activated Ras, which can lead to activation of AKT serine/threonine kinase and subsequent phosphorylation of proapoptotic protein Bad. Phosphorylated Bad is sequestrated and inactivated by cytosolic 14-3-3 proteins. This pathway may also block apoptosis in other ways. Akt also activates mTOR and it's downstream target the elongation factor eIF-4E. In at least one model expression of eIF-4E could block c-Myc induced apoptosis and cooperate with c-Myc in tumorigenesis. Anti-apoptotic proteins, such as Bcl-2 and Bcl-x_L, reside in the outer mitochondrial membrane and block cytochrome *c* release possibly through sequestration of Bax. The PTEN tumor suppressor acts as a negatively regulator of Akt activation.

c-Myc can also promote apoptosis through its effect on the expression of members of the Bcl-2 family, which are discussed elsewhere. For example, c-Myc represses expression of bcl-2 and bcl-xL, both of which are antiapoptotic proteins, but induces expression of proapoptotic members, such as Bim. This has the effect of permeabilizing the mitochondria to release cytochrome c and other proapoptotic factors, which ultimately lead to activation of caspases.

The Bcl-2 family proteins have recently emerged as fundamental regulators of mitochondrial outer membrane permeabilization (MOMP) necessary for cytochrome *c* release *(108)*. Although the precise mechanism involved in c-Myc-mediated cytochrome *c* release is presently unclear, it may involve activation of the proapoptotic protein Bax *(109,110)*, as well as potentially upregulating mitochondrial genes, including cytochrome *c*, by the transcription factor nuclear respiratory factor-1 (NRF-1), a potential c-Myc-target gene *(111)*. During apoptosis, Bax (normally an inactive monomer) can be induced to oligomerize and migrate from the cytoplasm to the mitochondria by various BH3-only proteins (those Bcl2 family members containing only the single Bcl2 homology-3 domain), such as Bid *(112–114)*. Once inserted into the outer mitochondrial membrane, Bax induces cytochrome c release by the creation or alteration of membrane pores *(115–117)*. Recent studies suggest that c-Myc might activate Bax at least in part by fostering ASK1 activation in the P38 signaling cascade *(118)*. Importantly, mutational inactivation of *Bax* is found to occur during evolution of human colon tumors *(119)*. Antiapoptotic proteins such as Bcl-2 and Bcl-X_L suppress apoptosis by blocking MOMP possibly through sequestration of activated Bax and so preventing cytochrome *c* release and/or binding to APAF-1 to prevent it from activating caspase-9. In this scenario, the balance of anti- and proapoptotic molecules present within a c-Myc-activated cell would determine whether it lives or dies.

The survival factor, IGF-1, has been shown to inhibit c-MYC-induced apoptosis in vitro by blocking cytochrome *c* release from the mitochondria *(120)*. Survival signals mediated via the IGF-1 receptor or activated Ras (*see* Fig. 4) can lead to activation of the AKT/PKB serine/threonine kinase *(121)*. Activated AKT then phosphorylates the pro-apoptotic BH3-only protein BAD resulting in its sequestration and inactivation by the cytosolic 14-3-3 proteins *(122)*. Referring back to the initial experiments in vitro, in which cells with deregulated expression of c-Myc die by apoptosis when grown in low serum, it becomes clear how growth factor signaling is able to regulate apoptosis. Given the importance of such growth factors in determining the survival or death of a cell, it is not surprising that elevated signaling through the IGF-1 pathway occurs in many tumors. Similarly, genetic mutations that activate the PI3K pathway dramatically collaborate with c-Myc during tumor progression.

Another possible mechanism that links c-Myc and apoptosis is through its indirect activation of P53 via $p19^{ARF}$ (ARF) *(123)*. The ARF tumor suppressor protein acts in a checkpoint that protects against unscheduled cellular proliferation in response to oncogenic signaling. Deregulated expression of c-Myc induces ARF expression and apoptosis through the ARF-Mdm2-p53 axis. However, ARF also prevents hyperproliferation and transformation caused by c-Myc and enhances c-Myc-induced apoptosis independently of p53 *(124)* (*see* Fig. 5).

The tumor suppressor, p53, is a master regulator of cell proliferation, and can induce apoptosis, senescence, or DNA repair in response to a variety of cellular stresses, including

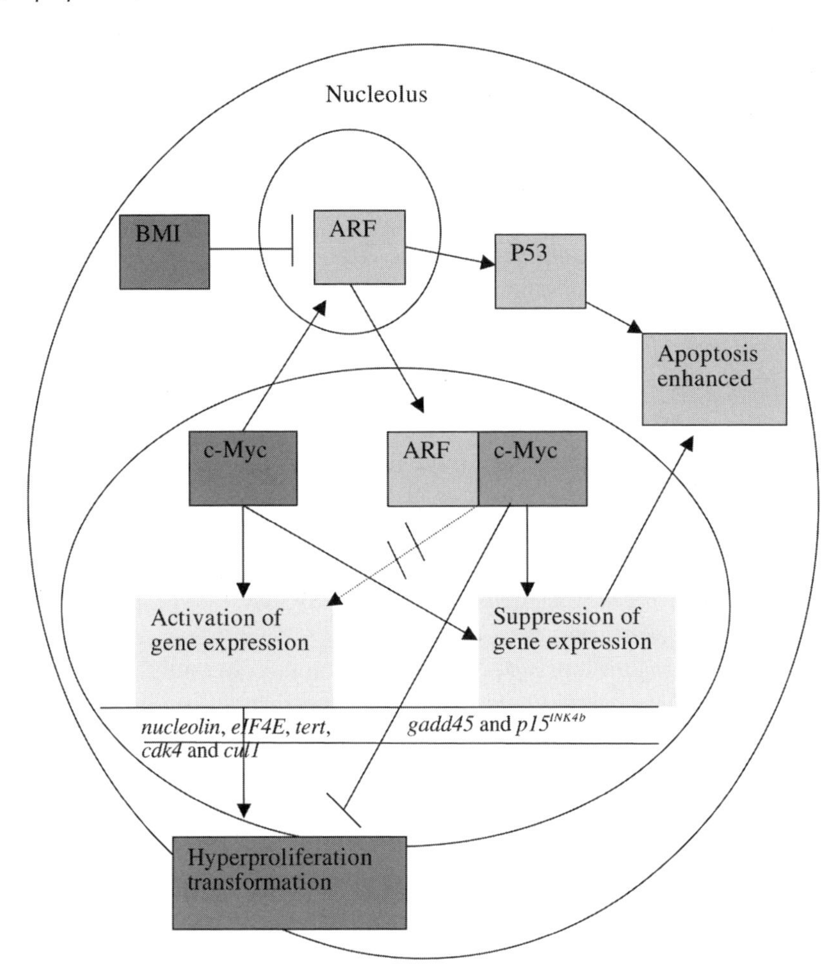

Fig. 5. ARF can promote c-Myc induced apoptosis directly or indirectly via p53. In the presence of c-Myc activation, ARF relocalizes from the nucleolus to the nucleus. ARF may bind to c-Myc thus preventing c-Myc transcriptional activation, but not the transcriptional repression. By this means, ARF may augment c-Myc induced apoptosis and block hyperproliferation- resisting tumorigenesis. ARF may also cooperate with c-Myc in inducing apoptosis via induction of p53. BMI may cooperate with c-Myc in tumors models in part via inhibiting ARF transcription.

DNA damage, hypoxia, and nutrient deprivation. The importance of $p19^{ARF}$ in c-MYC-induced apoptosis in vivo was underscored in genetically altered mice in which expression of *c-MYC* was targeted to B-lymphocytes. When these mice were cross-bred with another strain in which $p19^{ARF}$ was disrupted, Myc-induced lymphoma development was dramatically accelerated. This outcome, similar to that seen when p53 is disrupted, shows that c-MYC strongly collaborates with loss of p53 or $p19^{ARF}$ in murine lymphomagenesis presumably by inhibiting c-MYC-induced apoptosis (extrapolating from in vitro experiments). In a similar fashion, deregulated expression of *Bmi1* oncogene (which encodes a polycomb group protein), accelerates c-Myc-induced lymphomas by inhibiting expression from the *Ink4a* locus, which encodes Arf and $p16^{INK4A}$ (*see* Chapter 7). We now know that P53 activates many proapoptotic proteins, such as Bax, Apaf1, Noxa, and puma, which are involved in cytochrome *c* release from the mitochondria and/or the activation of caspases—enzymes responsible for the destruction of the cell.

Tiam1, a specific guanine nucleotide exchange factor of Rac1 binds to the c-Myc box II of c-Myc and modulates several of its biological functions. Overexpression of Tiam1 inhibits c-Myc apoptotic activity *(125)*. Recently the MBIII region has also been implicated in apoptosis.

It is clear that the favoured mechanism for c-Myc-induced apoptosis may be dictated by cell type as well as by tissue location and moreover modified by the presence or absence of additional mutations in other pro- and antiapoptotic genes.

2.6. c-Myc the Model "Super-Competitor"

Recent studies of the phenomenon of cell competition in *Drosophila* have produced some intriguing findings that may have implications in human cancer *(126–128)*. These recent studies in drosophila demonstrate that Myc may have another role in regulation of cell numbers beyond its ability to regulate expression of growth-promoting genes. The ability of Myc to induce apoptosis cell-autonomously has been extensively discussed already. However, it now seems that Myc may also induce apoptosis in neighboring cells—a non-cell-autonomous response. Thus, two recent papers in *Cell* have shown that cells with higher levels of Myc expression not only outgrow their neighbors with lower levels of Myc, perhaps not surprising, but directly lead to apoptosis in the cells with lower Myc expression. Cells with high levels of Myc expression can act as super-competitors that are capable of both out-growing and inducing death in nearby cells with lower Myc levels.

One likely explanation is that such cells are more efficient at accessing survival signals and/or growth factors in the microenvironment than cells with lower levels of Myc. Interestingly, other growth-promoting pathways that have been examined do not cause apoptosis in neighboring cells. Given that c-Myc is well-known to regulate multiple genes involved in biosynthesis cell growth and replication, it can be postulated that oncogenic *Myc* would enable cells to grow and clonally expand at the expense of their wild-type neighbors, which may even be killed as a consequence.

In order to further illuminate the complex interactions surrounding c-Myc and apoptosis, a role so sensitive to tissue location in vivo, we eagerly await the outcome of further studies in which apoptotic pathways are selectively manipulated in mouse mutants or by employing new pharmacological tools and phenotype determined in the all important context of the intact organism.

3. c-Myc, Evasion of Apoptosis and Cancer

Whereas regulated c-Myc expression occurs during normal cell proliferation in response to extracellular signals, such as ligation of growth factor receptors, constitutive or deregulated expression of c-Myc is often associated with tumor-derived cells. Unlike normal cells, tumor cells no longer require exogenous mitogenic stimulation from their tissue microenvironment in order to proliferate. This poses a danger to the host, as the normal restraints that exist in a given tissue to limit uncontrolled cell growth are disrupted.

Early in vitro experiments showed that constitutive overexpression of c-Myc can immortalize rat fibroblasts and prevent withdrawal from the cell cycle *(128,129)*. Although oncogenic activation of c-Myc alone causes uncontrolled proliferation in vitro, cellular transformation in vitro seemed to require additional oncogenic lesions *(130–135)*.

Deregulated c-Myc expression is often associated with aggressive, poorly differentiated tumors; however, given that most human tumors are quite advanced at the time of discovery—often possessing many genetic alterations—it is difficult to ascertain at which stage of tumor progression, oncogenic *c-Myc* activation occurred. This is an important point if we wish to establish the role of c-Myc in the initiation of a tumor, or indeed as a therapeutic target at later stages of progression. It is assumed from the majority of in vitro and in vivo data that the predominant role of deregulated c-Myc in tumor initiation/progression in vivo is through uncontrolled cell proliferation concomitant with loss of terminal differentiation (in most cases). Although this may be part of the picture, it is conceivable that there are other attributes afforded to c-Myc (e.g., the formation of new blood vasculature in the growing tissue mass) that are impossible to determine in the culture dish and in conventional transgenic systems.

Studies employing conventional c-Myc transgenic mice, in which the oncogene is constitutively expressed in a given cell type by means of a tissue-specific promoter, have supported the view that deregulated c-Myc, as an initial event, is important for the formation of certain cancers, albeit with a long latency *(94,95)*. However, it has been difficult to directly observe the proposed tumor suppressor function of c-Myc (apoptosis) because the oncoprotein is expressed through much of development and thereafter, and would be expected to obliterate the very cells in which it induced apoptosis.

Acceleration of tumorigenesis in transgenic mice is observed when apoptosis is suppressed. However, it is unclear whether exacerbation of c-Myc-induced tumorigenesis in these models is a direct consequence of antagonizing c-Myc-induced apoptosis or by nonspecific protection of cells from the apoptotic influence of genotoxic and other stresses. Indeed, the physiological relevance of c-Myc-induced apoptosis as a tumor suppressive mechanism in vivo has, until recently, been unclear. Moreover, as most human cancers are precipitated by somatic mutations that arise sporadically in adult tissues concerns were raised as to whether the early activation of the onco-transgene in mice accurately reflected the human disease.

These issues led to the development of regulatable transgenic mice in which oncogene expression is controlled within the target tissue. Although the widespread expression of given onco-transgenes (in all cells of the target tissue) does not precisely recapitulate the process during development of sporadic tumors, the ability to activate oncogenes at will in the adult provides a significant advance in the generation of physiological cancer models and an unprecedented potential to validate new drug targets by subsequently deactivating the oncoprotein (*see* Fig. 6).

A conditional transgenic expression system (c-MycER^{TAM}) was employed to investigate the effect of "switching on" c-Myc activation in distinct tissues of adult transgenic mice *(14,15,135–137)*. Directed expression of the conditional form of the c-Myc oncoprotein (c-MycER^{TAM}) to the suprabasal epidermis or pancreatic islet β-cells was achieved by placing *c-MycER^{TAM}* cDNA under the control of an involucrin promoter *(15,137)* or an insulin promoter *(14,137)*, respectively.

The differences in cell behavior in these two mouse models following c-Myc activation was striking. In the suprabasal epidermis *(15)*, c-Myc activation is sufficient to induce cell-cycle entry of postmitotic keratinocytes concomitant with impaired differentiation,

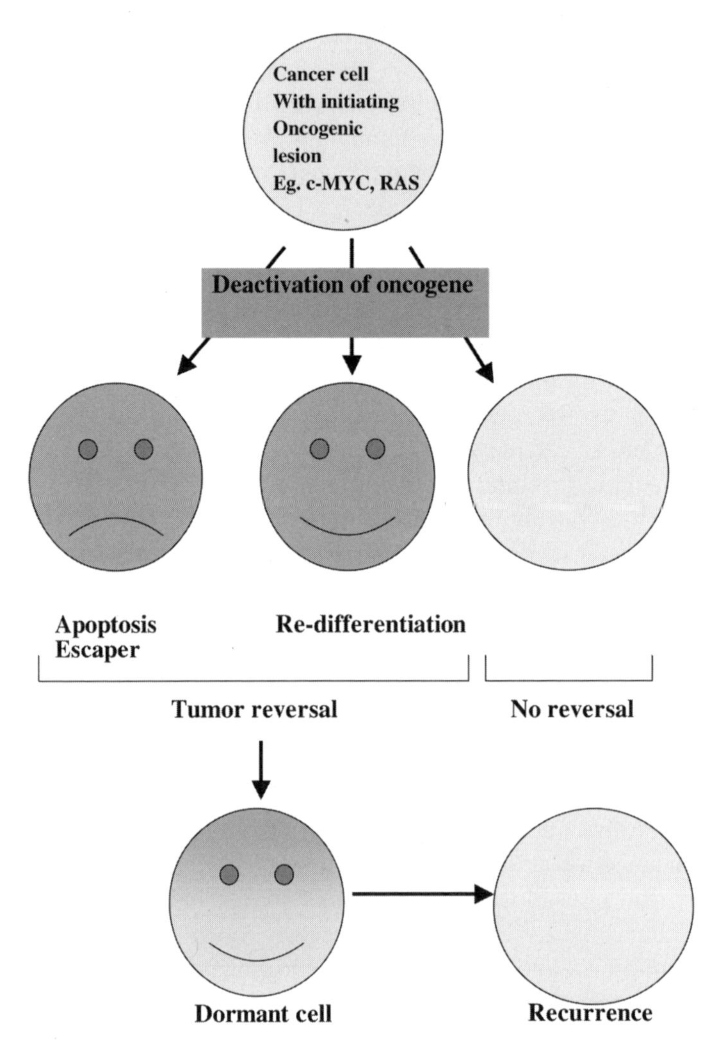

Fig. 6. Tumor regression following de-activation of initiating oncogenic lesion.Various transgenic mouse models of tumorigenesis have shown that subsequent deactivation of the initial oncogenic lesion leads to rapid and complete regression of most neoplastic lesions. Mechanisms of tumor regression include tumor cell apoptosis and/or differentiation and re-establishment of cell–cell contacts. Onset of vasculature collapse within tumor masses occurs rapidly which may also contribute to tumor cell apoptosis. In some cases, a subset of tumors escape reversal ("escapers") and thus are no longer maintained by the initiating oncogenic lesion. In these cases, it is likely that additional oncogenic lesions have occurred. Finally, a number of cells may remain dormant in an apparently regressed tumor that can after even along latency give rise to a recurrent tumor.

culminating in epidermal hyperplasia with areas of focal dysplasia and papillomatous lesions resembling the human premalignant skin lesions known as actinic keratosis—a precursor of squamous cell carcinoma. A notable feature was the widespread induction of angiogenesis, which argues against the necessity of a specific "angiogenic switch" in a subpopulation of tumor cells *(138)*. Instead, the angiogenesis observed may merely reflect the dramatic increase in the number of vascular endothelial growth factor (VEGF)-secreting keratinocytes as a consequence of c-Myc *(138,138a)*, although other angiogenic factors have yet to be investigated. Importantly, c-Myc has also been shown to be angiogenic in multiple cell types both through its repression of the angiogenesis inhibitor Thrombospondin-1 *(139–142)* and its induction of VEGF *(15,143,144)*. Recently, it has been shown that Myc may lie downstream of Ras signaling in the promotion of angiogenesis at least in part by inhibiting Thrombospondin 1 *(142)*. The key observation was the obvious lack of c-Myc-induced apoptosis in epidermal keratinocytes in vivo, despite the potent induction of apoptosis by c-Myc in vitro in isolated serum-deprived keratinocytes from the same transgenic mice. In this tissue proliferation is the predominant outcome over apoptosis. One can speculate that in this tissue the oncogenic potential of c-Myc is buffered by the continued differentiation and migration of keratinocytes to the skin surface where they become keratinized and then shed.

In contrast to skin, the predominant effect of activating c-Myc in pancreatic β-cells of adult transgenic mice *(14)* is apoptosis, and not proliferation. Thus, although activation of c-MycER^{TAM} is followed by widespread cell-cycle entry, in complete contrast to epidermal keratinocytes in vivo, c-Myc activation soon triggers rapid apoptosis of islet β-cells. Such β-cell apoptosis overwhelms any concomitant proliferation, resulting in islet involution and onset of diabetes within 9 d. The acute sensitivity to induction of apoptosis by c-Myc suggests that β-cells are only modestly buffered against cell death by survival signals or intrinsic antiapoptotic mechanisms, such as Bcl-2/Bcl-x_L expression, in vivo. The fact that the innate proapoptotic growth-suppressive potential of c-Myc dominates its growth-promoting action strongly indicates that, in this tissue, c-Myc acts as its own "tumor suppressor."

However, suppression of apoptosis (by coexpressing Bcl-X_L), exposes multiple innate oncogenic properties of c-Myc that are sufficient to trigger immediate carcinogenic progression *(14)*. These cellular aspects include immediate and sustained b-cell proliferation, which results in inexorable islet expansion concomitant with loss of differentiation (down-regulation of insulin and other differentiation markers), angiogenesis, loss of cell–cell contacts resulting from loss of E-cadherin, and local invasion of b-cells into surrounding exocrine pancreas. In contrast, with the RIP1-Tag2 model of b-cell carcinogenesis *(145)*, tumor progression is ubiquitous and rapid (2 wk) with multiple diverse aspects emerging concurrently. The dramatic and immediate oncogenic progression observed in the majority of islets in vivo strongly supported the at that time unorthodox notion that complex neoplastic phenomena involving the tumor cell and its interactions with normal surrounding tissues, may both be induced and maintained by a very limited platform of interdependent pleiotropic lesions.

3.1. Tumor Cell Apoptosis and Regression Following Oncogene Deactivation

Conditional systems to induce oncogenes in a tissue-specific and time-dependent manner also enable the study of tumor reversal, following deactivation of the oncogene (reviewed in refs. *146,147*). Given the importance of oncogene activation in human cancers, specific targeting of oncogenic pathways provides a potentially effective therapeutic strategy, which can be tested in such models. For example, targeting of the HER2/Neu receptor tyrosine kinase (which is overexpressed in up to 30% of primary human breast cancers) with the neutralizing antibody Trastuzumab, has been used successfully in clinical trials, in combination with other agents, to slow disease progression (*see* ref. *148* and references therein). Similarly, patients with chronic myelogenous leukaemia (CML) have been effectively treated with the ABL kinase inhibitor, imatinib (Gleevec), inducing clinical remission whilst in the CML phase *(149,150)*.

Previously, we have shown that sustained c-Myc inactivation in locally invasive pancreatic islet tumors (induced by c-Myc activation in β-cells on a background of Bcl-x_L overexpression) induced β-cell growth arrest and re-differentiation into mature β-cells, accompanied by the collapse of tumor vasculature and tumor cell mass resulting from apoptosis, despite the constitutive expression of Bcl-x_L in tumor cells *(14)*. Similarly, in premalignant skin epidermal tumors (papillomatosis) induced following activation of c-Myc alone, skin lesions completely regressed within 4 wk after sustained inactivation of c-Myc *(15)*. However, given the continual outward migration and shedding of growth-arrested and redifferentiated keratinocytes from the skin surface, it was not established whether this action alone was responsible for the removal of skin tumor cells or if apoptosis also played a part.

Tumor regression has now been observed in numerous other mouse cancer models (Table 1), following tagetting of various oncogenes. In fact, several studies, using conditional mouse models of various cancers, have shown unexpectedly, that inactivation of the initiating oncogene is sufficient not only for reversal of the primary tumor but also of invasive and metastatic lesions, many of which contain multiple genetic and epigenetic alterations *(14,15,151–169)*. The tumor regression observed in many of these tumor models following sustained oncogene inactivation, provides a powerful platform on which to build a deeper understanding of fundamental tumor biology and with which to preclinically evaluate novel therapeutics to target specific genes. However, as discussed later, there are some instances when more advanced tumors do not fully regress and further insight into the mechanisms of this divergence is required.

3.2. Epigentic Context, Apoptosis, and Tumor Regression

A recent study has shown that brief inactivation (10 d) of c-Myc was sufficient for the sustained regression of c-Myc induced invasive osteogenic sarcomas in transgenic mice *(170)*; subsequent re-activation of c-Myc led to extensive apoptosis rather than restoration of the neoplastic phenotype. Possible explanations for this outcome include changes in epigenetic context that may have occurred within the cell type, that is, presumably between the immature cell in which c-Myc was originally activated and the more differentiated cell resulting from subsequent (brief) inactivation of c-Myc. In this tumor model, although c-Myc expression is initiated in immature osteoblasts during embryogenesis, subsequent inactivation of c-Myc in osteogenic sarcoma cells induces differentiation into mature osteocytes. Therefore, reactivation of c-Myc now takes place in a different cellular context and induces apoptosis rather than neoplastic progression. However, irrespective of the actual underlying mechanisms, these intriguing findings suggest the novel possibility of employing transient inactivation of c-Myc as a therapeutic strategy in certain cancers, thus limiting potential toxic effects resulting from prolonged therapeutic inactivation *(146,147)*. However, to date this appears to be a unique feature of this model and sustained regression of tumors originating in different tissues and under differing circumstances following transient c-Myc inactivation has not been shown.

Thus, inactivating c-Myc transiently in skin keratinocytes or pancreatic β-cells does not lead to sustained regression. Rather, and in contrast to the osteogenic sarcoma model, reactivating c-Myc in islet tumors does not lead to accelerated β-cell apoptosis, but rather

Table 1
Use of Regulatable Transgenic Mouse Models of Cancer to Test Tumor Regression

Initial oncogenic lesion	Target tissue/cell type	Phenotype	De-activating initiating oncogene	Ref.
SV40 T antigen	Embryonic submandibular gland	(i) Atypical cells (4-wk-old) (ii) Transformed ductal cells (7-mo old)	(i) Full phenotype reversal (ii) Not reversed	*216*
H-Ras (+Ink4a−/−)	Melanocytes	Melanoma	Rapid tumor reversal Larger tumors (20%) not reversed	*217*
c-Myc	Embryonic hematopoietic cells	T cell lymphomas/ acute myeloid leukaemias (5-mo old)	Rapid tumor reversal 10% tumor relapse	*151*
c-Myc	Adult suprabasal epidermis (keratinocytes)	Papillomatosis (carcinoma *in situ*)	Rapid and complete tumor reversal	*15*
c-Myc	Adult suprabasal epidermis (keratinocytes)	Papillomatosis (carcinoma *in situ*)	Tumor recurrence after transient c-Myc inactivation	*137*
c-Myc	Adult suprabasal epidermis (keratinocytes)	Papillomatosis (carcinoma *in situ*)	Growth arrest after transient c-Myc inactivation	*138a*
Bcr-Abl1	B-cells	Acute B-cell leukaemia	Complete tumor reversal	*218*
c-Myc	Mammary epithelium	Invasive mammary adeno-carcinoma	Tumor reversal Subset not reversed (Ras activated)	*169*
K-Ras	Lung	Adeno-carcinoma	Tumor reversal	*219*
K-Ras (+p53−/− or Ink4a−/− or ARF−/−)	Lung	Adeno-carcinoma	Rapid and complete tumor reversal	*220*
c-Myc (+Bcl-xL)	Adult pancreatic islet β-cells	Invasive islet adeno-carcinoma	Rapid and complete tumor reversal	*14*

(*Continued*)

Table 1 (*Continued*)

Initial oncogenic lesion	Target tissue/cell type	Phenotype	De-activating initiating oncogene	Ref.
c-Myc (+Bcl-xL)	Adult pancreatic islet β-cells	Invasive islet adeno-carcinoma	Tumor recurrence after transient c-Myc inactivation	*137*
c-Myc	Embryonic osteocytes	Malignant osteogenic sarcoma	Rapid and complete tumor reversal even aftyer transient inactivation	*170*
c-Myc	Hepatocytes	Hepatocarcinoma	Tumor reversal, Dormant tumor cells give rise to recurrent tumors, when Myc is reactivated	*172*
c-Myc	Lymphocytes	T- and B-cell lymphoma	Reversal	*160*
Bcl2 (+ c-Myc)	Lymphocytes	Lymphoblastic leukaemia	Apoptosis and Reversal	*164*
Neu (and MMTV)	Mammary epithelium	Invasive mammary carcinoma	Essentially complete reversal even of metastatic lesions. Eventually, however tumors spontaneously recur	*221*
Wnt	Mammary epithelium	Invasive mammary carcinoma	Essentially complete reversal even of metastatic lesions. Eventually, however tumors spontaneously recur	*158*
Gli2	Basal keratinocytes	Basal cell carcinoma	Reversal	*222*

The potential for tumor regression following deactivation of the initial oncogenic lesion (c-MYC, SV40 T antigen, H-Ras, K-Ras, BCR-ABL, Neu, Wnt and Gli2) has been investigated using several regulatable transgenic mouse models of cancer. In general, the findings listed above have implications for therapy as they indicate that blocking oncogene function, even in advanced tumors, could lead to apoptosis or differentiation of tumor cells. Further to these findings, it was recently shown that transient, rather than sustained, inactivation of c-MYC is sufficient for full reversal of malignant osteogenic sarcoma in transgenic mice. If valid for other cancer models, transient inactivation of oncogenes could provide an effective cancer therapy limiting host cell toxicity. However, it has subsequently been shown in various other tissues, that such transient inactivation is not sufficient for tumor reversal (*see* text). In fact, in some tumors inactivation of the transgene is not followed by complete regression of all tumors or recurrence occurs spontaneously in regressed tumors, suggesting that mutations occur which can make the tumors independent of the initiating lesion.

restores the oncogenic properties of c-Myc, rapidly reinitiating β-cell proliferation, loss of differentiation, loss of E-cadherin, local invasion, and angiogenesis. This occurs despite the re-expression of differentiation markers and the loss of some of the newly acquired vasculature, occurring during the period of c-Myc inactivation. Moreover, as no new β-cells arise, during the period of c-Myc inactivation, replication is likely restored in those same cells that have previously experienced c-Myc activation. Similarly, in epidermis, reactivating c-Myc in suprabasal keratinocytes does not result in apoptosis, which remains confined to the shedding areas of parakeratosis at the skin surface, but restores the papillomatous phenotype, inducing cell proliferation and dysplasia *(137)*. These results are in line with a very elegant recent study in a different system by Shachaf and colleagues, who demonstrate that invasive c-Myc-induced hepatocellular carcinomas regress when *c-Myc* expression is turned off but, interestingly, some tumor cells remain "dormant" even for prolonged periods and contribute to cancer progression if *c-Myc* expression is subsequently reinitiated *(172)*. One of the most important aspects of this study was the use of a "lineage tracking" system that unambiguously allowed the researchers to conclude that tumor recurrence derived from cells that had previously been in the regressed cancer rather than de novo from previously normal cells.

Different types of cancers are prevalent in different age groups with the effects of oncogene activation dependent on the developmental stage of the target cell at that time. The biological consequences of activating c-Myc are clearly influenced not only by environment but also by developmental stage. This has been shown in two recent mouse tumor models, mammary gland and liver. C-Myc can inhibit postpartum lactation if activated within a specific 72-hour window during mid-pregnancy, whereas c-Myc activation either prior to or following this 72-h window does not *(161)*. In embryonic and neonatal mice, c-Myc overexpression in the liver immediately results in hyperproliferation and neoplasia, whereas in adult mice c-Myc overexpression induces cell growth and DNA replication but without mitotic cell division, and neoplasia is considerably delayed *(163)*.

Taken together, these findings suggest that a cautious approach is required in considering cancer therapies aimed at transient oncogene inactivation. First, a more comprehensive understanding of the genetic basis and environmental context of any individual tumor would be required in order to predict the likely success of such a treatment schedule. Second, at least under those circumstances where tumor cell differentiation and alteration of epigenetic context would not be predicted to reinstate apoptosis and no alternative mechanism exists for tumor cell removal, sustained inactivation of the offending oncogene would seem the desired therapeutic goal.

3.3. Tumor Regression—Cancer Cell "Escapers" and "Sleepers"

There are a few instances when more advanced tumors do not fully regress (Table 1) and, in these cases, further insight into the mechanisms of this divergence is required. Although it is not clear as to why some tumors escape dependence upon the c-Myc transgene, it has been reasoned that these tumors acquire genetic lesions that in some way substitute for the requirement for c-Myc. Importantly, tumor regression following down-regulation of *c-Myc* expression coincided with tumor cells differentiating to more mature lymphocytes with the restoration of normal host hematopoiesis, as well as a proportion of tumor cells being eliminated by apoptosis *(157)*. Reversal of c-Myc-induced invasive mammary adenocarcinomas has also been shown in a mouse model, although one subset of tumors failed to reverse and were found to carry additional mutations in

Ras (169). In fact, recent studies have demonstrated that c-Myc may cause genomic instability in some model systems in vivo *(173)* and in vitro *(174,175)*, which certainly in the long-term could contribute to c-Myc-induced neoplastic progression.

In general, the above findings have implications for therapy as they indicate that blocking oncogene function, even in advanced tumors, could lead to apoptosis or redifferentiation of tumor cells. The potential for tumor regression following deactivation of initial oncogenic lesions (c-Myc, SV40 T antigen, H-Ras, and K-Ras) have provided valuable information and hope for future development of candidate drug molecules (Table 1).

3.4. Activated c-Myc Expression in Human Tumors

Activation of c-Myc, as is the case in many human tumor cells, is defined as deregulation of the normal, highly controlled expression pattern of the *c-Myc* proto-oncogene. In normal dividing cells, c-Myc expression is maintained at a relatively constant intermediate level throughout the cell cycle, whereas in its oncogenic form, c-Myc might be constitutively expressed at levels ranging from moderate to very high *(176)*, and is nonresponsive to external signals *(4–6)*. Alternatively, the regulated pattern of c-Myc expression can remain intact, but exceed normal levels of expression for the given cell type.

c-Myc activation can occur by a wide range of direct and indirect mechanisms, such as stabilization of c-Myc mRNA transcripts or through enhanced initiation of translation resulting from mutation of the internal ribosomal entry site *(177,178)*, or indirectly via activation of upstream signaling cascade *(179–182)*. Historically, molecular pathologists have relied on the presence of gross chromosomal abnormalities, such as translocation or amplification, of the *c-Myc* locus to define activation of this oncogene in tumor cells (as is the case in Burkitt's lymphoma and neuroblastoma, respectively). However, this restrictive diagnostic criterion has resulted in an underestimate of the numbers of tumors with deregulated *c-Myc*.

The detection of c-Myc activation in tumor cells now relies on both elevated expression of *c-Myc* mRNA and genetic alterations of the *c-Myc* locus. Advances in technology allow us to evaluate expression levels of *c-Myc* mRNA in tumor cells (precisely excised using laser-capture microdissection) with real-time reverse transcription-polymerase chain reaction (RT-PCR) and expression profiling using microarray technology *(183,184)*. These approaches allow rapid, quantifiable, high-throughput screening of *c-Myc* activation in tumor cells. However, these techniques do not measure levels of c-Myc protein expression, which are largely assayed by immunohistochemical methods. Evidence now indicates that *c-Myc* activation can also occur through stabilization of the c-Myc protein. c-Myc proteins with mutations at or near T58, which abolish phosphorylation at this site, result in inefficient ubiquitination and a stabilization of the protein *(185,186)*.

The fact that c-Myc activation is present in a broad range of human cancers, and is often associated with poor prognosis of the disease, suggests that analysis of c-Myc deregulation in certain tumor types may be used as a diagnostic marker. Moreover, inactivating c-Myc, or downstream targets of c-Myc, may provide important therapeutic targets. Although still in its infancy, various approaches to target c-Myc are presently under investigation *(187,188)*.

4. c-Myc and Diabetes

Diabetes is a serious global health problem affecting 194 million people worldwide and this figure is expected to rise to 370 million by the year 2030. Current preventative and therapeutic strategies are manifestly inadequate, and there is a need for research to develop novel healthcare interventions to address this substantial biomedical challenge. Diabetes is caused by an absolute (type 1) or relative (type 2) deficiency of insulin-producing β cells.

Type 2 diabetes (T2D) constitutes between 85 and 95% of all diabetes and is caused by two distinct but overlapping processes, namely peripheral *insulin resistance* in muscle, fat, and liver and failure of pancreatic β-cells to adequately compensate for this resistance—"β-cell failure." β-cell failure encompasses impaired glucose sensing/insulin secretion and loss of β-cell mass (imbalances in β-cell apoptosis and renewal) and is now widely appreciated as the pre-eminent factor in T2D onset and disease progression in man *(189)*.

Comparatively little is understood about the causes of β-cell failure and much less is known about how to prevent or reverse it. Encouraging results with lifestyle intervention and metformin not withstanding, large studies such as the United Kingdom Prospective Diabetes Study (UKPDS) suggest currently available therapies do not prevent relentless progression of β-cell failure in T2D.

To progress in this key area we urgently need to develop a more complete picture of how key β-cell processes, such as replication, apoptosis, neogenesis, and differentiation, are controlled and how they become derailed in T2D. Moreover, better understanding of β-cell growth and regeneration mechanisms may allow new strategies in the treatment of T2D based on early limitation of β-cell damage and/or restoration of a functional β-cell mass.

4.1. c-Myc Apoptosis, β-Cell Failure and Diabetes

Recent studies by several laboratories, including our own, have provided support for the notion that c-Myc could provide a molecular link between insulin resistance (and accompanying adaptive β-cell growth/replication) and the two major potential contributors to β-cell failure: loss of function and sensitisation of β-cells to apoptosis. c-Myc is clearly of great relevance to β-cells; it is upregulated in and required for replication (G_1/S transition) in essentially all cell types, including β-cells, and is an important contributor to cell growth (size) even where replication is absent; c-Myc has been shown repeatedly to be expressed in β-cells exposed to hyperglycaemia, during increases in β-cell mass in vivo and in vitro and in β-cell-derived neoplasia. In fact, by implication c-Myc will be elevated in β-cells in insulin resistance as an inevitable accompaniment to the adaptive increases in β-cell mass and ultimately also in response to rising blood glucose *(189)*.

In animal models of diabetes, impaired glucose-stimulated insulin secretion (GSIS) is associated with loss of β-cell differentiation markers and of proteins involved in glucose stimulus-secretion coupling; accompanied by β-cell hypertrophy and increased expression of c-Myc *(190–193)*. Recent studies from our own laboratory and that of Susan Bonner-Weir and Gordon Weir have shown that c-Myc is a powerful trigger for β-cell apoptosis as well as loss of differentiation in rodent islets in vivo *(14,194)*. Importantly, c-Myc activity in β cells is enhanced by hyperglycaemia. Given, that c-Myc activation leads to reduced insulin content and also β-cell apoptosis in vivo, *(14)* induction of c-Myc by hyperglycemia may contribute to β-cell loss/dysfunction in diabetes (*see* Fig. 7). This is discussed in more detail below

During the progression to diabetes both functional defects and β-cell apoptosis contribute to defective insulin secretion *(189–191)* generally described as β-cell failure. Several potential contributors to this have been identified, including defective insulin/IGF/IRS2 signalling, activated c-Myc, and various extrinsic factors including glucose toxicity, lipotoxicity, various inflammatory cytokines and oxidative stress *(193)*. Importantly, recent studies suggest that replicating β-cells may be particularly vulnerable to apoptosis, thus indirectly implicating c-Myc *(193a)*.

The notion that c-Myc may be involved in β-cell failure is further supported by studies showing that β-cells also upregulate c-Myc in response to rising blood glucose levels. Even mild hyperglycaemia can adversely affect β-cell phenotype over time and result in progressive loss of β-cell differentiation as evidenced by reduced expression of key transcription factors (e.g., Pdx-1, BETA2/NeuroD) and genes involved in glucose sensing (GLUT-2 and glucokinase) *(192,192a)* On the other hand, genes normally suppressed in β-cells (c-Myc and various target genes such as lactate dehydrogenase-A) are increased. Recent studies have helped resolve this complicated interaction- ectopic activation of c-Myc in β-cells results in loss of differentiation markers and also in apoptosis. Thus, a likely model is that β-cells activate c-Myc in response to a need for adaptive growth or later in the process because of rising blood glucose. Elevated c-Myc may then contribute to loss of function and apoptosis of β-cells. Various processes, apart from c-Myc, have been described as underlying glucose toxicity and include altered calcium homeostasis, activation of the extrinsic pathway of apoptosis by cytokines, such as IL-1β or Fas/Fas-ligand interactions and reactive oxygen species. Intriguingly, c-Myc has been shown to trigger apoptosis through all of these pathways in other systems—including in particular activation of both apical caspases (caspase 8 and 9), Fas/Fas-ligand signalling and reactive oxygen species.

Targeting c-Myc-induced β-cell apoptosis and loss of differentiation would represent an attractive therapeutic option in diabetes. However, the means of preventing apoptosis may need to be carefully approached, given that suppressing c-Myc-induced apoptosis by overexpression of Bcl-X_L in mouse islets in vivo rapidly leads to c-Myc-induced beta cell neoplasia, described earlier *(14)*. Interestingly, a recent study suggests that the growth effect of Pax4 may be mediated by concurrent activation of both c-Myc and BclxL, and may thus play a role in normal expansion of β-cells *(195)*. By implication, it is deregulated and excessive expression of c-Myc and possibly BclxL that is oncogenic, whereas controlled expression, likely in response to normal growth-factor regulation is not.

Various downstream targets and processes of c-Myc may also contribute to β-cell growth, including cyclin D2 and it's partner CDK, CDK4 [196-200]. Moreover, inactivation of the CDKI, p27 is also able to promote beta cell growth.

5. Therapeutic Targeting of c-Myc

Given the central role of c-Myc in cell proliferation and death and, moreover, in the development and maintenance of many cancers, the potential therapeutic benefits of targeting aspects of c-Myc activity in human cancers are self-evident. Moreover, studies employing c-Myc-directed therapies may also finally help address a central question

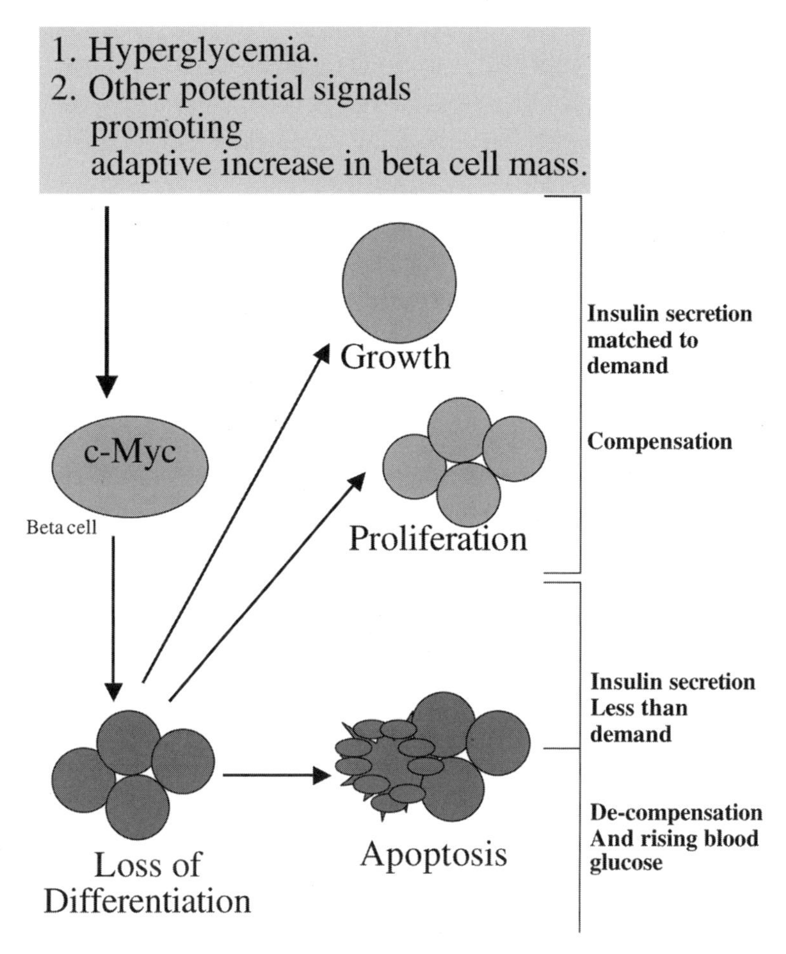

Fig. 7. Potential role of c-Myc in beta cell failure in diabetes.Type 2 diabetes arises or progresses when there is an imbalance between apoptosis and compensatory β-cell growth processes. Hyperglycaemia leads to c-Myc expression in insulin-secreting β-cells of the pancreatic islets. Although, as yet to be confirmed it is highly likely that c-Myc will be upregulated in β-cells induced to grow/replicate as part of normal adaptive growth (e.g., in response to pregnancy) or in response to insulin resistance. In animal models, following c-Myc activation β-cells replicate but undergo concomitant apoptosis. Moreover, even replicating β-cells that are prevented from undergoing apoptosis experience loss of differentiation as evidenced by reduced insulin content, and loss of markers such as Pdx1.

remaining unanswered in human cancer, namely can human tumors be reversed as readily as mouse equivalents by targeting an individual oncogenic lesion.

In order to realize the potential of targeting c-Myc in the treatment of cancer, drugs will have to be designed to manipulate specific c-Myc actions, such as cell-cycle entry, dedifferentiation, or apoptosis. It is likely that this will require the identification of drugable targets downstream of c-Myc activation. Despite recent advances in the identification of multiple c-Myc-target genes by employing powerful analytical techniques such as gene microarrays most of these studies have relied on the study of cell lines or tissues in vitro *(201–204)*, or during c-Myc-induced lymphomagenesis in chicken following transplantation of c-Myc infected embryonic bursal cells *(205)*. Interestingly, the latter study suggested that the level of c-Myc over-expression had a major impact on the numbers of genes regulated. Given the critical dependence of c-Myc-induced phenotype on tissue location in vivo we still do not know enough about the key divergence in c-Myc-targets under conditions, which favor the opposing potential outcomes of c-Myc-activation, namely either apoptosis or proliferation. However, there is considerable reason for optimism. Proof of hypothesis exists already, for example, at least in some situations, Cyclin D is essential for c-Myc-induced proliferation but not for c-Myc-induced apoptosis (in vitro), and the same may well apply to other genes/proteins. Studies in which the immediate and sequential evolution of changes in gene/protein expression can be tracked and correlated with the emergence of key cancer phenotypes in vivo are now required. To undertake this will involve the application of powerful new advances in conditional mutagenesis alongside analytical techniques such as microarrays, proteomics, and exciting new tools for the analysis of protein–protein interactions and putative functional protein networks. Some of these studies are on-going in our own laboratory and those of others. As it is likely that most heterogeneous and diverse cancers share a common and obligate mechanistic ancestry—deregulated proliferation and reduced cell death—targeting these pathways should lead to the collapse of the cancer platform. Therefore, the nature of the antiapoptotic mechanism would need to be deciphered in order to reinstate the apoptotic pathway. It seems increasingly likely that future cancer therapies will comprise a combination of specific drug targets as well as chemotherapy and/or radiotherapy. We can optimistically look forward to a future in which the mechanics of c-Myc-induced phenotypes have been elucidated and specific drugable targets in those processes identified.

5.1. Gene Therapy and c-Myc

Recent progress has led to the development of genetic therapy, which can be grouped under two general headings: gene therapy and antisense therapy. The second is to deliver to the target cells antisense oligodeoxynucleotides (ODNs) that can hybridize with mRNA and specifically inhibit the expression of pathogenic genes. With encouraging results from preclinical and clinical studies of ODNs in the past decade, significant progress has been made in developing antisense therapy, with the first antisense drug now being approved for clinical use. To date, phase I and II clinical trials have only been reported for Ras, Bcl-2, and H-Ras and for c-Myc only in the context of coronary heart disease interventions. For recent reviews of antisense approaches, the author recommends Wang et al. *(205)*.

5.2. c-Myc *Antisense Oligodeoxynucleotides and Cancer*

Many studies have now confirmed the utility of using ODNs to target *c-Myc* expression in cultured cells. Estrogen induces a fivefold increase in c-Myc protein expression within 90 min in steroid-deprived MCF-7 breast cancer cells. Exposure of MCF-7 breast cancer cells to a c-Myc ODN demonstrated up to 95% inhibition of c-Myc protein expression and inhibited estrogen-stimulated cell growth by up to 75% *(206)*.

Similar promising results have now also been shown in animal tumor models both as monotherapy and in combination with traditional chemotherapeutic agents. Continuous subcutaneous perfusion of ODNs to N-c-Myc in human neuroectodermal tumors grown in athymic mice in vivo led to a mean 50% reduction in tumor mass *(207)*.

Because many chemotherapeutic drugs kill tumor cells by inducing apoptosis, cellular resistance to apoptosis is thought to be a major factor limiting drug efficacy. Elucidation of those factors dictating whether a tumor will be sensitive or resistant to chemotherapy-induced apoptosis is a major goal in developing future new therapies for individual cancer patients. Chemotherapy resistance in lung cancer therapy has been found to correlate with amplification and overexpression of c-Myc. Using a Lewis lung syngeneic drug-resistant murine tumor model Knapp et al. *(208)* have shown that inhibition of c-Myc, using Morpholino oligomers (novel, stable, ODNs for inhibition of gene expression by an RNase H-independent mechanism), to mice previously given cisplatin led to a dramatic decrease in tumor growth rate. Interestingly using both agents concurrently was no more effective than using a single agent only, suggesting that timing of treatments may also be an important factor.

5.3. c-Myc *Antisense Oligodeoxynucleotides and Coronary Artery Restenosis*

Restenosis, resulting from neointimal hyperplasia, is the most important long-term limitation of stent implantation for coronary artery disease, occurring in 15 to 60% of patients. Over 1,500,000 percutaneous coronary interventions are performed annually. Growth and migration of vascular smooth-muscle cells result in neointimal proliferation after vascular injury and are the key mechanism of in-stent restenosis. The rationale of recent approaches to restenosis (e.g., brachytherapy and immunosuppressive agents) is based on the perceived similarities between tumor-cell growth and the benign proliferation of intimal hyperplasia. Several immunosuppressants have been tested for their potential to inhibit restenosis, with the novel strategy of administering the drug via a coated stent platform. Local drug delivery achieves higher tissue concentrations of drug without systemic effects, at a precise site and time. The first multicentre trials were with stents coated with sirolimus and there have now been several other studies showing the efficacy of sirolimus and also paclitaxel *(209)*. Given the known role of c-Myc in the proliferation of vascular smooth muscle cells, studies in animal models were undertaken and showed that restensosis after arterial injury or intervention might be profoundly reduced by administering antisense ODN directed against c-Myc. Intramural delivery of an advanced c-Myc neutrally charged antisense morpholino compound completely inhibited c-Myc expression and dramatically reduced neointimal formation in a dose dependent fashion in a porcine coronary stent restenosis model *(210)*. Sadly, a recent trial in patients designed to determine whether this approach could inhibit restenosis when given by local delivery immediately after coronary stent implantation has now been completed *(211)* and failed to demonstrate significant differences in angiographic restenosis rates.

5.4. *Alternative Approaches to Targeting c-Myc*

Various other approaches have been developed in order to modulate c-Myc activity. The application of RNA interference (RNAi) has been explored for some c-Myc-target genes in vitro (i.e., for the gene encoding mina53; c-Myc-induced nuclear antigen with a molecular mass of 53 kDa), with some success *(212)*. In a recent study, it has been shown that *c-Myc* (which bears the potential in its promoter to form a G-quadruplex) can be inhibited by the cationic porphyrin TMPyP4, which is able to bind to and stabilize G-quadruplexes *(213)*.

Indirectly, c-Myc may be targeted in several different ways, including through modulation of Max or members of the Mad family, which can regulate transcriptional activity of c-Myc, or through interference with signaling upstream of c-Myc expression, such as beta catenin. Mutations in the adenomatous polyposis coli (APC) gene, which initiate almost all human colon cancers, directly target the proto-oncogene, *c-Myc*, by elevating β-catenin/T-cell factor (TCF) signaling. Agents that are chemopreventive for colon cancer also stimulate β-catenin/TCF activity in vitro and can result in decreased c-Myc expression *(214)*.

Collier et al. *(215)* have shown that in rat hepatocytes in vitro the glucose-mediated induction of L-type pyruvate kinase and glucose-6-phosphatase mRNA levels were diminished by recombinant adenoviral vectors that interfere with c-Myc function through a dominant-negative Max protein. Interestingly, recent intriguing data suggest that over-expression of Mad, an antagonist of c-Myc transcriptional activity, may enhance some features of β-cell differentiation, including insulin secretion. Therefore, a future alternative approach to *c-Myc* antisense may be through the targeting of either the c Myc partner protein Max or antagonist Mad.

6. Future Directions

It is clear that c-Myc is an attractive treatment target in cancer and other diseases of hyperproliferation and also in diabetes. Compelling results from rodent models and cell-culture experiments suggest that future developments in c-Myc-targeted therapeutics hold great promise. However, with the important caveat that successful results in treating human diseases have yet to be demonstrated.

Intriguing paradoxes have been revealed by recent studies; for example, the demonstration that transient inactivation of c-Myc may be sufficient to arrest and reverse certain tumor models but not others, that in some cases seemingly complete regression is followed at some later stage by recrudescence from dormant cells that have escaped apoptosis, and finally that some cancer cells lose their dependence on c-Myc and fail to regress currently seem to complicate analysis, but on the other hand these features suggest that animal models may more completely model human cancer than has previously been appreciated.

Cells in metazoans must compete with one another for growth and survival signals. Intriguing studies with the *Drosophila* homolog of Myc (dMyc) activity have now shown that Myc levels may influence the ability of cells to compete. Cell competition may play a key role during *D. melanogaster* development. Recent studies have suggested that cells expressing higher levels of Myc than their neighbors may actually induce apoptosis in those cells with the lower levels of Myc- in other words such cells act as "super competitors" *(126–128)*. It is not known whether such a phenomenon also exists in vertebrates.

Another criticism that has been levied against mouse cancer models relates to the perceived numerical differences in genetic lesions required for cancer to develop *(223)*. Thus, immortalization of human fibroblasts required perturbation of six pathways-involving p53, pRb, PP2A, telomerase, Raf, and Ral-GEFs. Epithelial cells may require inactivation of the Rb/p16(INK4a) pathway and expression of the catalytic subunit of telomerase, hTERT, in combination with viral oncoproteins. However, it is possible that there are differences in the mutational requirements for "imortalization" and "transformation" of a cell in culture as compared to formation of a cancer cell in vivo. Moreover, it is also possible that some oncogenes, such as c-Myc, may take the place of multiple other potential lesions because of their diverse roles in promoting numerous cancer behaviors. It is interesting with this in mind that a very recent study has shown that c-Myc may be sufficient to transform a human prostate epithelial cell *(224)*; stabilizing telomere length through up-regulation of hTERT expression and overriding the accumulation of cell cycle inhibitors, such as p16(INK4a). However, these transformed cells still retained some features of normal cells.

Ultimately, we will no doubt resolve these issues, but it seems likely that simply targeting c-Myc or it's downstream mediators may only be part of the story and that timing and dosing strategies for such treatments, both alone and in combinations with other cancer treatments will likely prove equally important.

References

1. Vennstrom B, Sheiness D, Zabielski J, Bishop JM. Isolation and characterization of c-myc, a cellular homolog of the oncogene (v-myc) of avian myelocytomatosis virus strain 29. J Virol 1982;42:773–779.
2. Spencer CA, Groudine M. Control of c-myc regulation in normal and neoplastic cells. Adv Cancer Res 1991;6:1–48.
3. Dang CV. c-Myc target genes involved in cell growth, apoptosis, and metabolism. Mol Cell Biol 1999;19:1–11.
4. Nesbit C, Tersak J, Prochownik E. c-Myc oncogenes and human neoplastic disease. Oncogene 1999;18:3004–3016.
5. Schlagbauer-Wadl H, Griffioen M, Van Elsas A, et al. Influence of increased c-Myc expression on the growth characteristics of human melanoma. J Invest Dermatol 1999;112: 332–336.
6. Henriksson M, Luscher B. Proteins of the Myc network: essential regulators of cell growth and differentiation. Adv Cancer Res 1996;68:109–182.
7. Schmid P, Schulz WA, Hameister H. Dynamic expression pattern of the myc protooncogene in midgestation mouse embryos. Science 1989;243(4888):226–229.
8. Hirvonen H, Makela TP, Sandberg M, Kalimo H, Vuorio E, Alitalo K. Expression of the myc protooncogenes in developing human fetal brain. Oncogene 1990;5(12): 1787–1797.
9. Davis AC, Wims M, Spotts GD, Hann SR, Bradley A. A null c-myc mutation causes lethality before 10.5 days of gestation in homozygotes and reduced fertility in heterozygous female mice. Genes Dev 1993;7(4):671–682.
10. O'Connell BC, Cheung AF, Simkevich CP, et al. A Large Scale Genetic Analysis of c-Myc-regulated Gene Expression Patterns. J Biol Chem 2003;278(14):12,563–12,573.
11. Frye M, Gardner C, Li ER, Arnold I, Watt FM. Evidence that Myc activation depletes the epidermal stem cell compartment by modulating adhesive interactions with the local microenvironment. Development 2003;130(12):2793–2808.
12. Patel JH, Loboda AP, Showe MK, Showe LC, McMahon SB. Analysis of genomic targets reveals complex functions of c-Myc. Nat Rev Cancer 2004;4(7):562–568.
13. Fernandez PC, Frank SR, Wang L, Schroeder M, Liu S, Greene J, Cocito A, Amati B. Genomic targets of the human c-Myc protein. Genes Dev 2003;17(9):1115–1129.
14. Pelengaris S, Khan M, Evan GI. Suppression of Myc-induced apoptosis in beta cells exposes multiple oncogenic properties of Myc and triggers carcinogenic progression. Cell 2002;109(3):321–334.
15. Pelengaris S, Littlewood T, Khan M, Elia G, Evan G. Reversible activation of c-Myc in skin: induction of a complex neoplastic phenotype by a single oncogenic lesion. Mol Cell 1999;3(5):565–577.

16. Maddison K, Clarke AR.New approaches for modelling cancer mechanisms in the mouse. J Pathol 2005;205(2):181–193.
17. Baek SH, Kioussi C, Briata P, et al. Regulated subset of G1 growth-control genes in response to derepression by the Wnt pathway. Proc Natl Acad Sci USA 2003;100(6): 3245–3250.
18. Sasaki T, Suzuki H, Yagi K, et al. Lymphoid enhancer factor 1 makes cells resistant to transforming growth factor beta-induced repression of c-myc. Cancer Res 2003;63(4): 801–806.
19. Ramsey R, Barton A, Gonda T. Targeting c-Myb expression in human disease. EOTT 2003;7(2):235–248.
20. Amati B. Myc degradation: dancing with ubiquitin ligases. Proc Natl Acad Sci USA 2004;101(24):8843–8844.
21. David Dominguez-Sola & Riccardo Dalla-Favera. PINning down the c-Myc oncoprotein Nat Cell Biol 2004;6:288–289.
22. Welcker M, Orian A, Grim JA, Eisenman RN, Clurman BE. A nucleolar isoform of the Fbw7 ubiquitin ligase regulates c-Myc and cell size. Curr Biol 2004;14(20):1852–1857.
23. Yeh E, Cunningham M, Arnold H, et al. A signalling pathway controlling c-Myc degradation that impacts oncogenic transformation of human cells. Nat Cell Biol 2004;6(4): 308–318.
24. Patel JH, Du Y, Ard PG, et al. The c-Myc oncoprotein is a substrate of the acetyltransferases hGCN5/PCAF and TIP60. Mol Cell Biol 2004;24(24):10,826–10,834.
25. Tworkowski KA, Salghetti SE, Tansey WP. Stable and unstable pools of Myc protein exist in human cells. Oncogene 2002;21(55):8515–8520.
26. Blackwood EM, Eisenman RN. Max: a helix-loop-helix zipper protein that forms a sequence-specific DNA-binding complex with Myc. Science 1991;251(4998): 1211–1217.
27. Amati B, Brooks MW, Levy N, Littlewood TD, Evan GI, Land H. Oncogenic activity of the c-Myc protein requires dimerization with Max. Cell 1993;72(2):233–245.
28. Sakamuro D, Prendergast GC. New Myc-interacting proteins: a second Myc network emerges. Oncogene 1999;18(19):2942–2954.
29. Nilsson JA, Maclean KH, Keller UB, Pendeville H, Baudino TA, Cleveland JL. Mnt loss triggers Myc transcription targets, proliferation, apoptosis, and transformation. Mol Cell Biol 2004;24(4):1560–1569.
30. McMahon SB, Van Buskirk HA, Dugan KA, Copeland TD, Cole MD. The novel ATM-related protein TRRAP is an essential cofactor for the c-Myc and E2F oncoproteins. Cell 1998;94(3):363–374.
31. McMahon SB, Wood MA, Cole MD. The essential cofactor TRRAP recruits the histone acetyltransferase hGCN5 to c-Myc. Mol Cell Biol 2000;20(2):556–562.
32. Amati B, Frank SR, Donjerkovic D, Taubert S. Function of the c-Myc oncoprotein in chromatin remodeling and transcription. Biochim Biophys Acta 2001;1471(3): M135–M145.
33. Bouchard C, Dittrich O, Kiermaier A, et al. Regulation of cyclin D2 gene expression by the Myc/Max/Mad network: Myc- dependent TRRAP recruitment and histone acetylation at the cyclin D2 promoter. Genes Dev 2001;15(16):2042–2047.
34. Frank SR, Schroeder M, Fernandez P, Taubert S, Amati B. Binding of c-Myc to chromatin mediates mitogen-induced acetylation of histone H4 and gene activation. Genes Dev 2001;15(16):2069–2082.
35. Eberhardy SR, D'Cunha CA, Farnham PJ. Direct examination of histone acetylation on Myc target genes using chromatin immunoprecipitation. J Biol Chem 2000;275: 33,798–33,805.
36. Brenner C, Deplus R, Didelot C, et al. Myc represses transcription through recruitment of DNA methyltransferase corepressor. EMBO J 2005;24(2):336–346.

37. Herbst A, Hemann MT, Tworkowski KA, Salghetti SE, Lowe SW, Tansey WP. A conserved element in Myc that negatively regulates its proapoptotic activity. EMBO Rep 2005;6(2): 177–183.
38. Amati B, Alevizopoulos K, Vlach J. Myc and the cell cycle Front Bio sci 1998;3: D250–D268.
39. Eilers M. Control of cell proliferation by Myc family genes. Mol Cells 1999;9(1):1–6.
40. Amati B. Integrating Myc and TGF-beta signalling in cell-cycle control. Nat Cell Biol 2001;3(5):E112–E113.
41. Davis AC, Wims M, Spotts GD, Hann SR, Bradley A. A null c-myc mutation causes lethality before 10.5 days of gestation in homozygotes and reduced fertility in heterozygous female mice. Genes Dev 1993;7(4):671–682.
42. Mateyak MK, Obaya AJ, Adachi S, Sedivy JM. Phenotypes of c-Myc-deficient rat fibroblasts isolated by targeted homologous recombination. Cell Growth Differ 1997;8(10): 1039–1048.
43. Schorl C, Sedivy JM. Loss of Protooncogene c-Myc Function Impedes G(1) Phase Progression Both before and after the Restriction Point. Mol Biol Cell 2003;14(3):823–835.
44. Rosenwald IB, Rhoads DB, Callanan LD, Isselbacher KJ, Schmidt EV. Increased expression of eukaryotic translation initiation factors eIF-4E and eIF-2 alpha in response to growth induction by c-myc. Proc Natl Acad Sci USA 1993;90(13):6175–6178.
45. Jones RM, Branda J, Johnston KA, et al. An essential E box in the promoter of the gene encoding the mRNA cap-binding protein (eukaryotic initiation factor 4E) is a target for activation by c-myc. Mol Cell Biol 1996;16(9):4754–4764.
46. Coller HA, Grandori C, Tamayo P, et al. Expression analysis with oligonucleotide microarrays reveals that c-Myc regulates genes involved in growth, cell cycle, signaling, and adhesion. Proc Natl Acad Sci USA 2000;97(7):3260–3265.
47. Johnston LA, Prober DA, Edgar BA, Eisenman RN, Gallant P. Drosophila myc regulates cellular growth during development. Cell 1999;98(6):779–790.
48. Iritani BM, Eisenman RN. c-Myc enhances protein synthesis and cell size during B lymphocyte development. Proc Natl Acad Sci USA 1999;96(23):13,180–13,185.
49. Schuhmacher M, Staege MS, Pajic A, et al. Control of cell growth by c-Myc in the absence of cell division. Curr Biol 1999;9(21):1255–1258.
50. Beier R, Burgin A, Kiermaier A, et al. Induction of cyclin E-cdk2 kinase activity, E2F-dependent transcription and cell growth by Myc are genetically separable events. EMBO J 2000;19(21):5813–5823.
51. Gomez-Roman N, Grandori C, Eisenman RN, White RJ. Direct activation of RNA polymerase III transcription by c-Myc. Nature 2003;421(6920):290–294.
52. Arabi A, Wu S, Ridderstrale K, et al. c-Myc associates with ribosomal DNA and activates RNA polymerase I transcription. Nat Cell Biol 2005;7(3):303–310.
53. White RJ. RNA polymerases I and III, growth control and cancer. Nat Rev Mol Cell Biol 2005;6(1):69–78.
54. Steiner P, Philipp A, Lukas J, et al. Identification of a Myc-dependent step during the formation of active G1 cyclin-cdk complexes. EMBO J 1995;14(19):4814–4826.
55. Berns K, Hijmans EM, Bernards R. Repression of c-Myc responsive genes in cycling cells causes G1 arrest through reduction of cyclin E/CDK2 kinase activity. Oncogene 1997; 15(11):1347–1356.
56. Bouchard C, Thieke K, Maier A, et al. Direct induction of cyclin D2 by Myc contributes to cell cycle progression and sequestration of p27. EMBO J 1999;18(19):5321–5333.
57. Hermeking H, Rago C, Schuhmacher M, et al. Identification of CDK4 as a target of c-Myc. Proc Natl Acad Sci USA 2000;97(5):2229–2234.
58. O'Hagan RC, Ohh M, David G, et al. Myc-enhanced expression of Cul1 promotes ubiquitin-dependent proteolysis and cell cycle progression. Genes Dev 2000;14(17):2185–2191.

59. Perez-Roger I, Kim SH, Griffiths B, Sewing A, Land H. Cyclins D1 and D2 mediate myc-induced proliferation via sequestration of p27(Kip1) and p21(Cip1). EMBO J 1999; 18(19):5310–5320.
60. Muller D, Bouchard C, Rudolph B, et al. Cdk2-dependent phosphorylation of p27 facilitates its Myc-induced release from cyclin E/cdk2 complexes. Oncogene 1997; 15(21): 2561–2576.
61. Staller P, Peukert K, Kiermaier A, et al. Repression of p15INK4b expression by Myc through association with Miz-1. Nat Cell Biol 2001;3(4):392–399.
62. Herold S, Wanzel M, Beuger V, et al. Negative Regulation of the Mammalian UV Response by Myc through Association with Miz-1. Mol Cell 2002;10(3):509–521.
63. Weber A, Liu J, Collins I, Levens D. TFIIH operates through an expanded proximal promoter to fine-tune c-myc expression. Mol Cell Biol 2005;1:147–161.
64. Mao DY, Barsyte-Lovejoy D, Ho CS, Watson JD, Stojanova A, Penn LZ. Promoter-binding and repression of PDGFRB by c-Myc are separable activities. Nucleic Acids Res 2004; 32(11):3462–3468.
65. Li H, Wu X. Histone deacetylase inhibitor, Trichostatin A, activates p21WAF1/CIP1 expression through downregulation of c-myc and release of the repression of c-myc from the promoter in human cervical cancer cells. Biochem Biophys Res Commun 2004;324(2): 860–867.
66. Pawar SA, Szentirmay MN, Hermeking H, Sawadogo M. Evidence for a cancer-specific switch at the CDK4 promoter with loss of control by both USF and c-Myc. Oncogene 2004;23(36):6125–6135.
67. Barsyte-Lovejoy D, Mao DY, Penn LZ. c-Myc represses the proximal promoters of GADD45a and GADD153 by a post-RNA polymerase II recruitment mechanism. Oncogene 2004;23(19):3481–3486.
68. Pal S, Yun R, Datta A, et al. mSin3A/histone deacetylase 2- and PRMT5-containing Brg1 complex is involved in transcriptional repression of the Myc target gene cad. Mol Cell Biol 2003;23(21):7475–7487.
69. La Rocca SA, Crouch DH, Gillespie DA. c-Myc inhibits myogenic differentiation and myoD expression by a mechanism which can be dissociated from cell transformation. Oncogene 1994;9(12):3499–3508.
70. Ryan KM, Birnie GD. Cell-cycle progression is not essential for c-Myc to block differentiation. Oncogene 1997;14(23):2835–2343.
71. Grandori C, Cowley SM, James LP, Eisenman RN. The Myc/Max/Mad network and the transcriptional control of cell behavior. Annu Rev Cell Dev Biol 2000;16:653–699.
72. Foley KP, Eisenman RN. Two Mad tails: what the recent knockouts of Mad1 and Mxi1 tell us about the c-Myc/Max/Mad network. Biochim Biophys Acta 1999;1423(3): M37–M47.
73. Dean M, Levine RA, Campisi J. c-myc regulation during retinoic acid-induced differentiation of F9 cells is posttranscriptional and associated with growth arrest. Mol Cell Biol 1986;6(2):518–524.
74. Dotto GP, Gilman MZ, Maruyama M, Weinberg RA. c-myc and c-fos expression in differentiating mouse primary keratinocytes. EMBO J 1986;5(11):2853–2857.
75. Lemaitre JM, Buckle RS, Mechali M. c-Myc in the control of cell proliferation and embryonic development. Adv Cancer Res 1996;70:95–144.
76. Chin L, Schreiber-Agus N, Pellicer I, et al. Contrasting roles for Myc and Mad proteins in cellular growth and differentiation. Proc Natl Acad Sci USA 1995;92(18):8488–8492.
77. Hurlin PJ, Foley KP, Ayer DE, Eisenman RN, Hanahan D, Arbeit JM. Regulation of Myc and Mad during epidermal differentiation and HPV-associated tumorigenesis. Oncogene 1995;11(12):2487–2501.

78. Hurlin PJ, Queva C, Koskinen PJ, et al. Mad3 and Mad4: novel Max-interacting transcriptional repressors that suppress c-myc dependent transformation and are expressed during neural and epidermal differentiation. EMBO J 1995;14(22):5646–5659.
79. Gandarillas A, Watt FM. c-Myc promotes differentiation of human epidermal stem cells. Genes Dev 1997;11(21):2869–2882.
80. Arnold I, Watt FM. c-Myc activation in transgenic mouse epidermis results in mobilization of stem cells and differentiation of their progeny. Curr Biol 2001;11(8):558–568.
81. Iritani BM, Eisenman RN. c-Myc enhances protein synthesis and cell size during B lymphocyte development. Proc Natl Acad Sci USA 1999;96(23):13,180–13,185.
82. Ayer DE, Lawrence QA, Eisenman RN. Mad-Max transcriptional repression is mediated by ternary complex formation with mammalian homologs of yeast repressor Sin3. Cell 1995; 80(5):767–776.
83. Schreiber-Agus N, Chin L, Chen K, et al. An amino-terminal domain of Mxi1 mediates anti-Myc oncogenic activity and interacts with a homolog of the yeast transcriptional repressor SIN3. Cell 1995;80(5):777–786.
84. James L, Eisenman RN. Myc and Mad bHLHZ domains possess identical DNA-binding specificities but only partially overlapping functions in vivo. Proc Natl Acad Sci USA 2002;99(16):10,429–10,434.
85. Cartwright P, McLean C, Sheppard A, Rivett D, Jones K, Dalton S. LIF/STAT3 controls ES cell self-renewal and pluripotency by a Myc-dependent mechanism. Development 2005; 132(5):885–896.
86. Walkley CR, Fero ML, Chien WM, Purton LE, McArthur GA. Negative cell-cycle regulators cooperatively control self-renewal and differentiation of haematopoietic stem cells. Nat Cell Biol 2005;7(2):172–178.
87. Wilson A, Murphy MJ, Oskarsson T, et al. c-Myc controls the balance between hematopoietic stem cell self-renewal and differentiation. Genes Dev 2004;18(22):2747–2763.
88. Evan GI, Wyllie AH, Gilbert CS, et al. Induction of apoptosis in fibroblasts by c-myc protein. Cell 1992;69(1):119–128.
89. Askew DS, Ashmun RA, Simmons BC, Cleveland JL. Constitutive c-myc expression in an IL-3-dependent myeloid cell line suppresses cell cycle arrest and accelerates apoptosis. Oncogene 1991;6(10):1915–1922.
90. Debbas M, White E. Wild-type p53 mediates apoptosis by E1A, which is inhibited by E1B. Genes Dev 1993;7(4):546–554.
91. Qin XQ, Livingston DM, Kaelin WG, Jr, Adams PD. Deregulated transcription factor E2F-1 expression leads to S-phase entry and p53- mediated apoptosis. Proc Natl Acad Sci USA 1994;91(23):10,918–10,922.
92. Shan B, Lee WH. Deregulated expression of E2F-1 induces S-phase entry and leads to apoptosis. Mol Cell Biol 1994;14(12):8166–8173.
93. Wu X, Levine AJ. p53 and E2F-1 cooperate to mediate apoptosis. Proc Natl Acad Sci USA 1994;91(9):3602–3606.
94. Strasser A, Harris AW, Bath ML, Cory S. Novel primitive lymphoid tumours induced in transgenic mice by cooperation between myc and bcl-2. Nature 1990;348(6299): 331–333.
95. Adams JM, Harris AW, Pinkert CA, et al. The c-myc oncogene driven by immunoglobulin enhancers induces lymphoid malignancy in transgenic mice. Nature 1985;318(6046): 533–538.
96. Blyth K, Terry A, O'Hara M, et al. Synergy between a human c-myc transgene and p53 null genotype in murine thymic lymphomas: contrasting effects of homozygous and heterozygous p53 loss. Oncogene 1995;10(9):1717–1723.

97. Elson A, Deng C, Campos-Torres J, Donehower LA, Leder P. The MMTV/c-myc transgene and p53 null alleles collaborate to induce T-cell lymphomas, but not mammary carcinomas in transgenic mice. Oncogene 1995;11(1):181–190.
98. Jacobs JJ, Scheijen B, Voncken JW, Kieboom K, Berns A, van Lohuizen M. Bmi-1 collaborates with c-Myc in tumorigenesis by inhibiting c-Myc-induced apoptosis via INK4a/ARF. Genes Dev 1999;13(20):2678–2690.
99. Eischen CM, Weber JD, Roussel MF, Sherr CJ, Cleveland JL. Disruption of the ARF-Mdm2-p53 tumor suppressor pathway in Myc-induced lymphomagenesis. Genes Dev 1999;13(20):2658–2669.
100. Tanaka H, Matsumura I, Ezoe S, et al. E2F1 and c-Myc potentiate apoptosis through inhibition of NF-kappaB activity that facilitates MnSOD-mediated ROS elimination. Mol Cell 2002;9(5):1017–1029.
101. Vafa O, Wade M, Kern S, et al. c-Myc can induce DNA damage, increase reactive oxygen species, and mitigate p53 function: a mechanism for oncogene-induced genetic instability. Mol Cell 2002;9(5):1031–1044,
102. Hueber AO, Zornig M, Lyon D, Suda T, Nagata S, Evan GI. Requirement for the CD95 receptor-ligand pathway in c-Myc-induced apoptosis. Science 1997;278(5341): 1305–1309.
103. Klefstrom J, Vastrik I, Saksela E, Valle J, Eilers M, Alitalo K. c-Myc 103. induces cellular susceptibility to the cytotoxic action of TNF-alpha. EMBO J 1994; 13(22):5442–5450.
104. Lutz W, Fulda S, Jeremias I, Debatin KM, Schwab M. MycN and IFNgamma cooperate in apoptosis of human neuroblastoma cells. Oncogene 1998;17(3):339–346.
105. Ricci MS, Jin Z, Dews M, et al. Direct repression of FLIP expression by c-myc is a major determinant of TRAIL sensitivity. Mol Cell Biol 2004;24(19):8541–8555.
106. Juin P, Hueber AO, Littlewood T, Evan G. c-Myc-induced sensitization to apoptosis is mediated through cytochrome c release. Genes Dev 1999;13(11):1367–1381.
107. Acehan D, Jiang X, Morgan DG, Heuser JE, Wang X, Akey CW. Three-dimensional structure of the apoptosome: implications for assembly, procaspase-9 binding, and activation. Mol Cell 2002;9(2):423–432.
108. Martinou JC, Green DR. Breaking the mitochondrial barrier. Nat Rev Mol Cell Biol 2001;2(1):63–67.
109. Soucie EL, Annis MG, Sedivy J, et al. Myc potentiates apoptosis by stimulating Bax activity at the mitochondria. Mol Cell Biol 2001;21(14):4725–4736.
110. Juin P, Hunt A, Littlewood T, et al. c-Myc functionally cooperates with Bax to induce apoptosis. Mol Cell Biol 2002;22(17):6158–6169.
111. Morrish F, Giedt C, Hockenbery D. c-Myc apoptotic function is mediated by NRF-1 target genes. Genes Dev 2003;17(2):240–255.
112. Gross A, Jockel J, Wei MC, Korsmeyer SJ. Enforced dimerization of Bax results in its translocation, mitochondrial dysfunction and apoptosis. EMBO J 1998;17(14): 3878–3885.
113. Khaled AR, Kim K, Hofmeister R, Muegge K, Durum SK. Withdrawal of IL-7 induces Bax translocation from cytosol to mitochondria through a rise in intracellular pH. Proc Natl Acad Sci USA 1999;96(25):14,476–14,481.
114. Eskes R, Desagher S, Antonsson B, Martinou JC. Bid induces the oligomerization and insertion of Bax into the outer mitochondrial membrane. Mol Cell Biol 2000;20(3): 929–935.
115. Brenner C, Cadiou H, Vieira HL, et al. Bcl-2 and Bax regulate the channel activity of the mitochondrial adenine nucleotide translocator. Oncogene 2000;19(3):329–336.

116. Eskes R, Antonsson B, Osen-Sand A, et al. Bax-induced cytochrome C release from mitochondria is independent of the permeability transition pore but highly dependent on Mg2+ ions. J Cell Biol 1998;143(1):217–224.
117. Lowe SW, Cepero E, Evan G. Intrinsic tumour suppression. Nature 2004;432(7015): 307–315.
118. Desbiens KM, Deschesnes RG, Labrie MM, et al. c-Myc potentiates the mitochondrial pathway of apoptosis by acting upstream of Ask1 in the p38 signaling cascade. Biochem J 2003;372:631–641.
119. Ionov Y, Yamamoto H, Krajewski S, Reed JC, Perucho M. Mutational inactivation of the proapoptotic gene Bax confers selective advantage during tumor clonal evolution. Proc Natl Acad Sci USA 2000;97(20):10,872–10,877.
120. Kauffmann-Zeh A, Rodriguez-Viciana P, Ulrich E, et al. Suppression of c-Myc-induced apoptosis by Ras signalling through PI(3)K and PKB. Nature 1997;385(6616):544–548.
121. Zha J, Harada H, Yang E, Jockel J, Korsmeyer SJ. Serine phosphorylation of death agonist BAD in response to survival factor results in binding to 14-3-3 not Bcl-X(L) Cell 1996; 87(4):619–628.
122. Zindy F, Eischen CM, Randle DH, et al. Myc signaling via the ARF tumor suppressor regulates p53-dependent apoptosis and immortalization. Genes Dev 1998;12(15): 2424–2433.
123. Qi Y, Gregory MA, Li Z, Brousal JP, West K, Hann SR. p19ARF directly and differentially controls the functions of c-Myc independently of p53. Nature 2004;431(7009): 712–717.
124. Eischen CM, Alt JR, Wang P. Loss of one allele of ARF rescues Mdm2 haploinsufficiency effects on apoptosis and lymphoma development. Oncogene 2004;23(55): 8931–8940.
125. Otsuki Y, Tanaka M, Kamo T, Kitanaka C, Kuchino Y, Sugimura H. Guanine Nucleotide Exchange Factor, Tiam1, Directly Binds to c-Myc and Interferes with c-Myc-mediated Apoptosis in Rat-1 Fibroblasts. J Biol Chem 2003;278(7):5132–5140
126. Moreno E, Basler K. dMyc transforms cells into super-competitors. Cell 2004;117(1): 117–129.
127. de la Cova C, Abril M, Bellosta P, Gallant P, Johnston LA. Drosophila myc regulates organ size by inducing cell competition. Cell 2004;117(1):107–116.
128. Secombe J, Pierce SB, Eisenman RN. Myc: a weapon of mass destruction. Cell 2004; 117(2):153–156.
129. Land H, Parada LF, Weinberg RA. Tumorigenic conversion of primary embryo fibroblasts requires at least two cooperating oncogenes. Nature 1983;304(5927):596–602.
130. Land H, Chen AC, Morgenstern JP, Parada LF, Weinberg RA. Behavior of myc and ras oncogenes in transformation of rat embryo fibroblasts. Mol Cell Biol 1986;6(6): 1917–1925.
131. Lugo TG, Witte ON. The BCR-ABL oncogene transforms Rat-1 cells and cooperates with v-myc. Mol Cell Biol 1989;9(3):1263–1270.
132. Fanidi A, Harrington EA, Evan GI. Cooperative interaction between c-myc and bcl-2 proto-oncogenes. Nature 1992;359(6395):554–556.
133. Reed JC, Cuddy M, Haldar S, et al. Bcl2-mediated tumorigenicity of a human T-lymphoid cell line: synergy with c-Myc and inhibition by Bcl2 antisense. Proc Natl Acad Sci USA 1990;87(10):3660–3664.
134. Morgenbesser SD, DePinho RA. Use of transgenic mice to study myc family gene function in normal mammalian development and in cancer. Semin Cancer Biol 1994;5(1): 21–36.

135. Pelengaris S, Khan M, Evan G. c-Myc: more than just a matter of life and death. Nat Rev Cancer 2002;2:764–776.
136. Littlewood TD, Hancock DC, Danielian PS, Parker MG, Evan GI. A modified oestrogen receptor ligand-binding domain as an improved switch for the regulation of heterologous proteins. Nucleic Acids Res 1995;23(10):1686–1690.
137. Pelengaris S, Abouna S, Cheung L, Ifandi V, Zervou S, Khan M. Brief inactivation of c-Myc is not sufficient for sustained regression of c-Myc-induced tumours of pancreatic islets and skin epidermis. BMC Biology 2004;2(1):26.
138. Hanahan D, Folkman J. Patterns and emerging mechanisms of the angiogenic switch during tumorigenesis. Cell 1996;86(3):353–364.
138a. Flores I, Murphy DJ, Swigart LB, Knies U, Evan GI. Defining the temporal requirements for Myc in the progression and maintenance of skin neoplasia. Oncogene 2004;23(35): 5923–5930.
139. Janz A, Sevignani C, Kenyon K, Ngo CV, Thomas-Tikhonenko A. Activation of the myc oncoprotein leads to increased turnover of thrombospondin-1 mRNA. Nucleic Acids Res 2000;28(11):2268–2175.
140. Ngo CV, Gee M, Akhtar N, et al. An in vivo function for the transforming Myc protein: elicitation of the angiogenic phenotype. Cell Growth Differ 2000;11(4):201–210.
141. Brandvold KA, Neiman P, Ruddell A. Angiogenesis is an early event in the generation of myc-induced lymphomas. Oncogene 2000;19(23):2780–2785.
142. Watnick RS, Cheng YN, Rangarajan A, Ince TA, Weinberg RA. Ras modulates Myc activity to repress thrombospondin-1 expression and increase tumor angiogenesis. Cancer Cell 2003;3(3):219–231.
143. Okajima E, Thorgeirsson UP. Different regulation of vascular endothelial growth factor expression by the ERK and p38 kinase pathways in v-ras, v-raf, and v-myc transformed cells. Biochem Biophys Res Commun 2000;270(1):108–111.
144. Knies-Bamforth UE, Fox SB, Poulsom R, Evan GI, Harris AL. c-Myc interacts with hypoxia to induce angiogenesis in vivo by a vascular endothelial growth factor-dependent mechanism. Cancer Res 2004;64(18):6563–6570.
145. Christofori G, Naik P, Hanahan D. A second signal supplied by insulin-like growth factor II in oncogene-induced tumorigenesis. Nature 1994;369(6479):414–418.
146. Giuriato S, Rabin K, Fan AC, Shachaf CM, Felsher DW. Conditional animal models: a strategy to define when oncogenes will be effective targets to treat cancer. Semin Cancer Biol 2004;14:3–11.
147. Felsher DW. Cancer revoked: oncogenes as therapeutic targets. Nat Rev Cancer 2003;3: 375–380.
148. Slamon DJ, Leyland-Jones B, Shak S, et al. Use of chemotherapy plus a monoclonal antibody against HER2 for metastatic breast cancer that overexpresses HER2. N Engl J Med 2001;344:783–792.
149. Druker BJ, Talpaz M, Resta DJ, et al. Efficacy and safety of a specific inhibitor of the BCR-ABL tyrosine kinase in chronic myeloid leukemia. N Engl J Med 2001; 344: 1031–1037.
150. Kantarjian H, Sawyers C, Hochhaus A, et al. Hematologic and cytogenetic responses to imatinib mesylate in chronic myelogenous leukemia. N Engl J Med 2002;346:645–652.
151. Felsher DW, Bishop JM. Reversible tumorigenesis by c-Myc in hematopoietic lineages. Mol Cell 1999;4(2):199–207.
152. Chin L, Tam A, Pomerantz J, et al. Essential role for oncogenic Ras in tumour maintenance. Nature 1999;400:468–472.
153. Huettner CS, Zhang P, Van Etten RA, Tenen DG. Reversibility of acute B-cell leukaemia induced by BCR-ABL1. Nat Genet 2000;24:57–60.

154. Fisher GH, Wellen SL, Klimstra D, et al. Induction and apoptotic regression of lung adenocarcinomas by regulation of a K-Ras transgene in the presence and absence of tumour suppressor genes. Genes Dev 2001;15:3249–3262.
155. Wang R, Ferrell LD, Faouzi S, Maher JJ, Bishop JM. Activation of the Met receptor by cell attachment induces and sustains hepatocellular carcinomas in transgenic mice. J Cell Biol 2001;153:1023–1034.
156. Moody SE, Sarkisian CJ, Hahn KT, et al. Conditional activation of Neu in the mammary epithelium of transgenic mice results in reversible pulmonary metastasis. Cancer Cell 2002;2:451–461.
157. Karlsson A, Giuriato S, Tang F, Fung-Weier J, Levan G, Felsher DW. Genomically complex lymphomas undergo sustained tumour regression upon c-Myc inactivation unless they acquire novel chromosomal translocations. Blood 2003;101:2797–2803.
158. Gunther EJ, Moody SE, Belka GK, et al. Impact of p53 loss on reversal and recurrence of conditional Wnt-induced tumourigenesis. Genes Dev 2003;17:488–501.
159. Pao W, Klimstra DS, Fisher GH, Varmus HE. Use of avian retroviral vectors to introduce transcriptional regulators into mammalian cells for analyses of tumour maintenance. Proc Natl Acad Sci USA 2003;100:8764–8769.
160. Marinkovic D, Marinkovic T, Mahr B, Hess J, Wirth T. Reversible lymphomagenesis in conditionally c-Myc expressing mice. Int J Cancer 2004;110:336–342.
161. Blakely CM, Sintasath L, D'Cruz CM, et al. Developmental stage determines the effects of c-Myc in the mammary epithelium. Development. 2005;132(5):1147–1160.
162. Egle A, Harris AW, Bouillet P, Cory S. Bim is a suppressor of Myc-induced mouse B cell leukemia. Proc Natl Acad Sci USA 2004;101(16):6164–6169.
163. Beer S, Zetterberg A, Ihrie RA, et al. Developmental context determines latency of c-Myc-induced tumorigenesis. PLoS Biol 2004;2(11):e332.
164. Letai A, Sorcinelli MD, Beard C, Korsmeyer SJ. Antiapoptotic Bcl-2 is required for maintenance of a model leukemia. Cancer Cell 2004;6(3):241–249.
165. Cheung WC, Kim JS, Linden M, et al. Novel targeted deregulation of c-Myc cooperates with Bcl-X(L) to cause plasma cell neoplasms in mice. J Clin Invest 2004;113(12): 1763–1773.
166. Hemann MT, Zilfou JT, Zhao Z, Burgess DJ, Hannon GJ, Lowe SW. Suppression of tumorigenesis by the p53 target PUMA. Proc Natl Acad Sci USA 2004;101(25): 9333–9338.
167. Lewis BC, Klimstra DS, Varmus HE. The c-myc and PyMT oncogenes induce different tumor types in a somatic mouse model for pancreatic cancer. Genes Dev 2003;17(24): 3127–3138.
168. Ellwood-Yen K, Graeber TG, Wongvipat J, et al. Myc-driven murine prostate cancer shares molecular features with human prostate tumors. Cancer Cell 2003;4(3): 223–238.
169. D'Cruz CM, Gunther EJ, Boxer RB, et al. c-Myc induces mammary tumorigenesis by means of a preferred pathway involving spontaneous Kras2 mutations. Nat Med 2001; 7(2):235–239.
170. Jain M, Arvanitis C, Chu K, et al. Sustained loss of a neoplastic phenotype by brief inactivation of c-Myc. Science 2002;297(5578):102–104.
171. Pelengaris S, Khan M. The many faces of c-MYC. Arch Biochem Biophys 2003;416(2): 129–136.
172. Shachaf CM, Kopelman AM, Arvanitis C, et al. c-Myc inactivation uncovers pluripotent differentiation and tumour dormancy in hepatocellular cancer. Nature 2004; 431: 1112–1117.
173. Wu Y, Renard CA, Apiou F, et al. Recurrent allelic deletions at mouse chromosomes 4 and 14 in Myc-induced liver tumors. Oncogene 2002;21(10):1518–1526.

174. Felsher DW, Bishop JM. Transient excess of c-Myc activity can elicit genomic instability and tumorigenesis Proc Natl Acad Sci USA 1999;96(7):3940–3944.
175. Fest T, Mougey V, Dalstein V, et al. c-Myc overexpression in Ba/F3 cells simultaneously elicits genomic instability and apoptosis. Oncogene 2002; 21(19):2981–2990.
176. Oster SK, Ho CS, Soucie EL, Penn LZ. The myc oncogene: Marvelously. Complex. Adv Cancer Res 2002;84:81–154.
177. Bernasconi NL, Wormhoudt TA, Laird-Offringa IA. Post-transcriptional deregulation of myc genes in lung cancer cell lines. Am J Respir Cell Mol Biol 2000;23(4):560–565.
178. Chappell SA, LeQuesne JP, Paulin FE, et al. A mutation in the c-myc-IRES leads to enhanced internal ribosome entry in multiple myeloma: a novel mechanism of oncogene de-regulation. Oncogene 2000;19(38):4437–4440.
179. Barone MV, Courtneidge SA. Myc but not Fos rescue of PDGF signalling block caused by kinase-inactive. Src Nature 1995;378(6556):509–512.
180. Bowman T, Broome MA, Sinibaldi D, et al. Stat3-mediated Myc expression is required for Src transformation and PDGF-induced mitogenesis. Proc Natl Acad Sci USA 2001; 98(13):7319–7324.
181. Chiariello M, Marinissen MJ, Gutkind JS. Regulation of c-myc expression by PDGF through Rho GTPases. Nat Cell Biol 2001;3(6):580–586.
182. Kolligs FT, Kolligs B, Hajra KM, et al. gamma-catenin is regulated by the APC tumor suppressor and its oncogenic activity is distinct from that of beta-catenin. Genes Dev 2000; 14(11):1319–1331.
183. Huang H, Colella S, Kurrer M, Yonekawa Y, Kleihues P, Ohgaki H. Gene expression profiling of low-grade diffuse astrocytomas by cDNA arrays. Cancer Res 2000;60(24): 6868–6874.
184. Watatani M, Inui H, Nagayama K, et al. Identification of high-risk breast cancer patients from genetic changes of their tumors. Surg Today 2000;30(6):516–522.
185. Gregory MA, Hann SR. c-Myc proteolysis by the ubiquitin-proteasome pathway: stabilization of c-Myc in Burkitt's lymphoma cells. Mol Cell Biol 2000;20(7):2423–2435.
186. Salghetti SE, Kim SY, Tansey WP. Destruction of Myc by ubiquitin-mediated proteolysis: cancer-associated and transforming mutations stabilize Myc. EMBO J 1999;18(3): 717–726.
187. Sears R, Nuckolls F, Haura E, Taya Y, Tamai K, Nevins JR. Multiple Ras-dependent phosphorylation pathways regulate Myc protein stability. Genes Dev 2000;14(19):2501–2514.
188. Hahn WC, Weinberg RA. Modelling the molecular circuitry of cancer. Nat Rev Cancer 2002;2(5):331–341.
189. Butler AE, Janson J, Bonner-Weir S, Ritzel R, Rizza RA, Butler PC. Beta-cell deficit and increased beta-cell apoptosis in humans with type 2 diabetes. Diabetes 2003;52(1):102–110.
190. Deng et al. Structural and functional abnormalities in the islets isolated from type 2 diabetic subjects. Diabetes 2004;53(3):624–632.
191. Del Prato S, et al. Beta-cell mass plasticity in type 2 diabetes. Diabetes Obes Metab 2004;6(5):319–331.
192. Jonas JC, Sharma A, Hasenkamp W, Ilkova H, et al. Chronic Hyperglycaemia triggers loss of pancreatic β-cell differentiation in an animal model of diabetes. J Biol Chem 1999;20: 14,112–14,121.
192a. Ritzel RA, Butler PC. Replication increases beta-cell vulnerability to human islet amyloid polypeptide-induced apoptosis. Diabetes 2003;52(7):1701–1708.
193. Robertson RP, Harmon J, Tran PO, Poitout V. Beta-cell glucose toxicity, lipotoxicity, and chronic oxidative stress in type 2 diabetes. Diabetes 2001;53 Suppl 1:S119–S124.
193a. Ritzel RA, Butler PC. Replication increases beta-cell vulnerability to human islet amyloid polypeptide-induced apoptosis. Diabetes 2003;52(7):1701–1708.

194. Laybutt DR, Weir GC, Kaneto H, et al. Overexpression of c-Myc in beta-cells of transgenic mice causes proliferation and apoptosis, downregulation of insulin gene expression, and diabetes. Diabetes 2002;51(6):1793–1804.
195. Brun T, Franklin I, St-Onge L, et al. The diabetes-linked transcription factor PAX4 promotes β-cell proliferation and survival in rat and human islets. J Cell Biol 2004;167(6):1123–1135.
196. Hino S, Yamaoka T, Yamashita Y, Yamada T, Hata J, Itakura M. In vivo proliferation of differentiated pancreatic islet beta cells in transgenic mice expressing mutated cyclin-dependent kinase 4. Diabetologia 2004;47(10):1819–1830.
197. Uchida T, Nakamura T, Hashimoto N, et al. Deletion of Cdkn1b ameliorates hyperglycemia by maintaining compensatory hyperinsulinemia in diabetic mice. Nat Med 2005;11(2):175–182.
198. Marzo N, Mora C, Fabregat ME, et al. Pancreatic islets from cyclin-dependent kinase 4/R24C (Cdk4) knockin mice have significantly increased beta cell mass and are physiologically functional, indicating that Cdk4 is a potential target for pancreatic beta cell mass regeneration in Type 1 diabetes. Diabetologia 2004;47(4):686–694.
199. Zhang X, Gaspard JP, Mizukami Y, Li J, Graeme-Cook F, Chung DC. Overexpression of Cyclin D1 in Pancreatic β-Cells In Vivo Results in Islet Hyperplasia Without Hypoglycemia. Diabetes 2005;54(3):712–719.
200. Georgia S, Bhushan A. Beta cell replication is the primary mechanism for maintaining postnatal beta cell mass. J Clin Invest 2004;114(7):963–968.
201. Watson JD, Oster SK, Shago M, Khosravi F, Penn LZ. Identifying genes regulated in a Myc-dependent manner. J Biol Chem 2002;277:36,921–36,930.
202. Schumacher M, Kohlhuber F, Holzel M, et al. The transcriptional program of a human B cell line in response to Myc. Nucleic Acids Res 2001;29:397–406.
203. O'Hagan RC, Schreiber-Agus N, Chen K, et al. Gene-target recognition among members of the myc superfamily and implications for oncogenesis. Nat Genet 2000;24:113–119.
204. Neiman PE, Ruddell A, Jasoni C, et al. Analysis of gene expression during myc oncogene-induced lymphomagenesis in the bursa of Fabricius. Proc Natl Acad Sci USA 2001;98: 6378–6383.
205. Wang H, Prasad G, Buolamwini JK, Zhang R. Antisense anticancer oligonucleotide therapeutics. Curr Cancer Drug Targets 2001;1(3):177–196.
206. Watson PH, Pon RT, Shiu RP. Inhibition of c-myc expression by phosphorothioate antisense oligonucleotide identifies a critical role for c-myc in the growth of human breast cancer. Cancer Res 1991;51(15):3996–4000.
207. Whitesell L, Rosolen A, Neckers LM. In vivo modulation of N-myc expression by continuous perfusion with an antisense oligonucleotide. Antisense Res Dev 1991;1(4): 343–350.
208. Knapp DC, Mata JE, Reddy MT, Devi GR, Iversen PL. Resistance to chemotherapeutic drugs overcome by c-Myc inhibition in a Lewis lung carcinoma murine model. Anticancer Drugs 2003;14(1):39–47.
209. Fattori R, Piva T. Drug-eluting stents in vascular intervention. Lancet 2003;361(9353): 247–249.
210. Kipshidze NN, Kim HS, Iversen P, et al. Intramural coronary delivery of advanced antisense oligonucleotides reduces neointimal formation in the porcine stent restenosis model. J Am Coll Cardiol 2002;39(10):1686–1691.
211. Kutryk MJ, Foley DP, van den Brand M, et al. Trial Local intracoronary administration of antisense oligonucleotide against c-myc for the prevention of in-stent restenosis: results of the randomized investigation by the Thoraxcenter of antisense DNA using local delivery and IVUS after coronary stenting trial. J Am Coll Cardiol 2002; 39(2): 281–287.

212. Tsuneoka M, Koda Y, Soejima M, Teye K, Kimura H. A novel myc target gene, mina53, that is involved in cell proliferation. J Biol Chem 2002;277(38):35,450–35,459.
213. Grand CL, Han H, Munoz RM, et al. The cationic porphyrin TMPyP4 down-regulates c-Myc and human telomerase reverse transcriptase expression and inhibits tumor growth in vivo. Mol Cancer Ther 2003;2(2):208.
214. Wilson AJ, Velcich A, Arango D, et al. Novel detection and differential utilization of a c-myc transcriptional block in colon cancer chemoprevention. Cancer Res 2002;62(21): 6006–6010.
215. Collier JJ, Doan TT, Daniels MC, Schurr JR, Kolls JK, Scott DK. c-Myc Is Required for the Glucose-mediated Induction of Metabolic Enzyme Genes. J Biol Chem 2003;278(8): 6588–6595.
216. Ewald D, Li M, Efrat S, et al. Time-sensitive reversal of hyperplasia in transgenic mice expressing SV40 T antigen. Science 1996; 273(5280):1384–1386.
217. Chin L, Tam A, Pomerantz J, et al. Essential role for oncogenic Ras in tumour maintenance. Nature 1999;400(6743):468–472.
218. Huettner CS, Zhang P, Van Etten RA, Tenen DG. Reversibility of acute B-cell leukaemia induced by BCR-ABL1. Nat Genet 2000;24(1):57–60.
219. Jackson EL, Willis N, Mercer K, et al. Analysis of lung tumor initiation and progression using conditional expression of oncogenic K-ras. Genes Dev 2001;15(24):3243–3248.
220. Fisher GH, Wellen SL, Klimstra D, et al. Induction and apoptotic regression of lung adenocarcinomas by regulation of a K-Ras transgene in the presence and absence of tumor suppressor genes. Genes Dev 2001;15(24):3249–3262.
221. Hutchin ME, Kariapper MS, Grachtchouk M, et al. Sustained Hedgehog signaling is required for basal cell carcinoma proliferation and survival: conditional skin tumorigenesis recapitulates the hair growth cycle. Genes Dev 2005;19(2):214–223.
222. Boxer RB, Jang JW, Sintasath L, Chodosh LA. Lack of sustained regression of c-MYC-induced mammary adenocarcinomas following brief or prolonged MYC inactivation. Cancer Cell 2004;6(6):577–586.
223. Rangarajan A, Hong SJ, Gifford A, Weinberg RA. Species- and cell type-specific requirements for cellular transformation. Cancer Cell 2004;6(2):171–183.
224. Gil J, Kerai P, Lleonart M, et al. Immortalization of Primary Human Prostate Epithelial Cells by c-Myc. Cancer Res 2005;65(6):2179–2185.

7

Role of Lysophospholipids in Cell Growth and Survival

Xianjun Fang and Sarah Spiegel

Summary

Lysophospholipids are not only metabolites in membrane phospholipid synthesis, but also extracellular bioactive mediators of multiple biological processes. The best characterized of these are lysophosphatidic acid (LPA) and sphingosine-1-phosphate (S1P), which have both emerged as important regulators of cell growth and survival. Identification of LPA and S1P receptors in the past several years has led to the accumulation of evidence that most cellular responses to LPA and S1P are mediated through activation of their cognate G protein-coupled receptors (GPCR). There are at least three high-affinity receptors for LPA—named LPA_1, LPA_2 and LPA_3—and five receptor subtypes for S1P, designated $S1P_{1\text{-}5}$. The widespread expression of these receptors and their downstream G proteins of various classes allow activation of a variety of signal transduction pathways that are integrated to trigger diverse cellular responses to LPA or S1P.

This chapter reviews and puts into context current understandings of LPA and S1P receptors and their downstream signaling pathways involved in proliferative and survival responses to LPA and S1P. The significance of these biological responses in human health and disease, particularly potential roles of these lipid mediators in human carcinogenesis, is also discussed.

Key Words: Lysophosphatidic acid; sphingosine-1-phosphate; signaling; metabolism; apoptosis; cell growth and survival; cancer.

1. Lysophospholipid Metabolism

1.1. Lysophosphatidic Acid (see *Fig. 1*)

Although LPA is considered to be a key intermediate in *de novo* lipid synthesis in all cells formed by acylation of glycerol 3-phosphate catalyzed by glycerophosphate acyltransferase *(1)*, little is known of how its extracellular levels are regulated. LPA can be produced by activated platelets, adipocytes, fibroblasts, endothelial cells, and ovarian and prostate cancer cells from newly generated phosphatidic acid by the action of phospholipase D followed by phospholipase A1 or A2-mediated deacylation *(2–6)*. It was recently suggested, however, that the bulk of LPA arising from platelet activation results from the sequential cleavage of serum and membrane phospholipids to lysophospholipids by PLA1 and PLA2 secreted by platelets, followed by conversion to LPA by lysophospholipase D (LysoPLD) activity present in plasma *(7)* (*see* Fig. 2).

Plasma LysoPLD was recently purified and identified as autotaxin (ATX/NPP-2), a member of the nucleotide pyrophosphatase and phosphodiesterase family of exo- and ecto-enzymes *(8,9)*. ATX was originally identified as a tumor cell motility-stimulating factor *(10)*. Intriguingly, stimulation of tumor cell migration and invasion is also one of

From: *Apoptosis, Cell Signaling, and Human Diseases: Molecular Mechanisms, Volume 1*
Edited by R. Srivastava © Humana Press Inc., Totowa, NJ

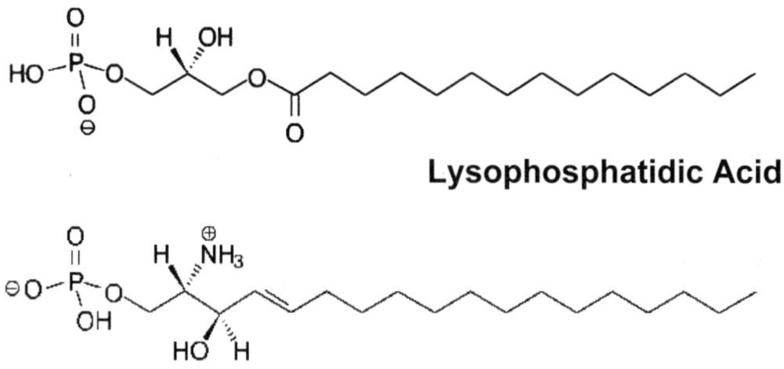

Lysophosphatidic Acid (LPA)

Sphingosine-1-phosphate (S1P)

Fig. 1. Structures of LPA and S1P.

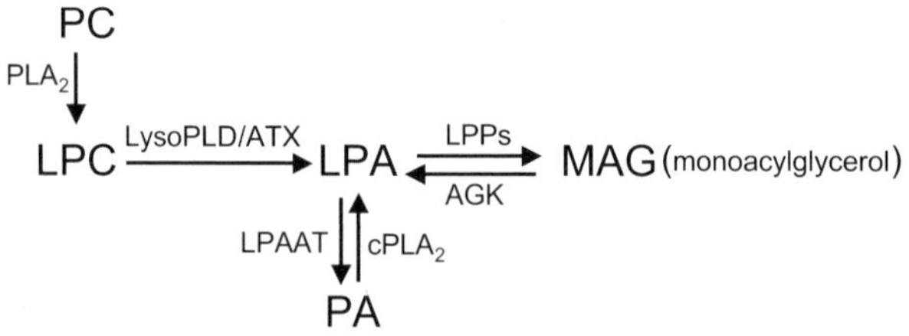

PC
PLA_2
LPC LysoPLD/ATX LPA LPPs AGK MAG (monoacylglycerol)
LPAAT $cPLA_2$
PA

Fig. 2. Major pathways for biosynthesis and metabolism of LPA.

the major biological functions of LPA *(11)*, suggesting that ATX promotes cell motility via formation and action of LPA.

Another potential pathway for synthesis of LPA is phosphorylation of monoacylglycerol by a specific lipid kinase activity first described more than 30 yr ago which has received scant attention following the characterization of the partially purified bovine brain enzyme in 1989 *(12)*. Recently, a novel lipid kinase that phosphorylates monoacylglycerol to form LPA, named AGK, has been identified *(13)*.

LPA can be acylated at the *sn*-2 hydroxyl group by acyl transferases, such as endophilin, to produce phosphatidic acid, another important lipid mediator *(14)*, deacylated by lysophospholipases to glycerol phosphate *(15)*, or dephosphorylated by lipid phosphate phosphohydrolases (LPPs), including LPP1/PAP2A, LPP3/PAP2B and LPP2/PAP2C, to generate monoacylglycerol *(16–19)*. LPPs are located at the cell membrane with their catalytic domains facing the extracellular media and degrade LPA associated with the external leaflet of the membrane. Overexpression of LPPs has been shown to antagonize cellular responses to LPA and reduce LPA concentrations in the medium *(18,19)*. In contrast, inhibition of platelet LPP sensitizes their responses to LPA and also increases thrombin-induced LPA production *(20)*. Overall, these data suggest that LPPs play a pivotal role in controlling extracellular levels of LPA and its functions.

1.2. Sphingosine-1-Phosphate (see *Fig. 1*)

Sphingosine, the backbone of all sphingolipids, is converted to S1P intracellularly by sphingosine kinase (SphK)-mediated phosphorylation *(21)*. Sphingosine is not an intermediate in the biosynthesis of sphingolipids, but rather is produced by ceramidase

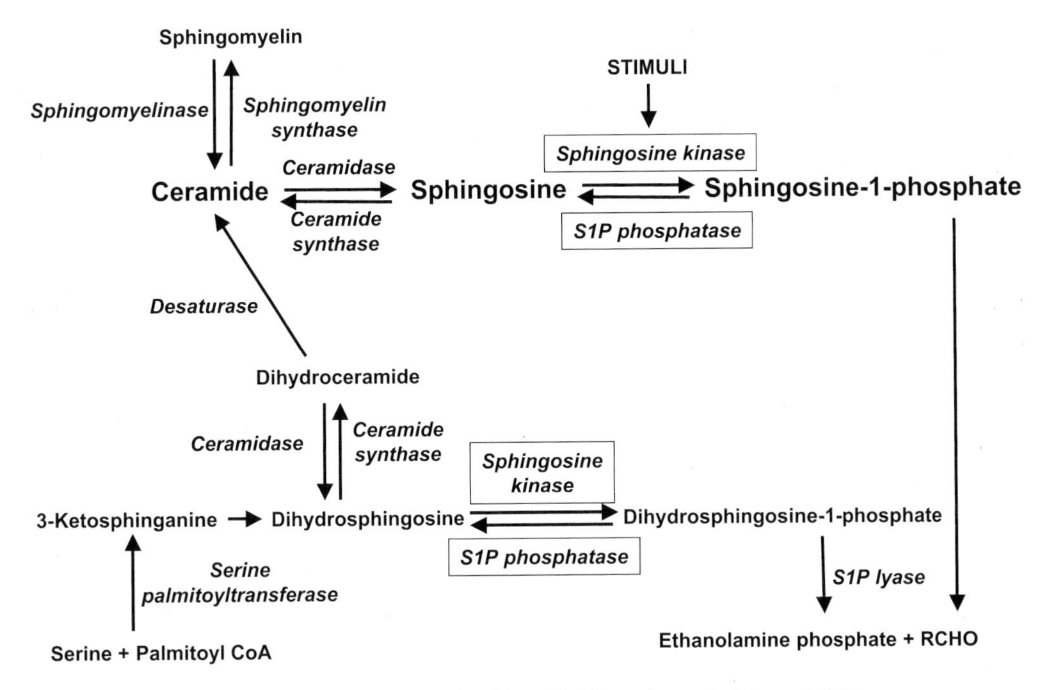

Fig. 3. Biosynthesis of sphingolipids and metabolism of S1P.

catalyzed degradation of ceramide, a fatty acid amide linked sphingoid base, of which sphingosine is the most predominant in mammalian cells. Ceramide is generated by a *de novo* biosynthetic pathway in which the intermediate sphinganine (dihydrosphingosine) is *N*-acylated to dihydroceramide prior to introduction of the 4–5 *trans* double bond characteristic of sphingosine *(22)* (*see* Fig. 3). Several different types of sphingomyelinases, which are activated by diverse cellular stresses, degrade membrane sphingomyelin to also produce ceramide. Intracellular levels of these interconvertible sphingolipid metabolites, ceramide, sphingosine, and S1P, are typically low and are determined by the balance between synthesis and degradation.

S1P can be reversibly degraded to sphingosine by dephosphorylation catalyzed by S1P phosphohydrolases. Two S1P-specific phosphohydrolases, SPP-1 and SPP-2, have been cloned *(23,24)* and both are located in the endoplasmic reticulum, suggesting that they are involved in dephosphorylation of intracellular S1P. An alternative degradation pathway is mediated by S1P lyase, a pyridoxal phosphate-dependent enzyme that cleaves S1P irreversibly to phosphoethanolamine and hexadecenal *(25)*.

Two isoforms of SphK (SphK1 and SphK2) are expressed in mammalian tissues and cells *(26,27)*. Database searches have identified homologous SphK genes in plants, insects, worms, and zebra fish, indicating that the S1P-generating mechanism is conserved throughout evolution *(21)*. There is a rapidly growing list of agonists that have been reported to increase SphK activity, including growth factor receptor tyrosine kinases *(28,29)*, ligands for GPCRs *(30,31)*, cross-linking of immunoglobulin receptors *(32–35)*, the protein kinase C (PKC) activator phorbol ester *(36)*, and inflammatory cytokines, such as tumor necrosis factor (TNF)-α *(37)*. So far, no agonists have been described that affect SphK2 activity. Although SphK2 is highly similar to SphK1 and has the same five conserved domains, it diverges in its central and N-terminal regions, and has a different developmental and tissue expression pattern *(27)*, suggesting that these two isoenzymes have distinct physiological functions. Indeed, whereas SphK1 promotes growth and survival *(38)*, SphK2 suppresses growth and enhances apoptosis *(39,40)*. It is still unclear whether SphK2 has redundant or antagonistic physiological functions to SphK1.

The mechanisms by which extracellular stimuli activate SphK1 and S1P production are beginning to be unraveled. A number of studies have addressed the involvement of intracellular calcium signaling in agonist-induced SphK1 activation *(31,41–43)*. In PDGF-stimulated cells, the phospholipase C (PLC) signaling pathway and calcium mobilization is required for SphK1 activation *(41)*. Consistent with the importance of calcium mobilization, SphK1 contains putative calcium/calmodulin binding domains *(26)*. A clue to the mechanism of SphK1 activation comes from the observation that the majority of SphK1 is found in the cytosol, whereas the hydrophobic substrate sphingosine is generated in membranes *(38,44)*. Growth factors and FcεRI crosslinking induce translocation of SphK1 to plasma membrane regions in various types of cells *(35, 45–47)*. This translocation could increase SphK1 activity, either through direct association with an activating membrane component, such as acidic phospholipids *(48)*, or by bringing it in close proximity to its substrate sphingosine. Although activation of PKC by phorbol ester *(47)* and vascular endothelial growth factor (VEGF) *(49)* induces phosphorylation of SphK1 and translocation to the plasma membrane, PKC cannot phosphorylate SphK1 directly, suggesting the participation of other protein kinases. In agreement, phosphorylation of overexpressed SphK1 on serine 225 by ERK2 induced by TNF-α not only increased its activity but was also necessary for its translocation from the cytosol to the plasma membrane *(50)*. Although a phosphorylation deficient SphK1 mutant retained full catalytic activity, it lost its ability to induce proliferation, form colonies in soft agar, and induce tumor formation in nude mice *(51)*. These results suggest that phosphorylation of SphK1 is essential for oncogenic signaling and is brought about by phosphorylation-induced translocation to the plasma membrane rather than enhanced catalytic activity. Another report suggested that SphK1 is translocated to membrane compartments enriched in phosphatidic acid (PA), produced by PLD. This translocation likely involves direct interaction of SphK1 with PA, as the purified enzyme has been shown to bind tightly to immobilized PA *(52)*.

An unresolved issue in S1P signaling is how S1P exits cells to serve as a ligand for multiple cell surface S1P receptors. The bulk of extracellular S1P seems to be secreted by unidentified trafficking mechanisms *(7)*. It remains unknown whether S1P is also generated extracellularly, for example, by secreted SphK1 *(53)*. In addition, ATX, the enzyme that produces extracellular LPA *(8,9)*, has been reported to be able to hydrolyze sphingosylphosphorylcholine (SPC) to S1P, although the catalytic activity towards SPC is much weaker than with lysophosphatidylcholine (LPC) *(54)*.

2. Signal Transduction of LPA and S1P

Research and interest in lysophospholipid mediators was greatly accelerated by the discovery of their specific cell surface receptors. Since the cloning of the first LPA receptor in 1996 *(55)*, sequence similarities have allowed rapid identification of multiple S1P receptors and additional LPA receptor subtypes. The four LPA receptors identified to date are LPA_1, LPA_2, LPA_3 *(56)*, and the newly identified LPA_4/GPR23

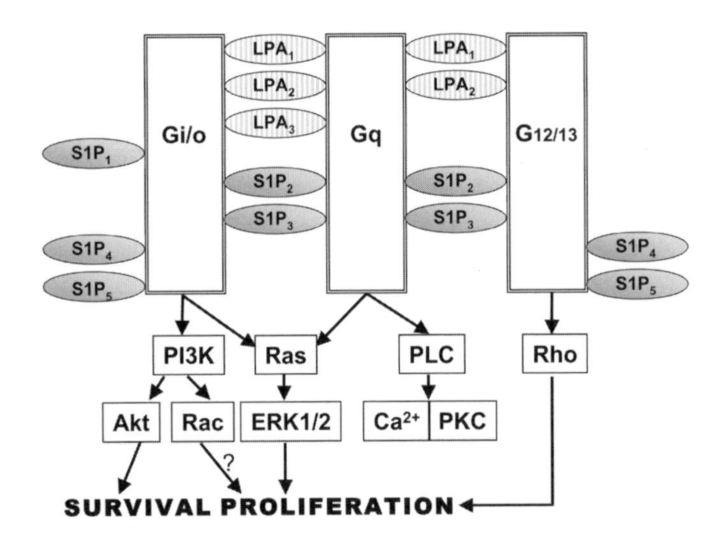

Fig. 4. LPA and S1P signaling important for cell growth and survival. A simplified scheme illustrating coupling of LPA and S1P receptors with different classes of G proteins, leading to activation or inhibition of downstream signaling involved in cell growth and survival.

receptor *(57)*, which is structurally distinct from other three LPA receptors. Whether LPA_4/GPR23 is a bona fide LPA receptor is still controversial. Five cognate GPCRs for S1P ($S1P_{1-5}$) have been identified and their signaling analyzed in detail *(58,59)*. It has recently been claimed that the orphan receptors GPR3, GPR6, GPR12 are authentic S1P receptors *(60)*, but this assignment needs further confirmation.

Whereas some other types of GPCRs show selectivity in interactions with particular G proteins, most of the LPA and S1P receptor subtypes appear to couple to multiple heterotrimeric G proteins (*see* Fig. 4). LPA_1, LPA_2, $S1P_2$, and $S1P_3$ couple to Gi, Gq, and G12/13 *(61)*. LPA_3 interacts with Gi and Gq but not G12/13 *(61)*. $S1P_1$ interacts exclusively with Gi, whereas $S1P_4$ and $S1P_5$ are capable of coupling to both Gi and G12/13 *(61)*.

LPA- or S1P-induced activation of Gi, which is sensitive to pertussis toxin, leads to inhibition of adenylyl cyclase (AC) and decreased cAMP, as well as activation of the Ras/MAPK cascade and the PI3 kinase-Akt pathway mediated by $\beta\gamma$ subunits released from Gi *(61–63)*. Decreased cAMP favors cell growth in many types of cells *(64,65)*. Thus, Gi-dependent inhibition of AC may contribute to proliferative responses to LPA and S1P. The MAPK and PI3K pathways downstream of the $\beta\gamma$ dimer are integral components of the proliferative and survival signaling networks as discussed below.

Gq binds to and activates PLCβ, which, in turn, catalyzes hydrolysis of phosphatidylinositol bisphosphate (PIP2) to diacylglycerol (DAG) and inositol trisphosphate (IP3) *(66)*. DAG activates PKCs and IP3 triggers intracellular calcium mobilization. Although activation of some Gq-coupled receptors or overexpression of constitutively active Gq can result in cell proliferation, activation of PLCβ, Ca^{2+} and PKC signaling pathways are generally insufficient to promote cell proliferation or survival *(63)*.

LPA- or S1P-induced activation of G12/13, which directly binds to the Rho-specific guanine nucleotide exchange factor p115RhoGEF, leads to Rho activation *(67,68)*. Rho signaling was once considered to be solely involved in remodeling of the actin cytoskeleton, ultimately controlling cell morphology and a variety of functions, such as cell motility, aggregation, polarity, and contraction *(69)*. It has become evident that Rho also regulates expression of genes associated with cell proliferation including c-jun, c-fos, and cyclin A *(70)*. S1P induces association of the scaffold protein CNK1 with p115RhoGEF, stimulating the JNK/MAPK cascade and thus controlling signaling specificity downstream of Rho *(71)*. A critical role for Rho in cell-cycle progression is further suggested by its ability to antagonize Ras and suppress expression of the cell-cycle inhibitors p21 and p27 *(72,73)*, thus facilitating entry into the S phase of the cell cycle.

The signaling pathways and ultimate biological consequences vary from one cell type to another depending on the spectrum and expression levels of the receptors and the specific G proteins present in the cell and possible "cross-talk" with signals mediated by other families of cell-surface receptors. In many cell types, treatment with either LPA or S1P can activate growth factor tyrosine kinase (TK) receptors, such as endothelial growth factor receptor (EGFR), platelet derived growth factor receptor (PDGFR), and c-met (the receptor for hepatocyte growth factor), in the absence of added growth factors, indicating that receptor TKs might be activated through intracellular receptor crosstalk (also known as transactivation) *(74–76)*. This cross-communication between different signaling systems is not only important for the growth promoting activity of LPA *(74)*, it also may be a clue to its pathophysiological role in prostate cancer *(75)*, head and neck squamous cell carcinoma *(77)*, and kidney and bladder cancers *(78)*.

Moreover, activation of receptor TKs or cross-linking of FcεRI can induce a reverse type of transactivation of S1P receptors that may be important for effects on cell movement *(35,45,79)*. According to this paradigm, stimulation of PDGFR activates and translocates SphK1 to the plasma membrane. This results in spatially restricted formation of S1P, that, in turn, activates $S1P_1$, resulting in activation and integration of downstream signals, such as the small GTPase Rac, important for protrusion of lamellipodia and forward movement *(45,79)*. However, ligation by S1P will suppress migration towards PDGF when $S1P_2$ is highly expressed *(80)*. These results revealed a role for receptor cross-communication in which activation of S1PRs by a receptor tyrosine kinase is critical for cell motility and shed light on their vital roles in vascular maturation *(81)* and angiogenesis *(82,83)*. Moreover, transactivation of S1PRs by FcεRI triggering is important for normal mast cell degranulation and chemotaxis towards antigen and might be involved in the movement of mast cells to sites of inflammation *(35)*. In addition, differential transactivation of S1PRs by nerve growth factor (NGF) regulates antagonistic signaling pathways that modulate neurite extension and might be important during development and differentiation of the central nervous system *(84)*.

3. LPA and S1P as Growth Factors

Most of the initial studies of biological effects mediated by exogenous LPA and S1P have centered on their mitogenic activity in fibroblasts. Both LPA and S1P, when exogenously applied to quiescent fibroblasts, stimulate cellular DNA synthesis and cell division *(85,86)*. Other related phospholipids, such as PA, LPC, sphingosine, and SPC have been also reported to have variable mitogenic effects on fibroblasts or other cell types *(85,87,88)*. Later studies suggested that the mitogenic activity of these LPA- and

S1P-related phospholipids most likely results from the contamination of the lipid preparations with LPA or S1P or conversion into LPA or S1P.

A common feature of LPA- and S1P-induced mitogenesis is the requirement of relatively high doses (μ*M*) of these ligands for significant mitogenic activity *(85,86,88)*. This requirement differentiates LPA and S1P from peptide growth factors such as EGF or PDGF, where nanomolar concentrations are usually sufficient to trigger pronounced DNA synthesis *(85,89)*. Because of the potential membrane lytic activity of high concentrations of LPA and S1P, it was initially argued that they might trigger cellular DNA synthesis through membrane perturbations. It is now clear that LPA is quickly degraded in culture by membrane-associated LPPs, thus limiting its functional concentrations at cell membranes *(16–19,90)*. A relatively high initial concentration of LPA appears to be critical for maintaining a threshold level of the ligand to elicit DNA synthesis. Recent studies from several groups demonstrated that the dose of LPA required for its functions could be markedly reduced by modifying its structure *(20,91)*. For example, replacement of the phosphate group of LPA with a phosphothionate has produced LPP-resistant LPA analogs with 10- to 100-fold increases in their potency to stimulate cell proliferation *(91)*.

Mitogenic effects of exogenous LPA and S1P have been observed in a wide range of cell types, including epithelial cells, endothelial cells, smooth muscle cells, astrocytes, renal mesangial cells, and various transformed cells *(62,63)*. Although LPA and S1P increase IP_3 formation and mobilize intracellular calcium in these cells, PKC activity or calcium signaling seems neither necessary nor sufficient for LPA- or S1P-induced cell proliferation *(63)*. As demonstrated in numerous systems, mitogenic responses to LPA or S1P are sensitive to pertussis toxin *(85,88,92,93)*, implicating Gi/o in transduction of signals from receptor activation to cell proliferation. An early study suggested that Gi-induced inhibition of AC and reduction of intracellular cAMP can mediate LPA-stimulated proliferation of Rat1 cells *(85)*. However, a general role for Gi inhibition of cAMP in LPA- or S1P-triggered cell proliferation seems unlikely because LPA still acts as a potent mitogen in some cell types, such as Swiss 3T3 cells and epithelial cells, where cAMP is a positive regulator of cell growth *(94,95)*.

Characterization of other signaling pathways downstream of Gi has revealed the importance of the Ras/ERK and PI3K/Akt pathways in LPA- and S1P-induced cell growth *(93,96–98)*. The universal requirement for ERK activation in cell proliferation likely reflects the necessity for the translocation of activated ERK to the nucleus where it enhances transcription of proliferation-associated genes. Of note, LPA induces biphasic activation of ERK in Rat1 cells, with a transient initial signal and subsequent sustained activation *(99)*. The sustained phase of ERK activation, which is sensitive to PTX, correlates well with the nuclear translocation of ERK and the initiation of cell proliferation.

The mechanism of the dependence on PI3K of LPA or S1P-induced cell proliferation is not fully understood. The most prominent downstream effector of PI3K is the serine/threonine kinase Akt, a key player in cell survival. Akt may contribute to cell growth by preventing apoptotic cell death or may directly modulate the mitogenic process. The lipid products of PI3K also activate other signaling molecules in addition to Akt. It was recently demonstrated that Tiam, a Rac-specific guanine nucleotide exchange factor (GEF), lies downstream of PI3K *(100,101)*. Besides its prominent role in lamellipodia protrusion and forward cell movement, Rac may also contribute to cell-cycle progression through its effector p21-activated kinase (PAK), superoxide production, and cyclin D1 expression *(102–104)*.

Which receptor subtypes are most intimately involved in LPA- or S1P-induced cell proliferation has not been fully established. Most cell types express more than one LPA or S1P receptor and LPA and S1P generally stimulate proliferation, making it difficult to link a proliferative response to a specific receptor. In theory, the receptor subtypes that couple to Gi and/or G12/13 may be capable of mediating growth responses to a certain degree. Consistent with its ability to activate Gi and G12/13, ectopic expression of the LPA_1 receptor in B103 neuronal cells sensitized LPA-induced DNA synthesis *(105)*. Similarly, the $S1P_2$ and $S1P_3$ receptors couple to both Gi and G12/13 and have been shown to mediate growth responses to S1P in hepatoma cells *(106)*. However, other LPA and S1P receptors have also been implicated in cell proliferation *(107–109)*.

It should be emphasized that studies of heterologous overexpression only indicate that a specific receptor has the ability to mediate growth responses but do not necessarily implicate them as physiological regulators. The most meaningful observations concerning the relevance of endogenous LPA receptors in cell proliferation were made recently with fibroblasts (MEFs) derived from specific LPA receptor-null mice *(109)*. The mitogenic response to LPA was only partially compromised in MEFs lacking either LPA_1 or LPA_2 *(109)*. However, LPA was not able to stimulate DNA synthesis in MEFs from LPA_1/LPA_2 double knockouts *(109)*, suggesting that both receptors contribute to the mitogenic effects of LPA. Similarly, the mitogenic effect of exogenous S1P was absent in MEFs devoid of functional S1PRs *(110)*. In normal physiological circumstances, the optimal response to LPA and S1P is likely determined by the relative expression of the endogenous receptors.

Interestingly, both LPA and S1P have been reported to be growth-inhibitory for certain cell types. For example, LPA inhibits growth of myeloma cells *(111)*. This antimitogenic action was proposed to be regulated by LPA-induced increases in intracellular cAMP, an antimitogenic signal for these cells. Further, both LPA and S1P inhibit proliferation of cells of epidermal origin. LPA treatment induced growth arrest of keratinocytes, probably acting through a Smad3-dependent mechanism, similar to the signaling pathways utilized by transforming growth factor (TGF)-β (112). Recent studies have shown that S1P induces rapid phosphorylation of Smad3 and its association with Smad4 leading to translocation of the complex into the nucleus and inhibition of keratinocyte growth *(112)*. Both S1PRs and TGFβ1R were essential, indicating a novel type of cross-talk between serine/threonine kinase receptors and S1PRs. Others have suggested however that the antimitogenic action of S1P in these cells is linked to the inhibition of cyclin D2 expression and associated induction of cyclin-dependent kinase inhibitors p21 and p27 *(113)*. Growth inhibitory activity of S1P was also observed in hepatic myofibroblasts *(92)*, rat hepatocytes *(114)*, and T-lymphocytes *(115)*, but the mechanisms involved have not been delineated.

4. LPA and S1P as Survival Factors

Like peptide growth factors, LPA and S1P also act as dual mediators of cell growth and survival. Frankel and Mills first recognized the survival-promoting activity of LPA based on their observation that LPA protected cisplatinum-treated ovarian cancer cells from apoptotic cell death *(116)*. Numerous studies have since been published that confirm the antiapoptotic activity of LPA in normal and transformed cells of diverse origins, including fibroblasts *(93)*, intestinal epithelial cells *(117)*, Schwann cells *(118,119)*, macrophages *(120)*, T-lymphocytes *(121)*, cardiac myocytes *(122)*, osteoblasts *(123)*, renal proximal tubular cells *(124)*, prostate cancer cells *(125)*, and primary chronic lymphocytic leukemia cells *(126)*. LPA also seems to be effective in preventing apoptosis induced by a wide variety of death inducing agents, including chemotherapeutic drugs *(116,117)*, radiation *(117)*, serum-deprivation *(93,124)*, detachment (anoikis) *(4)*, *Fas* ligand *(121,127)*, and activation of the death receptor TNF-related apoptosis-inducing ligand (TRAIL) *(128)*.

LPA administration also has cytoprotective effects in vivo. A study using a renal ischemia reperfusion injury model demonstrated that LPA inhibited apoptosis of tubular epithelial cells in a dose-dependent manner *(129)*. In irradiated mice, oral administration of LPA was shown to decrease the number of apoptotic cells found in intestinal crypts *(117)*. Moreover, rectally applied LPA in rats stimulates intestinal epithelial wound healing and diminishes mucosal damage *(130)*.

However, the underlying mechanisms responsible for this general protective action of LPA manifested in vitro and in vivo have not yet been fully defined. Some of these observations are consistent with a role for Gi-elicited signals in the antiapoptotic responses to LPA. PI3K, the most prominent survival regulator downstream of Gi, was reported to be required for LPA-dependent cell survival of Schwann cells *(118,119)*, macrophages *(120)*, prostate cancer cells *(125)*, and primary chronic lymphocytic leukemia cells *(126)*. Other studies showed synergism between the PI3K pathway and the Gi-linked ERK1/2 pathway *(119,131)*. In murine fibroblasts, however, Gi-mediated ERK1/2 activation plays a dominant role in LPA survival activity, whereas inhibition of PI3K has only minor effects *(93)*. In addition, in Schwann cells, G12/13-mediated activation of Rho GTPase may also participate in the cytoprotective effects of LPA *(132)*.

Other documented antiapoptotic processes mediated by LPA include suppression of proapoptotic proteins Bax in T-cells *(121)* and Bcl-2 in rat intestinal epithelial cells *(133)*, inhibition of caspase activity in ovarian cancer cells *(133,134)*, and prevention of cytochrome C release in intestinal epithelial cells (128). Notably, in many cell types, LPA potently induces activation of NF-κB and NF-κB-mediated transcription *(107,125,135–137)*. In light of the generally accepted prosurvival role for NF-κB, its activation by LPA may represent another signaling event related to the survival effect of LPA.

Interestingly, there are reports that in some types of cells, exogenous LPA treatment can promote apoptosis. For example, LPA enhances PMA-induced apoptosis of erythroblastic TF-1 cells by interfering with their adhesion *(138)*. In other cell types such as cortical neurons and airway smooth muscle cells *(139)*, low concentrations of LPA signal survival and/or mitogenesis whereas high levels of LPA lead to apoptosis.

It is now well accepted that S1P enhances survival and protects cells against ceramide-mediated cell death *(140–142)*. This antiapoptotic activity was dependent on activation of ERK1/2. S1P-mediated protection against apoptosis, like that of LPA, appears to be conferred primarily through Gi signaling cascades downstream of S1PRs. In a number of cell types, Gi-mediated activation of the PI3K-Akt cascade plays a primary role in the protective action of S1P *(123,128,143)*. In endothelial cells and melanoma cells, PI3K-Akt has been further linked to the regulation of expression or functional status of Bcl-2 family members such as Bax, Bcl-2, Bad, and Bim *(144,145)*. Gi-mediated activation of the ERK/MAPK pathway has been also found to be instrumental to S1P-enhanced cell survival *(106)*. For example, activation of ERK1/2 and concomitant suppression of JNK are required for S1P-mediated protection of murine fibroblasts from ceramide-induced apoptosis *(38,140,142)*. Similarly, activation of ERK1/2 by S1P protects melanocytes from UVB-mediated apoptosis *(146)*. Like LPA, S1P activates NF-κB *(147)*, which may also be a contributing factor to the cytoprotective effects of S1P.

5. S1P as an Intracellular Mediator of Growth and Survival

As discussed earlier, S1P is mainly generated intracellularly via the action of SphKs. In addition to the well-characterized receptor-dependent functionalities of S1P, several lines of evidence indicate that intracellularly produced S1P can also act within the cell, independent of S1P receptors. The biological effects of S1P, particularly its growth and survival-promoting activity, can be recapitulated by overexpression of SphK1 and intracellular S1P accumulation in the absence of detectable cell surface S1PRs or when specific S1PRs or downstream signaling are pharmacologically inhibited *(110)*. However, in the same cell, migratory responses to growth factors were dependent on the presence and function of S1P receptors, indicating pleiotropic cellular effects of S1P are mediated through both intracellular and extracellular routes *(110)*. Consistent with this postulated intracellular action, S1P was initially recognized as an intracellular regulator of cell growth. Furthermore, stimulation of mammalian cells with many growth factors, such as PDGF or VEGF, triggers activation of SphK1 and the subsequent accumulation of intracellular S1P. Substantial evidence exists that the intracellular production of S1P is essential for PDGF-induced cell proliferation and survival *(28,41,110,148)*. It was recently found that VEGF can also stimulate SphK1 and this has ramifications for its mitogenicity *(49)*. VEGF stimulation of PKC and consequent activation of SphK1 resulted in the conversion of sphingosine to S1P, and concomitant inhibition of Ras GTPase-activating proteins *(149)*, leading to activation of the ERK21/2 pathway and cell proliferation *(49)*. Several other lines of evidence further support the notion that S1P produced in response to growth factors also has an intracellular action independent of S1PR engagement: (1) dihydro-S1P, which is identical to S1P and only lacks the 4,5-*trans* double bond, binds to all of the S1PRs and activates them, yet does not mimic the effects of S1P on cell survival *(37,45,141,150)*; (2) microinjection of S1P, which elevates intracellular S1P, has been shown to mobilize calcium *(42)* and enhance proliferation and survival *(141,150,151)*; (3) SphK1 mediates VEGF-induced activation of Ras without involving S1PRs *(49)*; (4) yeast and plants do not possess S1PRs, yet phosphorylated long-chain sphingoid bases regulate environmental stress responses *(152,153)* and other important biological responses *(154,155)*. In contrast to the well-established receptor-mediated processes, the mechanisms for these intracellular actions of S1P are poorly understood. Identification of the protein targets or downstream effectors of S1P within the cell will provide insights into the mechanistic steps leading to S1P-induced cell proliferation and survival.

Whether LPA similar to S1P also has intracellular actions is not well established, as substantial evidence indicates that LPA acts essentially from outside of the cell through its cognate receptors and microinjection of LPA does not elicit similar cellular responses *(156)*. However, the mitogenic responses of some mammalian cells to LPA may be $LPA_{1\text{-}4}$ independent *(17)*. It has been suggested that LPA has a novel intracellular function as a high-affinity ligand for peroxisome proliferator-activated receptor (PPAR)γ *(157)*, a transcription factor that regulates genes controlling energy metabolism *(158)* and can exacerbate mammary gland tumor development *(159)*. Indeed, LPA induced proliferation of mast cells was attenuated by GW9662, a selective antagonist of PPARγ *(160)*. Moreover, LPA activates PPARγ in yeast which do not express LPARs *(158)*.

6. LPA and S1P in Health and Disease

Both LPA and S1P are secreted by activated platelets and hematopoietic cells at the platelet-endothelial cell interface. In humans, serum LPA levels increase within 1 to 24 h of blood clotting from approx 1 to 5–6 μ*M* *(161)*. S1P is abundantly stored within platelets which have high levels of SphK activity and low levels of S1P lyase activity *(162,163)*. S1P concentration in serum ranges from approx 0.5 to 0.8 μ*M* *(163)*. In vivo, LPA is rapidly and locally produced in response to injury, inflammation and stress conditions *(1)*. A number of studies in animals have shown that LPA promotes wound healing and tissue regeneration. For example, topically applied LPA promotes wound healing in skin *(164)*, whereas rectally administered LPA accelerates intestinal wound healing and reduces mucosal damages *(130)*. These protective effects of LPA have been ascribed largely to LPA's growth and survival-promoting activity towards epithelial and mesenchymal cells. In addition, LPA stimulates migration and contraction of these cells *(165,166)*, processes critical for normal wound healing.

Characterization of $S1P_1$ knockout mice has revealed a crucial role for S1P in angiogenesis and vascular maturation during development. The $S1P_1$ receptor null mice exhibit embryonic lethality at embryonic day (E)12.5–E14.5 resulting from defects in vascular smooth muscle migration *(81)*. Consistent with the phenotypes of the $S1P_1$ knockout mice, S1P stimulates proliferation, migration and differentiation of endothelial and smooth muscle cells in culture. In an in vitro wound healing assay, S1P stimulated closure of wounded monolayers of human umbilical vein endothelial cells and adult bovine aortic endothelial cells *(167)*. Thus S1P may contribute to vascular remodeling during tissue regeneration.

Both LPA and S1P have been implicated in the pathogenesis of malignant diseases. They potentially participate in multiple processes associated with cancer development and metastatic progression, including cell proliferation, evasion of apoptosis and chemotherapeutic killing, and tumor-cell migration and invasion. Accumulated evidence supports the notion that in many types of cancers, LPA or S1P production is deregulated and/or the expression of their corresponding receptors is upregulated, leading to aberrant signaling. In this regard, the role of LPA in ovarian cancer has been studied most extensively. High concentrations of multiple forms of LPA are found in ascites from ovarian cancer patients *(89,168)*.

As ascites fluid represents the in vivo microenvironment of tumor cells, high levels of LPA could affect the growth, survival, motility, and fate of ovarian tumor cells. Importantly, we have recently found that the LPA_2 and LPA_3 receptors are overexpressed in more than 50% of primary ovarian cancers and in established ovarian cancer cell lines compared with normal ovarian epithelial cells where only the LPA_1 receptor is present *(4,107)*. LPA potently induces ovarian cancer cell proliferation, survival and expression of oncogenic proteins, including uPA, VEGF, IL-6, and IL-8, whereas these responses in normal ovarian epithelial cells are absent or limited *(107,169–171)*. It is tempting to speculate that the acquired expression of LPA_2 and LPA_3 may be responsible for the differential responses to LPA of normal and malignant ovarian epithelial cells *(107,170)*.

In addition to ovarian cancer, overexpression of specific LPA receptors has been also found in colon cancer and colorectal cancer *(172,173)*, differentiated thyroid cancer *(174)*, and invasive ductal carcinomas *(175)*, indicating that amplification of LPA signaling through overexpression of LPARs may be a common oncogenic process. A recent study demonstrated that LPA_{1-3} are also expressed by primary breast cancer cells *(176)*. Ectopic overexpression of LPA_1 in MDA-BO2 breast cancer cells resulted in enhanced potential of these cells to metastasize to the bone in nude mice *(176)*. In contrast to ovarian cancer cells that may synthesize LPA, MDA-BO2 cells do not produce or secrete LPA. Instead, LPA was released from platelets that were aggregated and activated around the tumor cells. Therefore, the platelet antagonist Integrilin effectively blocked bone metastasis induced by overexpression of the LPA_1 receptor in MDA-BO2 cells *(176)*. This study implicates LPA as an endogenous factor in the regulation of metastatic progression of breast cancer.

A critical role for S1P signaling in oncogenesis is evidenced by the observation of the oncogenic effect of the S1P-producing enzyme SphK1 *(177)*. NIH 3T3 cells overexpressing SphK1 acquired the transformed phenotype as determined by focus formation, anchorage-independent growth in soft agar, and tumorigenicity in NOD/SCID mice *(51,177)*. Cellular SphK1 activity was found to underlie oncogenic Ras-induced transformation of NIH 3T3 cells *(177)*. We also demonstrated that SphK1 over-expression in estrogen-dependent MCF-7 breast cancer cells led to a more tumorigenic phenotype when implanted in nude mice *(178)*. Moreover, expression of a dominant-negative mutant SphK1 profoundly inhibited estrogen-mediated ERK1/2 activation and neoplastic cell growth *(179)*. These results suggest that SphK1 activation is important in transduction of estrogen-dependent mitogenic and carcinogenic action in human breast cancer cells. The ability of S1P to act in an autocrine or paracrine manner, combined with its actions on angiogenesis and vascular maturation that are also critical for tumor progression, suggest that SphK1 may not only protect tumors from apoptosis and enhance growth, it may also increase their vascularization. Because inhibitors of SphKs, such as *N,N*-dimethylsphingosine (DMS) and L-threo-dihydrosphingosine (known as safingol), induce apoptosis regardless of multidrug resistance expression *(180)*, they may provide a new strategy for treatment of anticancer drug-resistant cancers. In pilot clinical phase I trials with safingol, which also potentiated the tumor-inhibiting effect of doxorubicin in tumor-bearing animals, it was found that safingol can be given safely with doxorubicin *(181)*. Indeed, SphK inhibitors reduce gastric tumor growth *(182)* and mammary adenocarcinoma tumor growth in mice *(183)*. Because SphK1 is overexpressed in a variety of solid tumors, including breast, stomach, ovary, kidney, and lung, compared with normal tissues from the same patients *(183)*, anticancer therapeutics targeting this SphK isozyme might be useful adjuncts for management of many types of cancer.

As discussed above, S1P signaling through the $S1P_1$ receptor is required for stabilization of nascent blood vessels during embryonic development. Besides its growth- and survival-promoting activity, S1P may mediate tumor angiogenesis which is critical for the growth of tumors beyond a diffusion-limited size and for metastasis *(184)*. A recent study showed that $S1P_1$ expression is strongly induced in vessels within tumor xenografts *(185)*. Blocking its expression on neovessels with siRNA suppressed tumor angiogenesis and tumor growth in nude mice. These results provide support the involvement of S1P in tumor angiogenesis, further indication that S1P signaling deserves consideration as a therapeutic anticancer target.

As prototypic ligands of GPCRs, LPA and S1P have been the subjects of extensive research. Their potential roles in pathogenesis of human cancer are only beginning to be elucidated. Although the earliest work that implicated LPA and S1P in cancer focused on only a few cancer models, such as ovarian and breast cancer cells, more evidence is accumulating to support a general role for LPA and S1P in the development and progression of human malignancies. This can be explained by the fact that both LPA and S1P are general growth and survival factors with a broad range of target tissues and cell types. The relevance of LPA and S1P to abnormal growth control in cancers is further highlighted by the defects or abnormalities in LPA and S1P metabolism and their receptor expression/functions found in multiple types of human cancers. LPA and S1P (as small lipids) and their GPCRs are highly "drugable" targets. In fact, more than half of all drugs in current use target GPCRs. Future studies will no doubt uncover novel therapies intervening in production and/or receptor signal transduction pathways of LPA and S1P.

Acknowledgments

We apologize to those authors whose work could not be cited owing to space limitations. This work was supported by National Institutes of Health grants CA102196 (XF), GM43880 (SS), CA61774 (SS), and AI50094 (SS), and the Department of Defense grants W81XH-04-1-0103 (XF) and PC10392 (SS). The authors thank Dr. Sheldon Milstien for helpful comments.

References

1. Pages C, Simon MF, Valet P, Saulnier-Blache JS. Lysophosphatidic acid synthesis and release. Prostaglandins Other Lipid Mediat 2001;64:1–10.
2. van Dijk MC, Postma F, Hilkmann H, Jalink K, van Blitterswijk WJ, Moolenaar WH. Exogenous phospholipase D generates lysophosphatidic acid and activates Ras, Rho and Ca^{2+} signaling pathways. Curr Biol 1998;8:386–392.
3. Eder AM, Sasagawa T, Mao M, Aoki J, Mills GB. Constitutive and lysophosphatidic acid (LPA)-induced LPA production: role of phospholipase D and phospholipase A2. Clin Cancer Res 2000;6:2482–2491.
4. Fang X, Schummer M, Mao M, et al. Lysophosphatidic acid is a bioactive mediator in ovarian cancer. Biochim Biophys Acta 2002;1582:257–264.
5. Gesta S, Simon MF, Rey A, et al. Secretion of a lysophospholipase D activity by adipocytes: involvement in lysophosphatidic acid synthesis. J Lipid Res 2002;43:904–910.
6. Xie Y, Gibbs TC, Mukhin YV, Meier KE. Role for 18:1 lysophosphatidic acid as an autocrine mediator in prostate cancer cells. J Biol Chem 2002;277:32,516–32,526.

7. Sano T, Baker D, Virag T, et al. Multiple mechanisms linked to platelet activation result in lysophosphatidic acid and sphingosine 1-phosphate generation in blood. J Biol Chem 2002;277:21,197–21,206.
8. Umezu-Goto M, Kishi Y, Taira A, et al. Autotaxin has lysophospholipase D activity leading to tumor cell growth and motility by lysophosphatidic acid production. J Cell Biol 2002;158:227–233.
9. Tokumura A, Majima E, Kariya Y, et al. Identification of human plasma lysophospholipase D, a lysophosphatidic acid-producing enzyme, as autotaxin, a multifunctional phosphodiesterase. J Biol Chem 2002;277:39,436–39,442.
10. Stracke ML, Clair T, Liotta LA. Autotaxin, tumor motility-stimulating exophosphodiesterase. Adv Enzyme Regul 1997;37:135–144.
11. Imamura F, Horai T, Mukai M, Shinkai K, Sawada M, Akedo H. Induction of in vitro tumor cell invasion of cellular monolayers by lysophosphatidic acid or phospholipase D. Biochem Biophys Res Commun 1993;193:497–503.
12. Shim YH, Lin CH, Strickland KP. The purification and properties of monoacylglycerol kinase from bovine brain. Biochem Cell Biol 1989;67:233–241.
13. Bektas M, Payne SG, Liu H, Milstien S, Spiegel S. A novel acylglycerol kinase that produces lysophosphatidic acid modulates crosstalk with EGFR in prostate cancer cells. J Cell Biol 2005;169:801–811.
14. Reutens AT, Begley CG. Endophilin-1: a multifunctional protein. Int J Biochem Cell Biol 2002;34:1173–1177.
15. Thompson FJ, Clark MA. Purification of a lysophosphatidic acid-hydrolysing lysophospholipase from rat brain. Biochem J 1994;300:457–461.
16. Roberts R, Sciorra VA, Morris AJ. Human type 2 phosphatidic acid phosphohydrolases. Substrate specificity of the type 2a, 2b, and 2c enzymes and cell surface activity of the 2a isoform. J Biol Chem 1998;273:22,059–22,067.
17. Hooks SB, Santos WL, Im DS, Heise CE, Macdonald TL, Lynch KR. Lysophosphatidic acid induced mitogenesis is regulated by lipid phosphate phosphatases and is Edg-receptor independent. J Biol Chem 2001;276:4611–4621.
18. Sciorra VA, Morris AJ. Roles for lipid phosphate phosphatases in regulation of cellular signaling. Biochim Biophys Acta 2002;1582:45–51.
19. Brindley DN. Lipid phosphate phosphatases and related proteins: Signaling functions in development, cell division, and cancer. J Cell Biochem 2004;92:900–912.
20. Smyth SS, Sciorra VA, Sigal YJ, et al. Lipid phosphate phosphatases regulate lysophosphatidic acid production and signaling in platelets: studies using chemical inhibitors of lipid phosphate phosphatase activity. J Biol Chem 2003;278:43,214–43,223.
21. Maceyka M, Payne SG, Milstien S, Spiegel S. Sphingosine kinase, sphingosine-1-phosphate, and apoptosis. Biochim Biophys Acta 2002;1585:193–201.
22. Merrill AH, Jr. De novo sphingolipid biosynthesis: a necessary, but dangerous, pathway. J Biol Chem 2002;277:25,843–25,846.
23. Le Stunff H, Peterson C, Thornton R, Milstien S, Mandala SM, Spiegel S. Characterization of murine sphingosine-1-phosphate phosphohydrolase. J Biol Chem 2002;277:8920–8917.
24. Ogawa C, Kihara A, Gokoh M, Igarashi Y. Identification and characterization of a novel human sphingosine-1-phosphate phosphohydrolase, hSPP2. J Biol Chem 2003;278: 1268–1272.
25. Pyne S, Pyne NJ. Sphingosine 1-phosphate signalling in mammalian cells. Biochem J 2000;349:385–402.
26. Kohama T, Olivera A, Edsall L, Nagiec MM, Dickson R, Spiegel S. Molecular cloning and functional characterization of murine sphingosine kinase. J Biol Chem 1998;273: 23,722–23,728.

27. Liu H, Sugiura M, Nava VE, et al. Molecular cloning and functional characterization of a novel mammalian sphingosine kinase type 2 isoform. J Biol Chem 2000;275:19,513–19,520.
28. Olivera A, Spiegel S. Sphingosine-1-phosphate as a second messenger in cell proliferation induced by PDGF and FCS mitogens. Nature 1993;365:557–560.
29. Edsall LC, Pirianov GG, Spiegel S. Involvement of sphingosine 1-phosphate in nerve growth factor-mediated neuronal survival and differentiation. J Neurosci 1997;17:6952–6960.
30. Meyer zu Heringdorf D, Lass H, Alemany R, et al. Sphingosine kinase-mediated Ca^{2+} signalling by G-protein-coupled receptors. EMBO J 1998;17:2830–2837.
31. Alemany R, Sichelschmidt B, zu Heringdorf DM, Lass H, van Koppen CJ, Jakobs KH. Stimulation of sphingosine-1-phosphate formation by the P2Y(2) receptor in HL-60 cells: Ca(2+) requirement and implication in receptor-mediated Ca(2+) mobilization, but not MAP kinase activation. Mol Pharmacol 2000;58:491–497.
32. Melendez A, Floto RA, Gillooly DJ, Harnett MM, Allen JM. FcgRI coupling to phospholipase D initiates sphingosine kinase-mediated calcium mobilization and vesicular trafficking. J Biol Chem 1998;273:9393–9402.
33. Chuang FY, Sassaroli M, Unkeless JC. Convergence of Fc gamma receptor IIA and Fc gamma receptor IIIB signaling pathways in human neutrophils. J Immunol 2000;164: 350–360.
34. Prieschl EE, Csonga R, Novotny V, Kikuchi GE, Baumruker T. The balance between sphingosine and sphingosine-1-phosphate is decisive for mast cell activation after Fc epsilon receptor I triggering. J Exp Med 1999;190:1–8.
35. Jolly PS, Bektas M, Olivera A, et al. Transactivation of sphingosine-1-phosphate receptors by Fc(epsilon)RI triggering is required for normal mast cell degranulation and chemotaxis. J Exp Med 2004;199:959–970.
36. Mazurek N, Megidish T, Hakomori S-I, Igarashi Y. Regulatory effect of phorbol esters on sphingosine kinase in BALB/C 3T3 fibroblasts (variant A31): demonstration of cell type-specific response. Biochem Biophys Res Commun 1994;198:1–9.
37. Xia P, Gamble JR, Rye KA, et al. Tumor necrosis factor-a induces adhesion molecule expression through the sphingosine kinase pathway. Proc Natl Acad Sci USA 1998;95: 14,196–14,201.
38. Olivera A, Kohama T, Edsall LC, et al. Sphingosine kinase expression increases intracellular sphingosine-1-phosphate and promotes cell growth and survival. J Cell Biol 1999;147: 545–558.
39. Liu H, Toman RE, Goparaju S, et al. Sphingosine kinase type 2 is a putative BH3-Only protein that induces apoptosis. J Biol Chem 2003;278:40,330–40,336.
40. Igarashi N, Okada T, Hayashi S, Fujita T, Jahangeer S, Nakamura SI. Sphingosine kinase 2 is a nuclear protein and inhibits DNA synthesis. J Biol Chem 2003;278:46,832–46,839.
41. Olivera A, Edsall L, Poulton S, Kazlauskas A, Spiegel S. Platelet-derived growth factor-induced activation of sphingosine kinase requires phosphorylation of the PDGF receptor tyrosine residue responsible for binding of PLCgamma. FASEB J 1999;13: 1593–1600.
42. van Koppen CJ, Meyer zu Heringdorf D, Alemany R, Jakobs KH. Sphingosine kinase-mediated calcium signaling by muscarinic acetylcholine receptors. Life Sci 2001;68: 2535–2540.
43. Young KW, Willets JM, Parkinson MJ, et al. Ca^{2+}/calmodulin-dependent translocation of sphingosine kinase: role in plasma membrane relocation but not activation. Cell Calcium 2003;33:119–128.
44. Gijsbers S, Van der Hoeven G, Van Veldhoven PP. Subcellular study of sphingoid base phosphorylation in rat tissues: evidence for multiple sphingosine kinases. Biochim Biophys Acta 2001;1532:37–50.

45. Rosenfeldt HM, Hobson JP, Maceyka M, et al. EDG-1 links the PDGF receptor to Src and focal adhesion kinase activation leading to lamellipodia formation and cell migration. FASEB J 2001;15:2649–2659.
46. Melendez AJ, Ibrahim FB. Antisense knockdown of sphingosine kinase 1 in human macrophages inhibits c5a receptor-dependent signal transduction, Ca2+ signals, enzyme release, cytokine production, and chemotaxis. J Immunol 2004;173:1596–1603.
47. Johnson KR, Becker KP, Facchinetti MM, Hannun YA, Obeid LM. PKC-dependent activation of sphingosine kinase 1 and translocation to the plasma membrane. Extracellular release of sphingosine-1-phosphate induced by phorbol 12-myristate 13-acetate (PMA). J Biol Chem 2002;277:35,257–35,262.
48. Olivera A, Rosenthal J, Spiegel S. Effect of acidic phospholipids on sphingosine kinase. J Cell Biochem 1996;60:529–537.
49. Shu X, Wu W, Mosteller RD, Broek D. Sphingosine kinase mediates vascular endothelial growth factor-induced activation of ras and mitogen-activated protein kinases. Mol Cell Biol 2002;22:7758–7768.
50. Pitson SM, Moretti PA, Zebol JR, et al. Activation of sphingosine kinase 1 by ERK1/2-mediated phosphorylation. EMBO J 2003;22:5491–5500.
51. Pitson SM, Xia P, Leclercq TM, et al. Phosphorylation-dependent translocation of sphingosine kinase to the plasma membrane drives its oncogenic signalling. J Exp Med 2005; 201:49–54.
52. Delon C, Manifava M, Wood E, et al. Sphingosine kinase 1 is an intracellular effector of phosphatidic acid. J Biol Chem 2004;279:44,763–44,774.
53. Ancellin N, Colmont C, Su J, et al. Extracellular export of sphingosine kinase-1 enzyme: Sphingosine 1-phosphate generation and the induction of angiogenic vascular maturation. J Biol Chem 2002;277:6667–6675.
54. Clair T, Aoki J, Koh E, et al. Autotaxin hydrolyzes sphingosylphosphorylcholine to produce the regulator of migration, sphingosine-1-phosphate. Cancer Res 2003;63:5446–5453.
55. Hecht JH, Weiner JA, Post SR, Chun J. Ventricular zone gene-1 (vzg-1) encodes a lysophosphatidic acid receptor expressed in neurogenic regions of the developing cerebral cortex. J Cell Biol 1996;135:1071–1083.
56. Chun J, Goetzl EJ, Hla T, et al. International Union of Pharmacology. XXXIV. Lysophospholipid Receptor Nomenclature. Pharmacol Rev 2002;54:265–269.
57. Noguchi K, Ishii S, Shimizu T. Identification of p2y9/GPR23 as a novel G protein-coupled receptor for lysophosphatidic acid, structurally distant from the Edg family. J Biol Chem 2003;278:25,600–25,606.
58. Hla T, Lee MJ, Ancellin N, Paik JH, Kluk MJ. Lysophospholipids-receptor revelations. Science 2001;294:1875–1878.
59. Spiegel S, Milstien S. Sphingosine 1-phosphate, a key cell signaling molecule. J Biol Chem 2002;277:25,851–25,854.
60. Kostenis E. Novel clusters of receptors for sphingosine-1-phosphate, sphingosylphosphorylcholine, and (lyso)-phosphatidic acid: New receptors for "Old" ligands. J Cell Biochem 2004;92:923–936.
61. Anliker B, Chun J. Lysophospholipid G protein-coupled receptors. J Biol Chem 2004;279: 20,555–20,558.
62. Van Leeuwen FN, Olivo C, Grivell S, Giepmans BN, Collard JG, Moolenaar WH. Rac activation by lysophosphatidic acid LPA1 receptors through the guanine nucleotide exchange factor Tiam1. J Biol Chem 2003;278:400–406.
63. Radeff-Huang J, Seasholtz TM, Matteo RG, Brown JH. G protein mediated signaling pathways in lysophospholipid induced cell proliferation and survival. J Cell Biochem 2004; 92:949–966.

64. Heldin NE, Paulsson Y, Forsberg K, Heldin CH, Westermark B. Induction of cyclic AMP synthesis by forskolin is followed by a reduction in the expression of c-myc messenger RNA and inhibition of 3H-thymidine incorporation in human fibroblasts. J Cell Physiol 1989;138:17–23.
65. Magnaldo I, Pouyssegur, Paris S. Cyclic AMP inhibits mitogen-induced DNA synthesis in hamster fibroblasts, regardless of the signalling pathway involved. FEBS Lett 1989;245: 65–69.
66. Fukushima N, Chun J. The LPA receptors. Prostaglandins 2001;64:21–32.
67. Kranenburg O, Poland M, van Horck FP, Drechsel D, Hall A, Moolenaar WH. Activation of RhoA by lysophosphatidic acid and Galpha12/13 subunits in neuronal cells: induction of neurite retraction. Mol Biol Cell 1999;10:1851–1857.
68. Ren XD, Kiosses WB, Schwartz MA. Regulation of the small GTP-binding protein Rho by cell adhesion and the cytoskeleton. EMBO J 1999;18:578–585.
69. Bar-Sagi D, Hall A. Ras and Rho GTPases: a family reunion. Cell 2000;103:227–238.
70. Marinissen MJ, Chiariello M, Tanos T, Bernard O, Narumiya S, Gutkind JS. The small GTP-binding protein RhoA regulates c-jun by a ROCK-JNK signaling axis. Mol Cell 2004;14:29–41.
71. Jaffe AB, Hall A, Schmidt A. Association of CNK1 with Rho guanine nucleotide exchange factors controls signaling specificity downstream of Rho. Curr Biol 2005;15:405–412.
72. Olson MF, Paterson HF, Marshall CJ. Signals from Ras and Rho GTPases interact to regulate expression of p21Waf1/Cip1. Nature 1998;394:295–299.
73. Pruitt K, Der CJ. Ras and Rho regulation of the cell cycle and oncogenesis. Cancer Lett 2001;171:1–10.
74. Daub H, Weiss FU, Wallasch C, Ullrich A. Role of transactivation of the EGF receptor in signalling by G-protein-coupled receptors. Nature 1996;379:557–560.
75. Prenzel N, Zwick E, Daub H, et al. EGF receptor transactivation by G-protein-coupled receptors requires metalloproteinase cleavage of proHB-EGF. Nature 1999;402:884–888.
76. Baudhuin LM, Jiang Y, Zaslavsky A, Ishii I, Chun J, Xu Y. S1P3-mediated Akt activation and cross-talk with platelet-derived growth factor receptor (PDGFR). FASEB J 2004;18: 341–343.
77. Gschwind A, Prenzel N, Ullrich A. Lysophosphatidic acid-induced squamous cell carcinoma cell proliferation and motility involves epidermal growth factor receptor signal transactivation. Cancer Res 2002;62:6329–6336.
78. Schafer B, Gschwind A, Ullrich A. Multiple G-protein-coupled receptor signals converge on the epidermal growth factor receptor to promote migration and invasion. Oncogene 2004;23:991–999.
79. Hobson JP, Rosenfeldt HM, Barak LS, et al. Role of the sphingosine-1-phosphate receptor EDG-1 in PDGF-induced cell motility. Science 2001;291:1800–1803.
80. Goparaju K, Jolly PS, Watterson KR, et al. The S1P2 Receptor Negatively Regulates PDGF-Induced Motility and Proliferation. Mol Cell Biol 2005;25:4237–4249.
81. Liu Y, Wada R, Yamashita T, et al. Edg-1, the G protein-coupled receptor for sphingosine-1-phosphate, is essential for vascular maturation. J Clin Invest 2000;106:951–961.
82. Lee MJ, Thangada S, Claffey KP, et al. Vascular endothelial cell adherens junction assembly and morphogenesis induced by sphingosine-1-phosphate. Cell 1999;99:301–312.
83. Wang F, Van Brocklyn JR, Hobson JP, et al. Sphingosine 1-phosphate stimulates cell migration through a G(i)- coupled cell surface receptor. Potential involvement in angiogenesis. J Biol Chem 1999;274:35,343–35,350.
84. Toman RE, Payne SG, Watterson K, et al. Differential transactivation of sphingosine-1-phosphate receptors modulates nerve gowth factor-induced neurite extension. J Cell Biol 2004;166:381–392.

85. Van Corven EJ, Groenink A, Jalink K, Eicholtz T, Moolenaar WH. Lysophosphatidate-induced cell proliferation: Identification and dissection of signaling pathways mediated by G proteins. Cell 1989;59:45–54.
86. Zhang H, Desai NN, Olivera A, Seki T, Brooker G, Spiegel S. Sphingosine-1-phosphate, a novel lipid, involved in cellular proliferation. J Cell Biol 1991;114:155–167.
87. Desai NN, Spiegel S. Sphingosylphosphorylcholine is a remarkably potent mitogen for a variety of cell lines. Biochem Biophys Res Commun 1991;181:361–366.
88. van Corven EJ, van Rijswijk A, Jalink K, van der Bend RL, van Blitterswijk WJ, Moolenaar WH. Mitogenic action of lysophosphatidic acid and phosphatidic acid on fibroblasts. Dependence on acyl-chain length and inhibition by suramin. Biochem J 1992;281:163–169.
89. Fang X, Gaudette D, Furui T, et al. Lysophospholipid growth factors in the initiation, progression, metastases, and management of ovarian cancer. Ann NY Acad Sci 2000;905: 188–208.
90. Tanyi JL, Morris AJ, Wolf JK, et al. The human lipid phosphate phosphatase-3 decreases the growth, survival, and tumorigenesis of ovarian cancer cells: validation of the lysophosphatidic acid signaling cascade as a target for therapy in ovarian cancer. Cancer Res 2003;63: 1073–1082.
91. Hasegawa Y, Erickson JR, Goddard GJ, et al. Identification of a phosphothionate analogue of lysophosphatidic acid (LPA) as a selective agonist of the LPA3 receptor. J Biol Chem 2003;278:11,962–11,969.
92. Pebay A, Toutant M, Premont J, et al. Sphingosine-1-phosphate induces proliferation of astrocytes: regulation by intracellular signalling cascades. Eur J Neurosci 2001;13: 2067–2076.
93. Fang X, Yu S, LaPushin R, et al. Lysophosphatidic acid prevents apoptosis in fibroblasts via G(i)-protein-mediated activation of mitogen-activated protein kinase. Biochem J 2000;352:135–143.
94. Rozengurt E. Early signals in the mitogenic response. Science 1986;234:161–166.
95. Dumont JE, Jauniaux JC, Roger PP. The cyclic AMP-mediated stimulation of cell proliferation. Trends Biochem Sci 1989;14:67–71.
96. Cook SJ, McCormick F. Kinetic and biochemical correlation between sustained p44ERK1 (44 kDa extracellular signal-regulated kinase 1) activation and lysophosphatidic acid-stimulated DNA synthesis in Rat-1 cells. Biochem J 1996;320:237–245.
97. Van Brocklyn JR, Behbahani B, Lee NH. Homodimerization and heterodimerization of S1P/EDG sphingosine-1-phosphate receptors. Biochim Biophys Acta 2002;1582:89–93.
98. Kranenburg O, Moolenaar WH. Ras-MAP kinase signaling by lysophosphatidic acid and other G protein-coupled receptor agonists. Oncogene 2001;20:1540–1546.
99. Cook SJ, McCormick F. Inhibition by cAMP of Ras-dependent activation of Raf. Science 1993;262:1069–1072.
100. Fleming IN, Batty IH, Prescott AR, et al. Inositol phospholipids regulate the guanine-nucleotide-exchange factor Tiam1 by facilitating its binding to the plasma membrane and regulating GDP/GTP exchange on Rac1. Biochem J 2004;382:857–865.
101. Fleming IN, Gray A, Downes CP. Regulation of the Rac1-specific exchange factor Tiam1 involves both phosphoinositide 3-kinase-dependent and -independent components. Biochem J 2000;351:173–182.
102. Joyce D, Bouzahzah B, Fu M, et al. Integration of Rac-dependent regulation of cyclin D1 transcription through a nuclear factor-kappaB-dependent pathway. J Biol Chem 1999;274: 25,245–25,249.
103. Welsh CF, Roovers K, Villanueva J, Liu Y, Schwartz MA, Assoian RK. Timing of cyclin D1 expression within G1 phase is controlled by Rho. Nat Cell Biol 2001;3:950–957.
104. Babior BM. NADPH oxidase. Curr Opin Immunol 2004;16:42–47.

105. Fukushima N, Kimura Y, Chun J. A single receptor encoded by vzg-1/lpA1/edg-2 couples to G proteins and mediates multiple cellular responses to lysophosphatidic acid. Proc Natl Acad Sci USA 1998;95:6151–6156.
106. An S, Zheng Y, Bleu T. Sphingosine 1-phosphate-induced cell proliferation, survival, and related signaling events mediated by G protein-coupled receptors Edg3 and Edg5. J Biol Chem 2000;275:288–296.
107. Fang X, Yu S, Bast RC, et al. Mechanisms for lysophosphatidic acid-induced cytokine production in ovarian cancer cells. J Biol Chem 2004;279:9653–9661.
108. Katsuma S, Hada Y, Ueda T, et al. Signalling mechanisms in sphingosine 1-phosphate-promoted mesangial cell proliferation. Genes Cells 2002;7:1217–1230.
109. Contos JJ, Ishii I, Fukushima N, et al. Characterization of lpa(2) (Edg4) and lpa(1)/lpa(2) (Edg2/Edg4) lysophosphatidic acid receptor knockout mice: signaling deficits without obvious phenotypic abnormality attributable to lpa(2). Mol Cell Biol 2002;22: 6921–6929.
110. Olivera A, Rosenfeldt HM, Bektas M, et al. Sphingosine kinase type 1 Induces G12/13-mediated stress fiber formation yet promotes growth and survival independent of G protein coupled receptors. J Biol Chem 2003;278:46,452–46,460.
111. Tigyi G, Dyer DL, Miledi R. Lysophosphatidic acid possesses dual action in cell proliferation. Proc Natl Acad Sci USA 1994;91:1908–1912.
112. Sauer B, Vogler R, von Wenckstern H, et al. Involvement of Smad signaling in sphingosine 1-phosphate-mediated biological responses of keratinocytes. J Biol Chem 2004;279: 38,471–38,479.
113. Kim DS, Kim SY, Kleuser B, Schafer-Korting M, Kim KH, Park KC. Sphingosine-1-phosphate inhibits human keratinocyte proliferation via Akt/protein kinase B inactivation. Cell Signal 2004;16:89–95.
114. Ikeda H, Satoh H, Yanase M, et al. Antiproliferative property of sphingosine 1-phosphate in rat hepatocytes involves activation of Rho via Edg-5. Gastroenterology 2003;124: 459–469.
115. Jin Y, Knudsen E, Wang L, et al. Sphingosine 1-phosphate is a novel inhibitor of T-cell proliferation. Blood 2003;101:4909–4915.
116. Frankel A, Mills GB. Peptide and lipid growth factors decrease cis-diamminedichloroplatinum-induced cell death in human ovarian cancer cells. Clin Cancer Res 1996;2: 1307–1313.
117. Deng W, Balazs L, Wang DA, Van Middlesworth L, Tigyi G, Johnson LR. Lysophosphatidic acid protects and rescues intestinal epithelial cells from radiation- and chemotherapy-induced apoptosis. Gastroenterology 2002;123:206–216.
118. Weiner JA, Chun J. Schwann cell survival mediated by the signaling phospholipid lysophosphatidic acid. Proc Natl Acad Sci USA 1999;96:5233–5238.
119. Li Y, Gonzalez MI, Meinkoth JL, Field J, Kazanietz MG, Tennekoon GI. Lysophosphatidic acid promotes survival and differentiation of rat Schwann cells. J Biol Chem 2003;278: 9585–9591.
120. Koh JS, Lieberthal W, Heydrick S, Levine JS. Lysophosphatidic acid is a major serum non-cytokine survival factor for murine macrophages which acts via the phosphatidylinositol 3-kinase signaling pathway. J Clin Invest 1998;102:716–727.
121. Goetzl EJ, Kong Y, Mei B. Lysophosphatidic acid and sphingosine 1-phosphate protection of T cells from apoptosis in association with suppression of Bax. J Immunol 1999;162: 2049–2056.
122. Karliner JS, Honbo N, Summers K, Gray MO, Goetzl EJ. The lysophospholipids sphingosine-1-phosphate and lysophosphatidic acid enhance survival during hypoxia in neonatal rat cardiac myocytes. J Mol Cell Cardiol 2001;33:1713–1717.

123. Grey A, Chen Q, Callon K, Xu X, Reid IR, Cornish J. The phospholipids sphingosine-1-phosphate and lysophosphatidic acid prevent apoptosis in osteoblastic cells via a signaling pathway involving G(i) proteins and phosphatidylinositol-3 kinase. Endocrinology 2002; 143:4755–4763.
124. Levine JS, Koh JS, Triaca V, Lieberthal W. Lysophosphatidic acid: a novel growth and survival factor for renal proximal tubular cells. Am J Physiol 1997;273:F575–F585.
125. Raj GV, Sekula JA, Guo R, Madden JF, Daaka Y. Lysophosphatidic acid promotes survival of androgen-insensitive prostate cancer PC3 cells via activation of NF-kappaB. Prostate 2004;61:105–113.
126. Hu X, Haney N, Kropp D, Kabore AF, Johnston JB, Gibson SB. Lysophosphatidic acid (LPA) protects primary chronic lymphocytic leukemia cells from apoptosis through LPA receptor activation of the anti-apoptotic protein AKT/PKB. J Biol Chem 2005;280: 9498–9508.
127. Meng Y, Graves L, Do TV, So J, Fishman DA. Upregulation of FasL by LPA on ovarian cancer cell surface leads to apoptosis of activated lymphocytes. Gynecol Oncol 2004;95: 488–495.
128. Kang YC, Kim KM, Lee KS, et al. Serum bioactive lysophospholipids prevent TRAIL-induced apoptosis via PI3K/Akt-dependent cFLIP expression and Bad phosphorylation. Cell Death Differ 2004;11:1287–1298.
129. de Vries B, Matthijsen RA, van Bijnen AA, Wolfs TG, Buurman WA. Lysophosphatidic acid prevents renal ischemia-reperfusion injury by inhibition of apoptosis and complement activation. Am J Pathol 2003;163:47–56.
130. Sturm A, Dignass AU. Modulation of gastrointestinal wound repair and inflammation by phospholipids. Biochim Biophys Acta 2002;1582:282–288.
131. Sautin YY, Crawford JM, Svetlov SI. Enhancement of survival by LPA via Erk1/Erk2 and PI 3-kinase/Akt pathways in a murine hepatocyte cell line. Am J Physiol Cell Physiol 2001;281:C2010–C2019.
132. Weiner JA, Fukushima N, Contos JJ, Scherer SS, Chun J. Regulation of Schwann cell morphology and adhesion by receptor-mediated lysophosphatidic acid signaling. J Neurosci 2001;21:7069–7078.
133. Deng W, Wang DA, Gosmanova E, Johnson LR, Tigyi G. LPA protects intestinal epithelial cells from apoptosis by inhibiting the mitochondrial pathway. Am J Physiol Gastrointest Liver Physiol 2003;284:G821–G829.
134. Baudhuin LM, Cristina KL, Lu J, Xu Y. Akt activation induced by lysophosphatidic acid and sphingosine-1-phosphate requires both mitogen-activated protein kinase kinase and p38 mitogen-activated protein kinase and is cell-line specific. Mol Pharmacol 2002;62: 660–671.
135. Shahrestanifar M, Fan X, Manning DR. Lysophosphatidic acid activates NF-kappaB in fibroblasts. A requirement for multiple inputs. J Biol Chem 1999;274:3828–3833.
136. Palmetshofer A, Robson SC, Nehls V. Lysophosphatidic acid activates nuclear factor kappa B and induces proinflammatory gene expression in endothelial cells. Thromb Haemost 1999;82:1532–1537.
137. Lee H, Lin CI, Liao JJ, et al. Lysophospholipids increase ICAM-1 expression in HUVEC through a Gi- and NF-kappaB-dependent mechanism. Am J Physiol Cell Physiol 2004;287:C1657–1666.
138. Lai JM, Lu CY, Yang-Yen HF, Chang ZF. Lysophosphatidic acid promotes phorbol-ester-induced apoptosis in TF-1 cells by interfering with adhesion. Biochem J 2001;359: 227–233.
139. Ediger TL, Toews ML. Synergistic stimulation of airway smooth muscle cell mitogenesis. J Pharmacol Exp Ther 2000;294:1076–1082.

140. Cuvillier O, Pirianov G, Kleuser B, et al. Suppression of ceramide-mediated programmed cell death by sphingosine-1-phosphate. Nature 1996;381:800–803.
141. Morita Y, Perez GI, Paris F, et al. Oocyte apoptosis is suppressed by disruption of the acid sphingomyelinase gene or by sphingosine-1-phosphate therapy. Nature Med 2000;6: 1109–1114.
142. Edsall LC, Cuvillier O, Twitty S, Spiegel S, Milstien S. Sphingosine kinase expression regulates apoptosis and caspase activation in PC12 cells. J Neurochem 2001;76:1573–1584.
143. Morales-Ruiz M, Lee MJ, Zollner S, et al. Sphingosine 1-phosphate activates Akt, nitric oxide production, and chemotaxis through a Gi protein/phosphoinositide 3-kinase pathway in endothelial cells. J Biol Chem 2001;276:19,672–19,677.
144. Limaye VS, Li X, Hahn C, et al. Sphingosine kinase-1 enhances endothelial cell survival through a PECAM-1-dependent activation of PI-3K/Akt and regulation of Bcl-2 family members. Blood 2005;105:3169–3177.
145. Bektas M, Jolly PS, Muller C, Eberle J, Spiegel S, Geilen CC. Sphingosine kinase activity counteracts ceramide-mediated cell death in human melanoma cells: role of Bcl-2 expression. Oncogene 2005;24:178–187.
146. Kim DS, Kim SY, Lee JE, et al. Sphingosine-1-phosphate-induced ERK activation protects human melanocytes from UVB-induced apoptosis. Arch Pharm Res 2003;26:739–746.
147. Siehler S, Wang Y, Fan X, Windh RT, Manning DR. Sphingosine 1-phosphate activates nuclear factor-kappa B through Edg receptors. Activation through Edg-3 and Edg-5, but not Edg-1, in human embryonic kidney 293 cells. J Biol Chem 2001;276:48,733–48,739.
148. Rani CS, Berger A, Wu J, et al. Divergence in signal transduction pathways of PDGF and EGF receptors: Involvement of sphingosine-1-phosphate in PDGF but not EGF signaling. J Biol Chem 1997;272:10,777–10,783.
149. Wu W, Shu X, Hovsepyan H, Mosteller RD, Broek D. VEGF receptor expression and signaling in human bladder tumors. Oncogene 2003;22:3361–3370.
150. Van Brocklyn JR, Lee MJ, Menzeleev R, et al. Dual actions of sphingosine-1-phosphate: extracellular through the G_i-coupled orphan receptor edg-1 and intracellular to regulate proliferation and survival. J Cell Biol 1998;142:229–240.
151. Paris F, Perez GI, Fuks Z, et al. Sphingosine 1-phosphate preserves fertility in irradiated female mice without propagating genomic damage in offspring. Nature Med 2002;8: 901–902.
152. Mandala SM, Thornton R, Tu Z, et al. Sphingoid base 1-phosphate phosphatase: a key regulator of sphingolipid metabolism and stress response. Proc Natl Acad Sci USA 1998; 95:150–155.
153. Mao C, Saba JD, Obeid LM. The dihydrosphingosine-1-phosphate phosphatases of *Saccharomyces cerevisiae* are important regulators of cell proliferation and heat stress responses. Biochem J 1999;342:667–675.
154. Ng CK, Carr K, McAinsh MR, Powell B, Hetherington AM. Drought-induced guard cell signal transduction involves sphingosine-1-phosphate. Nature 2001;410:596–599.
155. Coursol S, Fan LM, Le Stunff H, Spiegel S, Gilroy S, Assmann SM. Sphingolipid signalling in Arabidopsis guard cells involves heterotrimeric G proteins. Nature 2003;423:651–654.
156. Fernhout BJ, Dijcks FA, Moolenaar WH, Ruigt GS. Lysophosphatidic acid induces inward currents in Xenopus laevis oocytes; evidence for an extracellular site of action. Eur J Pharmacol 1992;213:313–315.
157. Zhang C, Baker DL, Yasuda S, et al. Lysophosphatidic acid induces neointima formation through PPARγ activation. J Exp Med 2004;199:763–774.
158. McIntyre TM, Pontsler AV, Silva AR, et al. Identification of an intracellular receptor for lysophosphatidic acid (LPA): LPA is a transcellular PPARgamma agonist. Proc Natl Acad Sci USA 2003;100:131–136.

159. Saez E, Rosenfeld J, Livolsi A, et al. PPAR gamma signaling exacerbates mammary gland tumor development. Genes Dev 2004;18:528–540.
160. Bagga S, Price KS, Lin DA, Friend DS, Austen KF, Boyce JA. Lysophosphatidic acid accelerates the development of human mast cells. Blood 2004;104:4080–4087.
161. Baker DL, Desiderio DM, Miller DD, Tolley B, Tigyi GJ. Direct quantitative analysis of lysophosphatidic acid molecular species by stable isotope dilution electrospray ionization liquid chromatography-mass spectrometry. Anal Biochem 2001;292:287–295.
162. Yatomi Y, Ozaki Y, Ohmori T, Igarashi Y. Sphingosine 1-phosphate: synthesis and release. Prostaglandins Other Lipid Mediat 2001;64:107–122.
163. Okajima F. Plasma lipoproteins behave as carriers of extracellular sphingosine 1- phosphate: is this an atherogenic mediator or an anti-atherogenic mediator? Biochim Biophys Acta 2002;1582:132–137.
164. Balazs L, Okolicany J, Ferrebee M, Tolley B, Tigyi G. Topical application of the phospholipid growth factor lysophosphatidic acid promotes wound healing in vivo. Am J Physiol Regul Integr Comp Physiol 2001,280:R466–R472.
165. Abe M, Ho CH, Kamm KE, Grinnell F. Different molecular motors mediate platelet-derived growth factor and lysophosphatidic acid-stimulated floating collagen matrix contraction. J Biol Chem 2003;278:47,707–47,712.
166. van Leeuwen FN, Giepmans BN, van Meeteren LA, Moolenaar WH. Lysophosphatidic acid: mitogen and motility factor. Biochem Soc Trans 2003;31:1209–1212.
167. Lee H, Goetzl EJ, An S. Lysophosphatidic acid and sphingosine 1-phosphate stimulate endothelial cell wound healing. Am J Physiol Cell Physiol 2000;278:C612–C618.
168. Xu Y, Gaudette DC, Boynton JD, et al. Characterization of an ovarian cancer activating factor in ascites from ovarian cancer patients. Clin Cancer Res 1995;1:1223–1232.
169. Pustilnik TB, Estrella V, Wiener JR, et al. Lysophosphatidic acid induces urokinase secretion by ovarian cancer cells. Clin Cancer Res 1999;5:3704–3710.
170. Hu YL, Tee MK, Goetzl EJ, et al. Lysophosphatidic acid induction of vascular endothelial growth factor expression in human ovarian cancer cells. J Natl Cancer Inst 2001;93:762–768.
171. Schwartz BM, Hong G, Morrison BH, et al. Lysophospholipids increase interleukin-8 expression in ovarian cancer cells. Gynecol Oncol 2001;81:291–300.
172. Shida D, Watanabe T, Aoki J, et al. Aberrant expression of lysophosphatidic acid (LPA) receptors in human colorectal cancer. Lab Invest 2004;84:1352–1362.
173. Yun CC, Sun H, Wang D, et al. The LPA2 receptor mediates mitogenic signals in human colon cancer cells. Am J Physiol Cell Physiol 2005;289:C2–C11.
174. Schulte KM, Beyer A, Kohrer K, Oberhauser S, Roher HD. Lysophosphatidic acid, a novel lipid growth factor for human thyroid cells: over-expression of the high-affinity receptor edg4 in differentiated thyroid cancer. Int J Cancer 2001;92:249–256.
175. Kitayama J, Shida D, Sako A, et al. Over-expression of lysophosphatidic acid receptor-2 in human invasive ductal carcinoma. Breast Cancer Res 2004;6:R640–R646.
176. Boucharaba A, Serre CM, Gres S, et al. Platelet-derived lysophosphatidic acid supports the progression of osteolytic bone metastases in breast cancer. J Clin Invest 2004;114: 1714–1725.
177. Xia P, Gamble JR, Wang L, et al. An oncogenic role of sphingosine kinase. Curr Biol 2000; 10:1527–1530.
178. Nava VE, Hobson JP, Murthy S, Milstien S, Spiegel S. Sphingosine kinase type 1 promotes estrogen-dependent tumorigenesis of breast cancer MCF-7 cells. Exp Cell Res 2002;281: 115–127.
179. Sukocheva OA, Wang L, Albanese N, Pitson SM, Vadas MA, Xia P. Sphingosine kinase transmits estrogen signaling in human breast cancer cells. Mol Endocrinol 2003;17: 2002–2012.

180. Jendiroba DB, Klostergaard J, Keyhani A, Pagliaro L, Freireich EJ. Effective cytotoxicity against human leukemias and chemotherapy- resistant leukemia cell lines by N-N-dimethylsphingosine. Leuk Res 2002;26:301–310.
181. Schwartz GK, Ward D, Saltz L, et al. A pilot clinical/pharmacological study of the protein kinase C-specific inhibitor safingol alone and in combination with doxorubicin. Clin Cancer Res 1997;3:537–543.
182. Endo K, Igarashi Y, Nisar M, Zhou QH, Hakomori S. Cell membrane signaling as target in cancer therapy: inhibitory effect of N,N-dimethyl and N,N,N-trimethyl sphingosine derivatives on in vitro and in vivo growth of human tumor cells in nude mice. Cancer Res 1991;51: 1613–1618.
183. French KJ, Schrecengost RS, Lee BD, et al. Discovery and evaluation of inhibitors of human sphingosine kinase. Cancer Res 2003;63:5962–5969.
184. Risau W. Mechanisms of angiogenesis. Nature 1997;386:671–674.
185. Chae SS, Paik JH, Furneaux H, Hla T. Requirement for sphingosine 1-phosphate receptor-1 in tumor angiogenesis demonstrated by in vivo RNA interference. J Clin Invest 2004;114: 1082–1089.

8

Alternative Use of Signaling by the βGBP Cytokine in Cell Growth and Cancer Control

From Surveillance to Therapy

Livio Mallucci and Valerie Wells

Summary

βGBP is an antiproliferative cytokine produced by activated T-cells and endogenously released by somatic cells. In normal cells, βGBP negatively regulates the cell cycle; in cancer cells βGBP induces apoptosis. Mechanisms of action involve downregulation of PI3 kinase via p110 targeting. Downregulation of PI3 kinase by βGBP reflects on Ras-ERK signaling and Akt mediated signaling resulting in the delay of S/G2 transition in normal cells and activation of programmed cell death in cancer cells. The apoptotic response of cancer cells to βGBP takes place according to the molecular context of the cells. Where βGBP induces cell-cycle arrest, apoptosis relates to E2F-1 deregulation. In cancer cells characterized by strong mitogenic input and elevated Akt/PKB expression, loss of Akt protein deprives the cells of survival signaling. βGBP is an immunomolecule which can perform therapeutically what the native, endogenous protein would naturally perform in a surveillance role.

Key Words: Cancer therapy; apoptosis; E2F-1; PI3 kinase; Akt/PKB; βGBP; β-galactoside binding protein.

1. Signaling and Targeting

Within the context of cell proliferation, differentiation, and apoptosis, the fate of a cell is strictly determined by molecular events where an integrated cascade of signals connects molecules with phenotypes. Positive and negative regulatory cascades control transcriptional events and the dynamics of cell-cycle transitions *(1–3)*. Orderly cell replication depends on the balanced and time-programmed expression of protooncogenes, tumor suppressor genes, and associated events. Deregulation of this balance in favor of proliferation is a prime step toward tumorigenesis, where progression to malignancy depends on increased reliance on survival signaling as increasing mitogenic input—a consequence of multiple mutations—requires compensatory suppression of proapoptotic processes in order to allow uncontrolled expansion *(4–6)*. Alterations of the Ras-ERK and the PI3K/Akt pathways are key to successful tumorigenesis *(6–8)*. Knowledge of the integrated system of signals that extends from the cell surface to the nuclear transcriptional apparatus is therefore a necessary requirement to put into operation rational therapies that might negate tumor growth *(5)*. This system is governed at the initiation level by mitogens, growth factors, and by negative growth factors, which together determine the expression of effector molecules (which are basically the same in both normal and cancer cells but are expressed differently in measure and time in cancer cells). Exploiting the differential dependence on growth and survival signaling

From: *Apoptosis, Cell Signaling, and Human Diseases: Molecular Mechanisms, Volume 1*
Edited by R. Srivastava © Humana Press Inc., Totowa, NJ

between cancer and normal cells would create therapeutic opportunities to selectively eliminate cancers. Several strategies have been developed based on the identification of molecular lesions that are responsible for deregulated cell replication or molecular lesions that foster cell survival *(9–18)*. Drugs designed to affect identified targets, an approach emphasized within the past decade, are mostly based on inhibitors of the Ras-Raf-MEK-ERK pathway *(4,16,17,19–22)* and inhibitors of the PI3K/Akt pathway *(23–25)*, but it is now clear that these efforts have not been rewarded by satisfactory clinical success. A main obstacle in many cancers is the multiplicity of genetic lesions that, through different routes, promote replicative growth and protect cancer cells from apoptotic death, thus hampering the effectiveness of drugs directed at single targets. Other likely impediments are the ability of intracellular signaling to integrate and bypass individual blocks, the possible molecular promiscuity of the targeting compound, lack of cell specificity, and the toxicity implicit to the nature of the cell permeable low-molecular-weight pharmacological inhibitors used in these studies. Contrary to this scenario there is suggestion, based both on theoretical discrepancy between incidence of cancer and calculation of mutation probability *(26)* and on experimental evidence, that in the healthy organism a shift to hyperproliferation can be resolved by the activation of a molecular program which leads to the death of the aberrant cell (*see* Subheading 4.2.). The exploitation of the differential dependence on apoptosis between cells with deregulated growth and normal cells in the healthy organism can only be the prerogative of a naturally occurring molecule whose physiological function is to control cell replication by inducing regulatory changes to which normal and cancer cells respond differently *(26)*. This property, which implies surveillance, suggests that such a molecule could be used within a therapeutic context. We have identified one such molecule in the β-galactoside binding protein (βGBP) *(27)*, which we have cloned and found to be an antiproliferative cytokine *(28)* which negatively controls the cell cycle in normal cells while selectively activating apoptosis in cancer cells.

2. General Properties of βGBP

Produced by $CD8^+$ and $CD4^+$ activated T-cells (*see* Fig. 1) βGBP is a cytokine *(28)* with an important role in the silencing phase of T-cell immune response and in the switching off of T-lymphocyte effector functions *(29)*. βGBP is also released by somatic cells where its physiological autocrine function is to negatively control the cell cycle from S phase to G2/M transition *(27)*. βGBP is a 15-kDa monomeric globular molecule whose structural gene (Lgals 1) *(30)* is located in the sis/PDGFB homology region (mu-chromosome 15 E/Hu-chromosome 22 q12–q13) *(31)*, a syntenic group which undergoes deletions and translocations in a number of human tumors *(32–35)*. Previously classified as a lectin (galectin1), βGBP has no lectin properties. It has only one saccharide-binding site located in the region spanning from residue 64 to residue 76. At concentrations of about $10^{-4}M$ the βGBP molecules can be reversibly held in pairs by hydrophobic interactions becoming transient opportunistic dimers *(36)*, but in its native state βGBP is a monomer which exerts its function as a monomer. Hu-r-βGBP exerts its functions as a negative cell-cycle regulator and as an anticancer agent at concentrations of about $1–20 \times 10^{-9}M$ *(26,27,37–40)*, several orders of magnitude lower than the transient dimeric form. Hu-r-βGBP binds with high affinity (Kd $1.5 \times 10^{-10}M$) to about 5×10^4 cell surface specific receptors/cell through molecular domains other than those that link saccharide determinants *(27)*. This is demonstrated by sugar competition studies *(27,29)*, by a native form of the molecule where

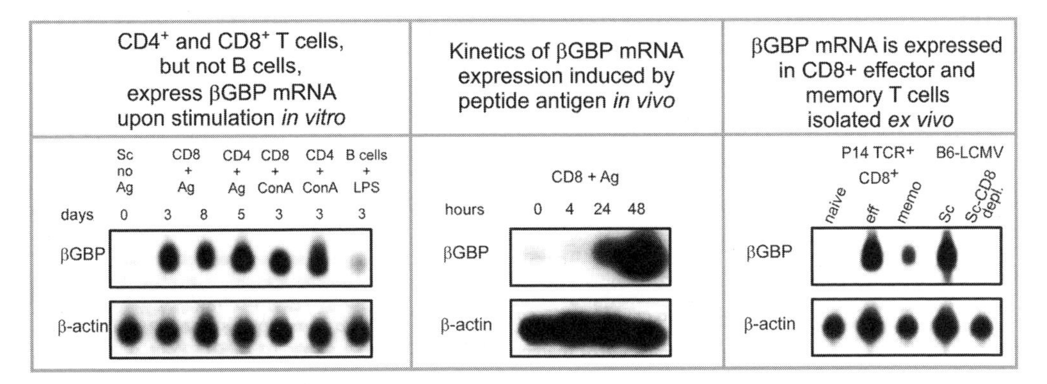

Fig. 1. Northern blot analysis of βGBP mRNA expression. *Left*: CD8+ and CD4+ T-and B-cells activated in vitro. *Lane 1*: spleen cells from naive C57BL/6 mice; *lanes 2* and *3*: CD8+ T-cells from P14 TCR-transgenic mice after peptide stimulation; *lane 4*: CD4+ T-cells from 2B4 TCR-transgenic mice 5 d after peptide stimulation; *lanes 5* and *6*: CD8+ and CD4+ T-cells from C57BL/6 mice 3 d after Con A activation; *lane 7*: B-cells activated with LPS. *Center*: CD8+ T-cells after peptide antigen treatment in vivo. P14 TCR-transgenic mice injected with 500 ug LCMV GP 33 peptide ip. *Lane 1*: no peptide; *lanes 2–4*: 4, 24, and 48 h after peptide treatment. *Right*: CD8+ T-cells ex vivo from P14 TCR-transgenic mice. *Lane 1*: uninfected; *lane 2*: 8 d after LCMV infection; *lane 3*: 4 wk after LCMV infection; *lane 4*: spleen cells from 8-d LCMV immune C57Bl/6 mice; *lane 5*: CD8+ T-cell-depleted spleen cells. Reprinted with permission from ref. *28*.

the saccharide-binding region is masked by a glycan complex and by the internalization of the saccharide-binding site when βGBP molecules are linked in a tetrameric complex *(41)*. These are all conditions which have no effect on the strictly dose related antiproliferative efficacy of the βGBP molecule which, via receptor interaction, negatively regulates the cell cycle in normal cells and selectively activate apoptosis in cancer cells *(26,27,37–40)*. Though undoubtedly originating within the lectin family, βGBP has evolved to become a cytokine making the term galectin a misnomer.

3. Role in Cell-Cycle Regulation

In normal cells, βGBP operates within the signaling context simplified in Fig. 2 to activate an S-phase checkpoint (*see* Fig. 3) which imposes a statutory cell-cycle pause before commitment to division. Downregulation of Ras-GTP loading by Hu-r-βGBP and consequent downregulation of the kinase sequence downstream of Ras in primary mouse embryo fibroblasts (MEF) results in reduction of cyclin-A levels, reduction of cyclin A-kinase activity, reduction of E2F function, and S/G2 cell-cycle arrest *(42)*. Negation of the Hu-r-βGBP by a neutralizing monoclonal antibody reverses S-phase arrest and promotes cell-cycle completion *(27)*. The importance of βGBP in the regulation of S/G2 transition is made evident by the fact that neutralization of the endogenous, constitutively released βGBP accelerates transition into G2/M, highlighting the existence of a programmed cell-cycle pause which may permit cells to correct errors which may cause genome instability and growth disorders. In addition to controlling an S-phase checkpoint, βGBP can hold cells in G0 by inhibiting the activation of the Ras-ERK module without affecting tyrosine kinase (TK) receptor phosphorylation. An analysis of the events involved shows that whereas Ras-GTP loading is inhibited in the presence of βGBP, Grb-2 recruitment to the receptor, Grb-2-SOS association and the ability of SOS to activate Ras in vitro are

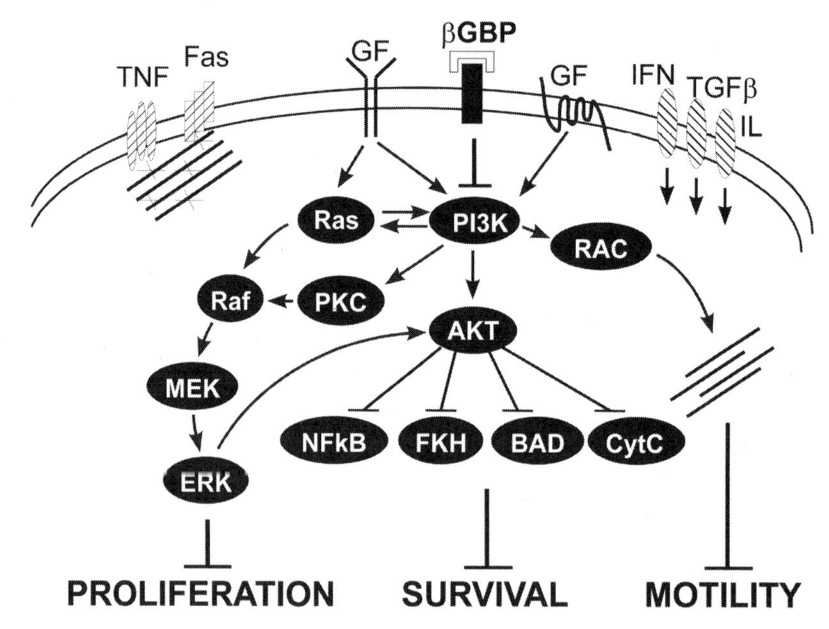

Fig. 2. Schematic model of principal signaling pathways initiating at the cell surface by ligand/receptor interaction and the resulting inhibitory effects of PI3K inhibition by βGBP induced signaling. Abbr: GF, growth factors; IFN, interferon α/β; TGF-β, transforming growth factor β; TNF, tumor necrosis factor; IL, interleukins.

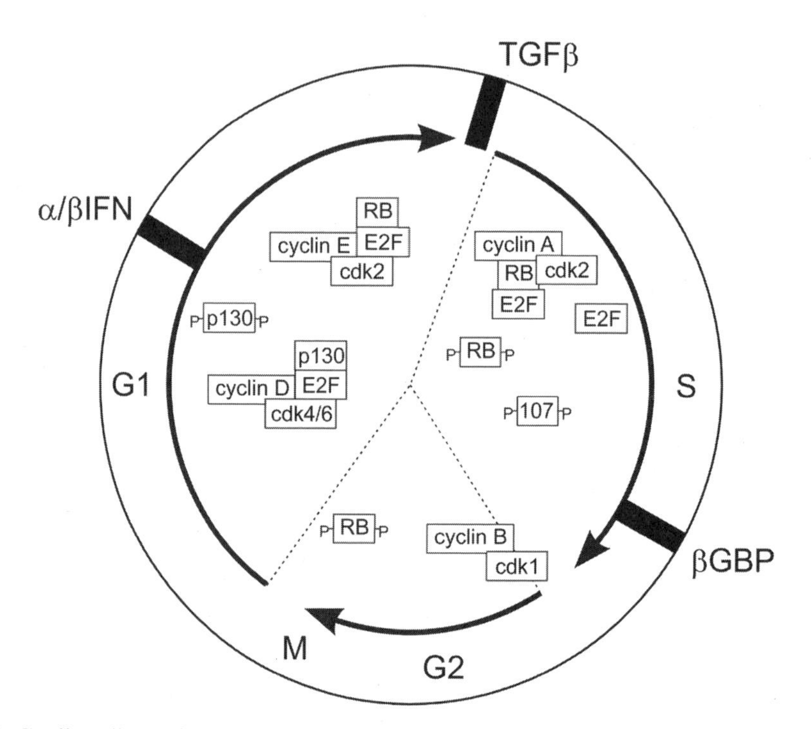

Fig. 3. Cyclin-cdks and E2F-RB/p107 associations in the course of regulatory events which control cell-cycle transitions and location of checkpoints (black bars) activated by IFN α/β *(96)*, TGF-β *(97,98)*, and βGBP *(27)*.

maintained. Also, Ras-GAP activity is not altered. Instead, lack of Ras-GTP loading in the cells treated with Hu-r-βGBP is a consequence of the inactivation of the p110 catalytic unit of phosphoinositide 3-OH kinase (PI3K) *(43)*. These responses make PI3K a prime target of βGBP induced signaling and position PI3K upstream of Ras.

4. Pathways to Apoptosis—Two Molecular Phenotypes

4.1. Molecular Phenotype I

The ability of βGBP to arrest normal cells in S phase, albeit reversibly because normal cells resume growth, extends to a number of tumor cells which, unlike normal cells where cell arrest is a temporary event, undergo programmed cell death. The mechanisms that can activate apoptosis in cancer cells are multifold. Apoptosis can be induced through the modulation of existing apoptosis regulatory proteins and through transcription and translation dependent changes to which normal and cancer cells respond differently *(44–50)*. The apoptotic response induced by βGBP conforms to two main molecular phenotypes. In the case of cancer cells arrested in S phase, ectopic transactivation of the E2F1 transcription factor, when highly expressed, has emerged as one mechanistic aspect *(26,51–56)*. E2F1 activity is required for entry into S phase *(57–60)*; E2F1 downregulation is necessary for S-phase progression and successful cell replication *(54,61)*. Transfection with E2F1 mutants defective in cyclin-A binding leads to S-phase arrest, continuous E2F1 transactivation and apoptosis *(53,54)*. E2F1 can promote apoptosis through p53 dependent and independent pathways *(62,63)* (*see* Fig. 4) and can force caspase enzymes to levels that may increase the probability that death inducing signals overcome the ability of inhibitor of apoptosis proteins (IAP) to impede the activation of programmed cell death *(64)*. The importance of E2F1 in the mediation of programmed cell death in cancer was indicated by the finding that loss of E2F1 function by dominant negative mutants in MBL 100 breast cancer cells inhibits apoptosis and induces tumors in severe combined immunodeficiency (SCID) mice *(65)*. By the enforcement of its regulatory functions, βGBP downregulates cyclin A-kinase activity, arrests cells in S phase, shifts E2F1 to a persistent transactivation mode, and activates apoptosis *(26,37–39)*. Cancer cells which respond to βGBP according to this pattern: molecular phenotype I (*see* Fig. 5) are cells where the ErbB2 oncoprotein receptor is expressed at low levels. They are typified by noninvasive MCF-7 breast cancer cells grown in the absence of estrogens, by p53 defective Ramos lymphoma cells and Ramos cells overexpressing Bcl-2 *(26)*.

4.2. Molecular Phenotype II

Unlike cancer cells which have low ErbB2 levels, in cancer cells where ErbB2 is overexpressed *(66–68)*, apoptosis in response to βGBP does not relate to E2F1 deregulation. A ligandless member of the ErbB/EGF family of TK receptors, ErbB2 strongly enhances Ras regulated Raf-MEK-ERK signaling by being constitutively active, by dimerizing with other ErbB members, by resisting endocytic degradation, and by returning to the cell surface *(69–71)*. ErbB2 downstream effects include cyclin D-kinase activation, activation of DNA synthesis and S-phase progression *(72–74)*. ErbB2 is also a strong activator of class IA phosphoinositide 3-kinase (PI3K) which consists of a 110-kDa catalytic subunit (α,β,δ) and a p85 regulatory/adapter subunit that plays roles in protein–protein interactions through a series of molecular domains *(75)*. PI3K catalyzes the production of phosphatidylinsitol-3,4,5 triphosphate (PIP3) which recruits to the cell membrane Akt/ protein

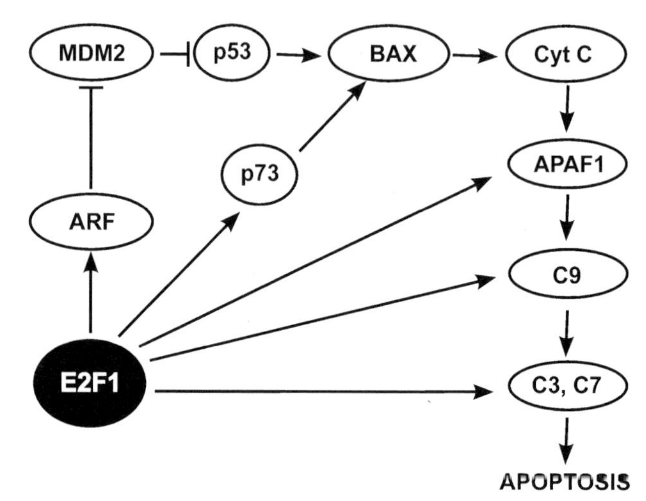

Fig. 4. A composite of pathways reported to be involved in E2F-1 activated apoptosis. Reprinted with permission from ref. *26*.

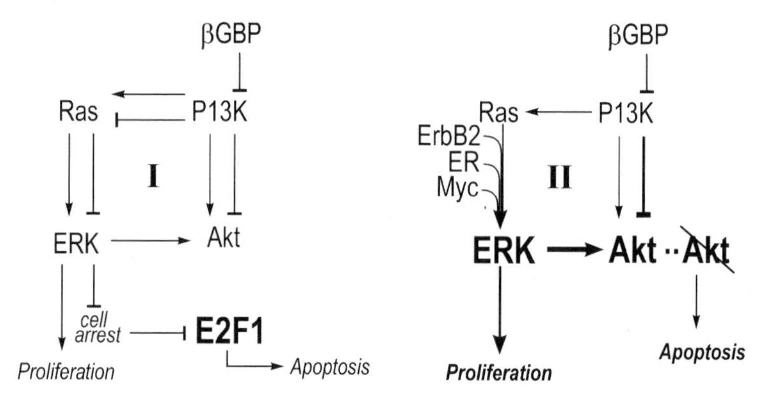

Fig. 5. Models of cancer cells' molecular phenotypes according to their response to treatment with Hu-r-βGBP. Molecular phenotype I is characterized by cancer cells which express high levels of E2F-1 and respond to βGBP by arresting in S phase. In these cells, apoptosis correlates to persistent ectopic E2F-1 transactivation. Molecular phenotype II is characterized by cancer cells where ERK activity is magnified by the converging signaling from multiple sources and PI3K/Akt activity is boosted. PI3K targeting by βGBP and consequent loss of PI3K/Akt signaling deprives the cells of survival capacity.

kinase B (PKB), a key regulator of cell survival *(76–78)*. PI3K also regulates a signaling network that mediates cell growth *(79)*, cell proliferation *(80,81)*, cytoskeletal organization *(81,82)*, cell motility and migration *(83–87)*, all processes of central importance to the evolution of tumorigenesis. Cells that overexpress ErbB2 are characterized by high-ERK levels, fast proliferation rates, and high levels of Akt expression *(71,88)*. In these cells, βGBP is unable to overcome the force of mitogenic signaling but, by targeting PI3K, βGBP abolishes Akt expression and the cancer cells' dependence on survival signaling *(42)*. Loss of Akt is followed after two to three replication cycles by massive cell death. A situation similar to that created by ErbB2 overexpression is also created by estrogen receptor (ER) overexpression, which boosts Ras-ERK signaling *(89,90)*, stimulates Myc activity, induces cyclin D-kinase activity *(89,90)*, stimulates S-phase entry *(91,92)*, and acutely

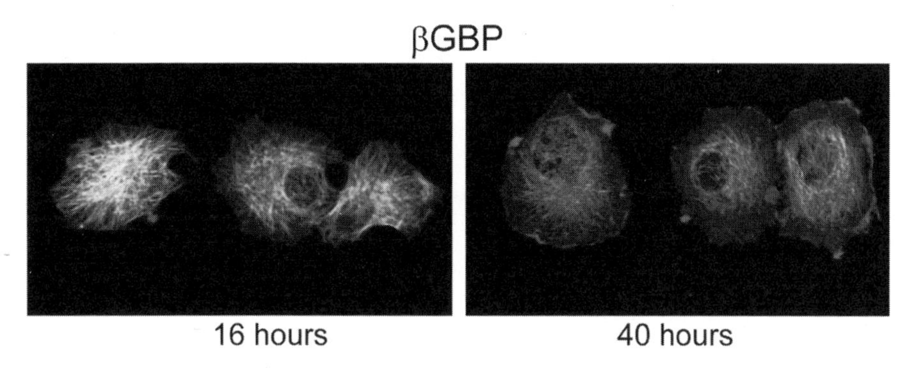

Fig. 6. MCF-7 breast cancer cells treated with βGBP 20 n*M* undergo cytoskeletal changes characteized by alteration of the tubulin network, loss of tubulin fluorescence, cell spreading, and increased visualization of actin at the edges of the cell membrane. (Photo courtesy of Daniel Zicha, Cancer Research UK.).

increases PI3K/Akt activity *(89)*. Cancer cells which respond to βGBP according to this molecular phenotype II pattern are typified by SKBR3 and BT474 breast cancer cells. Significantly, MCF 10A immortalized mammary ductal cells suffer little damage when exposed to βGBP; however, if their mitogenic imput is upregulated by increasing cAMP levels and ERK activity is boosted, their normal-like behavior changes to that of an aggressive phenotype characterized by faster proliferation rate and increased survival properties in terms of Akt phosphorylation. When challenged with Hu-r-βGBP, these cells incur Akt loss and respond with an intensity of apoptotic response similar to that of cells with ErbB2 overexpression. This experimental evidence signifies that where an increase of mitogenic signaling is the prime occurrence among events that lead to oncogenesis, nascent cancer cells could probably be eliminated in the healthy organism by the T-cell produced endogenous βGBP in a surveillance role.

The opening by βGBP of an Akt mediated apoptotic program as a consequence of strong mitogenic signaling whether enforced or endogenous (*see* Fig. 5), does not preclude a prime involvement of PI3K when apoptosis is the result of cell arrest and E2F1 deregulation. Further to the control of cell survival, PI3K plays a critical role in the regulation of S-phase entry via induction of cyclin D expression and activation of cdk 2 and cdk 4 and is necessary and sufficient for E2F activation *(93)*. Inhibition of PI3K results in elevation of $p27^{Kip}$ CDKI levels and cell-cycle arrest *(93,94)*. The inhibition of PI3K by βGBP adds also an extra dimension to the role that βGBP can play in the control of cancer. PI3K is critical to cytoskeleton dynamics and cell motility through the involvement of small GTPases with RAC as a primary regulator *(75,81,95)*. In MCF-7 breast cancer cells PI3K targeting by βGBP leads to loss of tubulin fluorescence, tubulin retraction, cell spreading, and accumulation of actin at the edges of the cell membrane (*see* Fig. 6). Loss of cytoskeletal function and motility is an important aspect in cancer control as it impedes cancer cells from leaving their site and metastasizing. Loss of cytoskeletal function can also cause clustering and activation of cell death receptors to induce apoptosis.

5. Therapeutic Prospects

Hu-r-βGBP has therapeutic efficacy on a wide range of cancer cells of both epithelial and mesenchymal origin while not harming normal cells including the cells of the

Table 1
Effect of Hu-r-βGBP 1–20 n*M* According to Cell Type
(From Mallucci et al. Biochem Pharmacol 2003;66:1563–1569)

Replication	Apoptosis	
Normal cells	Carcinomas	Lymphomas/leukemias
Luminal breast cells from mammoplasties examined in parallel with BT20 and T47D	**Breast** BT20, T47D, BT474, SKBR3, MCF-7	ST4, PF382, Jurkat, Raji, JM1, Daudi Ramos $p53^-$ Ramos $p53^-/Bcl\text{-}2^+$
Expanding T-cells from volunteers examined in parallel with ST4, PF382, and Jurkat	Tamoxifen resistant MCF7/TX9 Multidrug resistant MCF7/D40	U937 Topotecan resistant U937/RERC
Progenitor cells examined in clonogenic assays in parallel with cells from CML patients at diagnosis	**Ovarian** 2008 Cisplatin resistant 2008	**CML** K562, BV173 LAMA-84, KY01
	Prostate LNCaP, DU145, PC3	CML from patients at diagnosis
	Colon CCL233, SW480, SW620, LoVo, HT29 Colchicine resistant HT29/VMDR	
	Oral KB-3-1 Vinblastine resistant KB-3-1	

immune system (Table 1). The ability of βGBP to selectively kill cancer cells by opening opportunistic and alternative routes to apoptosis is a unique property not shared by conventional anticancer drugs nor by newly designed pharmacological inhibitors. As an effector molecule which enforces its function within a physiological context, βGBP induces regulatory changes to which normal and cancer cells respond differently because of their differential expression of molecular circuitries whose response to physiological challenges can harm cancer cells. These properties are of major therapeutic significance as they make unusable the acquired means of resisting therapeutic challenges that drug resistant cells have developed *(40)*. Hu-r-βGBP is as therapeutically effective in cancer cells which overexpress P-glycoprotein (MCF-7/D40, KB-V-1, HT-29), in cancer cells with increased DNA repair and metallothionine overexpression (2008/CP), or with altered expression or mutation of topoisomerase I (RERC) or II (MCF-7/D40, RERC), as it is against their parental counterparts (*see* Fig. 7). There are currently no therapeutic drugs in the clinic that combine

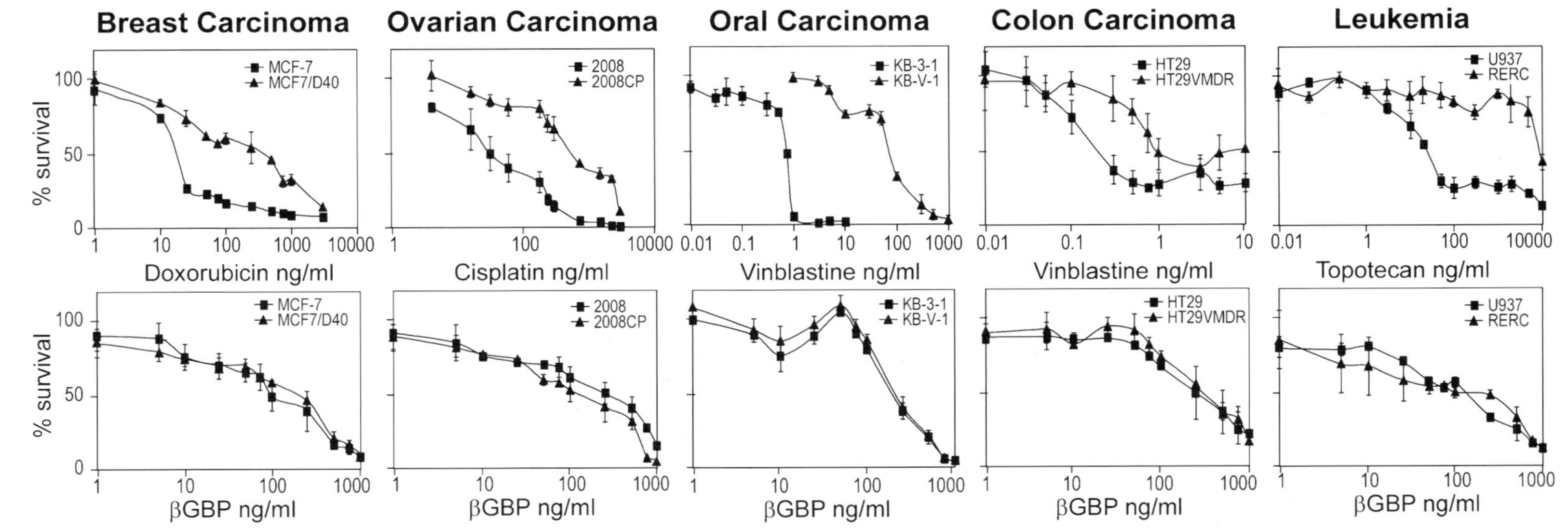

Fig. 7. βGBP bypasses drug resistance. Wild-type cells and their drug-resistant derivatives were treated with either the chemotherapeutic drugs that they were made resistant to or with Hu-r-βGBP and the deoxinucleotidyl transferase dUTP-biotin nick labeling assay performed 72 h post treatment. Reprinted with permission from ref. *40*.

the broad spectrum of antitumor activity with the ability to overcome multiple mechanisms of drug resistance as shown by βGBP.

6. Conclusion

βGBP is an immunomolecule and a conceivable new means by which the immune system may control malignancy. The ability of βGBP to perform therapeutically what it would perform naturally within a healthy organism offers a potentially safe and novel therapeutic approach to the control of cancer.

References

1. Hartwell LH, Weinert TA. Checkpoints: controls that ensure the order of cell cycle events. Science 1989;246:629–634.
2. Nurse P. Ordering S phase and M phase in the cell cycle. Cell 1994;79:547–550.
3. Elledge SJ. Cell cycle checkpoints: preventing an identity crisis. Science 1996;274: 1664–1672.
4. Johnstone RW, Ruefli AA, Lowe SW. Apoptosis: a link between cancer genetics and chemotherapy. Cell 2002;108:153–164.
5. Hahn WC, Weinberg RA. Modelling of the molecular circuitry of cancer. Nat Rev Cancer 2002;2:331–341.
6. Scherr CJ. Cancer cell cycles. Science 1996;274:1672–1677.
7. Downward J. Cell cycle: routine role for Ras. Curr Biol 1977;7:R258–R260.
8. Roberts EC, Shapiro PS, Stines-Nehareini et al. Distinct cell cycle timimg requirements for extracellular signal-regulated kinase and phosphoinositide-3-kinase signaling pathways in somatic cell mitosis. Mol Cell Biol 2002;22:7226–7241.
9. Hartwell LH, Szankasi P, Roberts CJ, et al. Integrating genetic approaches into the discovery of anticancer drugs. Science 1997;278:1064–1068.
10. Gibbs JB. Mechanism-based target identification and drug discovery in cancer research. Science 2000;287:1969–1973.
11. Mendelsohn L, Baselga J. The EGF receptor family as targets for cancer therapy. Oncogene 2000;19:6550–6565.
12. Sebti SM, Hamilton AD. Farnesyltransferase and geranylgeranyltransferase I inhibitors and cancer therapy: lessons from mechanisms and bench-to-bedside translational studies. Oncogene 2000;19:6584–6593.
13. Baell JB, Huang DC. Prospects for targeting the Bcl-2 family of oncoproteins to develop novel cytotoxic drugs. Biochem Pharmacol 2002;64:851–863.
14. Cox AD, Der CJ. Farnesyltransferase inhibitors: promises and realities. Curr Opin Pharmacol 2002;2:388–393.
15. Sawyers CL. Rational therapeutic intervention in cancer: kinases as drug targets. Curr Opin Genetic Dev 2002;12:111–115.
16. Singh SB, Lingham RB. Current progress in farnesyl protein transferase inhibitors. Current Opin Drug Discov Dev 2002;5:225–244.
17. Downward J. Targeting Ras signalling pathways in cancer therapy. Nat Rev Cancer 2003;3:11–22.
18. Reed JC. Apoptosis-targeted therapies for cancer. Cancer Cell 2003;3:17–22.
19. Levitzki A, Gazit A. Tyrosine kinase inhibition: an approach to drug development. Science 1995;267:1782–1788.
20. Shawver LK, Slamon D, Ulrich A. Tyrosine kinase inhibitors in cancer therapy. Cancer Cell 2002;1:117–123.
21. Martin S. Cell signalling and cancer. Cancer Cell 2003;4:167–174.

22. Areteaga CL, Baselga J. Tyrosine kinase inhibitors: why does the current process of clinical development not apply to them. Cancer Cell 2004;5:525–531.
23. Vivanco I, Sawyers CL. The phosphatidylinositol 3-kinase-Akt pathway in human cancer. Nat Rev Cancer 2002;2:489–501.
24. Luo J, Manning BD, Cantley LC. Targeting the PI3K-Akt pathway in human cancer: rationale and promise. Cancer Cell 2003;4:257–262.
25. Chang F, Lee JT, Navolanic PM, et al. Involvement of PI3K pathway in cell cycle progression, apoptosis and neoplastic transformation, a target for cancer therapy. Leukemia 2003; 176:590–603.
26. Mallucci L, Wells V, Danikas A, et al. Turning cell cycle controller genes into cancer drugs. A role for an antiproliferative cytokine (βGBP). Biochem Pharmacol 2003;66:1563–1569.
27. Wells V, Mallucci L. Identification of an autocrine negative growth factor: mouse beta-galactoside binding protein is a cytostatic factor and cell growth regulator. Cell 1991;64: 91–97.
28. Blaser C, Kaufman M, Muller C, et al. Beta-galactoside-binding protein secreted by activated T cells inhibits antigen-induced proliferation of T cells. Eur J Immunol 1998;28: 2311–2319.
29. Allione A, Wells V, Forni G, et al. β galactoside binding protein alters the cell cycle, upregulates expression of the α and β chains of the interferonγ receptor, and triggers IFNγ-mediated apoptosis of activated human T lymphocytes. J Immunol 1998;161:2114–2119.
30. Chiariotti L, Wells V, Bruni CB, et al. Structure and expression of the negative growth factor mouse β-galactoside binding protein gene. Biochim Biophys Acta 1991;1089:54–60.
31. Baldini A, Gress T, Patel K, et al. Mapping on human and mouse chromosomes of the gene for the beta-galactoside-binding protein, an autocrine negative growth factor. Genomics 1993;15:216–218.
32. Aurias A, Rimbaut C, Buffe D, et al. Translocation involving chromosome 22 in Ewing's sarcoma. A cytogenetic study of four fresh tumours. Cancer Genet Cytogenet 1984;12:21–25.
33. Turc-Carel C, Dal Cin P, Rao U, et al. Recurrent breakpoints at 9q31 and 22q12.2 in extraskeletal myzoid chondrosarcoma. Cancer Genet Cytogenet 1988;30:145–150.
34. Bridge JA, Borek DA, Neff JR, et al. Chromosomal abnormalities in clear cell sarcoma. Implications for histogenesis. Am J Clin Pathol 1990;93:26–31.
35. Rey JA, Bello MJ, de Campos JM, et al. Abnormalities of chromosome 22 in human brain tumors determined by combined cytogenetic and molecular genetic approaches. Cancer Genet Cytogenet 1993;66:1–10.
36. Cho M, Cummings RD. Galectin 1, β galactoside binding lectin in chinese hamster ovary cells. J Biol Chem 1995;270:5189–5206.
37. Mallucci L, Wells V. Negative control of cell proliferation. Growth arrest versus apoptosis. J Theor Med 1998;1:169–173.
38. Novelli F, Allione A, Wells V, et al. Negative cell cycle control of human T cells by beta-galactoside binding protein (beta-GBP): induction of programmed cell death in leukaemic cells. J Cell Physiol 1999;178:102–108.
39. Wells V, Davies D, Mallucci L. Cell cycle arrest and induction of apoptosis by beta galactoside binding protein (betaGBP) in human mammary cancer cells. A potential new approach to cancer control. Eur J Cancer 1999;35:978–983.
40. Ravatn R, Wells V, Nelson L, et al. Circumventing multidrug resistance by β-galactoside binding protein, an antiproliferative cytokine. Cancer Res 2005;65:1631–1634.
41. Wells V, Mallucci L. Molecular expression of the negative growth factor murine beta-galactoside binding protein (mGBP). Biochim Biophys Acta 1992;1121:239–244.
42. Mallucci L, Wells V. Potential role of the anti-proliferative cytokine β-galactosidase binding protein. Curr Opin Investig Drugs 2005;6:1228–1233.

43. Mallucci L, Wells V, Downward J. Control of G0/G1 transition by the betaGBP cytokine: inhibition of ras-MAPK signaling is preceded by loss of PI3K activity. In: Keystone Symposium: Cell response to Lipid Mediators. Keystone Symposia, Taos, NM: 2002:27.
44. Evan GI, Vousden KH. Proliferation, cell cycle and apoptosis in cancer. Nature 2001;411: 342–348.
45. Downward J. Ras signalling and apoptosis Curr Opin Genet Dev 1998;8:49–54.
46. Reed JC. Dysregulation of apoptosis in cancer. J Clin Oncol 1999;17:2941–2953.
47. Makin G, Hickman JA. Apoptosis and cancer chemotherapy. Cell Tissue Res 2000;301: 143–152.
48. Hanahan D, Weinberg RA. The hallmarks of cancer. Cell 2000;1900:57–70.
49. Letai A, Bassik MC, Walensky LD. Distint BH3 domains either sensitize or activate mitochondrial apoptosis, serving as prototype cancer therapeutics. Cancer Cell 2002;2:183–192.
50. Reed JC. Apoptosis based therapies. Nat Rev Drug Discov 2002;1:111–121.
51. Krek W, Ewen ME, Shirodkar S. Negative regulation of the growth-promoting transcription factor E2F-1 by a stably bound cyclin A dependent protein kinase. Cell 1994;78:161–172.
52. Helin K. Regulation of proliferation by the E2F transcription factors. Curr Opin Genet Dev 1998;8:28–35.
53. Shan B, Lee WH. Deregulated expression of E2F-1 induces S phase entry and leads to apoptosis. Mol Cell Biol 1994;14:8166–8173.
54. Krek W, Xu G, Livingston DM. Cyclin A-kinase regulation of E2F-1 DNA binding function underlies suppression of an S phase checkpoint. Cell 1995;83:1149–1158.
55. Fueyo J, Gomez-Manzano C, Yung WK, et al. Overexpression of E2F-1 in glioma triggers apoptosis and suppresses tumor growth in vitro and in vivo. Nat Med 1998;4:658–690.
56. Liu TJ, Wang M, Breau RL. Apoptosis induction by E2F-1 via adenoviral-mediated gene transfer results in growth suppression of head and squamous cell carcinoma cell lines. Cancer Gene Ther 1996;6:163–171.
57. Fagan R, Flint KJ, Jones N. Phosphorylation of E2F-1 modulates its interaction with the retinoblastoma gene product and the adenoviral E4 19kDa protein. Cell 1994;78:799–811.
58. Kitagawa M, Higashi H, Suzuki-Takahashi I, et al. Phosphorylation of E2F-1 by cyclinA-cdk2. Oncogene 1995;10:229–236.
59. Peeper DS, Keblusek P, Helin K, et al. Phosphorylation of specific cdk site in E2F-1 affects its electrophoretic mobility and promotes pRB-binding in vitro. Oncogene 1995;10:39–48.
60. Xu M, Sheppard KA, Peng CY, et al. Cyclin A/cdk2 binds directly to E2F-1 and inhibits the DNA binding activity of E2F-1/DP-1 by phosphorylation. Mol Cell Biol 1994;14: 8420–8431.
61. Stubbs M, Strachan G, Hall D. An early S phase checkpoint is regulated by the E2F-1 transcription factor. Bioch Biophys Res Comm 1999;258:77–80.
62. Moroni MC, Hickman ES, Denchi EL, et al. Apaf-1 is a transcriptional target for E2F and p53. Nat Cell Biol 2001;3:552–558.
63. Nahle Z, Polakoff J, Davuluri ME, et al. Direct coupling of the cell cycle and cell death machinery by E2F. Nat Cell Biol 2002;4:859–864.
64. Martins LM, Iaccarino I, Tenev T, et al. The serine protease Omi/HtrA2 regulates apoptosis by binding XIAP through a reaper-like motif. J Biol Chem 2002;277:1205–1213.
65. Bargou RC, Wagner C, Bommert K, et al. Blocking the transcription factor E2F/DP by dominant-negative mutants in a normal breast epithelial cell line efficiently inhibits apoptosis and induces tumor growth in SCID mice. J Exp Med 1996;183:1205–1213.
66. Brison O. Gene amplification and tumor progression. Biochim Biophys Acta 1993;1155:25–41.
67. Hynes NE. Amplification and overexpression of the erbB-2 gene in human tumors; its involvement in tumor development, significance as a prognostic factor, and potential as a target for cancer therapy. Semin Cancer Biol 1993;4:19–26.

68. Hynes NE, Stern DF. The biology of erbB-2/neu/HER-2 and its role in cancer. Biochim Biophys Acta 1994;1198:165–184.
69. Pinkas-Kramarski R, Soussan L, Waterman H, et al. Diversification of neu differentiation factor and epidermal growth factor signaling in combinatorial receptor interactions. EMBO J 1996;15:2452–2467.
70. Karunagaran D, Tzahar E, Beerli RR, et al. ErbB-2 is a common auxillary subunit of NDF and EGF receptors: implications for breast cancer. EMBO J 1996;15:254–264.
71. Brennan PJ, Kumagai T, Berezov A, et al. Her2/neu: mechanisms of dimerisation/oligomerisation. Oncogene 2000;19:6093–6101.
72. Graus-Porta D, Beerli RR, Daly JM, et al. ErbB-2, the preferred heterodimerisation partner of all ErbB receptors, is a mediator of lateral signaling. EMBO J 1997;16:1647–1655.
73. Blume-Jensen P, Hunter T. Oncogenic kinase signalling. Nature 2001;411;355–365.
74. Harari D, Yarden Y. Molecular mechanisms underlying erbB2/HER2 action in breast cancer. Oncogene 2000;19:6102–6114.
75. Vanhaesebroeck B, Leevers SJ, Almadi K, et al. Synthesis and function of 3-phosphorylated inositol lipids. Ann Rev Biochem 2001;70:535–602.
76. Marte BM, Downward J. Connecting PI3-kinase to cell survival and beyond. Trends Biochem Sci 1997;22:335–358.
77. Franke TF, Yang S, Chan TO, et al. The protein kinase encoded by the Akt proto-oncogene is a target of the PDGF-activated phosphatidylinositol 3-kinase. Cell 1995;81:727–736.
78. Dufner A, Thomas G. Ribosomal S6 kinase signalling and control of translation. Exp Cell Res 1999;253;100–109.
79. Wennstrom S, Downward J. Role of phosphoinositide-3-kinase in activation of Ras and mitogen-activated kinase by epidermal growth factor. Mol Cell Biol 1999;19:4279–4288.
80. Harbour JW, Dean DC. The Rb/E2F pathway; expanding roles and emerging paradigms. Genes Dev 2000;14:2393–2409.
81. Rodriguez-Vinciana P, Warne PH, Khwaja T, et al. Role of phosphoinositide-3-OH kinase in cell transformation and control of the actin cytoskeleton by Ras. Cell 1997;89:457–467.
82. Venkasteswarlu K, Oatley PB, Tavare JM, et al. Identification of centaurin-alpha 1 as a potential in vivo phosphatidyl 3,4,5-triphosphate-binding protein that is functionally homologous to the yeast ADP-ribosylation factor (ARF) GTPase-activating protein. Biochem J 1999;340:359–363.
83. Jin T, Zhang JT, Long Y, et al. Localisation of the G protein betagamma complex in living cells during chemotaxis. Science 2000;287:1034–1036.
84. Servant G, Weiner OD, Herzman P, et al. Polarisation of chemoattractant receptor signalling during neutrophil chemotaxis. Science 2000;287:1037–1040.
85. Hall A. Rho GTPases and the actin cytoskeleton. Science 1998;279:509–514.
86. Funamoto S, Meili R, Lee S. Spatial and temporal regulation of 3-phosphoinositides by PI3-kinase and PTEN mediated chemotaxis. Cell 2002;109:611–623.
87. Ridley AJ, Schwartz MA, Firtel RA. Cell migration: integrating signals from front to back. Science 2003;302:1704–1709.
88. Baselga J, Norton L. Focus on breast cancer. Cancer Cell 2000;1:319–322.
89. Castoria G, Migliaccio A, Bilancio A, et al. PI3-kinase in concert with Src promotes the S-phase entry of oestradiol-stimulated MCF-7 cells. EMBO J 2001;20:6050–6059.
90. van der Berg B, van Selm-Miltenberg AJ, Laat SW, et al. Direct effects of estrogen on c-fos and c-myc protooncogene expression and cellular proliferation in human breast cancer cells. Mol Cell Endocrinol 1989;64:223–228.
91. Migliaccio A, Di Domenico M, Castoria G, et al. Tyrosine kinase /p21^{Ras} /MAP-kinase Pathway activation by estradiol-receptor complex in MCF-7 cells. EMBO J 1996;15: 1292–1300.

92. Wang LH, Yang XY, Zhang X. Suppression of breast cancer by chemical modulation of vulnerable zinc fingers in estrogen receptor. Nat Med 2004;10:40–47.
93. Collado M, Medema RH, Garcia-Caol I, et al. Inhibition of the phosphoinositide 3-kinase pathway induces a senescence-like arrest mediated by p27^{kip1}. J Biol Chem 2000;275: 21,960–21,968.
94. Medema RH, Kops GJ, Bos JL, et al. AFX-like Forkhead transcription factors mediate cell cycle regulation by Ras and PKB through p27^{kip1}. Nature 2000;404:782–787.
95. Hawkins PT, Eguinoa A, Qui RG, et al. PDGF stimulates an increase in GTP-Rac via activation of phosphoinositide 3- kinase. Curr Biol 1995;5:393–403.
96. Wells V, Mallucci L. Cell cycle regulation (G1) by autocrine interferon and dissociation between autocrine interferon and 2′-5′olgigoadenylate synthetase expression. J Interferon Res 1988;8:793–802.
97. Laiho M, De Caprio JA, Ludlow JW, et al. Growth inhibition by TGF-beta linked to suppression in retinoblastoma protein phosphorylation. Cell 1990;62:175–185.
98. Zhang HS, Postigo AA, Dean DC. Active transcriptional repression by the Rb-E2F complex mediates G1 arrest triggered by p16^{INK4a}, TGF-beta and contact inhibition. Cell 1999; 97:53–61.

9

Control Nodes Linking the Regulatory Networks of the Cell Cycle and Apoptosis

Baltazar D. Aguda, Wee Kheng Yio, and Felicia Ng

Summary

Depending on the nature of extracellular stimuli and the ensuing intracellular signal transduction pathways, certain transcription factors are activated and subsequently determine the extent of expression of genes involved in cell proliferation, survival, and death. These factors are referred to as transcriptional control nodes because they permit the coordination of cell-cycle progression and the apoptosis program. This coordination is made possible by the existence of feedback loops in the regulatory networks of the entire system. A review of these networks for the following transcription factors is provided in this chapter: E2F, Myc, p53, and NF-κB.

Key Words: E2F; Myc; p53; NF-κB; cell cycle; apoptosis; regulatory networks; control nodes; modules.

1. Introduction

It is not surprising that the molecular regulatory networks of the cell–division cycle and apoptosis are tightly intertwined *(1)* because when the so-called cell cycle engine "overheats" in certain cells in an organism, a mechanism has to exist to get rid of these cells in order to steer a normal course of development or maintenance of the organism. Thus, it is quite reasonable to expect that at least one pathway leading to the initiation of the cell-suicide program is linked to at least one pathway leading to the initiation of the cell cycle. Without such link the coordination between the cell cycle and apoptosis machineries cannot be controlled. In this chapter, it is postulated that the key links are provided by certain transcription factors (TFs) controlling the expression of genes involved in the cell cycle and apoptosis. Activation of these TFs is initiated by extracellular signals transduced by intracellular pathways that directly cause the translocation of the TFs to the nucleus, binding to gene promoters and initiation of transcription. These transcriptional nodes are referred to in this chapter as "control nodes" because they permit the possibility of coordinating the cell cycle and apoptosis. The discussion is organized according to the general classification of signaling pathways and control nodes shown in Fig. 1 *(2)*. The pathways in this figure should be interpreted in a very general way; for example, type I signaling pathways activate type I control nodes which potentiate pathways promoting the cell cycle and apoptosis ("potentiate" here means that the pathway does not necessarily trigger activation). Note that the signaling pathways shown in Fig. 1 are only those that activate the control nodes but, in general, the discussion will also include activating and inhibiting signaling pathways and their interactions.

From: *Apoptosis, Cell Signaling, and Human Diseases: Molecular Mechanisms, Volume 1*
Edited by R. Srivastava © Humana Press Inc., Totowa, NJ

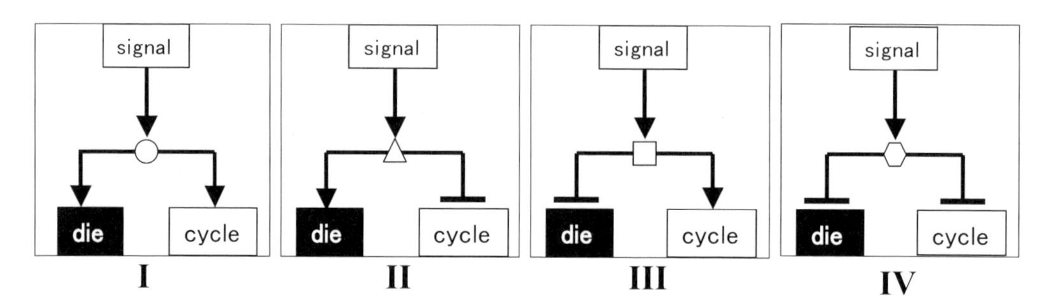

Fig. 1. Transcriptional nodes (symbolized by a circle, a triangle, a square, and a hexagon) are classified according to how they affect the cell cycle and apoptosis. Arrows represent "induction" whereas hammerheads represent "inhibition." Associated intracellular signalling pathways that activate these nodes are shown.

Furthermore, an important observation is that no case as shown in Fig. 1 would give rise to a mechanism in which the status of the cell-cycle module influences the status of the apoptosis module. Indeed, one can see that there should be at least a pathway from the cell-cycle module which feeds back to the transcriptional node or to any upstream signaling pathway that regulates the TF node activity; this feedback loop would generate the situation in which the TF control node activity is a function of the cell-cycle module status (i.e., the activity of some cell-cycle factor) and therefore the status of the apoptosis module would be an indirect function of the cell-cycle module status. Of course, the cell-cycle module can directly communicate to the apoptosis module by either regulating the pathways from the TF node to apoptosis and/or the apoptosis module itself.

The known molecular regulatory networks of the cell cycle and apoptosis are quite complex and can overlap. However, as suggested in Fig. 1, it will be assumed that these networks are separable and can be modularized by focusing on the primary enzymes that trigger these cellular processes, namely, the cyclin-dependent kinases (CDKs) for the cell cycle and the caspases for apoptosis. A discussion of the basic components and interactions in these modules will be provided in the next section. As will be shown, these modules represent tightly regulated interactions that communicate with the transcriptional nodes in both directions.

The control nodes discussed in depth in this chapter include the E2F, Myc, p53, and NF-κB transcription factors. E2F and Myc are combined into one node which will be referred to as the E2F-Myc node; E2F here specifically represents E2F-1 (a member of the E2F family) and Myc refers to c-Myc. These TFs are grouped together because of their synergistic interactions and the fact that they possess many common transcriptional targets. The E2F-Myc node represents case I in Fig. 1. The tumor suppressor function of the protein p53 is mainly attributed to the proapoptotic gene targets; in addition, p53 induces the expression of several genes whose protein products inhibit cell-cycle progression. Thus, p53 is an example of case II in Fig. 1. Case III is exemplified by the NF-κB family of dimeric transcription factors which is primarily known for its role in the immune response, but is now increasingly linked to carcinogenesis through pathways that connect to the cell cycle and apoptosis (case IV in Fig. 1 is not of interest in this chapter). Although there are more transcriptional nodes that can control the cell cycle and apoptosis programs, the above examples are deemed sufficient to illustrate some of the principles needed to understand the operation of transcriptional control nodes as they affect the cell cycle and apoptosis modules.

A detailed analysis of the E2F-Myc control node and its interactions with the signaling, cell cycle, and apoptosis modules illustrate some important concepts governing the operation of these networks. There are several coupled positive and negative feedback loops present in the regulatory networks, and these loops operate both at the level of a single module and at the level of module–module interactions. Positive feedback loops could lead to switching and bistable behavior whereas negative feedback loops can generate sustained oscillations. The structures of the regulatory networks of the other transcriptional nodes in terms of feedback loops are therefore discussed to demonstrate how sources of switching and other unstable behaviors are predicted. The intrinsic nonlinearities of these control networks make them very difficult to understand intuitively. Indeed, mathematical modeling and computer simulations are tools that would greatly aid in understanding these complex networks.

2. Cell Cycle and Apoptosis Modules

2.1. G1/S Module

Based on a kinetic analysis of the then prevailing consensus G1-S regulatory network of the mammalian cell cycle, Aguda and Tang *(3)* proposed that the G1 checkpoint called the "restriction point" is governed by the activation of cyclin E/CDK2 whose critical regulators include the phosphatase Cdc25A, and a CDK inhibitor (p27Kip1). This picture has not changed much as can be seen in a recent review of the field (for an example, *see* ref. *4*). However, recent experiments with mice knocked out of G1 cyclins and G1 CDKs suggest that the various mammalian cyclins and CDKs may take each others roles if necessary *(5,6)*.

According to the Aguda-Tang kinetic analysis, sharp switching behavior in CDK2 activity is predicted from the mutual-activation (between CDK2 and Cdc25A) and mutual-inhibition (between CDK2 and p27Kip1) as shown in Fig. 2. Cdc25A removes an inhibitory phosphate from CDK2 thereby promoting the kinase activity of the latter; in return, the phosphatase activity of Cdc25A is upregulated by its phosphorylation carried out by CDK2. Such positively coupled cycles of phosphorylation and dephosphorylation possess an intrinsic instability that generates switching behavior *(7)*; this switch is further sharpened when coupled to the mutual antagonistic interactions between p27Kip1 and CDK2. Several experimental observations are consistent with the Aguda-Tang model *(8–10)*.

In quiescent cells, members of the retinoblastoma protein (pRB) family bind and inhibit proliferative transcription factors such as the E2Fs which are essential for the expression of several S-phase genes. Growth factors stimulate the synthesis of the D-type cyclins which bind and activate CDK4 and CDK6; these kinases then phosphorylate and inactivate pRB subsequently freeing E2F to induce the expression of cyclin E, cyclin A, Cdc25A, several members of theorigin recognition complex (ORC). Cyclin E/CDK2 and cyclin A/CDK2 further contribute to the hyperphosphorylation of pRB (for review, *see* ref. *11*). The time lag in the activation of cyclin E/CDK2 after crossing the restriction point *(3,8)* could be explained by an induction time associated with the build up of the positive feedback loop involving pRB, E2F, and cyclin E/CDK2 (*see* Fig. 2).

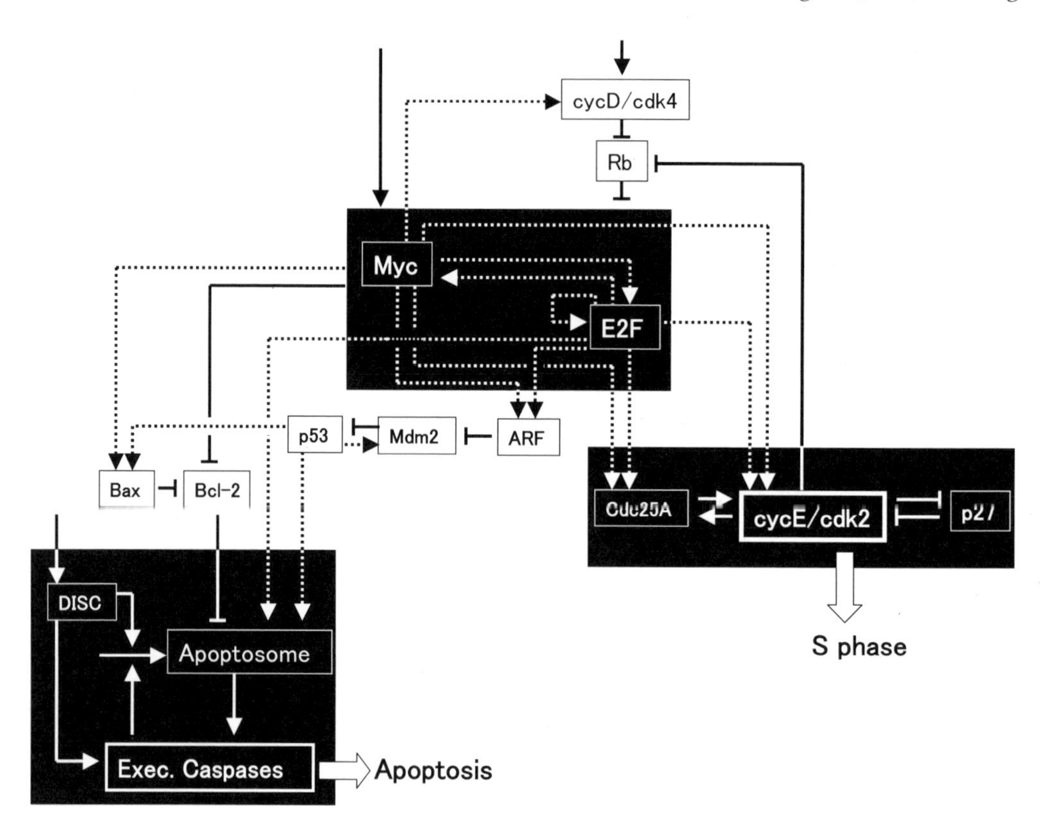

Fig. 2. The E2F-Myc node is the top-most black box which encapsulates the synergy between the E2F and Myc transcription factors. Dotted lines indicate transcriptional activation (with *arrow heads*). Regulation of the E2F-Myc node activity includes both Rb-dependent and Rb-independent pathways as represented by the two signaling routes shown. The right-most black box represents a minimal module for the regulatory network involved in CDK2 activation in S-phase entry. Solid lines mean posttranscriptional interactions with an arrow indicating "activation" and a hammerhead denoting "inhibition." The left-most black box represents the apoptosis module involving cascades of activation of the caspases.

2.2. Apoptosis Module

The apoptosis module involves cascades of activation of proteases of the caspase family (for a recent review, *see* ref. *12*). Members of this family include the initiator caspases (caspase-8, -9, -2, -10) and executioner caspases (caspase-3, -6, -7). The two major pathways that converge at active executioner caspases are summarized in Fig. 2 and in ref. *2*. The extrinsic or membrane receptor-mediated pathway involves a complex called death-inducing signaling complex (DISC) which forms upon ligand-receptor binding (e.g., tumor necrosis factor [TNF]-α and FasL ligands binding to corresponding receptors). For example, the accumulation of procaspase-8 in the DISC generates active caspase-8 molecules which are released from the complex to activate the executioner caspases. The extrinsic or mitochondrial pathway involves the formation of the "apoptosome," a complex composed of cytochrome *c*, Apaf-1, and procaspase-9. As a result of internal cellular stresses (e.g., DNA damage, reactive oxygen species [ROS], growth factor withdrawal) proapoptotic proteins (e.g., Bax, Bad) induce permeabilization of the mitochondria and subsequent release of cytochrome *c*. Formation of the apoptosome is followed by the activation of caspase-9 which promotes the activation of the executioner caspases. As shown in the apoptosis module in Fig. 2, the extrinsic pathway feeds into the intrinsic pathway (e.g., via activation of Bid by caspase-8; Bid induces release of cytochrome *c* from mitochondria); moreover, a positive feedback loop is shown from the executioner caspases to the step that forms the apoptosome (representing the observation that executioner caspases induce an increased rate of cytochrome *c* release from mitochondria *(13,14)*. This positive feedback loop is thought to be an essential mechanism for the irrevocable switch into apoptosis *(2)*.

3. The E2F-Myc Node

3.1. Regulation and Targets of the E2F-Myc Node

The E2F-Myc transcriptional node that induces expression of several genes involved in S-phase entry as well as genes involved in apoptosis is shown in Fig. 2. E2F dimerizes with DP proteins whereas Myc dimerizes with Max in order to carry out their respective transcriptional activities. E2F-1 has been shown to upregulate c-Myc expression *(15–17)* and c-Myc has been shown to induce transcription of E2F-1, E2F-2, and E2F-3 *(18,19)*. Thus, E2F-1 and c-Myc synergize to induce S-phase entry and apoptosis *(2,20)*.

A schematic representation of signal transduction pathways impinging on the E2F-Myc node is shown in Fig. 2. One involves a pathway from cyclin D/CDK to the retinoblastoma protein (pRB). Several studies have shown that the Ras family of small GTPases are involved in the upregulation of synthesis or stabilization of some D-type cyclins *(11)*. These cyclins bind and activate CDK4 or CDK6 which ultimately results to increased E2F transcriptional activity (because these CDKs phosphorylate pRB which then releases E2F). The significance of studying the pRB pathway is underlined by the fact that it is targeted for inactivation in at least 80% of sporadic human cancers *(21)*. Chau and Wang *(22)* recently reviewed evidence that pRB could be a key protein in the regulation of a cell's life and death decisions. This role of pRB is directly associated with its inhibition of the E2Fs. The interplay between mitogenic signals and death signals (e.g., TNF signaling) uses pRB as a key player; mitogenic signals inactivate pRB by phosphorylation via CDK activation, whereas death signals inactivate pRB by caspase-dependent degradation (for a review, *see* ref. 22).

A few of the major transcriptional targets of E2F-1 are shown in Fig. 2. These include S-phase genes such as cyclin E, cyclin A, Cdc25A, and several members of the prereplication protein complex formed prior to DNA replication (reviewed in ref. *11*). It is also interesting to note that there is a positive feedback loop involving the E2F-1 protein as a positive transcription factor for its corresponding gene *(23)*. Among the proapoptotic genes induced by E2F-1 are Apaf-1 (a member of the apoptosome complex) and various caspases *(24–26)*. Another E2F-1 target is ARF which indirectly stabilizes p53. E2F-1 can also promote apoptosis, independent of its transcriptional activity, by binding with the p65 subunit of NF-κB and inhibiting the latter's antiapoptotic functions *(27)*. In the presence of DNA damage, ATM, ATR, and Chk2 kinases have been shown to phosphorylate E2F-1 resulting in the stabilization of this substrate *(28)*.

Note that activation of cyclin D/CDK4/6 is not absolutely required for the activation of cyclin E/CDK2; this is because signaling pathways mediated by c-Myc can upregulate expression of proteins involved in CDK2 activation as well as in the downregulation of the CDK inhibitor p27Kip1 *(29)*. Santoni-Rugiu et al. *(29)* have demonstrated that Myc is involved in a G1/S-promoting mechanism that is parallel to the pRB/E2F pathway. Myc is an immediate-early gene that is expressed soon after a cell's exposure to growth factors leading to the activation of tyrosine kinases such as Src and JAK that upregulate Myc (reviewed in ref. *11*). Myc's transcriptional targets which are involved in S-phase entry include cyclin D2 *(30,31)*, CDK4 *(32,30)*, cdc25A *(33)*, E2F-2 *(18)*, and cyclin E *(34)*. Proapoptotic target genes include ARF and Bax as shown in Fig. 2. The pathway leading to apoptosis is also favored by the suppression of the protein and RNA levels of Bcl-2 by c-Myc as well as by E2F-1 *(35)*. Furthermore, it has been suggested *(36)* that c-Myc promotes apoptosis by repressing the expression of p21Cip1 which indirectly inhibits apoptosis via the mitochondrial pathway *(37)*.

3.2. Coordinating S Phase and Apoptosis

A major question of interest is how the E2F-Myc node coordinates the initiation of S phase and apoptosis; as an example of this coordination, E2F-1 first stimulates S-phase entry when its transcriptional activity is within some normal range but drives the apoptosis program when overexpressed. Looking at case I in Fig. 1, it is important to remember that the arrow from the E2F-Myc transcriptional node to apoptosis indicates that apoptosis is potentiated and not necessarily triggered; could it be that a trigger for apoptosis is generated by an over-stimulated cell cycle? Is it also possible that the cell-cycle mechanism itself generates a signal that somehow inhibits apoptosis so that the cell is not in danger of accidentally killing itself when everything else is normal? And how do the different cases in Fig. 1 interact to determine cell fate? These questions were addressed in a recent paper by Aguda and Algar *(2)*. It is proposed in this paper that the complexity of the network of molecular interactions can be reduced by subdividing the network into functional modules and analyzing module–module interactions. These modules and their interactions are shown in Fig. 3. The module–module interactions are labeled **a**, **b**, **b'**, and **c**. Interaction **a** includes positive feedback loops involving the transcriptional node and the S-phase module; examples would be the loop E2F-CyclinE/CDK2-RB-E2F and the autocatalytic loop involving E2F-1. Aguda and Algar *(2)* noted the importance of the negative feedback **b** from the cell-cycle module to the signaling pathway upstream of the transcriptional node; as the computer simulations in their paper demonstrate, this negative feedback permits a window or a range of transcriptional activities for normal cell cycles to exist by inhibiting apoptosis. The existence of a negative feedback loop involving the Ras/Raf/MEK/ERK pathway can be supported by experimental evidence *(38–40)* of multiple peaks of activities of Ras and/or MEK/ERK; these peaks look like damped oscillations which are indicative of negative feedback in a dynamical system. Another negative feedback is the interaction **b'** between the cell-cycle module and the transcriptional node; the phosphorylation of DP-1 (a partner of E2Fs) by CDK2 can turn off E2F-1 transcriptional activity *(41,42)*. Lastly, the feedforward type of interaction **c** in Fig. 3 is a direct way of triggering apoptosis when the cell cycle "overheats;" experimental evidence for this interaction has been reported *(43–45)*.

4. The p53 Node

The p53 protein is a DNA sequence-specific transcription factor that regulates the expression of certain genes associated with the induction of apoptosis, cell-cycle arrest, prevention of new blood vessel formation, and accelerated DNA repair. The induction

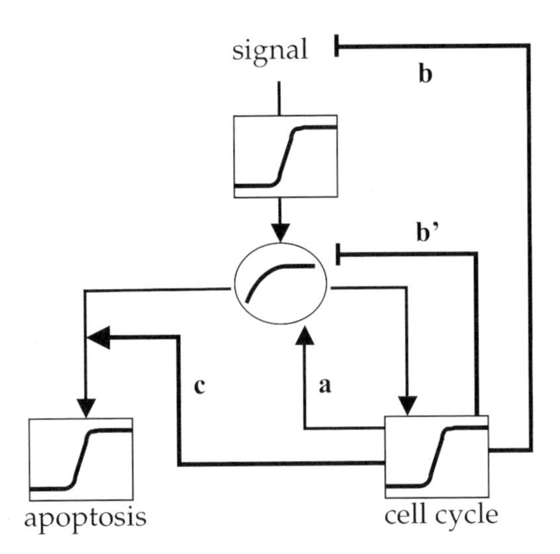

Fig. 3. Case I of Fig. 1 including module–module interactions. *Arrows* indicate activation whereas hammerheads indicate inhibition. This is the modular model analyzed by Aguda and Algar *(2)*. The signaling, apoptosis, and cell-cycle modules were assumed to possess ultrasensitive kinetics (as shown by the curves inside the rectangular boxes). The transcriptional node (circle) was not assumed to have that kinetics.

of apoptosis is widely accepted as the primary role of p53 in tumor suppression. The levels of p53 in normal cells are low but rapidly increases upon exposure to DNA-damaging stresses including ultraviolet light and intense oncogenic signaling. The p53 gene is mutated in about half of known human cancers whereas a majority of the remaining cases are caused by dysfunctional regulation of an otherwise normal p53 protein *(20,46–48)*.

The p53 control node shown in Fig. 4 symbolizes a set of active transcription factor complexes involving p53 that directly affect the expression of target genes, some of which are also shown in the figure. Signaling to the p53 node includes all upstream conditions and processes (e.g., exposure to ultraviolet light and presence of DNA damage) that cause elevated rates of p53 synthesis, covalent modifications of the protein, DNA binding and activation of p53-mediated transcription. Post-translational modifications (i.e., phosphorylation, acetylation, sumoylation) that regulate p53 activity are also considered processes signaling to the node. Many details of these regulatory pathways are already known, and comprehensive reviews of the current state of knowledge in the field are available *(46,49,50)* and will not be duplicated here. The primary aim here is to interpret the significance of these pathways in relation to their role in deciding whether the cell dies or survives. The interpretation is based on significance of the cycles present in the regulatory network. Figure 5 summarizes some of the positive and negative feedback loops in the p53 regulatory network. Harris and Levine *(51)* recently reviewed the various positive and negative feedback loops in the p53 regulatory network.

4.1. The p53-Mdm2-ARF Network and Negative Feedback Loops

As shown in Figs. 4 and 5, the regulation of p53 activity involves several coupled negative feeback loops; one of these is a two-cycle involving Mdm2 and p53, and another is a three-cycle involving p53, ARF, and Mdm2. In the 2-cycle, Mdm2 binds

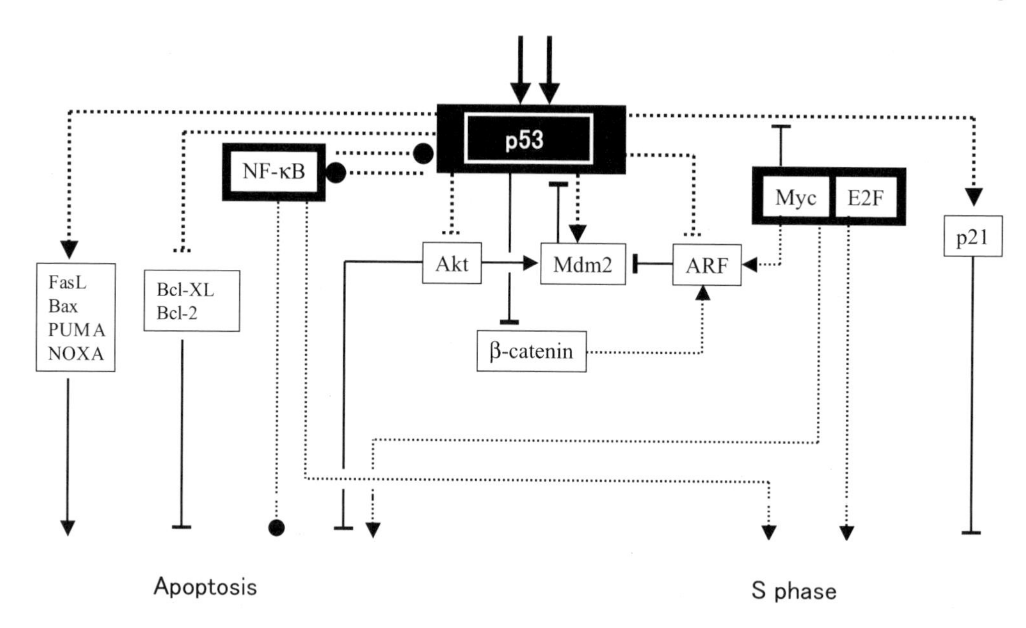

Fig. 4. The p53 transcriptional factor promotes expression of certain proapoptotic genes (e.g., FasL, Bax), of the CDK inhibitor p21, and represses transcription of certain antiapoptotic genes (e.g., Bcl-XL, Bcl-2). The core regulatory network involving Mdm2 and ARF is shown, as well as the links of this network to Akt and β-catenin. Interactions among the NF-κB and E2F-Myc nodes are discussed in the text. Dotted lines represent transcriptional processes while solid lines are post-translational interactions. Arrows mean "activation" and hammerheads mean "inhibition."

and inhibits p53 by blocking a transactivation domain of p53 and by promoting the ubiquitination of p53 leading to proteosomal degradation; on the other hand, p53 induces the expression of the *mdm*2 gene by binding to this genes promoter region. This negative feedback loop between p53 and Mdm2 has been offered as the origin for observed oscillations in p53 activity *(52,53)*.

The transcription of the p14/19/ARF gene is negatively regulated by p53 *(51)*. In turn, ARF binds to and represses the ubiquitin ligase activity of Mdm2 *(54)*. ARF is not expressed in normal proliferating cells but is rapidly expressed in response to many oncogenes such as c-Myc, E2F-1, Ras, and β-catenin; this is why ARF is considered as one of the most important "sensors for aberrant proliferative signaling" *(20)*.

As shown in Fig. 5, p53 induces transcription of SIAH-1, a ubiquitin ligase that targets β-catenin for degradation *(51)*. A negative feedback loop is thus formed by the pathway represented by the sequence p53-(SIAH-1)-(β-catenin)-ARF-Mdm2-p53. This links the β-catenin signaling pathway to the p53 network. Other negative feedback loops shown in Fig. 5 are p53-p21-CDK2-Mdm2-p53 and p53-(WIP-1)-p38MAPK. The gene p21Cip1, which produces a cyclin-dependent kinase (CDK) inhibitor, is one of the first identified transcriptional targets of p53. As shown in Fig. 5, CDK2 inhibits Mdm2 by phosphorylation; interestingly, this inhibitory phosphorylation is counteracted by the phosphatase PP2A in complex with cyclin G, which is rapidly transcribed after p53 activation in many cell types (reviewed in ref. *51*). This gives a picture of Mdm2 regulation by phosphorylation-dephosphorylation steps which are influenced by p53 via p21 (resulting to inhibition of the phosphorylation step) and via cyclin G (which enhances the dephosphorylation step).

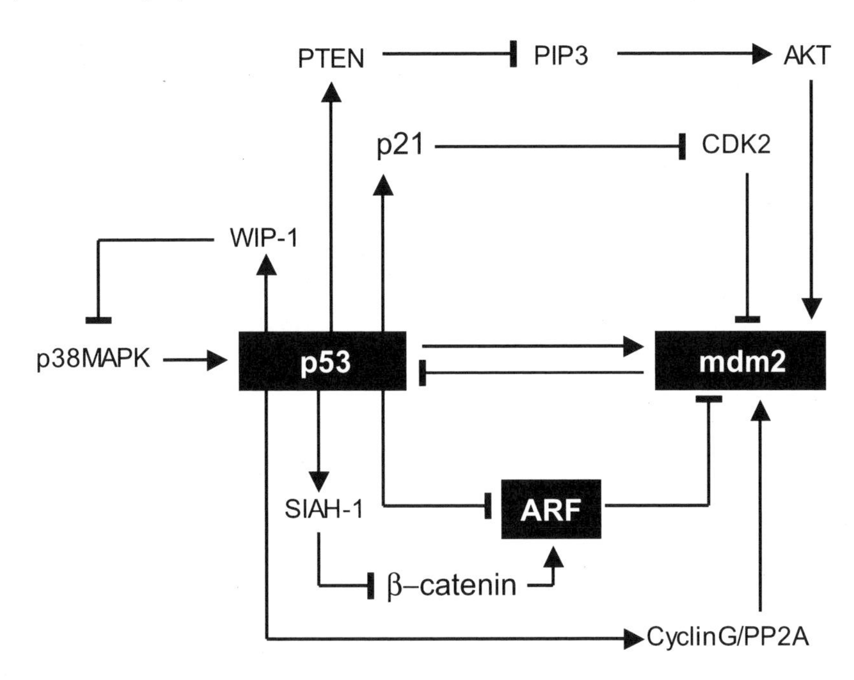

Fig. 5. Some of the feedback loops coupled with the p53-Mdm2-ARF core regulatory network as reviewed by Harris and Levine *(51)*. *See* text for details.

The Ras/MAPK signaling pathways are linked to the p53 regulatory network as shown in Fig. 5. Activation of p53 partly involves its phosphorylation by p38MAPK on certain serine residues (reviewed in ref. *51*). The protein p38MAPK itself is activated by phosphorylation which is reversed by the Wip-1 phosphatase.

4.2. Positive Feedback Loops in p53 Regulation

Shown in Fig. 5 is a positive feeback loop represented by the sequence p53-PTEN-PIP3-Akt-Mdm2-p53. PTEN is considered a tumor suppressor whose transcription is induced by p53 in some cell types (reviewed in ref. *51*). PTEN is a lipid phosphatase that dephosphorylates PIP3 (phosphatidyl inositol-3,4,5-triphosphate). PIP3 is required for the recruitment of Akt to the plasma membrane where it is phosphorylated and activated. In addition, p53 represses the catalytic subunit of PI3K (phosphatidylinositol 3-kinase) which catalyzes the formation of PIP3 *(55)*. Akt is a survival factor because it phosphorylates and subsequently inhibits proapoptotic proteins such as Bad and caspase-9. Akt also phosphorylates Mdm2 which subsequently translocates to the nucleus where it inhibits p53. The positive feedback loop between p53 and Akt is of the mutual antagonism kind: high p53 activity (apoptosis) means low Akt activity (low survival), and vice versa. Indeed, such mutual antagonism could provide a mechanism for switching between the two cell fates of survival and death.

The picture becomes more complicated when the participation of another important kinase, glycogen synthase kinase-3β (GSK-3β) is considered. The substrates of this kinase regulate cell proliferation and metabolism *(56,57)*. Shown in Fig. 6 are the interlocking pathways of p53, Akt, GSK-3β, and β-catenin. This figure shows various positive feedback loops in addition to that between Akt and p53 discussed in the preceding

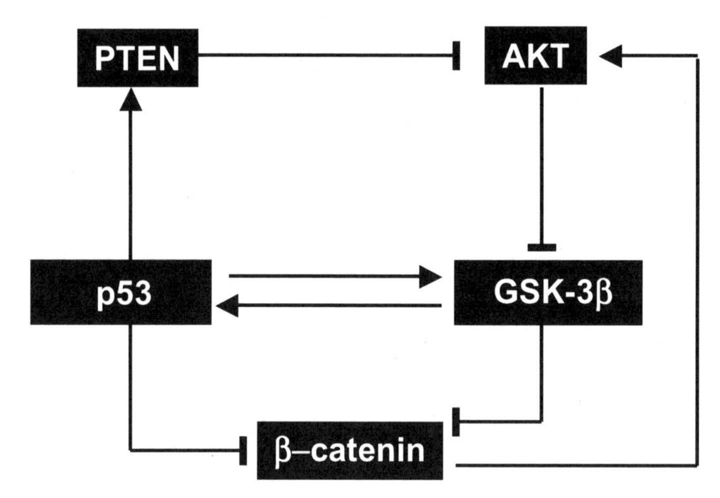

Fig. 6. Some of the positive feedback loops involving p53. Mutual antagonism between p53 (proapoptotic) and Akt (prosurvival), involving PTEN and GSK-3β, could provide a switching mechanism between cell survival and apoptosis. On the other hand, repression of β-catenin by p53 and GSK-3β could reduce β-catenin-Akt mediated tumorigenesis. *See* text for more details. Modified from ref. *48*.

paragraph. GSK-3β and p53 have been shown to interact directly *(58)* and that this interaction augments the activities of both proteins *(48)* (Fig. 6, double arrows). GSK-3β phosphorylates and subsequently inactivates proliferative factors such as cyclin D, c-Myc,and β-catenin, among others. True to its role as a survival factor, Akt can phosphorylate and inactivate GSK-3β as this kinase has been shown to trigger apoptosis when in excess *(59)*. The pathway of activation of Akt from β-catenin could involve increased expression of Wnt-induced secreted protein (WISP)-1(reviewed in refs. *4,60*).

The question remains regarding the role of the positive feedback loops involving p53. It was stated above that the positive feedback between Akt and p53 is a good candidate mechanism for a switch between survival and apoptosis. The positive feedback loop between GSK-3β and p53 could be a mechanism for favoring the apoptosis pathway. The positive loop (*see* Fig. 6) is another candidate for a switching mechanism between survival and death because it could be interpreted as a toggle switch between Akt (survival) and GSK-3β (apoptosis). The interactions among these positive loops (as well as between negative- and positive-feedback loops) are difficult to understand using mere intuitive reasoning and it would be interesting to see what computational modeling can contribute to future understanding of this complex network.

4.3. Are There Critical Determinants of p53-Mediated Decision on Life and Death?

If there is a critical determinant of a p53-mediated decision on life or death of a cell, it will surely depend on the cellular context such as the presence of survival signals and cell genotype *(48)*. For example, Myc has been suggested as a critical determinant between p53-mediated cell-cycle arrest and apoptosis *(47)*. Without Myc, cells exposed to ultraviolet light undergo growth arrest primarily as a result of the activation of the CDK inhibitor p21 *(61,62)*. Myc (via recruitment by Miz-1) effectively blocks p21 induction by p53 and tilts the decision towards apoptosis instead of growth arrest *(63)*. In human colorectal cancers, the balance between p21 and PUMA appears to be critical in deciding to arrest or die in response to exogenous p53 (reviewed in ref. *47*). Another possible critical determinant is E2F-1 which cooperates with p53 in various ways (e.g., by direct E2F1-p53 association), and by binding to adjacent promoter sites of proapoptotic genes as best illustrated by the case of Apaf-1 *(24)*. The authors of this chapter think that identification of critical determinants of p53-mediated life or death decisions will ultimately depend on understanding the feedback loops between p53-induced apoptotic pathways and survival pathways (e.g., the Akt pathway).

5. The NF-κB Node

5.1. Regulation and Targets of the NF-κB Node

NF-κB is a family of inducible transcription factors that are key responders to changes in cellular environment *(64)*. They play their most important role in the immune system by producing chemokines that trigger innate immune responses. The family consists of p50 (NF-κB1), p52 (NF-κB2), p65 (RelA), RelB, and c-Rel. These members form dimers in various combinations in order to carry out their transcriptional activities. The prototypical NF-κB complex is the heterodimer p50/p65. Each member protein contains a Rel-homology domain that is involved in binding to a DNA sequence motif called the κB site, in dimerization with other family members, and in binding to IκB which inhibits nuclear translocation of NF-κB. Some of the transcriptional targets of NF-κB are involved in cell proliferation (e.g., cyclin D1 and Myc), cell survival (e.g., Bcl-X_L, IAPs), angiogenesis (e.g., vascular endothelial growth factor [VEGF], IL-8), and the immune response (e.g., major histocompatibility complex [MHC]-1, MHC-II). Malfunction of NF-κB has been implicated in various hematological disorders, breast cancer, and many other cancers and human diseases *(65,66)*. Connections between inflammation and cancer have also been reviewed recently *(67,68)*.

The canonical pathway of NF-κB activation is initiated by upstream signals that activate a macromolecular complex called the "signalsome" *(69)*. The upstream signals include cytokines (e.g., TNFα, LPS, CD40L), reactive oxygen species, viral and bacterial products *(70)*. The "signalsome" is an IKK complex composed of IKKα, IKKβ, and IKKγ (NEMO). IKKβ is the principal kinase in the signalsome that phosphorylates IκB leading to the latter's degradation; free NF-κB then rapidly translocates to the nucleus.

Noncanonical pathways are initiated by agonists such as CD40L, BAFF, and RANKL that induce the activation of a protein kinase called NIK (NF-κB inducing kinase) *(64,70)*. NIK phosphorylates a dimer of IKKα which initiates proteolytic processing of p100 to p52. Subsequently, p52 along with its dimer partner (principally RelB) translocates to the nucleus.

Signaling to the NF-κB node involves both positive- and negative-feedback loops as shown in Fig. 7. The negative feedback is results from NF-κB inducing transcription of IκB which, in turn, inhibits NF-κB *(71–73)*; more specifically, inside the nucleus, the newly synthesized IκBα protein binds NF-κB resulting to the export of the complex out of the nucleus *(74,75)*. Interestingly, NF-κB is also known to upregulate the synthesis of proteins that activate NF-κB, thus forming a positive feedback loop (*see* Fig. 7); examples are the proteins activating upstream signals of the canonical pathway such as TRAF1/2 *(76)*, CD40 *(77)*, and CD40L *(78)*.

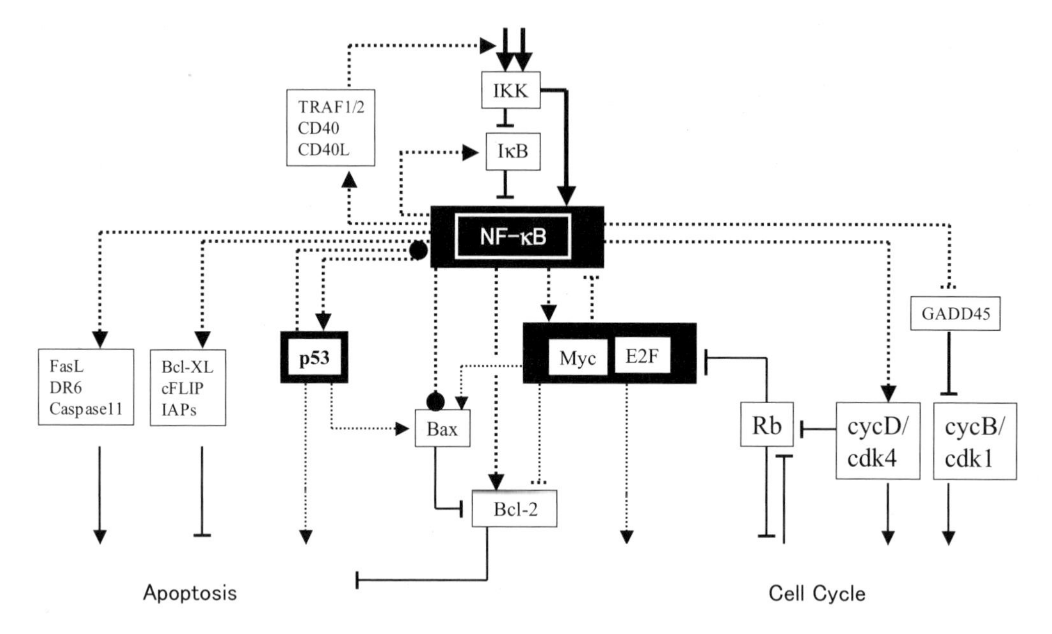

Fig. 7. Overview of the NF-κB control node. Upstream stimuli activate NF-κB via two groups of pathways. In the canonical pathway, activated IKK complex triggers degradation of IκB leading to activation of NF-κB; noncanonical pathways involve NIK activation of IKKα dimer to produce NF-κB heterodimers. NF-κB-induced transcription of IκB generates a negative feedback whereas NF-κB-induced transcription of TRAF1/2, CD40, and CD40L results in positive-feedback loops. As shown on the left-hand side of the diagram, NF-κB may induce expression of both antiapoptotic and apoptotic genes. The role of NF-κB in the cell cycle is exemplified by promoting the expression of genes such as cyclin D and gadd45 as shown on the right-hand side of the figure. See text for more details.

5.2. Antiapoptotic, Proapoptotic, and Proliferative Pathways From NF-κB

NF-κB protects the cell from apoptosis through induction of antiapoptotic genes such as Bcl-X_L *(79)*, Bfl-1/A1 *(80)*, cFLIP *(81)*, and IAPs *(76,82)*. These genes inhibit apoptosis in various ways. Examples include inhibitors of apoptosis (IAPs) inhibiting caspase-3, -6, -7 and -9; cFLIP inhibiting the apoptosome; and Bcl-2 inhibiting the release of cytochrome *c* from mitochondria.

On the other hand, NF-κB has been shown to target proapoptotic genes such as FasL *(83)*, DR6 *(84)*, and caspase-11 *(85)*. Characteristics of the activating stimulus of NF-κB may determine whether antiapoptotic or proapoptotic genes would be induced *(86)*. Interestingly, both the pro- and antiapoptotic functions of NF-κB could occur in the same cell *(87)*, and it may well be that the role of NF-κB in a cell's decision to live or die is the net result of these opposite apoptotic functions. What then determines the net effect of NF-κB on apoptosis? Differential regulation of apoptosis between different member proteins of the NF-κB family is possible (e.g., p65/p50 inhibits transcription of the Bax gene while p50/p50 upregulates it *[88]*). In addition, relationships between NF-κB family members and apoptotic p53 may play some role. There is evidence showing that NF-κB and p53 compete for the same coactivator, CBP/p300, effectively inhibiting each other *(89–93)*. On the other hand, it has been shown that p65/p50 induces p53 transcription *(94–96)* (later shown to require AP-1 and Myc/Max transcription factors *[97]*).

Also shown in Fig. 7 is the interaction between the NF-κB and the E2F-Myc nodes. NF-κB induces transcription of Myc whereas E2F-1 represses the activation of NF-κB by inhibiting the degradation of IκB *(98)* or by binding to the p65 subunit *(27)*, or through some other unknown proteins *(99)*. The influence of NF-κB on apoptosis and the cell cycle is therefore intertwined with that of the E2F-Myc transcriptional node as discussed in a previous section.

NF-κB regulates the cell cycle via induction of cyclin D1 *(100)* and c-Myc *(101)*, factors that promote G1-to-S progression. There is also evidence indicating the presence of NF-κB binding sites upstream of cyclin D2 *(102)* and cyclin D3 *(103)* genes. Studies *(104–106)* have shown that NF-κB represses GADD45 (-α, -γ), which in turn promotes cell survival by blocking JNK-induced apoptosis. Mak and Kultz *(107)* further showed that the various GADD45 isoforms (-α, -β, -γ) promote early stages of apoptosis but inhibit mitosis. A study by Taylor and Stark *(108)* also showed that p53 induces G2/M arrest partly through GADD45 proteins.

6. Conclusion

This chapter has presented a perspective for viewing the complexity of intracellular molecular networks associated with a cell's response to growth and death factors in its environment. These responses are coordinated by certain transcription factors which are referred to in this chapter as "control nodes" because each represents a meeting point of signalling pathways (inputs) and pathways (outputs) that diverge to the cell cycle and apoptosis machineries. The complexity of these molecular networks is reduced by subdividing them into modules and by assuming that the module–module interactions in a network are sufficient to understand the essential behaviour of the entire network.

Although only three transcriptional control nodes were discussed in this chapter there are others that can also affect both the apoptosis and cell-cycle programs, including, among others, the Smad family of transcription factors and FOXO *(4)*. The E2F-Myc node (primarily represented by the E2F-1 and c-Myc heterodimeric transcription factors) was discussed above to illustrate transcription factors that potentiate both the initiation of the cell cycle and the apoptosis programs. It may be surprising at first why such a possibility exists at all given that cell cycle and death are diametrically opposite cell fates; but, as discussed above, such a nodal link does provide a mechanism for the organism to carry out the fragile balance between cell proliferation and death (e.g., during its multicellular development). As explicitly analyzed by Aguda and Algar *(2)* through mathematical modelling, the module–module interactions allow a number of cell fates to be orchestrated sequentially—namely, quiescence, cell cycling, and death—as growth factor signalling intensities increase. The feedback loops in these modular interactions played crucial roles in the behavioural transitions of the system.

The second transcriptional node discussed was p53, a protein of high significance because of its tumor suppressor activity. As summarized in Fig. 5, the core p53-Mdm2-ARF regulatory module is linked to the Akt, p38MAPK, and β-catenin signaling pathways that involve feedback loops. It was pointed out that the mutual antagonism between p53 and Akt is a good candidate for a switching mechanism that toggles between cell death and survival. How this toggle switch is controlled by the associated pathways would be an interesting topic for kinetic modelling.

The NF-κB family of transcription factors was used as an example of a transcriptional control node that inhibits apoptosis but promotes the cell cycle. However, as discussed above, it is now known that NF-κB can also induce expression of proapoptotic genes, and it is still an open question as to how these opposing contributions to apoptosis are sorted out in a given cell. The positive- and negative-feedback loops involved in NF-κB signaling (*see* Fig. 7) have been the subject of recent mathematical modelling of temporal control of gene expression *(109,110)*.

Lastly, the three transcriptional nodes discussed here interact in various ways as shown in Figs. 2, 4, and 7. It is probable that only a subset of these interactions is realized for any given set of extracellular stimulus. The real challenge is to understand the logic of these interactions, their control, and to predict whether the cell proliferates, survives or die.

References

1. Evan GI, Vousden KH. Proliferation, cell cycle and apoptosis in cancer. Nature 2001;411: 342–348.
2. Aguda BD, Algar CK. A structural analysis of the qualitative networks regulating the cell cycle and apoptosis. Cell Cycle 2003;2:538–544.
3. Aguda BD, Tang Y. The kinetic origins of the restriction point in the mammalian cell cycle. Cell Prolif 1999;32:321–335.
4. Massagué J. G1 cell-cycle control and cancer. Nature 2004;432:298–306.
5. Tetsu O, McCormick F. Proliferation of cancer cells despite CDK2 inhibition. Cancer Cell 2003;3:233–245.
6. Malumbres M, Sotillo R, Santamaria D, et al. Mammalian cell cycles without the D-type cyclin-dependent kinases Cdk4 and Cdk6. Cell 2004;118:493–504.
7. Aguda BD. Instabilities in Phosphorylation-Dephosphorylation Cascades and Cell cycle Checkpoints. Oncogene 1999;18:2846–2851.
8. Ekholm SV, Zickert P, Reed SI, Zetterberg A. Accumulation of cyclin E is not a prerequisite for passage through the Restriction Point. Mol Cell Biol 2001;21:3256–3265.
9. Blomberg I, Hoffmann I. Ectopic expression of Cdc25A accelerates the G1/S transition and leads to premature activation of cyclin-E and cyclin-A-dependent kinases. Mol Cell Biol 1999;19:6183–6194.
10. Sandhu C, Donovan J, Bhattacharya N, Stampfer M, Worland P, Slingerland J. Reduction of Cdc25A contributes to cyclin E1-Cdk2 inhibition at senescence in human mammary epithelial cells. Oncogene 2000;19:5314–5323.
11. Aguda BD. Kick-starting the cell cycle: From growth-factor stimulation to initiation of DNA replication. Chaos 2001;11:269–276.
12. Boatright KM, Salvesen GS. Mechanisms of caspase activation. Curr Opin Cell Biol 2003;15:725–731.
13. Green D, Kroemer G. The central executioners of apoptosis: caspases or mitochondria? Trends Cell Biol 1998;8:267–271.
14. Kumar S, Vaux DL. A cinderella caspase takes center stage. Science 2002;297:1290–1291.
15. Hiebert SW, Lipp M, Nevins JR. E1A-dependent transactivation of the human MYC promoter is mediated by the E2F factor. Proc Natl Acad Sci USA 1989;86:3594–3598.
16. Thalmeier K, Synovzik H, Mertz R, Winnacker EL, Lipp M. Nuclear factor E2F mediates basic transcription and transactivation by E1A of the human MYC promoter. Genes Dev 1989;3:527–536.
17. Elliott MJ, Dong YB, Yang H, McMasters KM. E2F-1 up-regulates c-Myc and p14(ARF) and induces apoptosis in colon cancer cells. Clin Cancer Res 2001;7:3590–3597.
18. Sears R, Ohtani K, Nevins JR. Identification of positively and negatively acting elements regulating expression of the E2F2 gene in response to cell growth signals. Mol Cell Biol 1997;17:5227–5235.

19. Leone G, Sears R, Huang E, et al. Myc requires distinct E2F activities to induce S phase and apoptosis. Mol Cell 2001;8:105–113.
20. Lowe SW, Cepero E, Evan G. Intrinsic tumour suppression. Nature 2004;432:307–315.
21. Sherr CJ. Cancer cell cycles. Science 1996;274:1672–1677.
22. Chau BN, Wang JY. Coordinated regulation of Life and Death by RB. Nat Rev Cancer 2003;3:130–138.
23. Neuman E, Flemington EK, Sellers WR, Kaelin WG, Jr, Transcription of the E2F-1 gene is rendered cell cycle dependent by E2F DNA-binding sites within its promoter. Mol Cell Biol 1994;14:6607–6615.
24. Moroni MC, Hickman ES, Denchi EL, et al. Apaf-1 is a transcriptional target for E2F and p53. Nat Cell Biol 2001;3:552–558.
25. Nahle Z, Polakoff J, Davuluri RV, et al. Direct coupling of the cell cycle and cell death machinery by E2F. Nat Cell Biol 2002;4:859–864.
26. MacLachlan TK, El-Deiry WS. Apoptotic threshold is lowered by p53 transactivation of caspase-6. Proc Natl Acad Sci USA 2002;99:9492–9497.
27. Tanaka H, Matsumura I, Ezoe S, et al. E2F1 and c-Myc potentiate apoptosis through inhibition of NF-kappaB activity that facilitates MnSOD-mediated ROS elimination. Mol Cell 2002;9:1017–1029.
28. Stevens C, La Thangue NB. The emerging role of E2F-1 in the DNA damage response and checkpoint control. DNA Repair 2004;3:1071–1079.
29. Santoni-Rugiu E, Falck J, Mailand N, Bartek J, and Lukas J. Involvement of Myc activity in a G1/S-promoting mechanism parallel to the pRb/E2F pathway. Mol Cell Biol 2000;20: 3497–3509.
30. Coller HA, Grandori C, Tamayo P, et al. Expression analysis with oligonucleotide microarrays reveals that myc regulates genes involved in growth, cell cycle, signaling, and adhesion. Proc Natl Acad Sci USA 2000;97:3260–3265.
31. Bouchard C, Thieke K, Maier A, et al. Direct induction of cyclin D2 by Myc contributes to cell cycle progression and sequestration of p27. EMBO J 1999;18:5321–5333.
32. Hermeking H, Rago C, Schuhmacher M, et al. Identification of CDK4 as a target of *c*-Myc. Proc Natl Acad Sci USA 2000;97:2229–2234.
33. Galaktionov K, Chen X, Beach D. Cdc25 cell-cycle phosphatase as a target of *c*-myc. Nature 1996;382:511–517.
34. Perez-Roger I, Solomon DL, Sewing A, Land H. Myc activation of cyclin E/CDK2 kinase involves induction of cyclin E gene transcription and inhibition of p27Kip1 binding to newly formed complexes. Oncogene 1997;14:2373–2381.
35. Eischen CM, Packham G, Nip J, et al. Bcl-2 is an apoptotic target suppressed by both c-Myc and E2F-1. Oncogene 2001;20:6983–6993.
36. Vousden KH. Switching from life to death: The Miz-ing link between Myc and p53. Cancer Cell 2002;2:351–352.
37. Gartel AL, Tyner AL. The role of the cyclin-dependent kinase inhibitor p21 in apoptosis. Mol Cancer Ther 2002;1:639–649.
38. Meloche S. Cell cycle reentry of mammalian fibroblasts is accompanied by the sustained activation of p44mapk and p42mapk isoforms in the G1 phase and their inactivation at the G1/S transition. J Cell Physiol 1995;163:577–588.
39. Taylor SJ, Shalloway D. Cell cycle-dependent activation of Ras. Curr Biol 1996;6: 1621–1627.
40. Lee KY, Ladha MH, McMahon C, Ewen ME. The retinoblastoma protein is linked to the activation of Ras. Mol Cell Biol 1999;19:7724–7732.
41. Lees JA, Weinberg RA. Tossing monkey wrenches into the clock: New ways of treating cancer. Proc Natl Acad Sci USA 1999;96:4221–4223.

42. Krek W, Xu G, Livingston DM. Cyclin A-kinase regulation of E2F-1 DNA binding function underlies suppression of an S phase checkpoint. Cell 1995;83:1149–1158.
43. Konishi Y, Lehtinen M, Donovan N, Bonni A. Cdc2 Phosphorylation of BAD links the cell cycle to the cell death machinery. Mol Cell 2002;9:1005–1016.
44. Mendelsohn AR, Hamer JD, Wang ZB, Brent R. Cyclin D3 activates Caspase 2, connecting cell proliferation with cell death. Proc Natl Acad Sci USA 2002;99:6871–6876.
45. Sofer-Levi Y, Resnitzky D. Apoptosis induced by ectopic expression of cyclin D1 but not cyclin E. Oncogene 1996;13:2431–2437.
46. Vogelstein B, Lane D, Levine AJ. Surfing the p53 network. Nature 2000;408:307–310.
47. Haupt S, Berger M, Goldberg Z, Haupt Y. Apoptosis—the p53 network. J Cell Sci 2003; 116:4077–4085.
48. Oren M. Decision making by p53: life, death and cancer. Cell Death Differ 2003;10: 431–442.
49. Vousden KH, Prives C. P53 and prognosis: new insights and further complexity. Cell 2005; 120:7–10.
50. Bourdon JC, Laurenzi VD, Melino G, Lane D. p53: 25 years of research and more questions to answer. Cell Death Differ 2003;10:397–399.
51. Harris SL, Levine AJ. The p53 pathway: positive and negative feedback loops. Oncogene 2005;24:2899–2908.
52. Lev Bar-Or R, Maya R, Segel LA, Alon U, Levine AJ, Oren M. Generation of oscillations by the p53-Mdm2 feedback loop: a theoretical and experimental study. Proc Natl Acad Sci USA 2000;97:11,250–11,255.
53. Lahav G, Rosenfeld N, Sigal A, et al. Dynamics of the p53-Mdm2 feedback loop in individual cells. Nat Genet 2004;36:147–150.
54. Honda R, Yasuda H. Association of p19(ARF) with Mdm2 inhibits ubiquitin ligase activity of Mdm2 for tumor suppressor p53. Embo J 1999;18:22–27.
55. Singh B, Reddy PG, Goberdhan A, et al. p53 regulates cell survival by inhibiting PIK3CA in squamous cell. Genes Dev 2002;16:984–993.
56. Cohen P, Frame S. The renaissance of GSK3. Nat Rev Mol Cell Biol 2001;2:769–776.
57. Jope RS, Johnson GV. The glamour and gloom of glycogen synthase kinase-3. Trends Biochem Sci 2004;29:95–102.
58. Watcharasit P, Bijur GN, Zmijewski JW, et al. Direct, activating interaction between glycogen synthase kinase-3beta and. Proc Natl Acad Sci USA 2002;99:7951–7955.
59. Pap M, Cooper GM. Role of glycogen synthase kinase-3 in the phosphatidylinositol 3-Kinase/Akt cell survival pathway. J Biol Chem 1998;273:19,929–19,932.
60. Xu L, Corcoran RB, Welsh JW, Pennica D, Levine AJ. WISP-1 is a Wnt-1- and beta-catenin-responsive oncogene. Genes Dev 2000;14:585–595.
61. Sheen JH, Dickson RB. Overexpression of c-Myc alters G(1)/S arrest following ionizing radiation. Mol Cell Biol 2002;22:1819–1833.
62. Vafa O, Wade M, Kern S, et al. c-Myc can induce DNA damage, increase reactive oxygen species, and mitigate p53 function: a mechanism for oncogene-induced genetic instability. Mol Cell 2002;9:1031–1044.
63. Seoane J, Le HV, Massague J. Myc suppression of the p21(Cip1) Cdk inhibitor influences the outcome of the p53 response to DNA damage. Nature 2002;419:729–734.
64. Hayden MS, Ghosh S. Signaling to NF-kappaB. Genes Dev 2004;18:2195–2224.
65. Shishodia S, Aggarwal BB. Nuclear factor-kappaB: a friend or a foe in cancer? Biochem Pharmacol 2004;68:1071–1080.
66. Kumar A, Takada Y, Boriek AM, Aggarwal BB. Nuclear factor-kappaB: its role in health and disease. J Mol Med 2004;82:434–448.
67. Clevers H. At the crossroads of inflammation and cancer. Cell 2004;118:671–674.

68. Marx J. Cancer research. Inflammation and cancer: the link grows stronger. Science 2004;306:966–968.
69. Chen LF, Greene WC. Shaping the nuclear action of NF-kappaB. Nat Rev Mol Cell Biol 2004;5:392–401.
70. Chen F, Castranova V, Shi X. New insights into the role of nuclear factor-kappaB in cell growth regulation. Am J Pathol 2001;159:387–397.
71. Sun SC, Ganchi PA, Ballard DW, Greene WC. NF-kappaB controls expression of inhibitor I kappa B alpha: evidence for an inducible autoregulatory pathway. Science 1993;259: 1912–1915.
72. Beg AA, Finco TS, Nantermet PV, Baldwin AS, Jr. Tumor necrosis factor and interleukin-1 lead to phosphorylation and loss of I kappa B alpha: a mechanism for NF-kappaB activation. Mol Cell Biol 1993;13:3301–3310.
73. Brown K, Park S, Kanno T, Franzoso G, Siebenlist U. Mutual regulation of the transcriptional activator NF-kappaB and its inhibitor, I kappa B-alpha. Proc Natl Acad Sci USA 1993;90:2532–2536.
74. Arenzana-Seisdedos F, Thompson J, Rodriguez MS, Bachelerie F, Thomas D, Hay RT. Inducible nuclear expression of newly synthesized I kappa B alpha negatively regulates DNA-binding and transcriptional activities of NF-kappa B. Mol Cell Biol 1995;15: 2689–2696.
75. Arenzana-Seisdedos F, Turpin P, Rodriguez M, et al. Nuclear localization of I kappa B alpha promotes active transport of NF-kappa B from the nucleus to the cytoplasm. J Cell Sci 1997;110:369–378.
76. Wang CY, Mayo MW, Korneluk RG, Goeddel DV, Baldwin AS, Jr. NF-kappaB antiapoptosis: induction of TRAF1 and TRAF2 and c-IAP1 and c-IAP2 to suppress caspase-8 activation. Science 1998;281:1680–1683.
77. Hinz M, Loser P, Mathas S, Krappmann D, Dorken B, Scheidereit C. Constitutive NF-kappaB maintains high expression of a characteristic gene network, including CD40, CD86, and a set of antiapoptotic genes in Hodgkin/Reed-Sternberg cells. Blood 2001;97: 2798–2807.
78. Srahna M, Remacle JE, Annamalai K, et al. NF-kappaB is involved in the regulation of CD154 (CD40 ligand) expression in primary human T cells. Clin Exp Immunol 2001; 125:229–236.
79. Chen F, Demers LM, Vallyathan V, Lu Y, Castranova V, Shi X. Involvement of 5′-flanking kappaB-like sites within bcl-x gene in silica-induced Bcl-x expression. J Biol Chem 1999;274:35,591–35,595.
80. Zong WX, Edelstein LC, Chen C, Bash J, Gelinas C. The prosurvival Bcl-2 homolog Bfl-1/A1 is a direct transcriptional target of NF-kappaB that blocks TNFalpha-induced apoptosis. Genes Dev 1999;13:382–387.
81. Kreuz S, Siegmund D, Scheurich P, Wajant H. NF-kappaB inducers upregulate cFLIP, a cycloheximide-sensitive inhibitor of death receptor signaling. Mol Cell Biol 2001;21: 3964–3973.
82. Stehlik C, de Martin R, Kumabashiri I, Schmid JA, Binder BR, Lipp J. Nuclear factor (NF)-kappaB-regulated X-chromosome-linked iap gene expression protects endothelial cells from tumor necrosis factor alpha-induced apoptosis. J Exp Med 1998;188:211–216.
83. Matsui K, Fine A, Zhu B, Marshak-Rothstein A, Ju ST. Identification of two NF-kappa B sites in mouse CD95 ligand (Fas ligand) promoter: functional analysis in T cell hybridoma. J Immunol 1998;161:3469–3473.
84. Kasof GM, Lu JJ, Liu D, et al. Tumor necrosis factor-alpha induces the expression of DR6, a member of the TNF receptor family, through activation of NF-kappaB. Oncogene 2001;20:7965–7975.

85. Schauvliege R, Vanrobaeys J, Schotte P, Beyaert R. Caspase-11 gene expression in response to lipopolysaccharide and interferon-gamma requires nuclear factor-kappaB and signal transducer and activator of transcription (STAT) 1. J Biol Chem 2002;277: 41,624–41,630.
86. Kaltschmidt B, Kaltschmidt C, Hofmann TG, Hehner SP, Droge W, Schmitz ML. The pro- or anti-apoptotic function of NF-kappaB is determined by the nature of the apoptotic stimulus. Eur J Biochem 2000;267:3828–3835.
87. Bernard D, Monte D, Vandenbunder B, Abbadie C. The c-Rel transcription factor can both induce and inhibit apoptosis in the same cells via the upregulation of MnSOD. Oncogene 2002;21:4392–4402.
88. Grimm T, Schneider S, Naschberger E, et al. EBV latent membrane protein-1 protects B-cells from apoptosis by inhibition of BAX. Blood 2005;105:3263–3269.
89. Ravi R, Mookerjee B, van Hensbergen Y, et al. p53-mediated repression of nuclear factor-kappaB RelA via the transcriptional integrator p300. Cancer Res 1998;58:4531–4536.
90. Webster GA, Perkins ND. Transcriptional cross talk between NF-kappaB and p53. Mol Cell Biol 1999;19:3485–3495.
91. Wadgaonkar R, Phelps KM, Haque Z, Williams AJ, Silverman ES, Collins T. CREB-binding protein is a nuclear integrator of nuclear factor-kappaB and p53 signaling. J Biol Chem 1999;274:1879–1882.
92. Culmsee C, Siewe J, Junker V, et al. Reciprocal inhibition of p53 and nuclear factor-kappaB transcriptional activities determines cell survival or death in neurons. J Neurosci 2003;23:8586–8595.
93. Gu L, Findley HW, Zhou M. MDM2 induces NF-kappaB/p65 expression transcriptionally through Sp1-binding sites: a novel, p53-independent role of MDM2 in doxorubicin resistance in acute lymphoblastic leukemia. Blood 2002;99:3367–3375.
94. Wu H, Lozano G. NF-kappaB activation of p53. A potential mechanism for suppressing cell growth in response to stress. J Biol Chem 1994;269:20,067–20,074.
95. Hellin AC, Calmant P, Gielen J, Bours V, Merville MP. Nuclear factor - kappaB-dependent regulation of p53 gene expression induced by daunomycin genotoxic drug. Oncogene 1998;16:1187–1195.
96. Ryan KM, Ernst MK, Rice NR, Vousden KH. Role of NF-kappaB in p53-mediated programmed cell death. Nature 2000;404:892–897.
97. Kivch HC, Flaswinkel S, Rumpf H, Brockmann D, Esche H. Expression of human p53 requires synergistic activation of transcription from the p53 promoter by AP-1, NF-kappaB and Myc/Max. Oncogene 1999;18:2728–2738.
98. Chen M, Capps C, Willerson JT, Zoldhelyi P. E2F-1 regulates nuclear factor-kappaB activity and cell adhesion: potential antiinflammatory activity of the transcription factor E2F-1. Circulation 2002;106:2707–2713.
99. You Z, Madrid LV, Saims D, Sedivy J, Wang CY. c-Myc sensitizes cells to tumor necrosis factor-mediated apoptosis by inhibiting nuclear factor kappaB transactivation. J Biol Chem 2002;277:36,671–36,677.
100. Guttridge DC, Albanese C, Reuther JY, Pestell RG, Baldwin AS, Jr. NF-kappaB controls cell growth and differentiation through transcriptional regulation of cyclin D1. Mol Cell Biol 1999;19:5785–5799.
101. La Rosa FA, Pierce JW, Sonenshein GE. Differential regulation of the c-myc oncogene promoter by the NF-kappaB rel family of transcription factors. Mol Cell Biol 1994;14: 1039–1044.
102. Huang Y, Ohtani K, Iwanaga R, Matsumura Y, Nakamura M. Direct transactivation of the human cyclin D2 gene by the oncogene product Tax of human T-cell leukemia virus type I. Oncogene 2001;20:1094–1102.

103. Wang Z, Sicinski P, Weinberg RA, Zhang Y, Ravid K. Characterization of the mouse cyclin D3 gene: exon/intron organization and promoter activity. Genomics 1996;35:156–163.
104. De Smaele E, Zazzeroni F, Papa S, et al. Induction of gadd45beta by NF-kappaB downregulates pro-apoptotic JNK signalling. Nature 2001;414:308–313.
105. Jin R, De Smaele E, Zazzeroni F, et al. Regulation of the gadd45beta promoter by NF-kappaB. DNA Cell Biol 2002;21:491–503.
106. Papa S, Zazzeroni F, Bubici C, et al. Gadd45 beta mediates the NF-kappaB suppression of JNK signalling by targeting MKK7/JNKK2. Nat Cell Biol 2004;6:146–153.
107. Mak SK, Kultz D. Gadd45 proteins induce G2/M arrest and modulate apoptosis in kidney cells exposed to hyperosmotic stress. J Biol Chem 2004;279:39,075–39,084.
108. Taylor WR, Stark GR. Regulation of the G2/M transition by p53. Oncogene 2001;20: 1803–1815.
109. Hoffmann A, Levchenko A, Scott ML, Baltimore D. The IkappaB-NF-kappaB signaling module: temporal control and selective gene activation. Science 2002;298:1241–1245.
110. Lipniacki T, Paszek P, Brasier AR, Luxon B, Kimmel M. Mathematical model of NF-kappaB regulatory module. J Theor Biol 2004;228:195–215.

II

Molecular Basis of Disease Therapy

10

Regulation of NF-κB Function

Target for Drug Development

Daniel Sliva and Rakesh Srivastava

Summary

Nuclear factor-κB (NF-κB) is a family of transcription factors instrumental in a variety of physiological as well as pathophysiological conditions. Aberrant activation of NF-κB is responsible for the overexpression of different genes resulting in unregulated cell proliferation, survival, angiogenesis, cell adhesion, migration, and invasion that had been linked to cancer metastasis. Because the constitutively activated NF-κB had been demonstrated in a wide variety of cancers, its inhibition is a natural target for the development of new anticancer drugs. The activity of NF-κB is regulated by proteasome degradation, phosphorylation of NF-κB, and acetylation/deacetylation of histones. Therefore, compounds specifically suppressing proteasome activity, activities of kinases responsible for the phosphorylation of p65 subunit of NF-κB, and acetylation/deacetylation associated with NF-κB activation may be effective in the treatment in a wide variety of cancers.

Key Words: NF-κB; gene expression; cancer; cell growth; survival; angiogenesis; metastasis.

1. Introduction

In 1986, Sen and Baltimore *(1)* identified a nuclear factor, which bound to a sequence in the immunoglobulin κ-light-chain enhancer in B-cells stimulated with lipopolysaccharide. Although originally named nuclear factor-κB (NF-κB) in accordance with its discovery in B-cells, during the past 20 yr NF-κB has become a paradigm of transcription factor regulating of a variety of genes in numerous different cells and tissues. Because NF-κB is essential for immunity, cell proliferation, inflammation, and apoptosis, the activation of NF-κB has been linked to a variety of physiological as well as pathophysiological conditions and diseases.

NF-κB family of proteins consists of a group of structurally related and evolutionary conserved subunits that have been identified and cloned in mammalian cells *(2)*. These include Rel (c-Rel), RelA (p65), RelB, NF-κB1 (p50/p105), and NF-κB2 (p52/p100). The NF-κB subunits form homo- or heterodimers through their rel homology domain (RHD), which is also responsible for the DNA binding of NF-κB and interaction of NF-κB with IκB, the family of inhibitory proteins of NF-κB *(3)*. NF-κB usually exists in a latent state in the cytoplasm and its activation requires extracellular stimuli leading to the phosphorylation and subsequent proteasome-mediated degradation of inhibitory IκB proteins *(4)*. Finally, dimeric activated NF-κB translocates into the nucleus and controls the expression of target genes. However, constitutively active NF-κB was identified in a variety of cancers and linked to the oncogenesis and tumor resistance *(5)*.

From: *Apoptosis, Cell Signaling, and Human Diseases: Molecular Mechanisms, Volume 1*
Edited by R. Srivastava © Humana Press Inc., Totowa, NJ

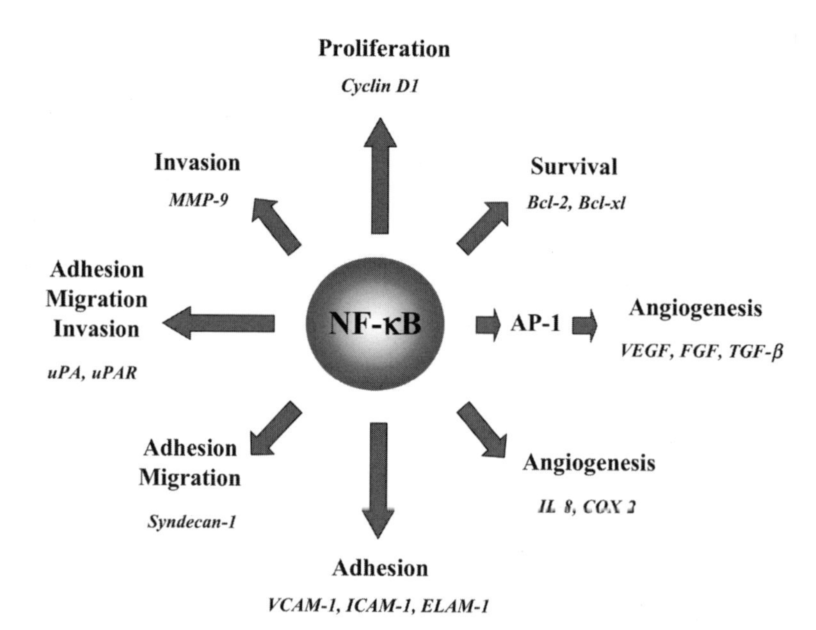

Fig. 1. NF-κB regulated genes involved in cancer growth and metastasis. Expression of genes which are crucial for the normal cell function were also identified in cancer growth and metastasis. These genes are responsible for cell proliferation, cell survival, angiogenesis, cell adhesion, migration, and invasion. However, the function of some gene products can be overlapping.

2. NF-κB and Oncogenesis

The involvement of NF-κB in cancer has been originally suggested because of the homology of mammalian c-Rel with highly oncogenic retroviral v-Rel, which induces aggressive tumors in chicken *(6,7)*. In mammalian cells, NF-κB also regulates expression of more than 150 genes and some of them were linked to cancer initiation, proliferation, angiogenesis, survival, and metastasis *(8)* (*see* Fig. 1). Interestingly, NF-κB activation was detected before tumor initiation, therefore connecting aberrant stimulation of expression of NF-κB-regulated genes prior to growth and progression of cancer *(9)*.

2.1. NF-κB and Cell Proliferation

Mitotic cellular division is a crucial feature in the growth of normal as well as cancer cells. The cellular division can be characterized as a cycling process where cells are proceeding from the resting stage (G_0) to DNA synthesis (S) and mitosis (M) stages of cell cycle. During the cell cycle the cells have to go through gap phases G_1 and G_2, which occur before and after DNA synthesis (S) phase. Cell division is conceived as a passage through a series of checkpoints, with the earliest point late in G1 when cells become irreversibly committed to DNA synthesis and division *(10,11)*. The checkpoints are controlled by a group of D cyclins, as well as cyclins E and A. Cyclin D1 acts in G_1 and promotes progression through the G_1–S phase of the cell cycle in mammalian cells. The cyclin-D1 protein is a regulatory subunit of holoenzyme, forming a complex with catalytic subunits of cyclin-dependent kinases (CDKs) cdk4 and cdk6, which phosphorylates and inactivates the retinoblastoma protein pRB

(10,11). The checkpoints of the cell cycle are usually deregulated in oncogenesis, and amplification or overexpression of cyclin D1 have been identified in the development of a subset of human cancers including cancers of breast *(12,13)*, prostate *(14)*, lung *(15)*, colon *(16)*, bladder *(17,18)*, ovary *(19)*, liver *(20)*, pancreas *(21)*, brain *(22)*, and esophagus *(23)*. Furthermore, overexpression of cyclin D1 has been identified in melanomas *(24)*, sarcomas *(25)*, parathyroid adenomas *(26)*, myelomas *(27)*, and myeloid leukemias *(28)*. In addition to the regulation of expression by NF-κB, the cyclin-D1 regulatory region also contains binding sites for activator protein (AP-1) (c-Jun/c-Fos) *(29)*, STAT5 *(30)*, E2F *(31)*, Sp-1/Sp-3 *(31)*, lymphoid enhancer factor-1 (LEF-1)/β-catenin *(32)*, and CREB/CREM proteins *(33)*. However, the constitutive activation of NF-κB (and AP-1) is probably the major factor responsible for the overexpression of cyclin D1 in a variety of cancers. The role of cyclin D1 in tumorigenesis was also confirmed by the overexpression of cyclin D1 in transgenic mice, which resulted in tumors of the mammary gland *(34)*.

2.2. NF-κB and Apoptosis

Apoptosis is a morphologically and biochemically specific form of cell death during physiological (e.g., embryological development and normal tissue turnover) as well as pathological conditions (including cancer, viral infections, autoimmune diseases, neurodegenerative disorders, and AIDS) *(35)*. In addition to the deregulation of cell proliferation, the suppression of cell death is crucial for the initiation and progression of a variety of cancers *(36)*. Apoptosis can be morphologically recognized by the blebbing of the plasma membrane—condensation of chromatin at the periphery of the nucleus—followed by disintegration of the cell into multiple membrane-enclosed vesicles, and nuclear fragmentation *(37)*. Biochemically, apoptosis can be the result of: (1) the activation of the classical apoptotic pathway through the mitochondrial cytochrome *c* and activation of proteases (caspases); (2) the necrotic programmed cell death mediated by the release of reactive oxygen species; or (3) the release of apoptosis inducing factors leading to the cell death without characteristic nuclear DNA fragmentation *(38,39)*. Caspases activation is controlled by the Bcl-2 family members, which are the key regulators that control release of cytochrome *c* and other apoptosis-promoting factors from mitochondria. Bcl-2 proteins have antiapoptotic as well as proapopototic functions, and the expression of antiapoptotic Bcl-2 and Bcl-xl was linked to a variety of chemotherapy resistant cancers *(40)*. For example, overexpression of Bcl-2 was associated with apoptosis loss in breast cancers and lymph nodes metastases *(41)*, Bcl-2 is frequently overexpressed in prostate cancer *(42)*, small-cell lung carcinoma *(43)*, colorectal adenomas and carcinomas *(44)*, cancers of bladder *(45)*, kidney *(46)*, and in B-cell chronic lymphocytic leukemia and non-Hodgkin's lymphoma *(47)*. Another antiapoptotic NF-κB regulated protein Bcl-xl, has been identified in invasive breast cancers *(48)*, and its expression correlated with increasing Gleason score and the presence of prostate cancer metastases *(49)*. In addition, expression of Bcl-xl was associated with the chemoresistance and chemotherapy failure in the treatment of ovarian cancer *(50)*. Bcl-xl overexpression was also identified in lung tumor cell lines *(51)*, colorectal cancers *(52)*, head and neck cancers *(53)*, metastatic malignant melanoma *(54)*, pancreatic cancer *(55)*, and lymphomas *(56)*. Therefore, small molecule ligands or antisense oligodeoxynucleotides for Bcl-2 and Bcl-xl, were recently suggested in preclinical testing to disrupt survival and chemoresistance properties of Bcl-2 and Bcl-xl in tumor cells *(57)*.

2.3. NF-κB and Angiogenesis

Angiogenesis, the formation of blood vessels by capillaries sprouting from preexisting vessels, is important for normal physiological processes such as embryogenesis, growth, and wound healing, as well as pathological processes such as tumor growth and metastasis *(58)*. Angiogenesis is a complex multistep process involving close orchestration of endothelial cells, the extracellular matrix, and angiogenic factors. Although vascular endothelial growth factor (VEGF) was originally the only growth factor proven to be specific for angiogenesis, other molecules have been implicated as positive regulators of angiogenesis, including fibroblast growth factor (FGF), transforming growth factor (TGF)-β, interleukin (IL)-8, cyclooxygenese (COX)-2, and others (for a review, *see* ref. 59). Expression of these angiogenic factors can be controlled by NF-κB indirectly (e.g., VEGF, FGF, TGF-β) or directly by NF-κB binding to the promoter region of specific genes (e.g., IL-8 and COX-2).

The up-regulation of VEGF expression has been detected in a variety of tumors including breast *(60,61)*, prostate *(62,63)*, lung *(64)*, gastrointestinal tract *(65)*, kidney and bladder *(66)*, brain *(67)*, ovary *(68)*, and thyroid *(69)*. In addition, increased levels of serum concentrations of VEGF were detected in solid tumors *(70–72)* as well as leukemia and non-Hodgkin's-lymphomas *(73)*. Although NF-κB controls expression of VEGF *(74)*, the VEGF promoter contains binding sites for AP-1, AP-2, Sp-1 *(75)*, hypoxia-inducible factor (HIF)-1α *(76)*, and STAT3 *(77)* but not for NF-κB. As recently demonstrated, NF-κB modulates the expression of VEGF indirectly through the induction of expression of c-Fos and AP-1 DNA-binding activity *(78)*.

FGF overexpression has been detected in cancers of breast *(79,80)*, prostate *(81)*, lung *(82)*, ovary *(83)*, brain *(84)*, thyroid *(85)*, as well as in melanomas *(86)* and non-Hodgkin's lymphoma *(87)*. In addition, TGF-β was overexpressed in cancers of breast *(88)*, prostate *(89)*, in lung cancer cells *(90)*, melanomas *(91)*, liver and pancreas tumors *(92,93)*, and Hodgkin's lymphoma *(94)*. Because the promoter region of FGF contains AP1, AP2, and Sp1 sites *(95)* and the promoter regions of TGF-β have Sp1, AP-1, EGR-1 binding sites *(96,97)*, it is possible that NF-κB can activate expression of FGF and TGF-β indirectly through the activation of AP-1 *(78,98)*.

The expression of other two factors implicated in angiogenesis, IL-8 and COX-2, is directly controlled by NF-κB *(99,100)*. In addition, their promoter regions contain binding sites for AP-1 and NF-IL6 for IL-8 *(99)*, and AP-1, NF-IL6, NFAT, and PEA3 for COX-2 *(100)*. Thus, IL-8 overexpression has been detected in cancers of breast *(101)*, prostate *(102)*, lung *(103)*, colon *(104)*, ovary *(105)*, melanoma *(106)*, thyroid *(107)*, brain *(108)*, pancreas *(109)*, leukemia *(110)*, and Hodgkin's disease *(111)*. Although the overexpression of COX-2 was originally described mainly in colorectal cancers *(112,113)*, COX-2 was also linked to the cancer of breast *(114,115)*, prostate *(116)*, lung *(117)*, bladder *(118)*, skin *(119)*, pancreas *(120)*, myeloid leukemia *(121)*, and clinical correlation between the COX-2 expression and prognostic factors in lymphoma patients was recently reported *(122)*.

2.4. NF-κB and Metastasis

Tumor invasion and metastasis are multifaceted processes involving cell adhesion, cell migration, and proteolytic degradation of tissue barriers *(123)*. Cell adhesion can be mediated through various cellular adhesion molecules, including vascular cell adhesion molecule (VCAM)-1, intercellular adhesion molecule (ICAM)-1, and endothelial-leukocyte adhesion molecule (ELAM)-1 (or E-selectin), which are regulated by NF-κB and AP-1 *(124–126)*. Significantly higher levels of VCAM-1 and ICAM-1 were detected in the patients with breast, ovarian, gastrointestinal and myeloma cancers *(127)*, gliomas *(128)*, pancreatic cancer *(129)*, and in patients with Hodgkin's disease *(130)*. VCAM-1 levels were elevated in patients with cancers of prostate *(131)*, colon *(131)*, and lymphoid and myeloid leukaemias *(132)*. ICAM-1 levels were elevated in sera from patients with melanoma *(130)* and expressed in pulmonary adenocarcinomas *(133)* and thyroid cancers *(134)*. Elevated levels of ELAM-1 were detected in the cancer of breast, ovary, gastrointestine *(127)*, prostate *(135)*, lung *(136)*, colon *(137)*, laryngeal carcinoma and oral carcinomas *(138)*, acute leukemia *(139)*, and Hodgkin's disease *(140)*. In addition to the typical cell adhesion molecules (VCAM-1, ICAM-1, and ELAM-1), other proteins were also implicated in cancer metastasis through cell adhesion, cell migration and cell invasion.

The expression of syndecan-1 and syndecan-4 (ryudocan), members of transmembrane heparan and chondroitin sulfate proteoglycans involved in cell–matrix signaling *(141)*, is also controlled by NF-κB *(142,143)*. Although expression of syndecan-1 inhibited invasion of lymphocytes *(144)*, overexpression of syndecan-4 decreased cell migration of Chinese hamster ovary (CHO) cells *(145)*, and the loss of syndecan expression was linked to tumor progression in a variety of cancers, other studies suggested tumor promoter function of syndecans *(146)*. Thus, increased expression syndecan-1 was detected in breast carcinomas with an aggressive phenotype and poor clinical behavior *(147)* and in hormone-refractory and metastatic prostate cancers *(148)*. In addition, enhanced syndecan-1 expression has been detected in ovarian *(149)*, pancreatic *(150)*, and gastric cancers *(151)*, and the serum levels of syndecan-1 correlated with increased tumor aggressiveness and poor outcome of lung cancer patients *(152)*. Syndecan-1 was also expressed in chronic lymphocytic leukaemia *(153)*, myeloma *(154)*, and Kaposi sarcoma-associated herpesvirus B-cell lymphomas *(155)*. Interestingly, overexpression of syndecan-4 was only detected in hepatocellular carcinomas and malignant mesotheliomas *(156,157)*. Therefore, the contradictory expression of syndecans in different cancers may reflect tissue and/or tumor stage-specific function, or may reflect the multiple functions of these integrin coreceptors *(146)*.

The urokinase-type plasminogen activator (uPA) is a serine protease that cleaves the extracellular matrix and stimulates the conversion of plasminogen to plasmin *(158)*. Plasmin can mediate cell invasion directly by degrading matrix proteins such as collagen IV, fibronectin, and laminin or indirectly by activating matrix metalloproteinases MMP-2, -3, and -9 and uPA *(159–162)*. Although originally identified for its proteolytic activity, uPA also binds to its receptor uPAR and form complexes with integrin receptors, which in turn modulate signaling pathways responsible for cytoskeleton reorganization resulting in cell adhesion and migration *(163)*. Expression of the uPA gene is controlled through the uPA promoter, which contains functional binding sites for the transcription factors AP-1, NF-κB, and PEA3 *(164,165)*, and expression of the uPAR gene is regulated by the transcription factors AP-1, AP-2, NF-κB, and Sp1 *(166)*. Both uPA and uPAR were identified in a variety of tumors and the correlation between uPA and uPAR overexpression and poor prognosis suggested *(167)*. Therefore, increased levels of uPA and/or uPAR in serum or tumors were identified in patients with breast *(168)*, prostate *(169)*, lung *(170)*, colon *(171)*, ovarian *(172)*, bladder *(173)*, endometrial *(174)*, head and neck cancers *(175)*, melanoma *(176)*, and acute leukemia *(177)*. Finally, inhibition of NF-κB resulted in the suppression of uPA secretion followed by inhibition of cell adhesion, migration, and invasion of highly metastatic breast cancer cells *(178,179)*.

Matrix metalloproteinases (MMPs) are a family of zinc-containing endopeptidases, which are responsible for the degradation of extracellular matrix during cell invasion in physiological (tissue remodeling during wound healing, embryonic development, and immune cell migration) as well as pathological processes as in cancer metastasis *(180)*. Especially, the expression of MMP-9 (type IV collagenase/gelatinase B), which is controlled by NF-κB and AP-1 *(181)*, has been correlated with metastasis in several systems. Plasma levels and/or expression of MMP-9 were elevated in cancers of breast *(182)*, prostate *(183)*, non-small cell lung cancer (NSCLC) *(184)*, colon *(182)*, bladder *(185)*, ovary *(186)*, kidney *(187)*, brain *(188)*, melanoma *(189)*, leukemia *(190)*, and lymphoma *(191)* and the levels of MMP-9 used as a prognostic factor of cancer progression.

3. Regulation of NF-κB

As summarized above, NF-κB controls the expression of genes involved in growth, survival, angiogenesis and metastasis in a variety of cancers (Table 1). Therefore, the activity of NF-κB is a natural target for the inhibition of aberrantly expressed proteins. The "classical pathway" of NF-κB activation, consisting of proteasome degradation of the natural inhibitor IκB and subsequent NF-κB dimerization, nuclear translocation and stimulation of gene transcription, is well recognized *(3,4)*. However, understanding how NF-κB is regulated will open more possibilities for the targeted modulation/inhibition of its activity.

3.1. Proteasome Degradation

The "classical" NF-κB activation pathway (for NF-κB subunits p50 and p65) is triggered by a variety of stimuli through specific receptors resulting in the activation of IκB-kinase (IKK) complex containing IKKα, IKKβ, and IKKγ (*see* Fig. 2). IKK activated complex in turn phosphorylates IκBα at Ser32 and Ser36 and is subsequently ubiquitinated and degraded by proteasome *(192)*. Alternative NF-κB activation pathway (for NF-κB subunits p52/p100 and RelB) employs NF-κB-inducing kinase (NIK), which activates an IKKα homodimer. Subsequently, IKKα phosphorylates p100 subunit of NF-κB, which is ubiquitinated and cleaved into a p52 subunit and forms the active NF-κB heterodimer with RelB *(192)*. Finally, the third pathway ("atypical") is independent of IKK, but still requires phosphorylation of IκBα by DNA damage induced p38-activated casein kinase 2 (CK2) followed by the IκBα degradation by proteasome *(193)*. The result of the activation of all three pathways is the translocation of NF-κB (p50/p65 or p52/RelB) complexes into the nucleus, where they activate gene expression. In addition, p65 in p50/p65 and RelB in p52/RelB must be phosphorylated for their optimal transcription activity *(194)*.

Table 1
Overexpression of NF-κB Regulated Genes in Cancer

	BCA	PrCA	LCA	PaCA	CCA	OCA	LEU
Proliferation							
Cyclin D1	+	+	+	+	+	+	+
Survival							
Bcl-2	+	+	+		+		+
Bcl-xl	+	+	+	+	+	+	
Angiogenesis							
VEGF	+	+	+		+	+	+
FGF	+	+	+			+	
TGF-β	+	+	+	+			
IL-8	+	+	+	+	+	+	+
COX-2	+	+	+	+			+
Adhesion							
VCAM-1	+	+		+	+	+	+
ICAM-1	+		+	+		+	
ELAM-1	+	+	+		+	+	+
Adhesion/Migration							
Syndecan-1	+	+	+	+	+		+
Adhesion/Migration/ Invasion							
uPA	+	+	+		+		+
uPAR	+	+	+		+	+	+
Invasion							
MMP-9	+	+	+		+	+	+

Note: NF-κB regulated genes that are overexpressed and/or have elevated serum levels elevated in leading cancer types. BCA, breast cancer; PrCA, prostate cancer; LCA, lung cancer; PaCA, pancreatic cancer; CCA, colon cancer; OCA, ovarian cancer; LEU, leukemia. References according to specific cancer may be found in the main text.

As mentioned previously, proteasome degradation of IκBα is a crucial step in the liberation of NF-κB and activation of gene transcription. The proteasome complex is a 26S multicatalytic protease composed of a 20S core catalytic complex with 14 different subunits and 19S regulatory complex that recognizes ubiquitinated proteins and regulate the activity of 20S *(195)*. Phosphorylated IκBα that will be degraded is first marked with ubiquitin chains by three subsequent reactions. Ubiquitin is activated by a ubiquitin-activating enzyme (E1), transferred to a ubiquitin-conjugating enzyme (E2), and finally ubiquitin-protein ligase (E3) transfers ubiquitin to lysine residues 21 and 22 of IκBα *(196)*. The ubiquitinated IκBα is then recognized by 19S and degraded by 20S complexes, respectively, resulting in the release of NF-κB *(197)*. Therefore, specific proteasome inhibitors were developed and screened for their antitumor activity *(198)*. One of them, bortezomib (Velcade), formerly known as PS-341, is the first proteasome inhibitor, which has been recently approved by the Food and Drug Administration for the treatment of newly diagnosed and relapsed/refractory multiple myeloma *(199)*. In addition, bortezomid has also recently shown significant preclinical and clinical activity in various hematologic and nonhematologic malignancies *(200–202)*.

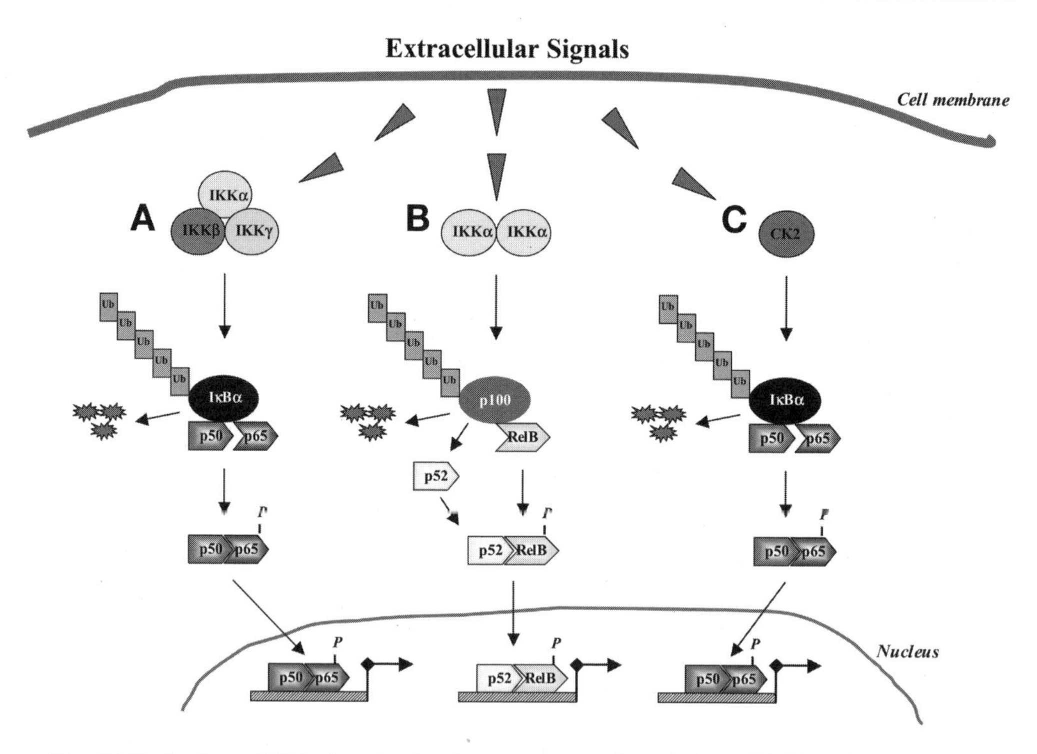

Fig. 2. Mechanism of NF-κB activation by proteasome degradation. **(A)** The "classical" NF-κB activation pathway through the activation of IKK complex (IKKα, IKKβ, IKKγ) results in the phosphorylation of IκBα and its ubiquitination and degradation leading to the liberation of NF-κB (p50 and p65) and NF-κB nuclear translocation. **(B)** The "alternative" NF-κB activation pathway through the activation of IKKα homodimers results in the phosphorylation of p100 protein and its proteolysis leading to the release of p52 which forms complex with RelB and translocates into nucleus. **(C)** The "atypical" NF-κB activation pathway through the activation of CK2 results in the degradation of IκBα and the liberation of NF-κB (p50 and p65) and NF-κB nuclear translocation.

3.2. Phosphorylation of NF-κB

In addition to the phosphorylation and subsequent degradation of IκBα, direct phosphorylation of NF-κB is also required for its optimal DNA-binding and transcription activation. Different protein kinases were identified to be necessary for the phosphorylation of specific sites in the NF-κB subunit p65 (*see* Fig. 3). Therefore, protein- kinase-A catalytic subunit (PKAc) phosphorylates p65 at Ser276 in the cytoplasm *(203)*, whereas mitogen- and stress-activated protein kinase-1 (MSKM1) phosphorylates the same site of p65 in the nucleus *(204)*. Protein kinase C-ζ (PKC-ζ) induces the activity of NF-κB through the phosphorylation of p65 at Ser311 *(205)* and casein kinase 2 (CK2) *(206)*, IKKα *(207)*, and Akt *(208)* phosphorylates p65 at Ser529. Interestingly, Ser536 in p65 was the target for the phosphorylation of a variety of kinases including IKKα *(207,209)*, IKKβ *(209)*, IKKε *(209)*, TRAF family member-associated (TANK)-binding kinase 1 (TBK1) *(209)*, serine/threonine kinases Akt *(208)*, ribosomal S6 kinase 1 (RSK1) *(210)*, and Bruton's tyrosine kinase (Btk) *(211)*. In addition, glycogen synthase kinase (GSK)-3β phosphorylated p65 on COOH-terminal transactivation domain, which contains four potential GSK-3 phosphorylation sites within amino acids 354-441 *(212)*.

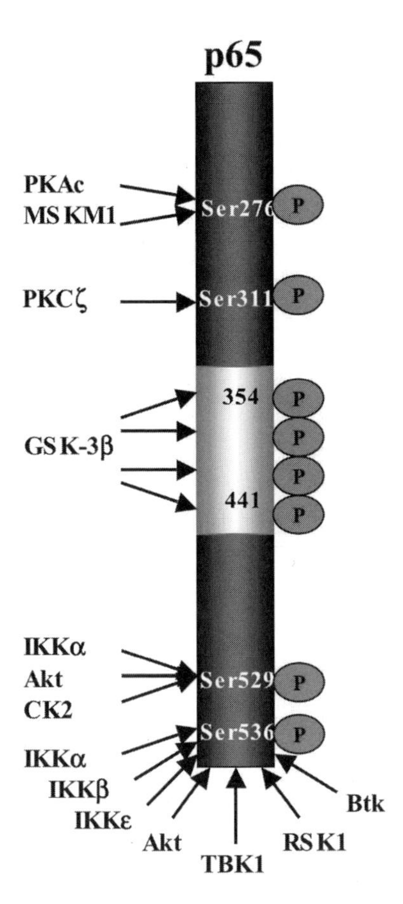

Fig. 3. Kinases phosphorylating p65 subunit of NF-κB. Direct phosphorylation of NF-κB enhances DNA-binding and transcription activation. Akt, serine/threonine kinase; Btk, Bruton's tyrosine kinase; CK2, casein kinase 2; GSK-3β, glycogen synthase kinase-3β; IKK, I-κB kinase; MSKM1, mitogen- and stress-activated protein kinase-1; PKAc, protein kinase A catalytic subunit; PKC-ζ, protein kinase C-ζ; RSK1, ribosomal S6 kinase; TBK-1, TRAF family member-associated (TANK)-binding kinase 1.

As demonstrated previously, different kinases are necessary for the phosphorylation of p65 and NF-κB activation. However, these data were obtained in different cell lines and with different biological stimuli. In addition, some of the phosphorylation sites might not be necessary for the full NF-κB activity in a particular system. Nevertheless, the kinases responsible for the phosphorylation of p65 are promising targets for the inhibition of aberrantly active NF-κB in cancer.

3.3. Acetylation and Deacetylation

Gene expression is a tightly regulated process requiring the fine-tuning of the transcription machinery through the establishment of an active transcription complex *(213)*. In its nonactive state, eukaryotic DNA is condensed into chromatin through the coiling of DNA on the surface of nucleosome resulting in the suppression of gene expression *(214)*. Transcription coactivators are able to uncoil the DNA from the nucleosome core (containing histone proteins H2A, H2B, H3 and H4) by acetylation of histones, therefore

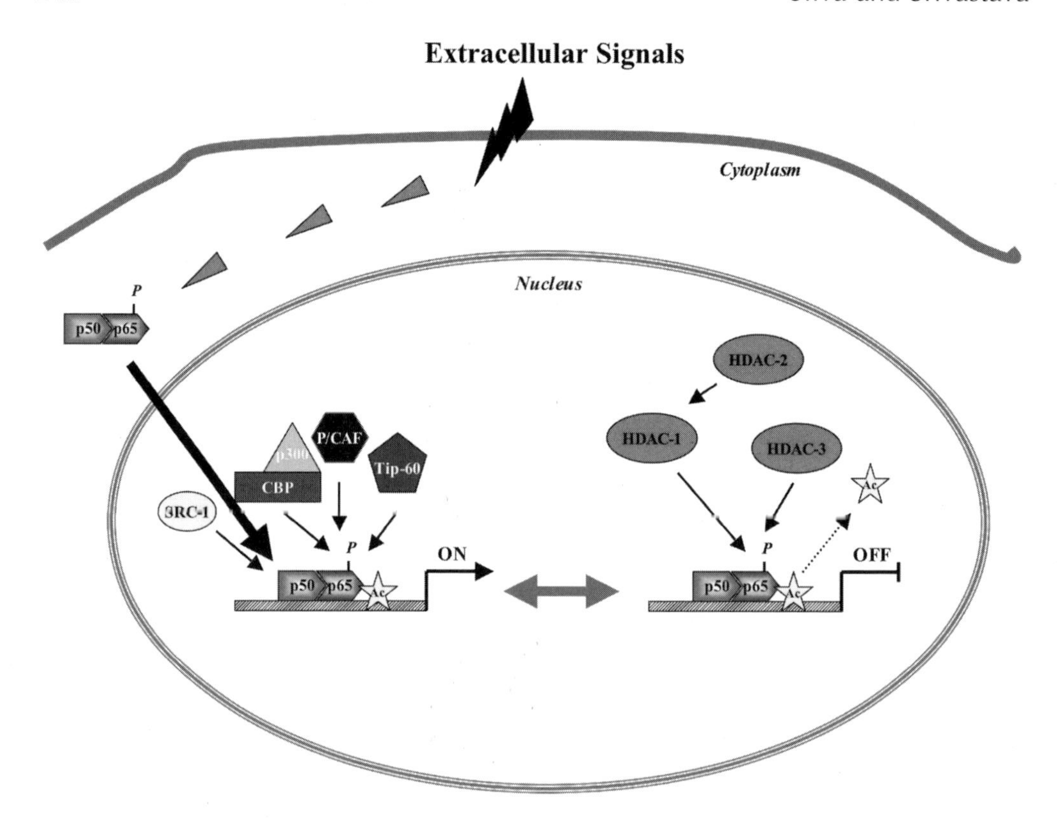

Fig. 4. Regulation of gene expression by acetylation/deacetylation. Direct interaction between HATs CBP/p300, p/CAF, Tip60, or SRC-1 with NF-κB results in histone acetylation and activation of gene expression. The interaction between HDACs and NF-κB results in histone deacetylation and suppression of gene expression. CBP/p300, cyclic AMP response element-CREB binding protein-CBP and p300; p/CAF, CBP-associated factor; SRC-1, steroid receptor coactivator-1.

opening and increasing the accessibility of DNA for the binding of transcription factors *(215)*. Because the histone acetylation is a reversible process, the regulation of gene transcription is controlled by coactivators—histone acetylases (HATs) and by corepressors—histone deacetylases (HDACs) (*see* Fig. 4).

Not surprisingly, the activation of transcription by NF-κB is regulated by acetylation and deacetylation. Therefore, histone acetylase CBP (cyclic AMP response element-CREB binding protein-CBP) and p300 (CBP/p300) *(216)*, steroid receptor coactivator-1 (SRC-1) *(217)*, CBP-associated factor (p/CAF) *(218)*, and Tip60 *(219)* potentiated the activation of NF-κB. Interestingly, phosphorylation of p65 at Ser276 enhanced the interaction of p65 with p300/CBP and stimulated the transcription activity of NF-κB *(220)*. However, NF-κB-induced activity by SRC-1 was the result of the interaction of p50 subunit of NF-κB with SRC-1 *(218)*.

NF-κB-dependent gene expression was repressed by histone deacetylases HDAC-1, HDAC-2 *(221)*, and HDAC-3 *(222)*. Although HDAC1 repressed NF-κB activity directly through its interaction with p65, HDAC2 activity on NF-κB was mediated indirectly through its association with HDAC1 *(221)*. As mentioned above, phosphorylation of p65 at Ser276 stimulated the transcription activity of NF-κB through p300/CBP interaction *(220)* and the same phosphorylation at Ser276 decreased the interaction of HDAC-1 with p65 *(223)*, suggesting that specific phosphorylation of p65 can determine if the activity of NF-κB will be regulated by coactivators or corepressors.

Although the importance of acetylation/deacetylation of histones was originally described for the activation of transcription mediated by NF-κB, HATs and HDACs can also directly acetylate/deacetylate p50 and p65 subunits of NF-κB *(224)*. Therefore, enhanced acetylation of p50 correlates with increased expression of COX-2 *(225)* and acetylated p65 was reversibly deacetylated by HDAC3, suggesting that HDAC3 can be an intranuclear molecular switch that controls the duration of the transcriptional response *(222,226)*. However, the transcription regulation through NF-κB is more complicated because acetylation of p65 can activate as well as suppress NF-κB-dependent gene expression *(222,227)*. Therefore, acetylation of p65 at lysines 218, 221, and 310 activates NF-κB by inhibiting IκBα binding to p65. Conversely, acetylation of p65 at lysines 122 and 123 suppresses NF-κB-dependent gene transcription by reducing DNA-binding to NF-κB *(227)*. Even though the exact molecular mechanism of acetylation/deacetylation-dependent transcriptional regulation is not fully elucidated, aberrant transcriptional repression by HDAC activity was associated with neoplastic transformation and HDAC inhibitors are considered as a promising new class of cancer therapeutic agents *(228)*.

However, HDAC inhibitor trichostatin A (TSA) demonstrated quite the opposite effects on the expression of NF-κB-regulated genes, which are involved in cancer proliferation, survival, and metastasis. Thus, TSA increased both basal and TNF-induced expression of IL-8 *(221)*, and TSA stimulated expression of Bcl-xl, MMP-9 and IL-8 in lung cancer cells *(229)*. In contrast, TSA down-regulated the expression of cyclin D1 by the inhibition of DNA binding of NF-κB/p65 independently of the acetylation status of p65 *(230)*. In addition, the inhibition of expression of NF-κB-regulated cyclin D1, COX-2, MMP-9 and Bcl-2 by indole-3-carbinol (natural compound with antitumor effects that is found in Brassica vegetables) in leukemia cells was associated with decreased acetylation of p65 *(231)*.

4. Summary

Significant progress has been made in our understanding of the mechanism of the activation of NF-κB during carcinogenesis, but many important aspects remain to be elucidated. Although the activation of NF-κB is vital for the proper function of the immune system, its aberrant activation results in inflammation and progression of cancer. Because constitutively activated NF-κB results in overexpression of genes involved in cancer cell proliferation, survival, angiogenesis, invasion, and metastasis, NF-κB is a characteristic signature for a wide variety of cancers. Thus, the inhibition of NF-κB is a rational target for the development of new drugs. Because the activity of NF-κB is regulated through different signaling mechanisms, their specific inhibition could result in the suppression of particular NF-κB-dependent gene expression in cancer cells. However, this inhibition should be specifically targeted only for cancer cells without affecting the regular immune function.

References

1. Sen R, Baltimore D. Inducibility of kappa immunoglobulin enhancer-binding protein Nf-kappa B by a posttranslational mechanism. Cell 1986;47:921–928.
2. Ghosh S, May MJ, Kopp EB. NF-κB and Rel proteins: evolutionary conserved mediators of immune responses. Annu Rev Immunol 1998;16:225–260.
3. Baldwin AS. The NF-κB and IκB: new discoveries and insights. Annu Rev Immunol 1996;14:649–683.
4. Karin M, Ben-Neriah Y. Phosphorylation meets ubiquitination: the control of NF-κB activity. Annu Rev Immunol 2000;18:621–663.
5. Mayo MW, Baldwin AS. The transcription factor NF-κB: control of oncogenesis and cancer therapy resistance. Biochem Biophys Acta 2000;1470:M55–M62.
6. Gilmore TD, Koedood M, Piffat K, White D. Rel/NF-kappaB/IkappaB proteins and cancer. Oncogene 1996;13:1567–1378.
7. Luque I, Gelinas C. Rel/NF-kappa B and I kappa B factors in oncogenesis. Semin Cancer Biol. 1997;8:103–111.
8. Pahl HL. Activators and target genes of Rel/NF-κB transcription factors. Oncogene 1999;18:6853–6866.
9. Kim DW, Sovak MA, Zanieski G, et al. Activation of NF-κB/Rel occurs early during neoplastic transformation of mammary cells. Carcinogenesis 2000;21:871–879.
10. Pestell RG, Albanese C, Reutens AT, Segall JE, Lee RJ, Arnold A. The cyclins and cyclin-dependent kinase inhibitors in hormonal regulation of proliferation and differentiation. Endocr Rev 1999;20:501–534.
11. Joyce D, Albanese C, Steer J, Fu M, Bouzahzah B, Pestell RG. NF-kappaB and cell-cycle regulation: the cyclin connection. Cytokine Growth Factor Rev 2001;12:73–90.
12. Gillett C, Fantl V, Smith R, et al. Amplification and overexpression of cyclin D1 in breast cancer detected by immunohistochemical staining. Cancer Res 1994;54:1812–1817.
13. McIntosh GG, Anderson JJ, Milton I, et al. Determination of the prognostic value of cyclin D1 overexpression in breast cancer. Oncogene 1995;11:885–891.
14. Wang L, Habuchi T, Mitsumori K, et al. Increased risk of prostate cancer associated with AA genotype of cyclin D1 gene A870G polymorphism. Int J Cancer 2003;103:116–120.
15. Reissmann PT, Koga H, Figlin RA, Holmes EC, Slamon DJ. Amplification and overexpression of the cyclin D1 and epidermal growth factor receptor genes in non-small-cell lung cancer. Lung Cancer Study Group. J Cancer Res Clin Oncol 1999;125:61–70.
16. Le Marchand L, Seifried A, Lum-Jones A, Donlon T, Wilkens LR. Association of the cyclin D1 A870G polymorphism with advanced colorectal cancer. JAMA 2003;290:2843–2848.
17. Bringuier PP, Tamimi Y, Schuuring E, Schalken J. Expression of cyclin D1 and EMS1 in bladder tumours; relationship with chromosome 11q13 amplification. Oncogene 1996;12:1747–1753.
18. Wang L, Habuchi T, Takahashi T, et al. Cyclin D1 gene polymorphism is associated with an increased risk of urinary bladder cancer. Carcinogenesis 2002;23:257–264.
19. Hung WC, Chai CY, Huang JS, Chuang LY. Expression of cyclin D1 and *C-Ki-ras* gene product in human epithelial ovarian tumors. Hum Pathol 1996;27:1324–1328.
20. Zhang YJ, Jiang W, Chen C, et al. Amplification and overexpression of cyclin D1 in human hepatocellular carcinoma. Biochem Biophys Res Commun 1993;196:1010–1016.
21. Gansauge S, Gansauge F, Ramadani M, et al. Overexpression of cyclin D1 in human pancreatic carcinoma is associated with poor prognosis. Cancer Res 1997;57:1634–1637.
22. Cavalla P, Dutto A, Piva R, Richiardi P, Grosso R, Schiffer D. Cyclin D1 expression in gliomas. Acta Neuropathol (Berl) 1998;95:131–135.

23. Jiang W, Zhang YJ, Kahn SM, et al. Altered expression of the cyclin D1 and retinoblastoma genes in human esophageal cancer. Proc Natl Acad Sci USA 1993;90:9026–9030.
24. Sauter ER, Yeo UC, von Stemm A, et al. Cyclin D1 is a candidate oncogene in cutaneous melanoma. Cancer Res 2002;62:3200–3206.
25. Kim SH, Lewis JJ, Brennan MF, Woodruff JM, Dudas M, Cordon-Cardo C. Overexpression of cyclin D1 is associated with poor prognosis in extremity soft-tissue sarcomas. Clin Cancer Res 1998;4:2377–2382.
26. Hemmer S, Wasenius VM, Haglund C, et al. Deletion of 11q23 and cyclin D1 overexpression are frequent aberrations in parathyroid adenomas. Am J Pathol 2001;158:1355–1362.
27. Bergsagel PL, Kuehl WM. Critical roles for immunoglobulin translocations and cyclin D dysregulation in multiple myeloma. Immunol Rev 2003;194:96–104.
28. Liu JH, Yen CC, Lin YC, et al. Overexpression of cyclin D1 in accelerated-phase chronic myeloid leukemia. Leuk Lymphoma. 2004;45:2419–2425.
29. Albanese C, Johnson J, Watanabe G, Eklund N, Vu D, Arnold A, Pestell RG. Transforming $p21^{ras}$ mutants and c-Ets-2 activate the cyclin D1 promoter through distinguishable regions. J Biol Chem 1995;270:23,589–23,597.
30. Matsumura I, Kitamura T, Wakao H, et al. Transcriptional regulation of cyclin D1 promoter by STAT5: its involvement in cytokine-dependent growth of hematopoietic cells. EMBO J 1999;18:101–111.
31. Watanabe G, Albanese C, Lee RJ, et al. Inhibition of cyclin D-kinase activity is associated with E2F-mediated inhibition of cyclin D1 promoter activity through E2F and Sp1. Mol Cell Biol 1998;18:3212–3222.
32. Shtutman M, Zhurinsky J, Simcha I, et al. The cyclin D-1 gene is a target of the β-catenin/LEF-1 pathway. Proc Natl Acad Sci USA 1999;96:5522–5527.
33. Watanabe G, Howe A, Lee RJ, et al. Induction of cyclin D1 by simian virus 40 small tumor antigen. Proc Natl Acad Sci USA 1996;93:12,861–12,866.
34. Wang TC, Cardiff RD, Zukerberg L, Lees E, Arnold A, Schmidt EV. Mammary hyperplasia and carcinoma in MMTV-cyclin D1 transgenic mice. Nature 1994;369:669–671.
35. Thomson CB. Apoptosis in the pathogenesis and treatment of disease. Science 1995;267: 1456–1462.
36. Evan GI, Vousden KH. Proliferation, cell cycle and apoptosis in cancer. Nature 2001;411: 342–348.
37. Arends MJ, Wyllie AH. Apoptosis: mechanisms and roles in pathology. Int Rev Exp Path 1991;32:223–254.
38. Kaufmann SH, Hengartner MO. Programmed cell death: alive and well in the new millennium. Trends Cell Biol 2001;11:526–534.
39. Leist M, Jaattela M. Four deaths and a funeral: from caspases to alternative mechanisms. Nat Rev Mol Cell Biol 2001;2:589–598.
40. Amundson SA, Myers TG, Scudiero D, Kitada S, Reed JC, Fornace AJ, Jr. An informatics approach identifying markers of chemosensitivity in human cancer cell lines. Cancer Res 2000;60:6101–6110.
41. Sierra A, Castellsague X, Tortola S, et al. Apoptosis loss and bcl-2 expression: key determinants of lymph node metastases in T1 breast cancer. Clin Cancer Res 1996;2:1887–1894.
42. Baltaci S, Orhan D, Ozer G, Tolunay O, Gogous O. Bcl-2 proto-oncogene expression in low- and high-grade prostatic intraepithelial neoplasia. BJU Int 2000;85:155–159.
43. Jiang SX, Sato Y, Kuwao S, Kameya T. Expression of *bcl*-2 oncogene protein is prevalent in small cell lung carcinomas. J Pathol 1995;177:135–138.
44. Sinicrope FA, Ruan S, Cleary KR, Lee JJ, Levin B. Bcl-2 and p53 oncoprotein expression during colorectal tumorigenesis. Cancer Res 1995;55:237–241.

45. Miyake H, Hanada N, Nakamura H, et al. Overexpression of Bcl-2 in bladder cancer cells inhibits apoptosis induced by cisplatin and adenoviral-mediated p53 gene transfer. Oncogene 1998;16:933–943.
46. Tomita Y, Bilim V, Kawasaki T, et al. Frequent expression of Bcl-2 in renal-cell carcinomas carrying wild-type p53. Int J Cancer 1996;66:322–325.
47. Zaja F, Di Loreto C, Amoroso V, et al. BCL-2 immunohistochemical evaluation in B-cell chronic lymphocytic leukemia and hairy cell leukemia before treatment with fludarabine and 2-chloro-deoxy-adenosine. Leuk Lymphoma 1998;28:567–572.
48. Olopade OI, Adeyanju MO, Safa AR, et al. Overexpression of BCL-x protein in primary breast cancer is associated with high tumor grade and nodal metastases. Cancer J Sci Am 1997;3:230–237.
49. Krajewski S, Krajewska M, Shabaik A, et al. Immunohistochemical analysis of *in vivo* patterns of Bcl-X expression. Cancer Res 1994;54:5501–5507.
50. Williams J, Lucas PC, Griffith KA, et al. Expression of Bcl-xL in ovarian carcinoma is associated with chemoresistance and recurrent disease. Gynecol Oncol 2005;96:287–295.
51. Reeve JG, Xiong J, Morgan J, Bleehen NM. Expression of apoptosis-regulatory genes in lung tumour cell lines: relationship to p53 expression and relevance to acquired drug resistance. Br J Cancer 1996;73:1193–1200.
52. Biroccio A, Benassi B, D'Agnano I, et al. c-Myb and Bcl-x overexpression predicts poor prognosis in colorectal cancer: clinical and experimental findings. Br J Cancer 1996;73: 1193–1200.
53. Pena JC, Thompson CB, Recant W, Vokes EE, Rudin CM. Bcl-xL and Bcl-2 expression in squamous cell carcinoma of the head and neck. Cancer 1999;85:164–170.
54. Tang L, Tron VA, Reed JC, et al. Expression of apoptosis regulators in cutaneous malignant melanoma. Clin Cancer Res 1998;4:1865–1871.
55. Friess H, Lu Z, Andren-Sandberg A, et al. Moderate activation of the apoptosis inhibitor Bcl-X_L worsens the prognosis in pancreatic cancer. Ann Surg 1998;228:780–787.
56. Xerri L, Parc P, Brousset P, et al. Predominant expression of the long isoform of Bcl-x (Bcl-xL) in human lymphomas. Br J Haematol 1996;92:900–906.
57. O'Neill J, Manion M, Schwartz P, Hockenbery DM. Promises and challenges of targeting Bcl-2 anti-apoptotic proteins for cancer therapy. Biochim Biophys Acta 2004;1705: 43–51.
58. Folkman J. Angiogenesis in cancer, vascular, rheumatoid and other disease, Nat Med 1995;1:27–31.
59. Yancopoulos GD, Davis S, Gale NW, Rudge JS, Wiegand SJ, Holash J. Vascular-specific growth factors and blood vessel formation. Nature 2000;407:242–248.
60. Brown LF, Berse B, Jackman RW, et al. Expression of vascular permeability factor (vascular endothelial growth factor) and its receptors in breast cancer. Hum Pathol 1995;26:86–91.
61. Yoshiji H, Gomez DE, Shibuya M, Thorgeirsson UP. Expression of vascular endothelial growth factor, its receptors, and other angiogenic factors in breast cancer. Cancer Res 1996; 56:2013–2016.
62. Harper ME, Glynne-Jones E, Goddard L, Thurston VJ, Griffiths K. Vascular endothelial growth factor (VEGF) expression in prostatic tumours and its relationship to neuroendocrine cells. Br J Cancer 1996;74:910–916.
63. Ferrer FA, Miller LJ, Andrawis RI, et al. Angiogenesis and prostate cancer: in vivo and in vitro expression of angiogenesis factors by prostate cancer cells. Urology 1998;51: 161–167.
64. Mattern J, Koomagi R, Volm M. Association of vascular endothelial growth factor expression with intratumoral microvessel density and tumour cell proliferation in human epidermoid lung carcinoma. Br J Cancer 1996;73:931–934.

65. Brown LF, Berse B, Jackman RW, et al. Expression of vascular permeability factor (vascular endothelial growth factor) and its receptors in adenocarcinomas of the gastrointestinal tract. Cancer Res 1993;53:4727–4735.
66. Brown LF, Berse B, Jackman RW, et al. Increased expression of vascular permeability factor (vascular endothelial growth factor) and its receptors in kidney and bladder carcinomas. Am J Pathol 1993;143:1255–1262.
67. Plate KH, Breier G, Weich HA, Risau W. Vascular endothelial growth factor is a potential tumour angiogenesis factor in human gliomas *in vivo*. Nature 1992:359:845–847.
68. Olson TA, Mohanraj D, Carson LF, Ramakrishnan S. Vascular permeability factor gene expression in normal and neoplastic human ovaries. Cancer Res 1994;54:276–280.
69. Viglietto G, Maglione D, Rambaldi M, et al. Upregulation of vascular endothelial growth factor (VEGF) and downregulation of placenta growth factor (PlGF) associated with malignancy in human thyroid tumors and cell lines. Oncogene 1995;11: 1569–1579.
70. Fujisaki K, Mitsuyama K, Toyonaga A, Matsuo K, Tanikawa K. Circulating vascular endothelial growth factor in patients with colorectal cancer. Am J Gastroenterol 1998;93: 249–252.
71. Salven P, Ruotsalaine NT, Mattson K, Joennsu H. High pretreatment serum level of vascular endothelial growth factor (VEGF) is associated with poor outcome in small cell lung cancer. Int J Cancer 1998;79:144–146.
72. Takano S, Yoshii Y, Kondo S, et al. Concentration of vascular endothelial growth factor in the serum and tumor tissue of brain tumor patients. Cancer Res 1996;56:2185–2190.
73. Moehler TM, Ho AD, Goldschmidt H, Barlogie B. Angiogenesis in hematologic malignancies. Crit Rev Oncol Hematol 2003;45:227–244.
74. Huang S, Robinson JB, Deguzman A, Bucana CD, Fidler IJ. Blockade of nuclear factor-kappaB signaling inhibits angiogenesis and tumorigenicity of human ovarian cancer cells by suppressing expression of vascular endothelial growth factor and interleukin 8. Cancer Res 2000;60:5334–5339.
75. Tischer E, Mitchell R, Hartman T, et al. The human gene for vascular endothelial growth factor. Multiple protein forms are encoded through alternative exon splicing. J Biol Chem 1991;2 66:11,947–11,954.
76. Forsythe JA, Jiang BH, Iyer NV, et al. Activation of vascular endothelial growth factor gene transcription by hypoxia-inducible factor 1. Mol Cell Biol 1996;16:4604–4613.
77. Wei D, Le X, Zheng L, et al. Stat3 activation regulates the expression of vascular endothelial growth factor and human pancreatic cancer angiogenesis and metastasis. Oncogene 2003; 22:319–329.
78. Fujioka S, Niu J, Schmidt C, et al. NF-κB and AP-1 connection: mechanism of NF-κB-dependent regulation of AP-1 activity. Mol Cell Biol 2004;24:7806–7819.
79. Penault-Llorca F, Bertucci F, Adelaide J, et al. Expression of FGF and FGF receptor genes in human breast cancer. Int J Cancer 1995;61:170–176.
80. Smith J, Yelland A, Baillie R, Coombes RC. Acidic and basic fibroblast growth factors in human breast tissue. Eur J Cancer 1994;30:496–503.
81. Giri D, Ropiquet F, Ittmann M. Alterations in expression of basic fibroblast growth factor (FGF) 2 and its receptor FGFR-1 in human prostate cancer. Clin Cancer Res 1999;5: 1063–1071.
82. Berger W, Setinek U, Mohr T, et al. Evidence for a role of FGF-2 and FGF receptors in the proliferation of non-small cell lung cancer cells. Int J Cancer 1999;83:415–423.
83. Fujimoto J, Ichigo S, Hori M, Hirose R, Sakaguchi H, Tamaya T. Expression of basic fibroblast growth factor and its mRNA in advanced ovarian cancers. Eur J Gynaecol Oncol 1997;18:349–352.

84. Ueba T, Takahashi JA, Fukumoto M, et al. Expression of fibroblast growth factor receptor-1 in human glioma and meningioma tissues. Neurosurgery 1994;34:221–225.
85. Thompson SD, Franklyn JA, Watkinson JC, Verhaeg JM, Sheppard MC, Eggo MC. Fibroblast growth factors 1 and 2 and fibroblast growth factor receptor 1 are elevated in thyroid hyperplasia. J Clin Endocrinol Metab 1998;83:1336–1341.
86. Albino AP, Davis BM, Nanus DM. Induction of growth factor RNA expression in human malignant melanoma: markers of transformation. Cancer Res 1991;51:4815–4820.
87. Bertolini F, Paolucci M, Peccatori F, et al. Angiogenic growth factors and endostatin in non-Hodgkin's lymphoma. Br J Haematol 1999;106:504–509.
88. Gorsch SM, Memoli VA, Stukel TA, Gold LI, Arrick BA. Immunohistochemical staining for transforming growth factor beta 1 associates with disease progression in human breast cancer. Cancer Res 1992;52:6949–6952.
89. Wikstrom P, Stattin P, Franck-Lissbrant I, Damber JE, Bergh A. Transforming growth factor beta1 is associated with angiogenesis, metastasis, and poor clinical outcome in prostate cancer. Prostate 1998;37:19–29.
90. Jakowlew SB, Mathias A, Chung P, Moody TW. Expression of transforming growth factor beta ligand and receptor messenger RNAs in lung cancer cell lines. Cell Growth Differen 1995;6:465–476.
91. Reed JA, McNutt NS, Prieto VG, Albino AP. Expression of transforming growth factor-beta 2 in malignant melanoma correlates with the depth of tumor invasion. Implications for tumor progression. Am J Pathol 1994;145:97–104.
92. Ito N, Kawata S, Tamura S, et al. Positive correlation of plasma transforming growth factor-beta 1 levels with tumor vascularity in hepatocellular carcinoma. Cancer Lett 1995;89:45–48.
93. Friess H, Yamanaka Y, Buchler M, et al. Enhanced expression of transforming growth factor beta isoforms in pancreatic cancer correlates with decreased survival. Gastroenterology 1993;105:1846–1856.
94. Kadin ME, Agnarsson BA, Ellingsworth LR, Newcom SR. Immunohistochemical evidence of a role for transforming growth factor beta in the pathogenesis of nodular sclerosing Hodgkin's disease. Am J Pathol 1990;136:1209–1214.
95. Payson RA, Chotani MA, Chiu IM. Regulation of a promoter of the fibroblast growth factor 1 gene in prostate and breast cancer cells. J Steroid Biochem Mol Biol 1998;66:93–103.
96. Kim SJ, Angel P, Lafyatis R, et al. Autoinduction of transforming growth factor beta is mediated by the AP-1 complex. Mol Cell Biol 1990;10:1492–1497.
97. Liu C, Calogero A, Ragona G, Adamson E, Mercola D. EGR-1, the reluctant suppression factor: EGR-1 is known to function in the regulation of growth, differentiation, and also has significant tumor suppressor activity and a mechanism involving the induction of TGF-beta1 is postulated to account for this suppressor activity. Crit Rev Oncog 1996;7:101–125.
98. Stanley G, Harvey K, Slivova V, Jiang J, Sliva D. Ganoderma lucidum suppresses angiogenesis through the inhibition of secretion of VEGF and TGF-beta1 from prostate cancer cells. Biochem Biophys Res Commun 2005;330:46–52.
99. Mukaida N, Okamoto S, Ishikawa Y, Matsushima K. Molecular mechanism of interleukin-8 gene expression. J Leukocyte Biol 1994;56:554–558.
100. Smith WL, DeWitt DL, Garavito RM. Cyclooxygenases: structural, cellular and molecular biology. Annu Rev Biochem 2000;69:145–182.
101. Green AR, Green VL, White MC, Speirs V. Expression of cytokine messenger RNA in normal and neoplastic human breast tissue: identification of interleukin-8 as a potential regulatory factor in breast tumors. Int J Cancer 1997;72:937–941.
102. Veltri RW, Miller MC, Zhao G, et al. Interleukin-8 serum levels in patients with benign prostatic hyperplasia and prostate cancer. Urology 1999;53:139–147.

103. Smith DR, Polverini PJ, Kunkel SL, et al. Inhibition of interleukin 8 attenuates angiogenesis in bronchogenic carcinoma. J Exp Med 1994;179:1409–1415.
104. Cuenca RE, Azizkhan RG, Haskill S. Characterization of GRO alpha, beta and gamma expression in human colonic tumours: potential significance of cytokine involvement. Surg Oncol 1992;1:417–422.
105. Ivarsson K, Runesson E, Sundfeldt K, et al. The chemotactic cytokine interleukin-8—a cyst fluid marker for malignant epithelial ovarian cancer? Gynecol Oncol 1998;71:420–423.
106. Scheibenbogen C, Mohler T, Haefele J, Hunstein W, Keilholz U. Serum interleukin-8 (IL-8) is elevated in patients with metastatic melanoma and correlates with tumour load. Melanoma Res 1995;5:179–181.
107. Yoshida M, Matsuzaki H, Sakata K, et al. Neutrophil chemotactic factors produced by a cell line from thyroid carcinoma. Cancer Res 1992;52:464–469.
108. Morita M, Kasahara T, Mukaida N, et al. Induction and regulation of IL-8 and MCAF production in human brain tumor cell lines and brain tumor tissues. Eur Cytokine Network 1993;4:351–358.
109. Shi Q, Abbruzzese J, Huang S, Fidler IJ, and Xie K. Constitutive and inducible interleukin-8 expression by hypoxia and acidosis renders human pancreatic cancer cells more tumorigenic and metastatic. Clin Cancer Res 1999;5:3711–3721.
110. Di Celle PF, Carbone A, Marchis D, et al. Cytokine gene expression in B-cell chronic lymphocytic leukemia: evidence of constitutive interleukin-8 (IL-8) mRNA expression and secretion of biologically active IL-8 protein. Blood 1994;84:220–228.
111. Gruss HJ, Brach MA, Drexler HG, Bonifer R, Mertelsmann RH, Herrmann F. Expression of cytokine genes, cytokine receptor genes, and transcription factors in cultured Hodgkin and Reed-Sternberg cells. Cancer Res 1992;52:3353–3360.
112. Eberhart CE, Coffey RJ, Radhika A, Giardiello FM, Ferrenbach S, DuBois RN. Up-regulation of cyclooxygenase 2 gene expression in human colorectal adenomas and adenocarcinomas. Gastroenterology 1994;107:1183–1188.
113. Kargman SL, O'Neill GP, Vickers PJ, Evans JF, Mancini JA, Jothy S. Expression of prostaglandin G/H synthase-1 and -2 protein in human colon cancer. Cancer Res 1995;55: 2556–2559.
114. Costa C, Soares R, Reis-Filho JS, Leitao D, Amendoeira I, Schmitt FC. Cyclooxygenase 2 expression is associated with angiogenesis and lymph node metastasis in human breast cancer. J Clin Pathol 2002;55:429–434.
115. Half E, Tang XM, Gwyn K, Sahin A, Wathen K, Sinicrope FA. Cyclooxygenase-2 expression in human breast cancers and adjacent ductal carcinoma in situ. Cancer Res 2002;62: 1676–1681.
116. Gupta S, Srivastava M, Ahmad N, Bostwick DG, Mukhtar H. Over-expression of cyclooxygenase-2 in human prostate adenocarcinoma. Prostate 2000;42:73–78.
117. Hasturk S, Kemp B, Kalapurakal SK, Kurie JM, Hong WK, Lee JS. Expression of cyclooxygenase-1 and cyclooxygenase-2 in bronchial epithelium and nonsmall cell lung carcinoma. Cancer 2002;94:1023–1031.
118. Komhoff M, Guan Y, Shappell HW, et al. Enhanced expression of cyclooxygenase-2 in high grade human transitional cell bladder carcinomas. Am J Pathol 2000;157:29–35.
119. Buckman SY, Gresham A, Hale P, Hruza G, Anast J, Masferrer J, Pentland AP. COX-2 expression is induced by UVB exposure in human skin: implications for the development of skin cancer. Carcinogenesis 1998;19:723–729.
120. Maitra A, Ashfag A, Gunn CR, et al. Cyclooxygenase 2 expression in pancreatic adenocarcinoma and pancreatic intraepithelial neoplasia: an immunohistochemical analysis with automated cellular imaging. Am J Clin Pathol 2002;194–201.

121. Zetterberg E, Lundberg LG, Palmblad J. Expression of cox-2, tie-2 and glycodelin by megakaryocytes in patients with chronic myeloid leukaemia and polycythaemia vera. Br J Haematol 2003;121:497–499.
122. Hazar B, Ergin M, Seyrek E, Erdogan S, Tuncer I, Hakverdi S. Cyclooxygenase-2 (Cox-2) expression in lymphomas. Leuk Lymphoma 2004;45:1395–1399.
123. Price JT, Bonovich MT, Kohn EC. The biochemistry of cancer dissemination. Crit Rev Biochem Mol Biol 1997;32:175–253.
124. Cybulsky MI, Fries JW, Williams AJ, et al. Gene structure, chromosomal location, and basis for alternative mRNA splicing of the human VCAM1 gene. Proc Natl Acad Sci USA 1991;88:7859–7863.
125. van de Stolpe A, Caldenhoven E, Stade BG, et al. 12-O-tetradecanoylphorbol-13-acetate- and tumor necrosis factor alpha-mediated induction of intercellular adhesion molecule-1 is inhibited by dexamethasone. Functional analysis of the human intercellular adhesion molecular-1 promoter. J Biol Chem 1994;269:6185–6192.
126. Montgomery KF, Osborn L, Hession C, et al. Activation of endothelial-leukocyte adhesion molecule 1 (ELAM-1) gene transcription. Proc Natl Acad Sci USA 1991;88:6523–6527.
127. Banks RE, Gearing AJ, Hemingway IK, Norfolk DR, Perren TJ, Selby PJ. Circulating intercellular adhesion molecule-1 (ICAM-1), E-selectin and vascular cell adhesion molecule-1 (VCAM-1) in human malignancies. Br J Cancer 1993;68:122–124.
128. Maenpaa A, Kovanen PE, Paetau A, Jaaskelainen J, Timonen T. Lymphocyte adhesion molecule ligands and extracellular matrix proteins in gliomas and normal brain: expression of VCAM-1 in gliomas. Acta Neuropathol (Berl). 1997;94:216–225.
129. Tempia-Caliera AA, Horvath LZ, Zimmermann A, et al. Adhesion molecules in human pancreatic cancer. J Surg Oncol 2002;79:93–100.
130. Christiansen I, Sundstrom C, Enblad G, Totterman TH. Soluble vascular cell adhesion molecule-1 (sVCAM-1) is an independent prognostic marker in Hodgkin's disease. Br J Haematol 1998;102:701–709.
131. Lynch DF, Jr, Hassen W, Clements MA, Schellhammer PF, Wright GL, Jr. Serum levels of endothelial and neural cell adhesion molecules in prostate cancer. Prostate 1997;32: 214–220.
132. Reuss-Borst MA, Ning Y, Klein G, Muller CA. The vascular cell adhesion molecule (VCAM-1) is expressed on a subset of lymphoid and myeloid leukaemias. Br J Haematol 1995;89:299–305.
133. Jiang Z, Woda BA, Savas L, Fraire AE. Expression of ICAM-1, VCAM-1, and LFA-1 in adenocarcinoma of the lung with observations on the expression of these adhesion molecules in non-neoplastic lung tissue. Mod Pathol 1998;11:1189–1192.
134. Pasieka Z, Stepien H, Komorowski J, Kolomecki K, Kuzdak K. Evaluation of the levels of bFGF, VEGF, sICAM-1, and sVCAM-1 in serum of patients with thyroid cancer. Recent Results Cancer Res 2003;162:189–194.
135. Bhaskar V, Law DA, Ibsen E, et al. E-selectin up-regulation allows for targeted drug delivery in prostate cancer. Cancer Res 2003;63:6387–6394.
136. Roselli M, Mineo TC, Martini F, et al. Soluble selectin levels in patients with lung cancer. Int J Biol Markers 2002;17:56–62.
137. Takahashi Y, Mai M, Watanabe M, Tokiwa M, Nishioka K. Relationship between serum ELAM-1 and metastasis among patients with colon cancer. Dis Colon Rectum 1998;41: 770–774.
138. Liu CM, Sheen TS, Ko JY, Shun CT. Circulating intercellular adhesion molecule 1 (ICAM-1), E-selectin and vascular cell adhesion molecule 1 (VCAM-1) in head and neck cancer. Br J Cancer 1999;79:360–362.
139. Sudhoff T, Wehmeier A, Kliche KO, et al. Levels of circulating endothelial adhesion molecules (sE-selectin and sVCAM-1) in adult patients with acute leukemia. Leukemia 1996;10:682–686.

140. Syrigos KN, Salgami E, Karayiannakis AJ, Katirtzoglou N, Sekara E, Roussou P. Prognostic significance of soluble adhesion molecules in Hodgkin's disease. Anticancer Res 2004;24:1243–1247.
141. Carey DJ. Syndecans: multifunctional cell-surface co-receptors. Biochem J 1997;327:1–16.
142. Hinkes MT, Goldberger OA, Neumann PE, Kokenyesi R, Bernfield M. Organization and promoter activity of the mouse syndecan-1 gene. J Biol Chem 1993;268:11,440–11,448.
143. Takagi A, Kojima T, Tsuzuki S, et al. Structural organization and promoter activity of the human ryudocan gene. J Biochem (Tokyo) 1996;119:979–984.
144. Liebersbach BF, Sanderson RD. Expression of syndecan-1 inhibits cell invasion into type I collagen. J Biol Chem 1994;269:20,013–20,019.
145. Longley RL, Woods A, Fleetwood A, Cowling GJ, Gallagher JT, Couchman JR. Control of morphology, cytoskeleton and migration by syndecan-4. J Cell Sci 1999;112: 3421–3431.
146. Beauvais DM, Rapraeger AC. Syndecans in tumor cell adhesion and signaling. Reprod Biol Endocrinol 2004;2:3 (doi: 10.1186/1477-7827-2-3. Published online 2004 January 7).
147. Barbareschi M, Maisonneuve P, Aldovini D, et al. High syndecan-1 expression in breast carcinoma is related to an aggressive phenotype and to poorer prognosis. Cancer 2003;98: 474–483.
148. Zellweger T, Ninck C, Bloch M, et al. Expression patterns of potential therapeutic targets in prostate cancer. Int J Cancer 2005;113:619–628.
149. Davies EJ, Blackhall FH, Shanks JH, et al. Distribution and clinical significance of heparan sulfate proteoglycans in ovarian cancer. Clin Cancer Res 2004;10:5178–5186.
150. Conejo JR, Kleeff J, Koliopanos A, et al. Syndecan-1 expression is up-regulated in pancreatic but not in other gastrointestinal cancers. Int J Cancer 2000;88:12–20.
151. Wiksten JP, Lundin J, Nordling S, et al. Epithelial and stromal syndecan-1 expression as predictor of outcome in patients with gastric cancer. Int J Cancer 2001;95:1–6.
152. Joensuu H, Anttonen A, Eriksson M, et al. Soluble syndecan-1 and serum basic fibroblast growth factor are new prognostic factors in lung cancer. Cancer Res 2002;62: 5210–5271.
153. Sebestyen A, Kovalszky I, Mihalik R, et al. Expression of syndecan-1 in human B cell chronic lymphocytic leukaemia. Eur J Cancer 1997;33:2273–2277.
154. Yang Y, Yaccoby S, Liu W, et al. Soluble syndecan-1 promotes growth of myeloma tumors in vivo. Blood 2002;100:610–617.
155. Deloose ST, Smit LA, Pals FT, Kersten MJ, van Noesel CJ, Pals ST. High incidence of Kaposi sarcoma-associated herpesvirus infection in HIV-related solid immunoblastic/plasmablastic diffuse large B-cell lymphoma. Leukemia 2005, in press.
156. Roskams T, De Vos R, David G, Van Damme B, Desmet V. Heparan sulphate proteoglycan expression in human primary liver tumours. J Pathol 1998;185:290–297.
157. Gulyas M, Hjerpe A. Proteoglycans and WT1 as markers for distinguishing adenocarcinoma, epithelioid mesothelioma, and benign mesothelium. J Pathol 2003;199:479–487.
158. Blasi F. uPA, uPAR, PAI-I: A key intersection in proteolysis, adhesion and chemotaxis highways? Immunol Today 1997;18:415–417.
159. Sheela S, Barrett JC. Degradation of type IV collagen by neoplastic human skin fibroblasts. Carcinogenesis 1982;3:363–369.
160. Danø K, Andreasen PA, Grondahl-Hansen J, Kristensen P, Nielsen LS, Skriver L. Plasminogen activators, tissue degradation, and cancer. Adv Cancer Res 1985;44:139–266.
161. Murphy G, Willenbrock F, Crabbe T, et al. Regulation of matrix metalloproteinase activity. Ann NY Acad Sci 1994;732:31–41.
162. Schmitt M, Janicke F, Moniwa N, Chuckolowski N, Pache L, Graeff H. Tumor-associated urokinase-type plasminogen activator: biological and clinical significance. Biol Chem Hoppe-Seyler 1992;373:611–622.

163. Blasi F, Carmeliet P. uPAR: a versatile signalling orchestrator. Nat Rev Mol Cell Biol 2002;3:932–943.
164. Nerlov C, Rorth P, Blasi F, Johnsen M. Essential AP-1 and PEA3 binding elements in the human urokinase enhancer display cell type-specific activity. Oncogene 1991;6:1583–1593.
165. Lengyel E, Gum R, Stepp E, Juarez J, Wang H, Boyd D. Regulation of urokinase-type plasminogen activator expression by an ERK1-dependent signaling pathway in a squamous cell carcinoma cell line. J Cell Biochem 1996;61:430–438.
166. Wang Y. The role and regulation of urokinase-type plasminogen activator receptor gene expression in cancer invasion and metastasis. Med Res Rev 2001;21:146–170.
167. Andreasen PA, Kjøller L, Christensen L, Duffy MJ. The urokinase-type plasminogen activator system in cancer metastasis: a review. Int J Cancer 1997;72:1–22.
168. Duggan C, Maguire T, McDermott E, O'Higgins N, Fennelly JJ, Duffy MJ. Urokinase plasminogen activator and urokinase plasminogen activator receptor in breast cancer. Int J Cancer 1995;61:597–600.
169. Miyake H, Hara I, Yamanaka K, Gohji K, Arakawa S, Kamidono S. Elevation of serum levels of urokinase-type plasminogen activator and its receptor is associated with disease progression and prognosis in patients with prostate cancer. Prostate 1999;39:123–129.
170. Pedersen H, Brunner N, Francis D, et al. Prognostic impact of urokinase, urokinase receptor, and type 1 plasminogen activator inhibitor in squamous and large cell lung cancer tissue. Cancer Res 1994;54:4671–4675.
171. Berger DH. Plasmin/plasminogen system in colorectal cancer. World J Surg 2002;26: 767–771.
172. Sier CF, Stephens R, Bizik J, et al. The level of urokinase-type plasminogen activator receptor is increased in serum of ovarian cancer patients. Cancer Res 1998;58:1843–1849.
173. Seddighzadeh M, Steineck G, Larsson P, et al. Expression of UPA and UPAR is associated with the clinical course of urinary bladder neoplasms. Int J Cancer 2002;99:721–726.
174. Tecimer C, Doering DL, Goldsmith LJ, Meyer JS, Abdulhay G, Wittliff JL. Clinical relevance of urokinase-type plasminogen activator, its receptor, and its inhibitor type 1 in endometrial cancer. Gynecol Oncol 2001;80:48–55.
175. Schmidt M, Hoppe F. Increased levels of urokinase receptor in plasma of head and neck squamous cell carcinoma patients, Acta Otolaryngol 1999;119:949–953.
176. Stabuc B, Markovic J, Bartenjev I, Vrhovec I, Medved U, Kocijancic B. Urokinase-type plasminogen activator and plasminogen activator inhibitor type 1 and type 2 in stage I malignant melanoma. Oncol Rep 2003;10:635–639.
177. Scherrer A, Wohlwend A, Kruithof EK, Vassalli JD, Sappino AP. Plasminogen activation in human acute leukaemias. Br J Haematol 1999;105:920–927.
178. Sliva D, Rizzo MT, English D. Phosphatidylinositol 3-Kinase and NF-κB regulate motility of invasive MDA-MB-231 human breast cancer cells by the secretion of urokinase-type plasminogen activator (uPA). J Biol Chem 2002;277:3150–3157.
179. Slivova V, Zaloga G, DeMichele SJ, et al. Green tea polyphenols modulate secretion of urokinase plasminogen activator (uPA) and inhibit invasive behavior of breast cancer cells. Nutrition and Cancer 2005, in press.
180. Matrisian LM. The matrix-degrading metalloproteinases. Bioessays 1992;14:455–463.
181. He C. Molecular mechanism of transcriptional activation of human gelatinase B by proximal promoter. Cancer Lett 1996;106:185–191.
182. Zucker S, Lysik RM, Zarrabi MH, Moll U. M(r) 92,000 type IV collagenase is increased in plasma of patients with colon cancer and breast cancer. Cancer Res 1993;53:140–146.
183. Hamdy FC, Fadlon EJ, Cottam D, et al. Matrix metalloproteinase 9 expression in primary human prostatic adenocarcinoma and benign prostatic hyperplasia. Br J Cancer 1994;69:177–182.

184. Iizasa T, Fujisawa T, Suzuki M, et al. Elevated levels of circulating plasma matrix metalloproteinase 9 in non-small cell lung cancer patients. Clin Cancer Res 1999;5:149–153.
185. Davies B, Waxman J, Wasan H, et al. Levels of matrix metalloproteases in bladder cancer correlate with tumor grade and invasion. Cancer Res 1993;53:5365–5369.
186. Takemura M, Azuma C, Kimura T, Kanai T, Saji F, Tanizawa O. Type-IV collagenase and tissue inhibitor of metalloproteinase in ovarian cancer tissues. Int J Gynaecol Obstet 1994;46:303–309.
187. Lein M, Jung K, Laube C, et al. Matrix-metalloproteinases and their inhibitors in plasma and tumor tissue of patients with renal cell carcinoma. Int J Cancer 2000;85:801–804.
188. Kachra Z, Beaulieu E, Delbecchi L, et al. Expression of matrix metalloproteinases and their inhibitors in human brain tumors. Clin Exp Metastasis 1999;17:555–566.
189. Bodey B, Bodey B, Jr, Siegel SE, Kaiser HE. Matrix metalloproteinase expression in malignant melanomas: tumor-extracellular matrix interactions in invasion and metastasis. In Vivo 2001;15:57–64.
190. Matsuzaki A, Janowska-Wieczorek A. Unstimulated human acute myelogenous leukemia blasts secrete matrix metalloproteinases. J Cancer Res Clin Oncol 1997;123:100–106.
191. Kossakowska AE, Urbanski SJ, Huchcroft SA, Edwards DR. Relationship between the clinical aggressiveness of large cell immunoblastic lymphomas and expression of 92 kDa gelatinase (type IV collagenase) and tissue inhibitor of metalloproteinases-1 (TIMP-1) RNAs. Oncol Res 1992;4:233–240.
192. Ghosh S, Karin M. Missing pieces in the NF-κB puzzle. Cell 2002;108:S81–S96.
193. Kato T, Jr, Delhase M, Hoffmann A, Karin M. CK2 Is a C-Terminal IkappaB Kinase Responsible for NF-kappaB Activation during the UV Response. Mol Cell 2003;12: 829–839.
194. Viatour P, Merville MP, Bours V, Chariot A. Phosphorylation of NF-kappaB and IkappaB proteins: implications in cancer and inflammation. Trends Biochem Sci 2005; 30:43–52.
195. Baumeister W, Lupas A. The proteasome. Curr Opin Struct Biol 1997;7:273–278.
196. Scheffner M, Smith S, Stefan Jentsch S. The ubiquitin-conjugation system. In: Ubiquitin and the biology of the cell. Peters JM, Harris JR, Finley D. eds. New York, Plenum Press, 1998.
197. Scherer DC, Brockman JA, Chen Z, Maniatis T, Ballard DW. Signal-induced degradation of I kappa B alpha requires site-specific ubiquitination. Proc Natl Acad Sci USA 1995;92:11,259–11,263.
198. Adams J, Palombella VJ, Sausville EA, et al. Proteasome inhibitors: a novel class of potent and effective antitumor agents. Cancer Res 1999;59:2615–2622.
199. Chauhan D, Hideshima T, Anderson KC. Proteasome inhibition in multiple myeloma: therapeutic implication. Annu Rev Pharmacol Toxicol 2005;45:465–476.
200. Adams J. The proteasome as a novel target for the treatment of breast cancer. Breast Dis 2002;15:61–70.
201. Price N, Dreicer R. Phase I/II trial of bortezomib plus docetaxel in patients with advanced androgen-independent prostate cancer. Clin Prostate Cancer 2004;3:141–143.
202. Rajkumar SV, Richardson PG, Hideshima T, Anderson KC. Proteasome inhibition as a novel therapeutic target in human cancer. J Clin Oncol 2005;23:630–639.
203. Zhong H, SuYang H, Erdjument-Bromage H, Tempst P, Ghosh S. The transcriptional activity of NF-kappaB is regulated by the IkappaB-associated PKAc subunit through a cyclic AMP-independent mechanism. Cell 1997;89:413–424.
204. Vermeulen L, De Wilde G, Van Damme P, Vanden Berghe W, Haegeman G. Transcriptional activation of the NF-kappaB p65 subunit by mitogen- and stress-activated protein kinase-1 (MSK1). EMBO J 2003;22:1313–1324.

205. Duran A, Diaz-Meco MT, Moscat J. Essential role of RelA Ser311 phosphorylation by zetaPKC in NF-kappaB transcriptional activation. EMBO J 2003;22:3910–3918.
206. Bird TA, Schooley K, Dower SK, Hagen H, Virca GD. Activation of nuclear transcription factor NF-kappaB by interleukin-1 is accompanied by casein kinase II-mediated phosphorylation of the p65 subunit. J Biol Chem 1997;272:32,606–32,612.
207. O'Mahony AM, Montano M, Van Beneden K, Chen LF, Greene WC. Human T-cell lymphotropic virus type 1 tax induction of biologically Active NF-kappaB requires IkappaB kinase-1-mediated phosphorylation of RelA/p65. J Biol Chem 2004;279:18,137–18,145.
208. Madrid LV, Mayo MW, Reuther JY, Baldwin AS, Jr. Akt stimulates the transactivation potential of the RelA/p65 Subunit of NF-kappa B through utilization of the Ikappa B kinase and activation of the mitogen-activated protein kinase p38. J Biol Chem 2001;276:18,934–18,940.
209. Buss H, Dorrie A, Schmitz ML, Hoffmann E, Resch K, Kracht M. Constitutive and interleukin-1-inducible phosphorylation of p65 NF-κB at serine 536 is mediated by multiple protein kinases including IκB kinase (IKK)-α, IKKβ, IKKε, TRAF family member-associated (TANK)-binding kinase 1 (TBK1), and an unknown kinase and couples p65 to TATA-binding protein-associated factor II31-mediated interleukin-8 transcription. J Biol Chem 2004;279:55,633–55,643.
210. Bohuslav J, Chen LF, Kwon H, Mu Y, Greene WC. p53 induces NF-kappaB activation by an IkappaB kinase-independent mechanism involving phosphorylation of p65 by ribosomal S6 kinase 1. J Biol Chem 2004;279:26,115–26,125.
211. Doyle SL, Jefferies CA, O'Neill LA. Bruton's tyrosine kinase is involved in p65-mediated transactivation and phosphorylation of p65 on serine 536 during NF-kappa B activation by LPS. J Biol Chem 2005, in press.
212. Schwabe RF, Brenner DA. Role of glycogen synthase kinase-3 in TNF-alpha-induced NF-kappaB activation and apoptosis in hepatocytes. Am J Physiol Gastrointest Liver Physiol 2002;283:G204–G211.
213. Sternglanz R. Histone acetylation: a gateway to transcriptional activation. Trends Biol Sci 1996;21:357–358.
214. Wu W. Chromatin remodeling and the control of gene expression. J Biol Chem 1997; 272:28,171–28,174.
215. Imhof A, Wolffe AP. Transcription: gene control by targeted histone acetylation. Curr Biol 1998;8:R422–R424.
216. Gerritsen ME, Williams AJ, Neish AS, Moore S, Shi Y, Collins T. CREB-binding protein/p300 are transcriptional coactivators of p65. Proc Natl Acad Sci USA 1997;94: 2927–2932.
217. Na SY, Lee SK, Han SJ, Choi HS, Im SY, Lee JW. Steroid receptor coactivator-1 interacts with the p50 subunit and coactivates nuclear factor kappaB-mediated transactivations. J Biol Chem 1998;273:10,831–10,834.
218. Sheppard KA, Rose DW, Haque ZK, et al. Transcriptional activation by NF-kappaB requires multiple coactivators. Mol Cell Biol 1999;19:6367–6378.
219. Baek SH, Ohgi KA, Rose DW, Koo EH, Glass CK, Rosenfeld MG. Exchange of N-CoR corepressor and Tip60 coactivator complexes links gene expression by NF-κB and β-amyloid precursor protein. Cell 2002;110:55–67.
220. Zhong H, Voll RE, Ghosh S. Phosphorylation of NF-kappa B p65 by PKA stimulates transcriptional activity by promoting a novel bivalent interaction with the coactivator CBP/p300. Mol Cell 1998;1:661–671.
221. Ashburner BP, Westerheide SD, Baldwin AS, Jr. The p65 (RelA) subunit of NF-kappaB interacts with the histone deacetylase (HDAC) corepressors HDAC-1 and HDAC-2 to negatively regulate gene expression. Mol Cell Biol 2001;21;7065–7077.

222. Chen LW, Fischle W, Verdin E, Greene WC. Duration of nuclear NF-kappaB action regulated by reversible acetylation. Science 2001;293:1653–1657.
223. Zhong H, May MJ, Jimi E, Ghosh S. The phosphorylation status of nuclear NF-kappa B determines its association with CBP/p300 or HDAC-1. Mol Cell 2002;9:625–636.
224. Quivy V, Van Lint C. Regulation at multiple levels of NF-kappaB-mediated transactivation by protein acetylation. Biochem Pharmacol 2004;68:1221–1229.
225. Deng WG, Zhu Y, Wu KK. Up-regulation of p300 binding and p50 acetylation in tumor necrosis factor-α-induced cyclooxygenase-2 promoter activation. J Biol Chem 2003;278: 4770–4777.
226. Kiernan R, Bres V, Ng RW, et al. Post-activation turn-off of NF-κB-dependent transcription is regulated by acetylation of p65. J Biol Chem 2003;278:2758–2766.
227. Chen LF, Mu Y, Greene WC. Acetylation of RelA at discrete sites regulates distinct nuclear functions of NF-κB. EMBO J 2002;21:6539–6548.
228. Johnstone RW, Licht JD. Histone deacetylase inhibitors in cancer therapy: is transcription the primary target? Cancer Cell 2003;4:13–18.
229. Mayo MW, Denlinger CE, Broad RM, et al. Ineffectiveness of histone deacetylase inhibitors to induce apoptosis involves the transcriptional activation of NF-kappa B through the Akt pathway. J Biol Chem 2003;278:18,980–18,989.
230. Hu J, Colburn NH. Histone deacetylase inhibition down-regulates cyclin D1 transcription by inhibiting nuclear factor-kappaB/p65 DNA binding. Mol Cancer Res 2005;3:100–109.
231. Takada Y, Andreeff M, Aggarwal BB. Indole-3-carbinol suppresses NF-κ B and IκB αkinase activation causing inhibition of expression of NF-κB-regulated antiapoptotic and metastatic gene products and enhancement of apoptosis in myeloid and leukemia cells. Blood 2005; in press.

11

5-Fluorouracil

Molecular Mechanisms of Cell Death

Daniel B. Longley and Patrick G. Johnston

Summary

5-Fluorouracil (5-FU) has been the mainstay of colorectal cancer treatment for over 40 years. However, response rates for 5-FU in advanced colorectal cancer are modest. Although combining 5-FU with the newer chemotherapeutic agents oxaliplatin and irinotecan has improved response rates, new therapeutic strategies are necessary. Understanding the molecular mechanism by which tumors become resistant to 5-FU is needed if drug resistance is to be overcome. Tumor drug resistance is often due to insufficient chemotherapy-induced cell death. In this chapter, we describe the mechanisms of action of 5-FU, focusing on 5-FU-induced cell death. In the future, strategies aimed at increasing the effectiveness of 5-FU may target the cell death and cell survival pathways that are activated by this drug.

Key Words: 5-Fluorouracil; apoptosis; death receptors; growth factor receptors; Bcl-2.

1. Introduction

5-Fluorouracil (FU) is widely used in the treatment of a range of cancers including breast and cancers of the aerodigestive tract, but has had the greatest impact in colorectal cancer. 5-FU-based chemotherapy improves overall and disease-free survival of patients with resected stage III colorectal cancer *(1)*. Nonetheless, response rates for 5-FU-based chemotherapy as a first-line treatment for advanced colorectal cancer are only between 10 and 15% *(2)*. Combination of 5-FU with newer chemotherapies, such as irinotecan and oxaliplatin, has improved the response rates for advanced colorectal cancer to between 40 and 50% *(3,4)*. However, despite these improvements, new therapeutic strategies are urgently needed. Understanding the mechanisms by which 5-FU causes cell death and by which tumors become resistant to 5-FU is an essential step toward overcoming that resistance. This chapter discusses the mechanisms of action of 5-FUand highlights important determinants of drug sensitivity, focusing, in particular, on regulation of apoptosis in response to 5-FU, as we believe that many future therapeutic strategies will be aimed at enhancing the effectiveness of 5-FU by modulating cell death and cell survival pathways.

2. Mechanism of Action of 5-Fluorouracil

5-Fluorouracil (FU) is converted intracellularly to several active metabolites: fluorodeoxyuridine monophosphate (FdUMP), fluorodeoxyuridine triphosphate (FdUTP), and fluorouridine triphosphate (FUTP) (*see* Fig. 1). The active metabolites of 5-FU disrupt

From: *Apoptosis, Cell Signaling, and Human Diseases: Molecular Mechanisms, Volume 1*
Edited by R. Srivastava © Humana Press Inc., Totowa, NJ

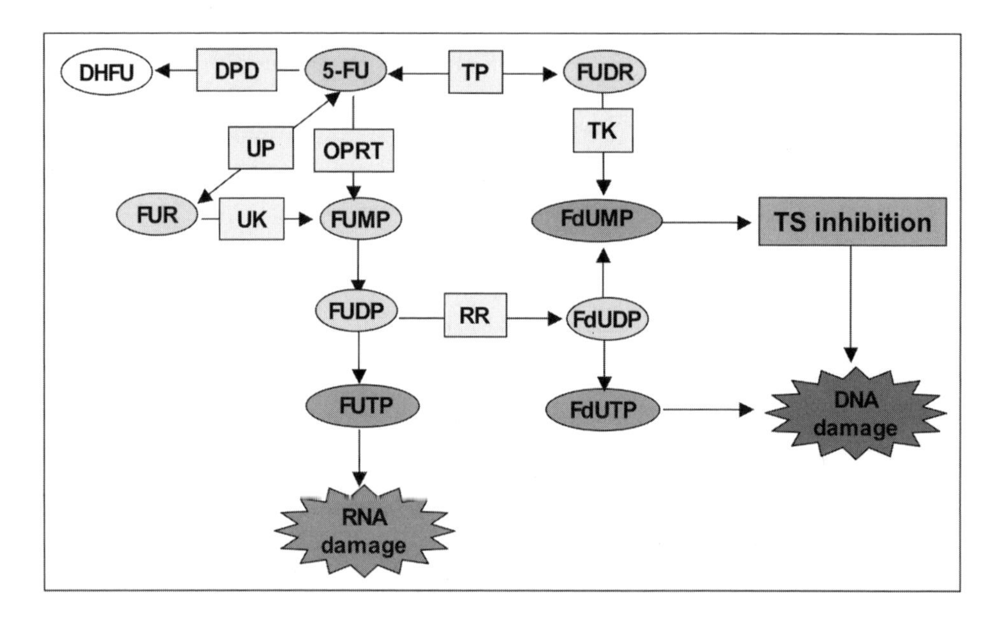

Fig. 1. 5-Fluorouracil (5-FU) is converted to three major active metabolites: (1) fluorodeoxyuridine monophosphate (FdUMP), (2) fluorodeoxyuridine triphosphate (FdUTP), and (3) fluorouridine triphosphate (FUTP). The main mechanism of 5-FU activation is conversion to fluorouridine monophosphate (FUMP) either directly by orotate phosphoribosyl transferase (OPRT), or indirectly via fluorouridine (FUR) through the sequential action of uridine phosphorylase (UP) and uridine kinase (UK). FUMP is then phosporylated to fluorouridine diphosphate (FUDP), which can be either further phosphorylated to the active metabolite fluorouridine triphosphate (FUTP), or converted to fluorodeoxyuridine diphosphate (FdUDP) by ribonucleotide reductase (RR). In turn, FdUDP can either be phosphorylated or dephosphorylated to generate the active metabolites FdUTP and FdUMP respectively. An alternative activation pathway involves the thymidine phosphorylase catalyzed conversion of 5-FU to fluorodeoxyuridine (FUDR), which is then phosphorylated by thymidine kinase (TK) to the thymidylate synthase (TS) inhibitor, FdUMP. Dihydropyrimidine dehydrogenase (DPD)-mediated conversion of 5-FU to dihydrofluorouracil (DHFU) is the rate-limiting step of 5-FU catabolism in normal and tumor cells.

RNA synthesis (FUTP), inhibit the action of thymidylate synthase (TS)—a nucleotide synthetic enzyme (FdUMP)—and can also be directly misincorporated into DNA (FdUTP). The rate-limiting enzyme in 5-FU catabolism is dihydropyrimidine dehydrogenase (DPD), which converts 5-FU to dihydrofluorouracil (DHFU) *(5)*. Over 80% of administered 5-FU is normally catabolized primarily in the liver, where DPD is abundantly expressed *(5)*.

2.1. TS Inhibition

TS catalyses the reductive methylation of deoxyuridine monophosphate (dUMP) to deoxythymidine monophosphate (dTMP) with the reduced folate 5,10-methylenetetrahydrofolate (CH_2THF) as the methyl donor. This reaction provides the sole *de novo* source of thymidylate, which is necessary for DNA replication and repair. TS contains a nucleotide-binding site and a binding site for CH_2THF. The 5-FU metabolite FdUMP binds to the nucleotide-binding site of TS, forming a stable ternary complex with the enzyme and CH_2THF which blocks binding of the normal substrate dUMP, thereby inhibiting dTMP synthesis *(6,7)*. Inhibition of thymidylate synthesis causes disruption of nucleotide levels that results in DNA damage.

2.2. DNA Damage

The capacity of a cancer cell to repair DNA can determine resistance to chemotherapeutic drugs that induce DNA damage such as 5-FU. The response to DNA damage is either repair or cell death and therefore has a profound effect on tumor chemosensitivity and chemoresistance. Signaling pathways have evolved to arrest the cell cycle following DNA damage to allow more time for DNA repair. Only when repair is incomplete (e.g., when the DNA damage is too extensive) will cells undergo apoptosis. The exact molecular mechanisms that mediate events downstream of TS inhibition have not been fully elucidated. Depletion of dTMP results in subsequent depletion of deoxythymidine triphosphate (dTTP), which induces perturbations in the levels of the other deoxynucleotides (e.g., dATP, dGTP, and dCTP) through various feedback mechanisms *(8)*. Deoxynucleotide pool imbalances (in particular the dATP/dTTP ratio) are thought to severely disrupt DNA synthesis and repair, resulting in lethal DNA damage *(9,10)*. TS inhibition can also result in accumulation of deoxyuridine triphosphate (dUTP), as conversion of deoxyuridine monophosphate (dUMP) to deoxythymidine monophosphate (dTMP) is blocked *(11)*. Both dUTP and another 5-FU metabolite fluorodeoxyuridine triphosphate (FdUTP) can be misincorporated into DNA. Repair of uracil- and 5-FU-containing DNA is mediated by the base excision repair enzyme uracil-DNA-glycosylase (UDG) *(12)*. However, this repair mechanism is futile in the presence of high (F)dUTP/dTTP ratios, and only results in further false nucleotide incorporation. These futile cycles of misincorporation, excision and repair eventually lead to DNA strand breaks. DNA damage resulting from dUTP misincorporation is highly dependent on the levels of the pyrophosphate dUTPase, which limits intracellular accumulation of dUTP *(13,14)*. Increased dUTPase expression has been associated with resistance to TS inhibitors *(13–16)*.

2.3. RNA Misincorporation

The 5-FU metabolite FUTP is extensively incorporated into RNA, disrupting normal RNA processing and function. Significant correlations between 5-FU misincorporation into RNA and loss of clonogenic potential have been demonstrated in human colon and breast cancer cell lines *(17,18)*. Ribosomes are ribonucleoprotein complexes that play an essential role in the translation of messenger RNA (mRNA) into protein. The RNA component of ribosomes (rRNA) makes up approx 60% of the complex. rRNA is transcribed by RNA polymerase I and is then processed to generate mature rRNA species. 5-FU misincorporation has been found to inhibit the processing of pre-rRNA into mature rRNA *(19,20)*. Furthermore, 5-FU has been shown to disrupt post-transcriptional modification of transfer RNAs (tRNAs), which also play an essential role in mRNA translation *(21,22)*. Small nuclear RNAs (snRNAs) play key roles in the splicing of pre-mRNA (also known as heterogenous nuclear RNA [hnRNA]) into mature mRNA. 5-FU can disrupt the assembly and activity of snRNA/protein complexes, thereby inhibiting splicing of pre-mRNA *(23,24)*. rRNA, tRNA, and snRNA contain the modified base pseudouridine, and 5-FU has been shown to inhibit the posttranscriptional conversion of uridine to pseudouridine in these RNA species *(23,25)*. These in vitro studies indicate that 5-FU misincorporation can potentially disrupt many aspects of RNA processing, leading to profound effects on cellular metabolism and viability.

3. Mechanisms of Resistance to 5-Fluorouracil

3.1. Drug Activation/Inactivation

Mechanisms that inactivate drugs can diminish the amount of free drug available to bind to its intracellular target. More than 80% of 5-FU is normally catabolized by dihydropyrimidine dehydrogenase (DPD), primarily in the liver *(5)*. In vitro studies have demonstrated that DPD overexpression in cancer cell lines confers resistance to 5-FU *(26)*. Furthermore, high levels of DPD mRNA expression in colorectal tumors have been shown to correlate with resistance to 5-FU *(27)*, presumably reflecting greater DPD-mediated degradation of 5-FU in these tumors. The activation of 5-FU to its active metabolites is complex, and the levels of 5-FU activating enzymes such as thymidine phosphorylase (TP), uridine phosphorylase (UP), and orotate phosphoribosyl transferase (OPRT) (*see* Fig. 2) have been associated with 5-FU sensitivity *(28–30)*.

3.2. TS Expression

Alterations in expression levels or mutation of a chemotherapeutic drug target can have a major impact on drug resistance. As mentioned above, the 5-FU metabolite fluorodeoxyuridine monophosphate (FdUMP) is a potent inhibitor of TS, and it is the inhibition of TS that is believed to be the primary anticancer activity of 5-FU *(31)*. Numerous preclinical studies have demonstrated that TS expression is a key determinant of 5-FU sensitivity *(32)*. Furthermore, immunohistochemical and reverse-transcription PCR studies have shown improved response rates to 5-FU-based chemotherapy in patients with low-tumor TS expression *(33,34)*. More recently, genotyping studies have found that patients homozygous for a particular polymorphism in the *TS* promoter (TSER3/TSER3) that increases TS expression are less likely to respond to 5-FU-based chemotherapy than patients who are heterozygous (TSER2/TSER3), or homozygous for the alternative polymorphism (TSER2/TSER2) *(35)*. Collectively, these studies indicate that high-TS expression correlates with increased 5-FU resistance.

Treatment with 5-FU has been shown to acutely induce TS expression in both cell lines and tumors *(36,37)*. This induction of TS seems to be to the result of inhibition of a negative feedback mechanism in which ligand-free TS binds to its own mRNA and inhibits its own translation *(38)*. When stably bound by FdUMP, TS can no longer bind its own mRNA and suppress translation, resulting in increased protein expression. This constitutes a potentially important resistance mechanism, as acute increases in TS would facilitate recovery of enzyme activity. In vitro studies have demonstrated that acute increases in TS expression induce resistance to TS inhibitors such as 5-FU and the antifolate drugs tomudex (TDX) and multitargeted antifolate (MTA, Alimta) *(39,40)*.

3.3. DNA Repair

Meyers et al. found that restoration of hMLH1 in MMR-deficient HCT116 colon cancer cells renders them more sensitive to 5-FU, suggesting that MMR deficient cells are more resistant to 5-FU *(41)*. However, the MSI phenotype has been associated with excellent survival in colorectal cancer patients receiving adjuvant 5-FU-based chemotherapy *(42)*. These apparently contradictory findings may be to the result of intrinsic biological

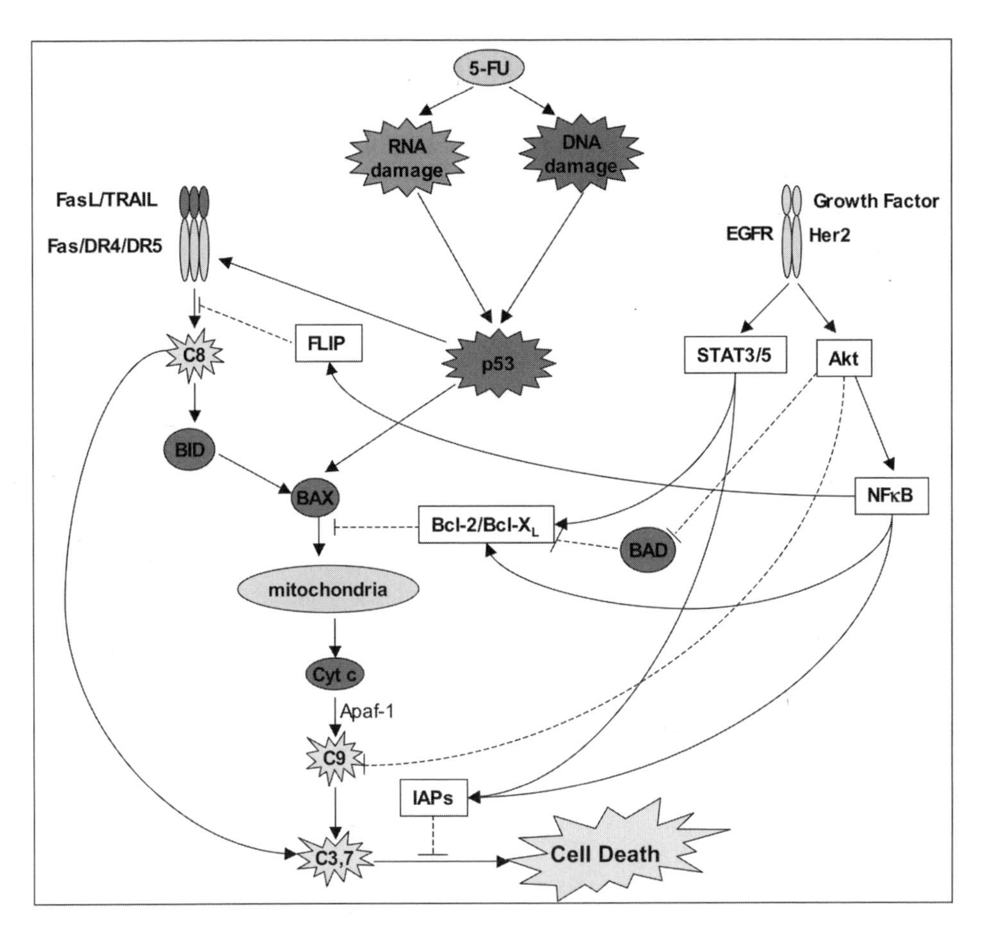

Fig. 2. Interactions between proapoptotic and prosurvival signal transduction pathways that may play important roles in determining response to 5-FU. *Abbr:* Apaf-1, apoptotic protease activating factor; C3, caspase 3; C7, caspase 7; C8, caspase 8; C9, caspase 9; cyt c, cytochrome *c*.

differences between MSI-positive and MSI-negative colorectal tumors, with MSI-positive (MMR deficient) tumors being less aggressive. For example, it has been suggested that MSI may lead to an increased anti-tumor immune response *(43)*. In addition, it has been demonstrated that the majority of MSI-positive tumors express wild type p53 *(44)*, which is an important determinant of 5-FU sensitivity (discussed later).

3.4. p53

As already mentioned, there is a critical balance between cell-cycle arrest (i.e., promoting DNA repair and survival) and cell death following chemotherapy. The tumor suppressor protein p53 plays a central role in regulation of cell-cycle arrest and cell death *(45)*. The gene encoding p53, *TP53*, is the most frequently mutated gene in human cancers, with about 50% of all tumors estimated to carry a mutation *(46)*. DNA damage results in activation of upstream kinases such as ATM (Ataxia-telangiectasia mutated), ATR (ATM and Rad-3 related), and DNA-PK (DNA-dependent protein kinase), which can directly or indirectly activate p53 *(47)*. Phosphorylation of p53 by upstream kinases inhibits its negative regulation by MDM-2, which targets p53 for ubiquitin-mediated degradation *(47)*. The role of p53 in determining cell fate following DNA damage has been attributed to its role as a transcription factor. p53 transcriptionally up-regulates genes such as those encoding $p21^{WAF\text{-}1/CIP\text{-}1}$ and GADD45, which induce cell-cycle arrest in response to DNA damage *(48,49)*. However, depending on the cellular context, p53 can trigger elimination of the damaged cells by promoting apoptosis through the up-regulation of proapoptotic genes such as *Bax, NOXA, TRAIL-R2 (DR5),* and *Fas (CD95/Apo-1)* *(50–53)*. Both p53-induced cell-cycle arrest and apoptosis act to maintain genomic integrity and prevent damaged DNA being passed on to daughter cells. In addition, we have demonstrated that p53 and p53-target genes are activated in response to RNA-directed 5-FU cytotoxicity *(40)*.

A number of experimental reports have indicated that lack of functional p53 contributes to drug resistance, and this has been attributed to an inability to undergo p53-mediated apoptosis. For example, in vitro studies have reported that loss of p53 function reduces chemosensitivity to 5-FU *(40,54)*. Furthermore, a number of clinical studies have found that p53 overexpression (a surrogate marker for p53 mutation) correlated with resistance to 5-FU-based chemotherapy *(42,55,56)*, although other studies failed to find a correlation *(57)*. Such conflicting findings may be due at least in part to the fact that p53 overexpression does not actually reflect *TP53* mutation in as many as 30 to 40% of cases *(58)*. In addition, gain-of-function p53 mutations have been described that actively contribute to transformation and drug resistance *(59)*. Indeed, a recent study has suggested that certain p53 mutants may increase dUTPase expression, resulting in 5-FU resistance *(15)*. So, 5-FU chemosensitivity may be dependent on the particular *TP53* genotype.

4. Induction of Apoptosis

The ultimate goal of cytotoxic chemotherapies is to induce cell death in tumor cells. The onset of apoptosis is regulated by multiple intra- and extracellular signals, amplification of these signals by second messengers, and activation of the effectors of apoptosis—the caspases. There are two main pathways for activation of caspases: the intrinsic pathway regulated by Bcl-2 proteins and the extrinsic pathway regulated by members of the tumor necrosis factor (TNF) receptor superfamily *(60)*. Activation of proximal caspases in these pathways leads to activation of downstream effector caspases, most importantly caspase 3 and 7. These executioner caspases cleave a cassette of cellular substrates to bring about the morphological and biochemical changes that characterize apoptosis, including chromatin condensation and nuclear fragmentation, membrane blebbing, and cell shrinkage *(60)*. Eventually, the cell breaks into small membrane-bound fragments (apoptotic bodies) that are cleared by phagocytosis without causing an inflammatory response.

Activation of the intrinsic apoptotic pathway is regulated by the Bcl-2 family of proteins. Although these proteins share some homology, some Bcl-2 family members, such as Bax and Bid, promote apoptosis whereas others, such as Bcl-X_L and Bcl-2 itself, are antiapoptotic *(60,61)*. Not surprisingly, the role of Bcl-2 family members in regulating response to chemotherapy has been extensively studied. Several in vitro studies have demonstrated that antisense targeting of Bcl-2 or Bcl-X_L sensitizes cancer cells to 5-FU *(62–64)*, whereas overexpression of Bax has been found to sensitize cancer cells to 5-FU-induced apoptosis *(65)*. It has also been demonstrated that loss of Bax expression decreases sensitivity to 5-FU *(66)*. Furthermore, Violette et al. found that high levels of

Bcl-2 and Bcl-X_L coupled to low levels of Bax correlated with resistance to 5-FU in colon cancer cell lines *(67)*. In addition, Bid null mouse embryonic fibroblasts were found to be more resistant to 5-FU, and Bid overexpression sensitized hepatocellular carcinoma cells to 5-FU *(68,69)*.

Oblimersen (G3139, Genasense) is an antisense oligonucleotide designed to specifically downregulate Bcl-2 that has been shown to synergize with cytotoxic drugs in preclinical models *(70)*. Various clinical trials are currently evaluating the antitumor potential of combining oblimersen with chemotherapy. In terms of the clinical predictive value of Bcl-2 proteins, several studies have shown no correlation between Bcl-2 expression and response to 5-FU-containing chemotherapeutic regimens *(71,72)*. Similarly, Paradiso et al. found no correlation between Bax expression and response to 5-FU-based chemotherapy in patients with advanced colorectal cancer, whereas Sjostrom et al. found no correlation between response to 5-FU-based chemotherapy and Bcl-2, Bcl-X_L, and Bax expression in advanced breast cancers *(73,74)*. So, although Bcl-2 family members undoubtedly play important roles during chemotherapy-induced apoptosis, their usefulness as molecular predictive markers of response appears to be limited.

The extrinsic apoptotic pathway is regulated by cell surface "death receptors" of the TNF-receptor family. These includeFas (CD95/APO-1), DR4 (TNF-related apoptosis-inducing ligand receptor 1, TRAIL-R1), and DR5 (TRAIL-R2). Houghton et al. demonstrated that apoptosis of colon cancer cells in response to thymidylate deficiency was mediated via the Fas pathway, suggesting an important role for this apoptotic pathway following 5-FU-mediated inhibition of TS *(75)*. Furthermore, in vitro studies have shown that targeting death receptors with recombinant death ligands or agonistic antibodies can enhance 5-FU-induced apoptosis *(51,76–78)*. Efficient induction of Fas-mediated apoptosis in response to agonistic Fas antibodies and recombinant FasL has been demonstrated in vivo; however, systemic treatment with Fas-targeted agents has been shown to cause severe liver damage *(79)*. Most preclinical studies are now focused on local administration of rFasL, or the use of FasL-expressing vectors as gene therapy *(79)*. However, Ichikawa et al. have successfully developed a nonhepatotoxic agonistic Fas antibody *(80)*, suggesting that it is possible to develop less toxic Fas antibodies.

A clinical study by Backus et al. found that 5-FU treatment up-regulated Fas expression in colorectal tumors *(81)*, suggesting that Fas is a clinically important mediator of response to chemotherapy. In several cancers, down-regulation of Fas and up-regulation of FasL have been demonstrated to occur during disease progression *(82–85)*. It has been postulated that tumor FasL induces apoptosis of Fas-sensitive immune effector cells, thereby inhibiting the antitumor immune response *(86)*. A recent study found that Fas-negative/FasL-positive breast cancer tumors had significantly shorter DFS and OS following adjuvant chemotherapy *(87)*. However, a number of clinical studies have failed to show any relationship between Fas/FasL expression and response to 5-FU-based chemotherapy *(74,88)*.

Targeting the TNF-related apoptosis-inducing ligand (TRAIL) receptors may prove to be a more promising therapeutic approach than targeting Fas, as TRAIL has been shown to exert marked anticancer activity without systemic toxicity in mice *(89)*. Wang et al., recently found that RNAi-mediated silencing of DR5 increased resistance to 5-FU in colon cancer cells *(90)*. Furthermore, several in vitro studies have demonstrated synergy between rTRAIL and 5-FU (and other chemotherapies) in a number of different cancers *(91–93)*. Furthermore, a study by Naka et al. using fresh surgical specimens taken from human colon tumors implanted in severe combined immunodeficiency (SCID) mice found that combining rTRAIL with 5-FU (or irinotecan) produced a greatly enhanced antitumor effect as compared with either agent alone *(94)*. Clinical trials are currently examining rTRAIL and TRAIL receptor-targeted antibodies as anticancer agents.

Apoptosis mediated by both Fas and DR4/DR5 can be inhibited by cytoplasmic factors, most notably c-FLIP (a FADD-like interleukin-1β-converting enzyme-inhibitory protein), which binds to the death-induced signal complex (DISC) and inhibits caspase 8 activation *(95)*. c-FLIP overexpression has been found to inhibit death ligand-induced apoptosis in a number of in vitro studies *(96–98)*. Furthermore, Ganten et al. identified down-regulation of c-FLIP as the mechanism of 5-FU-mediated sensitization to rTRAIL in hepatocellular carcinoma cells *(99)*. Recently, we have found that RNAi targeting of c-FLIP dramatically sensitized a panel of colon cancer cell lines to 5-FU (and irinotecan and oxaliplatin) in the absence of cotreatment with any death ligand, suggesting that c-FLIP plays an important role in regulating colon cancer cell chemosensitivity *(99a)*. Interestingly, c-FLIP has been found to be overexpressed in a high percentage of colonic and gastric carcinomas *(100,101)*, however the significance of c-FLIP overexpression for clinical drug resistance has not yet been studied.

Death receptor-mediated apoptosis can also be inhibited by decoy receptors: decoy receptor 3 (DcR3) in the case of Fas, and DcR1 and DcR2 in the case of the TRAIL receptors *(89)*. These decoy receptors bind to FasL/TRAIL, but lack the intracellular domains necessary for DISC formation and therefore inhibit death receptor-mediated apoptosis. A recent study in stage II and III colorectal cancer found that the *DcR3* gene was amplified in 63% of cases and that the DcR3 protein was overexpressed in 73% of cases *(102)*. Although, adjuvant chemotherapy was found to be significantly more beneficial in patients with normal *DcR3* gene copy number in this study, DcR3 protein expression was not associated with the effectiveness of adjuvant 5-FU-based chemotherapy. So, the clinical significance of decoy receptor expression for drug resistance has yet to be demonstrated; however, decoy receptors may prove to be important clinical determinants of sensitivity to agents targeted directly against death receptors, such as rTRAIL, which may eventually be used in combination with 5-FU.

5. Prosurvival Signaling

Protein tyrosine kinases (PTKs) can have an important impact on drug resistance through their regulation of antiapoptotic signal transduction pathways *(103)*. Overexpression and oncogenic mutations of many PTKs have been described in human cancers *(103)*. Some of the best-characterized PTKs are the epidermal growth factor receptor (EGFR) family, which are comprised of EGFR (ErbB1, Her1), Her2 (ErbB2, Neu), Her3 (ErbB3), and Her4 (ErbB4) *(104,105)*. Binding of growth factors such as epidermal growth factor (EGF), transforming growth factor-α (TGF-α), and heregulins result in homo- and heterodimerisation of EGFR, Her3, and Her4, with the preferred binding partner being Her2, for which there is no known ligand *(105)*. Receptor dimerisation leads to cross autophosphorylation of key tyrosine residues in the receptor cytoplasmic domain, which creates docking sites for downstream signal transducers. Antiapoptotic downstream signals activated by these receptor tyrosine kinases include the phosphatidylinositol 3-kinase (PI3K)/Akt (protein kinase B, PKB) pathway and the signal transducers and activators of transcription (STAT) pathway, specifically STAT3 and STAT5 *(103,104)*. Numerous reports have described EGFR overexpression or mutation in various human cancers, indicating an important role for this receptor in cancer *(103,106)*.

EGFR and Her2 have long been recognized as potential drug targets, and specific inhibitors of these receptors have entered the clinical arena *(106,107)*. The Her2 inhibitor herceptin (trastuzamab) is used to treat the 20 to 40% of breast cancers that are Her2-positive *(108)*. The EGFR tyrosine kinase inhibitors gefitinib (Iressa, ZD1839) is currently being tested in clinical trials in a number of solid tumors and is approved for use in the treatment of advanced non-small cell lung cancer *(107)*. Cetuximab (C225), another EGFR inhibitor, is in phase III trials in combination with chemotherapy in advanced colorectal cancer *(107,109)*. In vitro and xenograft studies have found that gefitinib enhances the cytotoxic effects of a variety of chemotherapies including 5-FU *(110–113)*. We have found that 5-FU treatment enhances phosphorylation on EGFR tyrosine residue 1068 (a surrogate marker for EGFR activation) in a panel of colorectal cancer cell lines *(113a)*. Furthermore, this up-regulation appears to be the molecular basis of the synergy between gefitinib and 5-FU in these cell lines.

STAT proteins transmit cytoplasmic signals from cytokine and growth factor receptors to the nucleus, where they activate transcription of a diverse set of target genes *(114)*. Persistent activation of STATs, in particular STAT3 and STAT5, has been demonstrated in a large number of cancers, and this often occurs as a result of constitutive activation of an upstream tyrosine kinase *(114)*. Phosphorylated STAT proteins form activated dimers that translocate to the nucleus and up-regulate target gene transcription. STAT3 and STAT5 have been shown to regulate Bcl-X_L expression and apoptosis in a wide range of tumor cells *(115–120)*. A study by Real et al. demonstrated that STAT3-dependent overexpression of Bcl-2 inhibited chemotherapy-induced apoptosis in breast cancer cells *(121)*, whereas Masuda et al. demonstrated that inhibition of STAT3 enhanced the sensitivity of head and neck cancer cells to 5-FU *(122)*. Given its antiapoptotic activity and the frequency of its activation in cancer, STAT3 is a particularly attractive therapeutic target.

The transcription factor nuclear factor-κB (NF-κB) is a key regulator of oncogenesis through its promotion of proliferation and inhibition of apoptosis *(123)*. NF-κB exerts its antiapoptotic effects by up-regulating a number of antiapoptotic proteins, including inhibitors of apoptosis proteins (IAPs), TNF-receptor associated factors (TRAFs), c-FLIP, Bcl-2, Bfl-1 (A1), and Bcl-X_L *(123)*. NF-κB has been connected with multiple pathways involved in oncogenesis, including cell-cycle regulation and apoptosis *(124,125)*. Chuang et al. demonstrated that a wide range of cytotoxic drugs including 5-FU activated NF-κB in a panel of cancer cell lines *(126)*. This suggests that NF-κB activation is a general feature of cancer cell response to chemotherapy. Whether constitutively or inducibly activated, NF-κB appears to be a critical determinant of drug resistance, with NF-κB activation blunting the ability of chemotherapy to induce cell death *(125)*. Increased NF-κB activity in patients with oesophageal cancer has been correlated with reduced response to neoadjuvant 5-FU-based chemotherapy and radiation therapy *(127)*. Inhibiting NF-κB signaling may prove to be an effective strategy to enhance 5-FU-induced apoptosis in a range of cancers. NF-κB is believed to be a major target for proteasome inhibitors, as proteasome inhibition prevents degradation of the NF-κB inhibitor IκB *(128,129)*. Clinical trials with proteasome inhibitors such as bortezomib are underway.

6. Conclusion

In the past, attempts to modulate 5-FU cytotoxicity have focused primarily on increasing activation, decreasing degradation, and enhancing TS inhibition *(32)*. The development of new proteomic and genomic technologies now enables characterization of the downstream signaling pathways involved in regulating tumor cell response to chemotherapy moreso than ever before. This will facilitate the future development of rational combined chemotherapy regimens designed to maximize drug activity. Newer therapies that target apoptotic pathways are likely to be used increasingly in combination with cytotoxic drugs to enhance chemotherapy activity. In the future, the ability to predict response to cytotoxic drugs and to modulate this response with targeted therapies will permit selection of the best combined treatment for an individual patient.

References

1. IMPACT, Efficacy of adjuvant fluorouracil and folinic acid in colon cancer. International Multicentre Pooled Analysis of Colon Cancer Trials (IMPACT) investigators. Lancet 1995; 345(8955):939–944.
2. Johnston PG, Kaye S. Capecitabine: a novel agent for the treatment of solid tumors. Anticancer Drugs 2001;12(8):639–646.
3. Giacchetti S, et al. Phase III multicenter randomized trial of oxaliplatin added to chronomodulated fluorouracil-leucovorin as first-line treatment of metastatic colorectal cancer. J Clin Oncol 2000;18(1):136–147.
4. Douillard JY, et al. Irinotecan combined with fluorouracil compared with fluorouracil alone as first-line treatment for metastatic colorectal cancer: a multicentre randomised trial. Lancet 2000;355(9209):1041–1047.
5. Diasio RB, Harris BE. Clinical pharmacology of 5-fluorouracil. Clin Pharmacokinet 1989;16(4):215–237.
6. Sommer H, Santi DV. Purification and amino acid analysis of an active site peptide from thymidylate synthetase containing covalently bound 5-fluoro-2′-deoxyuridylate and methylenetetrahydrofolate. Biochem Biophys Res Commun 1974;57(3):689–695.
7. Santi DV, McHenry CD, Sommer H. Mechanism of interaction of thymidylate synthetase with 5-fluorodeoxyuridylate. Biochemistry 1974;13(3):471–481.
8. Jackson RC, Grindley GB. The biochemical basis for methotrexate cytotoxicity. In: Biochemistry, Molecular Actions, and Synthetic Design. Sirotnak FM, et al., eds. New York, Academic, 1984:289–315.
9. Houghton JA, Tillman DM, Harwood FG. Ratio of 2′-deoxyadenosine-5′-triphosphate/thymidine-5′-triphosphate influences the commitment of human colon carcinoma cells to thymineless death. Clin Cancer Res 1995;1(7):723–730.
10. Yoshioka A, et al. Deoxyribonucleoside triphosphate imbalance. 5-Fluorodeoxyuridine-induced DNA double strand breaks in mouse FM3A cells and the mechanism of cell death. J Biol Chem 1987;262(17):8235–8241.
11. Aherne GW, et al. Immunoreactive dUMP and TTP pools as an index of thymidylate synthase inhibition; effect of tomudex (ZD1694) and a nonpolyglutamated quinazoline antifolate (CB30900) in L1210 mouse leukaemia cells. Biochem Pharmacol 1996;51(10): 1293–1301.
12. Lindahl T. An N-glycosidase from Escherichia coli that releases free uracil from DNA containing deaminated cytosine residues. Proc Natl Acad Sci USA 1974;71(9):3649–3653.
13. Ladner RD. The role of dUTPase and uracil-DNA repair in cancer chemotherapy. Curr Protein Pept Sci 2001;2(4):361–370.

14. Webley SD, et al. Deoxyuridine triphosphatase (dUTPase) expression and sensitivity to the thymidylate synthase (TS) inhibitor ZD9331. Br J Cancer 2000;83(6):792–799.
15. Pugacheva EN, et al. Novel gain of function activity of p53 mutants: activation of the dUTPase gene expression leading to resistance to 5-fluorouracil. Oncogene 2002;21(30): 4595–4600.
16. Webley SD, et al. The ability to accumulate deoxyuridine triphosphate and cellular response to thymidylate synthase (TS) inhibition. Br J Cancer 2001;85(3):446–452.
17. Kufe DW, Major PP. 5-Fluorouracil incorporation into human breast carcinoma RNA correlates with cytotoxicity. J Biol Chem 1981;256(19):9802–9805.
18. Glazer RI, Lloyd LS. Association of cell lethality with incorporation of 5-fluorouracil and 5-fluorouridine into nuclear RNA in human colon carcinoma cells in culture. Mol Pharmacol 1982;21(2):468–473.
19. Kanamaru R, et al. The inhibitory effects of 5-fluorouracil on the metabolism of preribosomal and ribosomal RNA in L-1210 cells in vitro. Cancer Chemother Pharmacol 1986;17(1): 43–46.
20. Ghoshal K, Jacob ST. Specific inhibition of pre-ribosomal RNA processing in extracts from the lymphosarcoma cells treated with 5-fluorouracil. Cancer Res 1994;54(3):632–636.
21. Santi DV, Hardy LW. Catalytic mechanism and inhibition of tRNA (uracil-5-)methyltransferase: evidence for covalent catalysis. Biochemistry 1987;26(26):8599–8606.
22. Randerath K, et al. Specific effects of 5-fluoropyrimidines and 5-azapyrimidines on modification of the 5 position of pyrimidines, in particular the synthesis of 5-methyluracil and 5-methylcytosine in nucleic acids. Recent Results Cancer Res 1983;84:283–297.
23. Patton JR. Ribonucleoprotein particle assembly and modification of U2 small nuclear RNA containing 5-fluorouridine. Biochemistry 1993;32(34):8939–8944.
24. Doong SL, Dolnick BJ. 5-Fluorouracil substitution alters pre-mRNA splicing in vitro. J Biol Chem 1988;263(9):4467–4673.
25. Samuelsson T. Interactions of transfer RNA pseudouridine synthases with RNAs substituted with fluorouracil. Nucleic Acids Res 1991;19(22):6139–6144.
26. Takebe N, et al. Retroviral transduction of human dihydropyrimidine dehydrogenase cDNA confers resistance to 5-fluorouracil in murine hematopoietic progenitor cells and human CD34+-enriched peripheral blood progenitor cells. Cancer Gene Ther 2001;8(12): 966–973.
27. Salonga D, et al. Colorectal tumors responding to 5-fluorouracil have low gene expression levels of dihydropyrimidine dehydrogenase, thymidylate synthase, and thymidine phosphorylase. Clin Cancer Res 2000;6(4):1322–1327.
28. Schwartz PM, et al. Role of uridine phosphorylase in the anabolism of 5-fluorouracil. Biochem Pharmacol 1985;34(19):3585–3589.
29. Houghton JA, Houghton PJ. Elucidation of pathways of 5-fluorouracil metabolism in xenografts of human colorectal adenocarcinoma. Eur J Cancer Clin Oncol 1983;19(6): 807–815.
30. Evrard A, et al. Increased cytotoxicity and bystander effect of 5-fluorouracil and 5-deoxy-5-fluorouridine in human colorectal cancer cells transfected with thymidine phosphorylase. Br J Cancer 1999;80(11):1726–1733.
31. Peters GJ, Kohne CH. Fluoropyrimidines as antifolate drugs. Antifolate drugs in cancer therapy, ed. A.L. Jackman. 1999: Humana Press,101–145.
32. Longley DB, Harkin DP, Johnston PG. 5-fluorouracil: mechanisms of action and clinical strategies. Nat Rev Cancer 2003;3(5):330–338.
33. Johnston PG, et al. Thymidylate synthase gene and protein expression correlate and are associated with response to 5-fluorouracil in human colorectal and gastric tumors. Cancer Res 1995;55(7):1407–1412.

34. Lenz HJ, et al. p53 point mutations and thymidylate synthase messenger RNA levels in disseminated colorectal cancer: an analysis of response and survival. Clin Cancer Res 1998;4(5):1243–1250.
35. Marsh S, McLeod HL. Thymidylate synthase pharmacogenomics in colorectal cancer. Clinical Colorectal Cancer 2001;1:175–178.
36. Chu E, et al. Regulation of thymidylate synthase in human colon cancer cells treated with 5-fluorouracil and interferon-gamma. Mol Pharmacol 1993;43(4):527–533.
37. Swain SM, et al. Fluorouracil and high-dose leucovorin in previously treated patients with metastatic breast cancer. J Clin Oncol 1989;7(7):890–899.
38. Chu E, et al. Identification of a thymidylate synthase ribonucleoprotein complex in human colon cancer cells. Mol Cell Biol 1994;14(1):207–213.
39. Longley DB, et al. Characterization of a thymidylate synthase (TS)-inducible cell line: a model system for studying sensitivity to TS- and non-TS-targeted chemotherapies. Clin Cancer Res 2001;7(11):3533–3539.
40. Longley DB, et al. The role of thymidylate synthase induction in modulating p53-regulated gene expression in response to 5-fluorouracil and antifolates. Cancer Res 2002; 62(9):2644–2649.
41. Meyers M, et al. Role of the hMLH1 DNA mismatch repair protein in fluoropyrimidine-mediated cell death and cell cycle responses. Cancer Res 2001;61(13):5193–5201.
42. Elsaleh H, et al. P53 alteration and microsatellite instability have predictive value for survival benefit from chemotherapy in stage III colorectal carcinoma. Clin Cancer Res 2001;7(5):1343–1349.
43. Banerjea A, et al. Colorectal cancers with microsatellite instability display mRNA expression signatures characteristic of increased immunogenicity. Mol Cancer 2004;3(1):21.
44. Mori S, Ogata Y, Shirouzu K. Biological features of sporadic colorectal carcinoma with high-frequency microsatellite instability: special reference to tumor proliferation and apoptosis. Int J Clin Oncol 2004;9(4):322–329.
45. Vogelstein B, Lane D, Levine AJ. Surfing the p53 network. Nature 2000;408(6810):307–310.
46. Levine AJ. p53, the cellular gatekeeper for growth and division. Cell 1997;88(3):323–331.
47. Ljungman M. Dial 9-1-1 for p53: mechanisms of p53 activation by cellular stress. Neoplasia 2000;2(3):208–225.
48. Dotto GP. p21(WAF1/Cip1): more than a break to the cell cycle? Biochim Biophys Acta 2000;1471(1):M43–M56.
49. Zhan Q, et al. Tumor suppressor p53 can participate in transcriptional induction of the GADD45 promoter in the absence of direct DNA binding. Mol Cell Biol 1998;18(5): 2768–2778.
50. Miyashita T, et al. Tumor suppressor p53 is a regulator of bcl-2 and bax gene expression in vitro and in vivo. Oncogene 1994;9(6):1799–1805.
51. Petak I, Tillman DM, Houghton JA. p53 dependence of Fas induction and acute apoptosis in response to 5-fluorouracil-leucovorin in human colon carcinoma cell lines. Clin Cancer Res 2000;6(11):4432–4341.
52. Schuler M, Green DR. Mechanisms of p53-dependent apoptosis. Biochem Soc Trans 2001;29(Pt 6):684–688.
53. Yu J, et al. Identification and classification of p53-regulated genes. Proc Natl Acad Sci USA 1999;96(25):14,517–14,522.
54. Bunz F, et al. Disruption of p53 in human cancer cells alters the responses to therapeutic agents. J Clin Invest 1999;104(3):263–269.
55. Liang JT, et al. P53 overexpression predicts poor chemosensitivity to high-dose 5-fluorouracil plus leucovorin chemotherapy for stage IV colorectal cancers after palliative bowel resection. Int J Cancer 2002;97(4):451–457.

56. Ahnen DJ, et al. Ki-ras mutation and p53 overexpression predict the clinical behavior of colorectal cancer: a Southwest Oncology Group study. Cancer Res 1998;58(6):1149–1158.
57. Paradiso A, et al. Thymidylate synthase and p53 primary tumour expression as predictive factors for advanced colorectal cancer patients. Br J Cancer 2000;82(3):560–567.
58. Sjogren S, et al. The p53 gene in breast cancer: prognostic value of complementary DNA sequencing versus immunohistochemistry. J Natl Cancer Inst 1996;88(3-4):173–182.
59. van Oijen MG, Slootweg PJ. Gain-of-function mutations in the tumor suppressor gene p53. Clin Cancer Res 2000;6(6):2138–2145.
60. Hengartner MO. The biochemistry of apoptosis. Nature 2000;407(6805):770–776.
61. Reed JC. Bcl-2 family proteins. Oncogene 1998;17(25):3225–3236.
62. Nita ME, et al. Bcl-X(L) antisense sensitizes human colon cancer cell line to 5-fluorouracil. Jpn J Cancer Res 2000;91(8):825–832.
63. Yang JH, et al. Chemosensitization of breast carcinoma cells with the use of bcl-2 antisense oligodeoxynucleotide. Breast 2004;13(3):227–231.
64. Kim R, et al. Effect of Bcl-2 antisense oligonucleotide on drug-sensitivity in association with apoptosis in undifferentiated thyroid carcinoma. Int J Mol Med 2003;11(6):799–804.
65. Xu ZW, et al. Overexpression of Bax sensitizes human pancreatic cancer cells to apoptosis induced by chemotherapeutic agents. Cancer Chemother Pharmacol 2002;49(6): 504–510.
66. Zhang L, et al. Role of BAX in the apoptotic response to anticancer agents. Science 2000; 290(5493):989–992.
67. Violette S, et al. Resistance of colon cancer cells to long-term 5-fluorouracil exposure is correlated to the relative level of Bcl-2 and Bcl-X(L) in addition to Bax and p53 status. Int J Cancer 2002;98(4):498–504.
68. Sax JK, et al. BID regulation by p53 contributes to chemosensitivity. Nat Cell Biol 2002; 4(11):842–849.
69. Miao J, et al. Bid sensitizes apoptosis induced by chemotherapeutic drugs in hepatocellular carcinoma. Int J Oncol 2004;25(3):651–659.
70. Herbst RS, Frankel SR. Oblimersen sodium (Genasense bcl-2 antisense oligonucleotide): a rational therapeutic to enhance apoptosis in therapy of lung cancer. Clin Cancer Res 2004;10(12 Pt 2):4245s–4248s.
71. Colleoni M, et al. Prediction of response to primary chemotherapy for operable breast cancer. Eur J Cancer 1999;35(4):574–579.
72. Bottini A, et al. p53 but not bcl-2 immunostaining is predictive of poor clinical complete response to primary chemotherapy in breast cancer patients. Clin Cancer Res 2000; 6(7):2751–2758.
73. Paradiso A, et al. Expression of apoptosis-related markers and clinical outcome in patients with advanced colorectal cancer. Br J Cancer 2001;84(5):651–658.
74. Sjostrom J, et al. The predictive value of bcl-2, bax, bcl-xL, bag-1, fas, and fasL for chemotherapy response in advanced breast cancer. Clin Cancer Res 2002;8(3):811–816.
75. Houghton JA, Harwood FG, Tillman DM. Thymineless death in colon carcinoma cells is mediated via fas signaling. Proc Natl Acad Sci USA 1997;94(15):8144–8149.
76. Longley DB, et al. The roles of thymidylate synthase and p53 in regulating Fas-mediated apoptosis in response to antimetabolites. Clin Cancer Res 2004;10(10):3562–3571.
77. Fan QL, et al. Synergistic antitumor activity of TRAIL combined with chemotherapeutic agents in A549 cell lines in vitro and in vivo. Cancer Chemother Pharmacol, 2004.
78. Wu XX, Ogawa O, Kakehi Y. TRAIL and chemotherapeutic drugs in cancer therapy. Vitam Horm 2004;67:365–383.
79. Timmer T, de Vries EG, de Jong S. Fas receptor-mediated apoptosis: a clinical application? J Pathol 2002;196(2):125–134.

80. Ichikawa K, et al. A novel murine anti-human Fas mAb which mitigates lymphadenopathy without hepatotoxicity. Int Immunol 2000;12(4):555–562.
81. Backus HH, et al. 5-Fluorouracil induced Fas upregulation associated with apoptosis in liver metastases of colorectal cancer patients. Ann Oncol 2001;12(2):209–216.
82. Mottolese M, et al. Prognostic relevance of altered Fas (CD95)-system in human breast cancer. Int J Cancer 2000;89(2):127–132.
83. von Reyher U, et al. Colon carcinoma cells use different mechanisms to escape CD95-mediated apoptosis. Cancer Res 1998;58(3):526–534.
84. Niehans GA, et al. Human lung carcinomas express Fas ligand. Cancer Res 1997;57(6): 1007–1012.
85. Gratas C, et al. Up-regulation of Fas (APO-1/CD95) ligand and down-regulation of Fas expression in human esophageal cancer. Cancer Res 1998;58(10):2057–2062.
86. O'Connell J, et al. Resistance to Fas (APO-1/CD95)-mediated apoptosis and expression of Fas ligand in esophageal cancer: the Fas counterattack. Dis Esophagus 1999;12(2):83–89.
87. Botti C, et al. Altered expression of FAS system is related to adverse clinical outcome in stage I-II breast cancer patients treated with adjuvant anthracycline-based chemotherapy. Clin Cancer Res 2004;10(4):1360–1365.
88. Bezulier K, et al. Fas/FasL expression in tumor biopsies: a prognostic response factor to fluoropyrimidines? J Clin Pharm Ther 2003;28(5):403–408.
89. Ashkenazi A. Targeting death and decoy receptors of the tumour-necrosis factor superfamily. Nat Rev Cancer 2002;2(6):420–430.
90. Wang S, El-Deiry WS. Inducible silencing of KILLER/DR5 in vivo promotes bioluminescent colon tumor xenograft growth and confers resistance to chemotherapeutic agent 5-fluorouracil. Cancer Res 2004;64(18):6666–6672.
91. Shimoyama S, et al. Supra-additive antitumor activity of 5FU with tumor necrosis factor-related apoptosis-inducing ligand on gastric and colon cancers in vitro. Int J Oncol 2002; 21(3):643–648.
92. Mizutani Y, et al. Potentiation of the sensitivity of renal cell carcinoma cells to TRAIL-mediated apoptosis by subtoxic concentrations of 5-fluorouracil. Eur J Cancer 2002;38 (1):167–176.
93. von Haefen C, et al. Multidomain Bcl-2 homolog Bax but not Bak mediates synergistic induction of apoptosis by TRAIL and 5-FU through the mitochondrial apoptosis pathway. Oncogene 2004;23(50):8320–8332.
94. Naka T, et al. Effects of tumor necrosis factor-related apoptosis-inducing ligand alone and in combination with chemotherapeutic agents on patients' colon tumors grown in SCID mice. Cancer Res 2002;62(20):5800–5806.
95. Krueger A, et al. FLICE-inhibitory proteins: regulators of death receptor-mediated apoptosis. Mol Cell Biol 2001;21(24):8247–8254.
96. Srinivasula SM, et al. FLAME-1, a novel FADD-like anti-apoptotic molecule that regulates Fas/TNFR1-induced apoptosis. J Biol Chem 1997;272(30):18,542-18,545.
97. Irmler M, et al. Inhibition of death receptor signals by cellular FLIP. Nature 1997; 388(6638):190–195.
98. Hu S, et al. I-FLICE, a novel inhibitor of tumor necrosis factor receptor-1- and CD-95-induced apoptosis. J Biol Chem 1997;272(28):17,255–17,257.
99. Ganten TM, et al. Enhanced caspase-8 recruitment to and activation at the DISC is critical for sensitisation of human hepatocellular carcinoma cells to TRAIL-induced apoptosis by chemotherapeutic drugs. Cell Death Differ 2004;11 Suppl 1:S86–S96.
99a. Longley DB, et al. c-FLIP inhibits chemothrapy-induced colorectal cancer cell death. Oncogene 2006;25(6):838–848.
100. Ryu BK, et al. Increased expression of cFLIP(L) in colonic adenocarcinoma. J Pathol 2001;194(1):15–19.

101. Zhou XD, et al. Overexpression of cellular FLICE-inhibitory protein (FLIP) in gastric adenocarcinoma. Clin Sci (Lond) 2004;106(4):397–405.
102. Mild G, et al. DCR3 locus is a predictive marker for 5-fluorouracil-based adjuvant chemotherapy in colorectal cancer. Int J Cancer 2002;102(3):254–257.
103. Blume-Jensen P, Hunter T. Oncogenic kinase signalling. Nature 2001;411(6835): 355–365.
104. Jorissen RN, et al. Epidermal growth factor receptor: mechanisms of activation and signalling. Exp Cell Res 2003;284(1):31–53.
105. Olayioye MA, et al. The ErbB signaling network: receptor heterodimerization in development and cancer. Embo J 2000;19(13):3159–3167.
106. Herbst RS. Review of epidermal growth factor receptor biology. Int J Radiat Oncol Biol Phys 2004;59(2 Suppl):21–26.
107. Gschwind A, Fischer OM, Ullrich A. The discovery of receptor tyrosine kinases: targets for cancer therapy. Nat Rev Cancer 2004;4(5):361–370.
108. Ross JS, et al. HER-2/neu testing in breast cancer. Am J Clin Pathol 2003;120 Suppl:S53–S71.
109. Cunningham D, et al. Cetuximab monotherapy and cetuximab plus irinotecan in irinotecan-refractory metastatic colorectal cancer. N Engl J Med 2004;351(4):337–345.
110. Ciardiello F, et al. Antitumor effect and potentiation of cytotoxic drugs activity in human cancer cells by ZD-1839 (Iressa), an epidermal growth factor receptor-selective tyrosine kinase inhibitor. Clin Cancer Res 2000;6(5):2053–2063.
111. Koizumi F, et al. Synergistic interaction between the EGFR tyrosine kinase inhibitor gefitinib ("Iressa") and the DNA topoisomerase I inhibitor CPT-11 (irinotecan) in human colorectal cancer cells. Int J Cancer 2004;108(3):464–472.
112. Magne N, et al. Sequence-dependent effects of ZD1839 ('Iressa') in combination with cytotoxic treatment in human head and neck cancer. Br J Cancer 2002;86(5):819–827.
113. Sirotnak FM, et al. Efficacy of cytotoxic agents against human tumor xenografts is markedly enhanced by coadministration of ZD1839 (Iressa), an inhibitor of EGFR tyrosine kinase. Clin Cancer Res 2000;6(12):4885–4892.
113a. Van Schaeybroeck S, et al. Epidermal growth factor receptor activity determines response of colorectal cancer cells to gefitinib alone and in combination with chemotherapy. Clin Cancer Res 2005;11(20):7480–7489.
114. Yu H, Jove R. The STATs of cancer—new molecular targets come of age. Nat Rev Cancer 2004;4(2):97–105.
115. Bromberg JF, et al. Stat3 as an oncogene. Cell, 1999;98(3):295–303.
116. Niu G, et al. Roles of activated Src and Stat3 signaling in melanoma tumor cell growth. Oncogene 2002;21(46):7001–7010.
117. Horita M, et al. Blockade of the Bcr-Abl kinase activity induces apoptosis of chronic myelogenous leukemia cells by suppressing signal transducer and activator of transcription 5-dependent expression of Bcl-xL. J Exp Med 2000;191(6):977–984.
118. Grandis JR, et al. Constitutive activation of Stat3 signaling abrogates apoptosis in squamous cell carcinogenesis in vivo. Proc Natl Acad Sci USA 2000;97(8):4227–4232.
119. Gesbert F, Griffin JD. Bcr/Abl activates transcription of the Bcl-X gene through STAT5. Blood 2000;96(6):2269–2276.
120. Zamo A, et al. Anaplastic lymphoma kinase (ALK) activates Stat3 and protects hematopoietic cells from cell death. Oncogene 2002;21(7):1038–1047.
121. Real PJ, et al. Resistance to chemotherapy via Stat3-dependent overexpression of Bcl-2 in metastatic breast cancer cells. Oncogene 2002;21(50):7611–7618.
122. Masuda M, et al. The roles of JNK1 and Stat3 in the response of head and neck cancer cell lines to combined treatment with all-trans-retinoic acid and 5-fluorouracil. Jpn J Cancer Res 2002;93(3):329–339.

123. Lin A, Karin M. NF-kappaB in cancer: a marked target. Semin Cancer Biol 2003; 13(2): 107–114.
124. Karin M, et al. NF-kappaB in cancer: from innocent bystander to major culprit. Nat Rev Cancer 2002;2(4):301–310.
125. Baldwin AS. Control of oncogenesis and cancer therapy resistance by the transcription factor NF-kappaB. J Clin Invest 2001;107(3):241–246.
126. Chuang SE, et al. Basal levels and patterns of anticancer drug-induced activation of nuclear factor-kappaB (NF-kappaB), and its attenuation by tamoxifen, dexamethasone, and curcumin in carcinoma cells. Biochem Pharmacol 2002;63(9):1709–1716.
127. Abdel-Latif MM, et al. NF-kappaB activation in esophageal adenocarcinoma: relationship to Barrett's metaplasia, survival, and response to neoadjuvant chemoradiotherapy. Ann Surg 2004;239(4):491–500.
128. Cusack JC. Rationale for the treatment of solid tumors with the proteasome inhibitor bortezomib. Cancer Treat Rev 2003;29 Suppl 1:21–31.
129. Richardson P. Clinical update: proteasome inhibitors in hematologic malignancies. Cancer Treat Rev 2003;29 Suppl 1:33–39.

12

Apoptosis-Inducing Cellular Vehicles for Cancer Gene Therapy

Endothelial and Neural Progenitors

Gergely Jarmy,* Jiwu Wei,* Klaus-Michael Debatin, and Christian Beltinger

Summary

Endothelial progenitor cells (EPCs) and neural progenitor cells (NPCs) are promising for cancer therapy because they specifically target tumors. They have the capacity to home to, invade, migrate within, and incorporate into tumor structures. They are easily expanded and can be armed with therapeutic payloads protected within the progenitor cells. Once in the tumor, armed progenitors can be triggered to induce apoptosis in surrounding tumor cells. Pro- and antiapoptotic mechanisms are pivotal to effectively kill tumor cells while simultaneously protecting the cellular vehicles from premature demise. Increasing the ratio of tumor cell apoptosis to progenitor apoptosis will be crucial among other efforts to enhance the efficacy of endothelial and neural progenitor cells to a level sufficient for clinical application.

Key Words: Apoptosis; endothelial progenitor cells; neural progenitor cells; cellular vehicle; cancer; gene therapy.

1. Cellular Vehicles for Gene Therapy of Tumors

Conventional tumor therapy by surgery, chemotherapy, and radiation often fails or is associated with severe side effects. Improved therapies are therefore needed. This may be achieved by improved targeting, that is, by confining the therapy-induced cytotoxicity to the tumor cells. Recent developments in tumor targeting have led to small molecules, such as receptor tyrosine kinase (TK) inhibitors, aimed at signaling pathways pivotal for tumors, to antibodies targeted at surface molecules, and to targeted gene therapy. However, the degree of targeting is often inefficient: the targets of small molecules may not be as tumor-specific as required and large antibodies may not penetrate the depth of a tumor mass. The success of gene therapy for treatment of cancer, in particular, depends on delivering genes specifically to tumor cells. For this, both viral and nonviral vectors have been used. Despite significant advances, gene therapy approaches using these vectors have been hampered by imprecise tumor targeting leading to unexpected side effects, low-viral titers, rapid degradation by the immune system, nonspecific adhesion, poor transduction efficiency, and low-level transgene expression. Moreover, these vectors have been mostly administered loco-regionally by direct intratumoral injection. Because metastases or inaccessible tumors account for a significant proportion of morbidity and mortality in cancer patients, there is a need for vectors that can be

---

*The first two authors contributed equally to this work.

From: *Apoptosis, Cell Signaling, and Human Diseases: Molecular Mechanisms, Volume 1*
Edited by R. Srivastava © Humana Press Inc., Totowa, NJ

administered systemically to target these tumor manifestations. To overcome the limitations of viral vectors, mammalian cells have been proposed as potential gene delivery vehicles. In principle, cellular vehicles offer several advantages. They may be able to home to tumors and mediate tumor cell death while, ideally, being protected against apoptosis induced by their therapeutic payload. If chosen appropriately, they should not evoke an immune response. Furthermore, some cell types intended as vehicles may be readily available, easy to isolate noninvasively, expandable ex vivo, and receptive to genetic manipulation. Not surprisingly, a variety of cellular vehicles have been investigated. In a pioneering clinical study genetically marked autologous tumor-infiltrating lymphocytes, expanded and activated in vitro, were used as vehicles for tumor-specific immunotherapy *(1)*. Other cell types, including fibroblasts *(3–6)*, macrophages *(7)*, endothelial cells (ECs) *(8)*, natural killer (NK) cells, dendritic cells, and autologous or allogeneic tumor cells *(9)* have been evaluated as cellular vehicles for tumor gene therapy. However, modest tumor-specific homing, limited ex vivo expandability, difficult genetic manipulation, or poor tolerance have hindered clinical translation of these vehicles. By virtue of their intrinsic properties, adult and embryonic stem and progenitor cells promise to overcome these problems. This chapters focuses on two well-characterized progenitors, endothelial cell (EPC) and neural progenitor cells (NPC) as potential cellular vehicles to target two of the most therapy-recalcitrant cancer manifestations: metastases and brain tumors.

2. Endothelial Progenitors for Systemic Apoptosis-Inducing Cancer Gene Therapy

2.1. The Physiological Role of Endothelial Progenitors—Vasculogenesis and Angiogenesis

During development, the vasculature is formed by vasculogenesis, that is the *in situ* differentiation of ECs from their precursors and their subsequent organization into a primary capillary network. Vasculogenesis during development requires EPCs that are derived from a common precursor of both the hematopoietic system and the vascular system called the hemangioblast *(10)*. The crucial role of vascular endothelial growth factor receptor (VEGFR)2 and VEGF in embryonic vasculogenesis is shown by the phenotype of both VEGFR2 and VEGF knock-out mice who die in utero as a result of lack of endothelial and blood cells *(11,12)*. In adult life new vessels are formed during wound healing, the endometrial cycle and the growth of tumors. This postnatal vessel formation has for a long time been thought to occur exclusively by sprouting from existing vessels and was termed angiogenesis. Recent work has shown that adult EPCs are recruited from the bone marrow to form new vessels in adults during both physiological and pathological conditions *(13,14)*. Adult EPCs have been characterized as highly proliferative cells, having properties similar to those of embryonic angioblasts. In response to angiogenic stimuli they are mobilized from the bone marrow and after differentiation into mature ECs they functionally incorporate into vessels. These findings have suggested that vasculogenesis is not restricted to embryogenesis but also plays an important role in adults. The discovery of EPCs has led to the concept that angiogenesis—the remodeling or sprouting of preexisting blood vessels—and vasculogenesis may constitute complementary mechanisms for postnatal neo-vascularization.

To what extent and under which conditions bone marrow derived endothelial progenitors contribute to new vessel formation still has to be elucidated. Recent preclinical and clinical

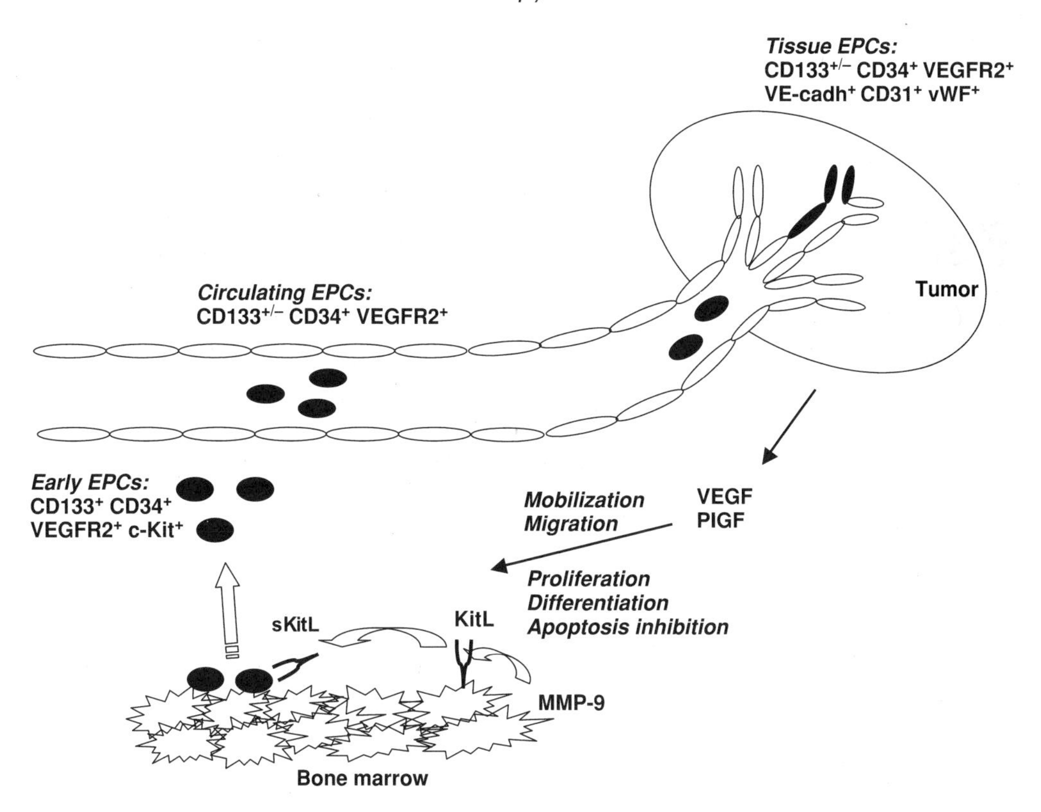

Fig. 1. Markers of EPCs and mechanisms of their recruitment to tumors. EPCs are recruited to tumors by secreted VEGF and PlGF which mobilize EPCs from the bone marrow. These cytokines also have proliferative, antiapoptotic, differentiating, and migratory effects on EPCs. *See* Subheading 2.2. for details on mobilization mechanisms. EPCs change surface marker expression when they mature from early EPCs to circulating and finally to tissue EPCs.

studies have shown that transplantation of ex vivo expanded EPCs can be successfully used to promote tissue vascularization after ischemic events *(13)*.

Adult EPCs are characterized and isolated by cell surface markers (*see* Fig. 1). In general, early functional EPCs are characterized by the combined expression of the cell surface markers CD133 and VEGFR2 *(16,17)*. CD133 is an early stem cell marker with as yet poorly understood functions. Whereas VEGFR2 is expressed on both mature ECs and their progenitors, CD133 distinguishes "true" EPCs from mature ECs sloughed from the injured vessel wall and from cells of myelomonocytic origin, which are able to transdifferentiate into endothelial-like cells. Additional markers associated with EPCs include CD34, CD31, VE-cadherin, von Willebrand factor (vWF), CD105, UEA, lectin staining, and uptake of acetylated LDL *(13,18)*. $CD34^-/CD14^+$ mononuclear cells have the potential to acquire an endothelial phenotype and to coexpress endothelial lineage markers. Infusion of these cells appears to be effective for re-endothelialization after vascular injury *(19)*.

2.2. The Role of EPCs in Tumor Angiogenesis

There is evidence for an important role of bone marrow-derived cells in the neovascularization of tumors. Increased numbers of EPCs have been detected in the circulation of tumor-bearing animals as well as in cancer patients *(20)*. Importantly, transplanted EPCs incorporated functionally into tumor-associated vessels of implanted tumors in different animal models. Studies have taken advantage of transgenic mouse models and transplantation techniques, in which sex-mismatched or genetically marked bone marrow-derived cells were used to distinguish between tumor-induced vasculogenesis and angiogenesis. Asahara et al. transplanted tumor-bearing mice with transgenic bone marrow-derived cells, which were designed to constitutively express a genetic marker under the transcriptional regulation of different endothelial-specific promoters. He demonstrated that neovasculature of the developing tumors comprised marker gene-expressing cells *(21)*. Similar results were obtained in wild type mice transplanted with bone marrow cells expressing green fluorescent protein (GFP) or β-galactosidase or lacking sphingomyelinase *(22,23)*. Lyden and colleagues used Id-mutant mice that did not grow tumors as a result of defective angiogenesis to determine the importance of recruiting EPCs for tumor vasculogenesis. Following transplantation with wild-type bone marrow endothelial and hematopoietic progenitors from the bone marrow significantly contributed to neovascularization and thus restored tumor growth *(24)*. The majority of the newly formed tumor vessels were derived from the genetically marked donor, mainly during the early stages of tumor growth. In contrast to these studies, no contribution of bone marrow-derived cells to tumorigenesis was found by Göthert et al. *(25)*. Another study also did not find an involvement of bone marrow-derived EPCs in tumorigenesis, but defined a subset of bone marrow-derived cells of hematopoietic origin that contributed indirectly to tumor neovascularization *(26)*. Similarly, Rajantie and colleagues failed to detect a contribution of EPCs to tumor vasculogenesis, but provided evidence that bone marrow-derived cells contribute to the formation of periendothelial vascular mural cells *(27)*. These partially conflicting data suggest that whereas bone marow-derived cells take part in the cellular make-up of tumor vessels, their relative contribution to tumor ECs, periendothelial cells and nonstructural but proangiogenic cells has yet to be determined.

Mobilization and recruitment of EPCs to the tumor vasculature are regulated by a variety of growth factors, cytokines, and surface receptors. Under steady state conditions the number of circulating EPCs in the peripheral blood is low. However, in response to angiogenic stimuli, EPCs are rapidly mobilized from the bone marrow. This process is associated with elevated levels of VEGF in the plasma, suggesting that VEGF is the most critical factor promoting mobilization of EPCs *(20,28)*. VEGF *(29,30)* and placental growth factor (PlGF), another member of the VEGF family, are released by many tumors. These angiogenic factors induce expression of matrix metalloproteinase (MMP)-9 in the bone marrow microenvironment, which in turn activates the release of soluble Kit ligand (sKitL). Increased levels of sKitL then promote the proliferation and mobilization of stem and progenitor cells into the peripheral circulation *(30)*. Furthermore, VEGF interacts with its receptors, VEGF receptor-1 (VEGFR1, also termed Flt-1), and VEGFR2 (also termed Flk-1or kinase insert domain receptor, KDR), expressed both on endothelial and hematopoietic stem cells, thereby promoting further maturation of EPCs and the recruitment of these cells to the sites of neovascularization. PlGF—which signals primarily through VEGFR1—also contributes to vessel growth in tumors by recruiting BM derived cells *(31)*. The crucial role of these growth factors in tumor vasculogenesis was further confirmed by studies, in which either the growth factors or their receptors were inhibited. Inhibition of VEGF production or function by different classes of VEGF antagonists including neutralizing antibodies, soluble receptors, and compounds that block VEGF receptor signaling resulted in the inhibition of tumor growth *(29,32)*. Inhibition of either VEGFR1 or VEGFR2 resulted in vascular malfunction, but could only partially block tumor growth. However, simultaneous inhibition of both receptors led to impaired mobilization of bone marrow-derived cells to the tumor vasculature and was highly effective in retarding tumor growth, indicating that signaling through both receptors is required for tumor vasculogenesis *(31,33,34)*. Emerging evidence suggests that stromal-derived factor 1 (SDF-1/CXCL12) in synergy with VEGF may contribute to the recruitment of EPCs to tumors *(33)*. Little is known about signals that direct incorporation of EPCs into tumor-associated vessels.

2.3. Apoptotic and Antiapoptotic Mechanisms in EPCs

Similar to other cellular vehicles apoptosis of EPCs has to be curtailed during ex vivo expansion and homing to the tumor as well as after activation of its proapoptotic payload. On the other hand, apoptosis of the targeted tumor has to be induced, while normal tissue must be spared.

There is only scant information about the pro- and antiapoptotic mechanisms operating in EPCs. Much relevant information, however, can be gleaned from apoptosis occurring during vessel development and from apoptosis in mature ECs. Angiogenic growth factors not only stimulate EC proliferation and migration, but also concomitantly inhibit apoptosis, thus providing a survival advantage to ECs. VEGF and the VEGF receptor Fkl-1 play a pivotal role in maintaining EC integrity as demonstrated by targeted disruption of these genes *(12,36,37)*. The antiapoptotic effect of VEGF and also of angiopoietin-1 (Ang-1; via activation of its receptor Tie-2) is mainly mediated through activation of the survival promoting PI3K/Akt signaling pathway *(38,39)*. Activation of Akt stimulates the phosphorylation of proapoptotic Bad promoting its sequestration and thus decreasing apoptosis *(40)*. Moreover, Akt activates the endothelial nitric oxide synthase (eNOS). Enhanced synthesis of endothelial NO promotes EC survival by inhibiting the cysteine protease activity of the effector caspases *(41)*. Akt is also known to activate mTOR. mTOR-mediated effects appear to play a role in adult EPCs, because rapamycin-mediated inhibition of mTOR, the target of rapamycin, increases apoptosis and decreases proliferation and differentiation of adult EPCs *(42)*. VEGF has been also shown to mediate the upregulation of antiapoptotic proteins such as survivin, XIAP and Bcl-2 *(43–45)*, the importance of the latter being highlighted by its targeted disruption leading to attenuation of vascular development and of pathological angiogenesis *(46)*. VEGF-activated Akt also mediates down-regulation of p38 mitogen-activated protein kinase (MAPK)-dependent apoptosis *(47)*. Intercellular contacts such as mediated by VE-cadherin *(48)* also promote EC survival as does CD31 (PECAM-1), which inhibits Bax-dependent mitochondria-mediated apoptosis *(49)*. Cellular receptors interacting with the extracellular matrix can emanate antiapoptotic signals. Thus, aυβ3, the receptor for several ligands anchored in the extracellular matrix, triggers antiapoptotic NF-κB signaling once ligated *(50)*. This activity may form part of the dependence receptor nature of aυβ3. The receptor acts antiapoptotic when ligated to ECM-bound ligands by activating the Src, PI3K, and Erk pathways and by increasing Bcl-2, c-FLIP, and IAP protein activity. The receptor switches to proapoptotic activity involving p53 when triggered by ligands not bound to the ECM *(51)*. This occurs during detachment of therapeutic EPCs from their culture matrix and during homing until the EPCs have integrated into the tumor and are protected against apoptosis again *(52)*. Constitutive overexpression of human telomerase reverse transcriptase (hTERT) rescues adult EPCs from starvation-induced apoptosis, enhances their proliferative capacity, and keeps them in an undifferentiated state *(53)*. Compared with human umbilical vein ECs, adult EPCs derived from blood express significantly higher levels of the antioxidative enzymes MnSOD, catalase, and glutathione peroxidase leading to decreased oxidative stress-induced apoptosis *(54)*. Thus, adult therapeutic EPCs may possess a degree of intrinsic protection against radical producing proapoptotic payloads.

With so many antiapoptotic mechanisms in place, strong stimuli are necessary to induce apoptosis in ECs. Such stimuli comprise, among others, lipopolysaccharides and endostatin *(55,56)*, whose proapoptotic effects are mediated by the mitochondrial pathway. In contrast, death receptor-mediated apoptosis plays a limited role in EC that may be secondary to the high expression of cFLIP *(57)* and activation of NF-κB.

From these findings it can be surmised that the default mode of adult ECs is antiapoptotic. This is reflected by the low physiological turnover of ECs in the adult. Marked changes, such as infection, wound healing, hormone withdrawal during the uterine cycle, or tumor growth are required to increase turnover and associated apoptosis of ECs. In addition to these intrinsic antiapoptotic mechanisms, ex vivo expanded EPCs are subjected to antiapoptotic growth factors such as VEGF or bFGF and antiapoptotic extracellular matrices such as collagen and fibronectin. In concert, these mechanisms allow for ex vivo expansion of therapeutic EPCs with minimal spontaneous apoptosis. There are, however, two major proapoptotic insults to EPCs used for tumor therapy. One occurs when EPCs remain in a suspended state following disruption from the matrix of the culture vessel and during homing until they have integrated into the tumor (vessels). As discussed above, the disruption of antiapoptotic integrin signaling renders them susceptible to apoptosis during transit. Because the second major proapoptotic insult in the life of therapeutic EPCs, the activation of its payload, occurs during or shortly after this vulnerable phase, the therapeutic effect of the EPC may be impaired. Increasing protection against apoptosis without diminishing homing and tumor killing efficiency is thus an important goal of further research.

EC apoptosis plays a crucial role in tumor vasculature, because survival of the EC within the tumor vasculature influences the survival of the tumor. Therefore, inhibition of endothelial survival factors or the activation of the EC apoptotic machinery could be an attractive antitumor strategy. This is highlighted by recent work proposing endothelial apoptosis as a key determinant of the tumor response to ionizing radiation treatment. Tumors transplanted into mice deficient in the proapoptotic molecules Bax or acid sphingomyelinase were more resistant to radiotherapy than the same tumors in wild-type animals *(23)*.

2.4. EPCs for Tumor Therapy

The involvement of endothelial progenitors in tumor vasculogenesis provides the rationale to use these cells for therapeutic targeting of the tumor vasculature. Growth and metastasis of the majority of tumors critically depends on the formation of new blood vessels *(58)*. Because new vessel formation in adults is restricted to wound healing and to the female ovulatory cycle, it has been put forward, that the natural tropism of circulating EPCs could be exploited to deliver and selectively express genes at tumor sites. The use of endothelial progenitors as potential cellular vehicles for tumor gene therapy requires the isolation and ex vivo expansion of the cells, followed by their genetic manipulation and subsequent reinfusion into the patients.

2.4.1. Sources of EPCs

EPCs have been derived from mouse embryos, mouse or human embryonic stem (ES) cells, fetal liver, human umbilical cord blood, postnatal bone marrow, and peripheral blood. Embryonic EPCs derived from ES cells have been shown to differentiate into mature ECs and, following transplantation, to contribute to vasculogenesis also in the adult *(59)*. Embryonic EPCs have an unlimited proliferative capacity and are easy to manipulate genetically. In a proof of principle study, the potential of these cells for systemic cancer gene therapy has been explored *(60)*. Given the potential immunogenicity of differentiated ES-cell derivatives and ethical considerations limiting their generation, there is, however, a need for other sources of EPCs.

Umbilical cord blood contains a large number of stem and progenitor cells. In contrast to progenitors generated from adult bone marrow, progenitors derived from cord blood have distinctive proliferative advantages, including the capacity to form a greater number of colonies, a higher cell-cycle rate, and longer telomeres. EPCs generated from human cord blood have been shown to achieve more than 100 population doublings, as opposed to those generated from the mononuclear fraction of adult peripheral blood that stopped dividing after 20 to 30 population doublings *(61)*. Moreover, in contrast with bone marrow, cord blood can be obtained noninvasively. Studies have shown, that transplantation of cord blood-derived EPCs efficiently augments postnatal neovascularization in vivo *(62)*.

Adult endothelial progenitors have been initially isolated from CD34-positive peripheral blood mononuclear cells *(13)*. Since their identification a number of studies have reported more defined methods of isolation and culture. However, because most of the studies have focused on promoting the differentiation of isolated EPCs into mature, terminally differentiated ECs, conditions required for the long-term proliferation, and maintenance of undifferentiated EPCs still have to be elucidated. Recently, Reyes and colleagues identified a unique cell population from postnatal bone marrow that has extensive proliferation potential and the capacity to differentiate into different cell types of the mesodermal lineage *(63)*. These nonendothelial stem cells, termed multipotent adult progenitor cells (MAPCs), were able to differentiate into ECs in the presence of VEGF in vitro and to contribute to vasculogenesis in vivo. Remarkably, intravenous injection of undifferentiated MAPCs resulted in new vessel formation in transplanted tumors, indicating that MAPCs can differentiate into ECs in vivo, in response to local clues. The authors proposed MAPCs as the precursors of EPCs and as a promising source of ECs for therapeutic approaches.

2.4.2. In Vitro Expansion of EPCs

EPCs have been isolated from mononuclear cells by antibodies against surface markers like CD133 or VEGFR2—coupled to magnetic microbeads—or by adherence culture of unfractionated peripheral blood mononuclear cells (PBMCs). Isolated EPCs are grown on fibronectin or collagen-coated dishes in the presence of growth factors, such as VEGF, basic fibroblast growth factor (bFGF), and insulin-like growth factor (IGF) and can be differentiated into mature adherent ECs *(13,18)*. Upon endothelial differentiation EPCs downregulate CD133 expression and start to express mature endothelial-specific markers, such as VEGFR2, VE-cadherin, platelet EC adhesion molecule (PECAM-1, CD31), von Willebrand factor (vWF), and CD105.

The success of therapeutic approaches depends on the isolation of a sufficient number of EPCs for expansion. As described previously, the number of EPCs in the bone marrow and in the circulation is low. However, clinical studies have demonstrated that the number of circulating EPCs can be significantly increased by injecting chemotactic factors, such as VEGF, PlGF, and granulocyte-macrophage colony-stimulating factor (GM-CSF) *(64)*. Drugs such as statins can also be used to mobilize larger number of EPCs for in vitro manipulation *(65)*. Lin and colleagues described a reliable method for isolating a virtually unlimited number of bone marrow-derived EPCs *(66)*. After long-term culture of PBMCs late outgrowth ECs with high proliferative capacity were generated. In contrast, vessel wall-derived, or circulating mature ECs gave rise to early outgrowth ECs with limited proliferation potential, thereby providing a functional assay to discriminate bone marrow-derived EPCs from contaminating ECs. Late outgrowth cells, termed as blood outgrowth endothelial cells (BOECs) could be passaged up to 30 times.

2.4.3. Genetic Manipulation of EPCs

Ex vivo expanded EPCs derived from different sources are easily amenable to genetic manipulation. Transduction of EPCs and their putative precursors with retrovirus *(22,67)* and lentivirus *(26)* vectors have been shown to be feasible and resulted in long-term transgene expression in mice. These vectors mediate stable genomic integration of the transgene, thus allowing for long-term transgene amplification. This may be necessary for several therapeutic approaches using EPCs as cellular vehicles. Moreover, replication-defective retro- and lentiviral vectors do not express any viral proteins, thereby decreasing the chance of immune clearance of the transduced EPC following readministration to the host.

EPCs or other bone marrow-derived cells may also be manipulated using nonintegrating vectors, such as those based on adenovirus and herpes simplex virus (HSV). Such vectors are advantageous if an antitumor effect can be achieved within a short interval.

2.4.4. EPCs as Cellular Vehicles for Tumor Gene Therapy

Recent advances in the biology of EPCs have suggested the feasibility of their use as cellular vehicles. As described previously, EPCs can be easily isolated and ex vivo expanded in quantities required for therapeutic approaches. Cells can be efficiently transduced with therapeutic payloads using viral vectors. What renders EPCs attractive cellular vehicles for systemic tumor gene therapy is their capacity to localize not only into the tumor proper but also into the tumor's vasculature.

Tumor vessels are promising targets for anticancer therapeutics for several reasons. The strict dependence of the majority of tumors on recruiting new blood vessels to grow and metastasize provides a common target for the treatment of different tumor types irrespective of their antigenic and genetic make-up. Furthermore, in adults—with the exception of wound healing and the uterine cycle—neovascularization is restricted to tumor growth. In addition, tumor vessels are accessible to circulating EPCs. This might eliminate some of the major obstacles associated with targeting of solid tumors, such as poor tissue penetration. It has to be noted, however, that systemically administered EPCs incorporate into only a minority of tumor vessels *(21,22,63,67)*, which limits the ability of therapeutic EPCs to destroy tumor vessels. However, because EPCs have been shown to home also to the tumor proper, a second venue to attack the tumor is open. Therefore, as far as the ability to target tumors as a whole is concerned, the relative contribution of EPCs to the formation of tumor vessels is a somewhat moot point.

2.4.5. EPS Can be Fitted With Apoptosis Effectors

2.4.5.1. Suicide Gene Bearing EPCs

Antitumor suicide gene therapy is one of the emerging strategies against cancer. Suicide genes, which encode nonmammalian enzymes, are capable of converting nontoxic prodrugs into highly toxic metabolites. The successful eradication of the tumor depends on the ability of the suicide gene-expressing target cells to exert cytotoxicity on neighbouring tumor cells, thereby providing a potent bystander effect *(68)*. Several groups have investigated the feasibility of using suicide gene-expressing bone marrow cells and EPCs as gene-targeting vectors to mediate systemic antitumor responses.

$CD34^+$ cells transduced by a TK-encoding HSV vector have been shown to migrate to a skin autograft in a nonhuman primate model and to mediate accelerated regression of the graft following ganciclovir (GCV) treatment *(69)*. Genetically modified CD34-positive cells have also been tested by the same group for the ability to exert a strong cytotoxic bystander effect *(70)*. Transduced cells expressed the TK gene with high efficiency, became sensitive to the effect of GCV, and mediated a strong bystander killing of both tumor cells and proliferating ECs in vitro.

Bone marrow-derived endothelial progenitor-like cells transplanted into tumor-bearing mice incorporated into the angiogenic vasculature of the growing tumors *(67)*. EPCs contributed to between 10 and 25% of the tumor-associated vasculature in sublethally irradiated mice, whereas the number of donor EPCs was far smaller in the tumor vessels of nonirradiated mice. This observation underlined the hypothesis that systemically administered EPCs might have to compete with endogenous EPCs, thus suppression of endogenous EPCs by irradiation may be necessary for optimal donor EPC incorporation. When the EPCs were modified to express the TK gene, they induced a strong antitumor effect without any systemic toxicity after GCV treatment, pointing to the feasibility of employing EPCs as angiogenesis-selective cancer-targeting vectors.

DePalma and co-workers transplanted bone-marrow progenitors expressing the TK gene under the control of an endothelial-specific promoter/enhancer (Tie2) *(26)*. As described above, these cells homed specifically to tumors and interacted with vascular ECs. The EPCs and tumor bystander cells were selectively eliminated following GCV treatment, resulting in slower tumor growth without systemic toxicity.

More recently it has been shown that murine embryonic EPCs administered systemically preferentially home to hypoxic areas of lung metastases *(60)*. Because embryonic EPCs do not express MHC I proteins and are not killed by nonactivated NK cells, they are able to evade immunological rejection. When modified to express the cytosine deaminase gene, these embryonic EPCs exerted a bystander effect on tumor cells upon administration of 5-FC, thus prolonging the life of tumor bearing mice.

2.4.5.2. Antiangiogenic EPCs

An increasing number of clinical trials (phase I, II, and III) have demonstrated the efficacy of antiangiogenic compounds for the treatment of cancer *(71)*. Gene-mediated delivery of these agents has been proposed to have potential benefits such as generation of high-drug concentrations limited to the tumor. This notion was supported by recent clinical data showing that transplantation of unfractionated bone marrow cells engineered to express an angiogenesis inhibitor—a truncated form of VEGFR2—resulted in long-term expression of the therapeutic protein in the tumor in mice and in reduced tumor growth *(22)*.

2.4.5.3. EPCs as Packaging Cells for Viral Vectors

EPCs, as other cellular vehicles, can be engineered to produce viral vectors *(72,73)*. This elegant method avoids systemic exposure to viral vectors and protects them from immune inactivation, nonspecific adhesion, and other causes of particle loss during transit to the tumor. Using such vector-producing cells appears to be an attractive strategy if the incorporation efficiency of the particular cell carrier is low. Each cell can release multiple virus particles at the tumor site, thereby amplifying the therapeutic effect.

2.4.6. Clinical Translation

Because of the availability and ease of isolation, expansion and manipulation, and because of immunological tolerance, EPCs derived from autologous blood will most likely be the first choice as the source of therapeutic EPCs. However, several points have to be addressed before autologous EPC-based tumor gene therapies can enter clinical trials. First, to yield an adequate therapeutic effect a sufficient number of these cellular carriers has to reach the tumor site. As discussed above, the number of endogenous EPCs recruited from the bone marrow into tumors appears to be low, as is the number of these EPCs actually incorporating into tumor vessels. The same holds true for ex vivo expanded autologous EPCs in preclinical tumor models. Thus, data on the tumor homing efficacy of systemically injected autologous EPCs in patients are required but are not yet available. Furthermore, a better understanding is needed as to what degree of differentiation is required for ex vivo expanded autologous therapeutic EPCs to achieve optimal tumor homing and whether genetic manipulation of the EPCs is necessary to increase homing. In addition, it has to be investigated whether myelosuppression to suppress recruitment of endogenous EPCs that potentially compete with therapeutic EPCs may prove beneficial.

Second, recruitment of EPCs depends on the tumor microenvironment, which may vary not only between tumor entities but also within a given tumor. An obvious intratumoral variable is the degree of hypoxia that leads to differential recruitment of EPCs predominantly into hypoxic areas *(60)*. More investigations are required to define tumor entities or tumor states that are dependent on the recruitment of EPCs and to delineate the molecular mechanisms governing EPC recruitment to tumors.

Third, although it appears that EPCs are partially protected against apoptosis, further research is needed to elucidate how to protect them better against their proapoptotic payload without sacrificing the reliability of the elimination of EPCs at the end of therapy. Final elimination is needed to avoid an ongoing proangiogenic effect of surviving EPCs on the tumor and to preclude the remote possibility of malignant transformation of the highly proliferative EPCs.

The reward of these efforts may be a cellular vehicle that provides efficient, safe, and specific systemic delivery of highly proapoptotic effectors into tumors.

3. Neural Progenitors for Brain Tumor Gene Therapy

3.1. Gene Therapy of Brain Tumors

Brain tumors are hard to treat. In particular the infiltrative nature of many of these tumors, such as glioblastoma multiforme, often makes it impossible to achieve complete surgical resection or to apply curative radiotherapy without risking severe side effects. Brain tumors are also often chemoresistant. Thus, novel therapies are urgently needed for these tumors.

Brain tumors often have characteristic genetic alterations in the form of activated oncogenes or disrupted tumor suppressor genes. Gene therapy holds the promise of augmenting conventional brain tumor therapy by specifically targeting these genetic alterations. However, such approaches require a degree of gene transfer and targeting efficiency impossible to achieve with current vector systems. Therefore strategies have been developed which are independent of the genes specifically altered in brain tumors allowing more leeway in gene transfer efficiency and targeting. The effectors of these strategies include suicide/prodrug systems, toxins, cytokines, and angiogenesis inhibitors. But even such spatially broad-acting effectors require a degree of targeting efficiency hard to obtain. It is thus not surprising that cellular vehicles with brain tumor homing abilities, such as neural stem or progenitor cells, have recently received increasing attention.

3.2. The Physiological Role of Neural Progenitor Cells

During the embryonic and early postnatal period, a large number of cells in the central nervous system (CNS) are born. These neurons are derived from self-renewing multipotent neural stem and progenitor cells that are highly migratory. For a long time the adult brain had been considered to be postmitotic. Recently, however, neural progenitor cells (NPCs) have been shown also to exist in the adult CNS *(74,75)*. In the adult brain of rodents these NPCs are found within the subventricular zone *(76)*, the dentate gyrus of the hippocampus *(77)*, the septum and striatum *(78)*, the spinal cord *(79)*, the neocortex and the optic nerve *(80)*. Adult neurogenesis has also been found in nonhuman primates *(81)* and in humans *(82)*.

3.3. NPCs for Brain Tumor Therapy

3.3.1. NPCs Home to Brain Tumors

NPCs from various sources and from various species are highly migratory when responding to environmental cues during development, following injury, or in the presence of brain tumors. NPCs from nonhuman primates can migrate over a distance of more than 2 cm *(83)*. Similarly, human fetal neural stem cells migrate extensively when injected into the brain of immunodeficient neonatal mice *(84)*. Human fetal NPCs have also been shown to migrate towards ischemic areas in the rat cerebral cortex with migrating NPCs expressing the neuroblast marker doublecortin, whereas the majority of NPCs that have stopped migration at the lesion border express the early neuronal marker beta-3-tubulin *(85)*.

What has kindled the interest of oncologists is the marked homing efficiency of NPCs to tumors. Aboody et al. showed that human and murine NPCs home specifically to gliomas implanted in mice even when the NPCs are injected into the hemisphere contralateral to the tumor *(86)*. In addition, NPCs appeared to "chase down" individual tumor cells "escaping" from the main tumor mass into normal brain tissue. A similar tropism of NPCs towards both the main tumor mass and tumor satellites of experimental glioblastomas was demonstrated by Ehtesham et al. *(87)*. Both glial-restricted progenitors and ES cell-derived neural stem cells have been shown to migrate towards experimental gliomas *(88)* as have been neural stem-like cells expanded from human adult bone marrow *(89)*. Furthermore, NPCs homed to brain tumor even when they were administered outside the brain by intravenous injection *(90)*. The migratory tropism of NPCs to experimental brain tumors has been monitored noninvasively in vivo using luciferase-labeled NPCs and bioluminescence imaging *(91)*. Systemically administered bone marrow cells labeled with superparamagnetic iron oxide nanoparticles could be detected in the tumor vessels of mouse glioma by in vivo magnetic resonance imaging *(92)*. Such techniques allow NPC-based therapies to be monitored for safety and efficacy.

The mechanisms underlying the migration and homing of NPCs have only begun to be elucidated. Thus, CXCR4-bearing NPCs have been shown to migrate to SDF-1a/CXCL12 expressing ischemic brain areas *(93)*. Information about other interactions may be gleaned from the migration of glioma cells. Because these cells express similar receptors as NPCs they may be targets for the same migration-inducing chemoattractants.

3.3.2. NPCs Can be Isolated From a Variety of Sources and Expanded Ex Vivo

NPCs can be obtained or derived from two sources: either from nonregional cells such as embryonic stem cells and bone marrow cells or from regional tissue stem cells. NPCs can be isolated and propagated from many neural tissues at any stage after the formation of neuroectoderm (i.e., from embryonic, fetal, or adult CNS). After isolation, NPCs are grown in serum-free medium as clonally derived proliferating floating aggregates, so-called neurospheres. NPCs can be expanded using mitogens such as fibroblast growth factor 2 (FGF-2) and epidermal growth factor (EGF) and cytokines including leukemia inhibitory factor. NPCs may also be expanded through introduction of an immortalizing gene. Expanded NPCs can be differentiated into neurons, astrocytes and oligodendrocytes *(77)* which are capable of integrating into neural structures *(84,94–97)*. This raises the vision of NPCs being used not only as cellular vehicles for brain tumor therapy, but also to mitigate neural loss caused by the tumor or its therapy.

Adult NPCs have been derived from the brain of adult humans *(98)*. Availability of autologous NPCs from the adult human brain is limited because of the invasive nature of their procurement and the dearth of dispensable brain tissue. Supply of brain tissues from cadavers and fetuses is also limited and carries the risk of immunological rejection.

NPCs can be generated from ES cells of various species, including rodents *(99,100)* nonhuman primates *(101,102)*, and humans *(103)*. Although this source of NPCs is potentially unlimited, immunogenicity, the potential for malignant transformation, and ethical considerations may limit their use as cellular vehicles.

Immortalized NPC cell lines have been generated from mouse *(104)*, rat *(105)*, and human *(106–108)*. Although immortalized NPC lines are useful for experimental purposes, the potential malignant transformation precludes their clinical use.

Progenitor cells that are more easily accessible, such as hematopoietic stem cells from the bone marrow, may substitute progenitor cells that are difficult to access for autologous grafting, such as brain-derived NPCs. Several groups have demonstrated by immunohistochemistry that bone marrow cells can be differentiated into neurons and glial cells in vitro and in vivo *(109–111)*. Although this concept of "transdifferentiation" has been challenged *(112)*, such cells may still be useful as autologous cellular vehicles for brain tumor gene therapy, because experimental brain tumors were successfully treated using neural stem-like cells expanded from human adult bone marrow and transduced with platelet factor 4 (PF4), an antiangiogenic protein *(89)*.

For brain tumor gene therapy employing NPCs it will be important to determine not only from which source NPC should be derived from, but also which stage of differentiation is to be used. For example, neural stem cells derived from neurospheres from the subventricular zone may have a higher capacity for expansion but a decreased ability to migrate compared to more differentiated NPCs from the same area *(113)*.

3.3.3. Apoptotic and Antiapoptotic Mechanisms in NPCs

Apoptotic mechanisms are operational in gene therapy using neural progenitors in several aspects. First, spontaneous apoptosis occurring in NPCs has to be controlled if a sufficient number of NPCs is to be efficiently generated in vitro and delivered in vivo. Second, NPCs acting as cellular vehicles have to be protected from apoptosis while homing to the tumor. Third, NPCs should be resistant to apoptosis induced by their proapoptotic payload. Fourth, apoptosis is a crucial mechanism of cell death of the cancer cells targeted by therapeutic NPCs. Finally, normal brain tissue should be resistant to apoptotic insults induced by NPCs.

Despite the obvious importance of apoptosis in NPC gene therapy little is known about it. Indirect clues can be gained by extrapolating from developmental apoptosis during brain development. During this process an excess of neural progenitors is born requiring widespread apoptosis to form the definitive neural structures. The molecular mechanisms underlying this process have just begun to be elucidated. Developmental apoptosis in embryonic NPCs depends on Apaf1, caspase-9, and caspase-3, but is independent of Bax and Bcl-X_l *(114–118)*. Notch signaling reduces the embryonic NPC pool by inducing p53-dependent apoptosis in embryonic NPCs *(119)*. Notch signaling also induces embryonic NPCs to differentiate *(120,121)*. Embryonic NPCs are protected against death receptor mediated apoptosis since caspase-8 is absent and because the death receptor inhibitory protein PED/PEA-15 is highly expressed in embryonic NPCs *(122)*. An interesting mechanism of controlling the number of embryonic NPCs is the asymmetric distribution of proapoptotic molecules such as prostate apoptosis response (PAR)-4 and endogenous ceramide into only one of the daughter cells during mitosis *(123)*. Using a ceramide analog, this mechanism can be exploited to induce selective apoptosis in nondifferentiated pluripotent stem cells, thus enriching ES cell-derived NPCs *(124)*. Taken together, these data show that apoptosis of embryonic NPCs proceeds via the intrinsic apoptosis pathway with caspase-3 being the pivotal executioner. Spontaneous apoptosis of therapeutic embryonic NPCs may follow the same pathway. In contrast, apoptosis of embryonic NPCs induced by DNA-damaging agents and γ-irradiation depends on p53 and caspase-9, but neither on Bax nor caspase-3 *(125)*. Rendering embryonic NPC deficient in p53 and caspase-9 may thus induce protection of therapeutic embryonic NPC against apoptosis-inducing DNA-damaging payloads.

In adult NPCs Bax and Bak increase apoptosis and differentiation of NPCs without decreasing their production or proliferation *(126,127)*. Protection against receptor-mediated apoptosis is provided by the absence of caspase-8 and overexpression of PED/PEA-15 *(122)*. Thus, similar to embryonic NPCs, the intrinsic apoptosis pathway appears to be pivotal in programmed cell death of adult NPCs. Unlike in embryonic NPCs, however, Bax and Bak play important roles. Inhibiting the intrinsic or the common execution pathway of apoptosis should protect NPCs from apoptosis and thus increase their therapeutic value. Haploinsufficiency for PTEN in adult NPCs increases phosphorylation of AKT, rendering NPCs more resistant to oxidative stress-induced apoptosis *(128)*. Interestingly, adult NPCs are exquisitely sensitive to γ-irradiation-induced apoptosis *(129)*, which is associated with ROS production and may involve p53 *(130)*. Similar apoptosis regulating mechanisms may be operational in spontaneous apoptosis of adult therapeutic NPCs.

3.3.4. NPCs are Partially Protected From Immune Responses

Several factors contribute to a relative immune privilege of NPCs within the brain. The brain *per se* is to some extent an immune-privileged site. Although leaky in pathological states such as malignancies, the blood–brain barrier (BBB) does donate some degree of immune protection by virtue of limiting transgression of lymphocytes. The brain also has few lymphatics and brain tumors often secrete immunosuppressive proteins such as tumor growth factor (TGF)-β. In addition, NPCs themselves have features causing immunoprotection. EGF-responsive NPCs from postnatal day 1 mice do not express major hisotcompatibility complex (MHC) I and II antigens and survive for a prolonged period of time at a nonimmune-privileged site in an allogeneic recipient *(131)*. Although differentiation was not reported to increase MHC I and II, an increase was reported after exposure to γ-interferon. Taken together, these data raise the possibility that NPCs can be employed even in an allogeneic setting, which would markedly increase the utility of ES cell-derived NPCs.

3.3.5. NPCs Can be Genetically Modified With Apoptosis Effectors

3.3.5.1. Suicide Gene Bearing NPCs

One promising brain tumor therapy is augmentation of the specific cytotoxicity of anticancer drugs by selectively increasing their concentration within the tumor through on-site conversion from nontoxic prodrugs. This gene-directed enzyme/prodrug system is also called suicide gene therapy. Given the low gene transfer efficiency in vivo this strategy would be futile if all cells of a tumor had to be transfected. What makes suicide gene therapy an option is the so-called "bystander effect," that is, the killing of untransfected cells surrounding the transfected target cells. Many suicide gene/prodrug systems have been investigated in preclinical trials. One of the first and most widely used systems is the herpes simplex virus thymidine kinase (TK) gene with the prodrug GCV. HSV-TK phosphorylates nontoxic GCV to intermediates that are metabolized to a toxic form, which is incorporated into replicating DNA causing apoptotic cell death *(132)*. TK/GCV-mediated apoptosis is characterized by CD95-ligand-independent CD95 aggregation leading to the formation of a Fas-associated death domain protein (FADD) and caspase-8-containing, death-inducing signaling complex (DISC) *(133)*. Mitochondria amplify TK/GCV-induced apoptosis by regulating p53 accumulation and the effector phase of apoptosis *(134)*. TK/GCV-induced apoptosis of tumor cells is augmented by TNF-related apoptosis-inducing ligand (TRAIL) depending both on caspase activation and on mitochondrial apoptogenic function *(135)*. Although gene therapy of glioma using TK/GCV has, with few exceptions *(5,6,136)*, failed in the clinic as a result of low transduction efficacy *(2,3,137)*, this problem may be circumvented by using the highly migratory NPCs expressing HSV-TK, as suggested by a preliminary in vitro study *(138)*. TK/GCV, however, depends on the presence of intercellular gap junctions, which may limit the killing effect of NPCs to tumor cells in the immediate vicinity.

Another commonly used suicide gene prodrug system is cytosine deaminase/5-fluorocytosine (CD/5-FC). CD deaminates the nontoxic pyrimidine 5-fluorocytosine to the cytotoxic 5-fluorouracil *(139)*. Aboody and colleagues were the first to combine an enzyme/prodrug system and NPCs for brain tumor gene therapy. Murine NPCs expressing CD migrated to tumors and microscopic satellites and significantly reduced the tumor mass upon administration of the prodrug *(86)*. Immortalized CD-expressing NPCs coinjected with glioma cells into the brain of rodents significantly decreased tumor size after application of 5-FC *(140)*. 5-FU diffuses out of the cell harboring CD and into neighbouring tumor cells independent of gap junctions, thus making this suicide gene system more versatile and probably also more efficient than TK/GCV.

3.3.5.2. Cytokine Secreting NPCs

Malignant gliomas are able to overcome host immune defences through a variety of mechanisms. Most clinical immunotherapy protocols have manipulated the immune response in an attempt to direct it against cancer cells. One approach is stimulation with cytokines to enhance recognition and thus eradication of the tumor cells. Interleukin (IL)-2, IL-4, IL-12, and GM-CSF have been used for this purpose in animal models.

NPCs that migrate to the tumor and secrete tumoractive cytokines constitute an attractive therapy modality. Mouse primary neural progenitor cells transduced with the IL-4 gene and injected into established syngeneic brain glioblastomas elicited a tumor-directed T-cell response with a strong antitumor effect *(141)*. This effect was more pronounced than the effect of retrovirus-mediated in vivo transfer of IL-4. Similarly, NPCs derived from fetal forebrain infected with an adenoviral vector carrying the gene for IL-12 prolonged survival of intracranial GL26 glioma-bearing mice *(142)*. This was associated with enhanced T-cell infiltration into tumor microsatellites and long-term antitumor immunity.

3.3.5.3. NPCs Shedding Oncolytic Viruses

An increasing number of oncolytic virus vectors have been developed lately for cancer therapy. HSV, one of the first oncolytic viruses to be used for this purpose, can be genetically engineered to replicate and spread selectively in tumor cells while sparing normal neurons *(143)*. Other oncolytic viruses, such as adenovirus and measles virus, have also been shown to be effective in lysing brain tumor cells. However, delivery of viral vectors to tumors in the brain remains a challenge, especially systemic delivery. To meet this challenge, NPCs have been employed as packaging cells. These deliver

HSV-1 mutants that only replicate in dividing (i.e., tumor) cells into established glioma in the mouse *(144)*. Replication of the mutant HSV-1 was blocked in the NPCs prior to implantation into the tumor by chemical growth arrest. NPCs were observed to migrate extensively throughout the tumor and into the surrounding parenchyma, with appropriate gene expression of HSV-1. Thus, NPC-mediated viral vector delivery has the potential to increase the reach of viral vectors within the brain and to magnify the efficacy of viral-mediated gene transfer into diffusely growing brain tumor cells.

3.3.5.4. NPCs Expressing TRAIL

TRAIL is a member of the tumor necrosis factor protein superfamily. It selectively induces apoptosis in a variety of neoplastic cells without evident toxicity to normal cells.

Murine fetal brain NPCs adenovirally infected to release TRAIL migrate to glioblastoma xenografted into the brain of mice, track down their satellites and, by inducing significant apoptosis of both the main tumor and its satellites, cause a marked inhibition of tumor growth while sparing normal brain tissue *(87)*. Importantly, the NPCs are not killed by the TRAIL they produce, highlighting the preferential killing of malignant cells by TRAIL. Using TRAIL as an apoptosis effector thus has the important advantage of leaving the cellular vehicle unharmed. Similar, NPCs differentiated from adult mouse bone marrow and transduced with TRAIL have been shown to migrate to glioblastoma xenografts and induce apoptosis in the tumors, thus prolonging the life of the tumor-bearing mice *(89)*.

3.3.5.5. Antiangiogenic NPCs

Brain tumors, like other cancers, require angiogenesis to grow. Angiogenesis can be inhibited via diffusible factors. NPCs that home into brain tumors are good candidates for delivering such antiangiogenic factors within or surrounding the tumor mass and its satellites. Human adult bone marrow-derived neural stem-like cells modified to secrete platelet factor 4 (PF4) homed to glioblastoma xenografts, reduced their microvascular density, and prolonged the life of the mice *(89)*.

3.3.5.6. Intrinsic Antitumor Effects of NPCs

Interestingly and for unknown reasons even NPCs that have not been engineered to express cytotoxic or immunomodulatory effectors have a growth inhibiting effect on rodent glioma leading to prolongation of the life of glioma bearing rodents *(145)*.

3.3.6. Clinical Translation

Whereas proof of principle has been achieved using NPCs as cellular vehicles for brain tumor gene therapy, clinical relevance still has to be proven. Human brain tumors differ in several important aspects from the rodent mouse tumor models employed in the proof of principle studies. First, the human brain is many times larger than the rodent brain. This translates into much greater distances the therapeutic NPC must cover. Second, the large rise of human tumors also challenges both the expandability and the infiltrative ability of NPCs. Third, human brain tumors are often infiltrative, a feature often not recapitulated in rodent brain tumor models. Fourth, in experimental tumors, especially in those of rodent origin, the proportion of cycling cells is higher than in patient tumors. Cycling tumor cells are usually easier to treat than quiescent ones. Fifth, most experimental tumors are naive and thus more susceptible to therapy, whereas many patients who might be considered for NPC-based therapy will be heavily pre-treated. Finally, neural stem cells, and to a lesser degree progenitor cells, share certain characteristics with tumor cells such as self-renewal, migration, and maybe apoptosis resistance. Thus, these cells may be prone to malignant transformation *(146)* and it has to be assured that therapeutic NPCs are reliably killed after they have fulfilled their therapeutic function. These caveats notwithstanding, NPCs do constitute a very promising means to deliver therapeutic genes including proapoptotic ones to infiltrative growing brain tumors.

Acknowledgments

This work was supported by grants of the DFG and the Wilhelm Lander-Liftung (to CB).

References

1. Kasid A, Morecki S, Aebersold P, et al. Human gene transfer: characterization of human tumor-infiltrating lymphocytes as vehicles for retroviral-mediated gene transfer in man. Proc Natl Acad Sci USA 1990;87:473–477.
2. Lang FF, Bruner JM, Fuller GN, et al. Phase I trial of adenovirus-mediated p53 gene therapy for recurrent glioma: biological and clinical results. J Clin Oncol 2003;21:2508–2518.
3. Packer RJ, Raffel C, Villablanca JG, et al. Treatment of progressive or recurrent pediatric malignant supratentorial brain tumors with herpes simplex virus thymidine kinase gene vector-producer cells followed by intravenous ganciclovir administration. J Neurosurg 2000;92:249–254.
4. Ram Z, Culver KW, Oshiro EM, et al. Therapy of malignant brain tumors by intratumoral implantation of retroviral vector-producing cells. Nat Med 1997;3:1354–1361.
5. Prados MD, McDermott M, Chang SM, et al. Treatment of progressive or recurrent glioblastoma multiforme in adults with herpes simplex virus thymidine kinase gene vector-producer cells followed by intravenous ganciclovir administration: a phase I/II multi-institutional trial. J Neurooncol 2003;65:269–278.
6. Valery CA, Seilhean D, Boyer O, et al. Long-term survival after gene therapy for a recurrent glioblastoma. Neurology 2002;58:1109–1112.
7. Griffiths L, Binley K, Iqball S, et al. The macrophage—a novel system to deliver gene therapy to pathological hypoxia. Gene Ther 2000;7:255–262.
8. Ojeifo JO, Lee HR, Rezza P, Su N, Zwiebel JA. Endothelial cell-based systemic gene therapy of metastatic melanoma. Cancer Gene Ther 2001;8:636–648.
9. Fabre JW. The allogeneic response and tumor immunity. Nat Med 2001;7:649–652.
10. Choi K, Kennedy M, Kazarov A, Papadimitriou JC, Keller G. A common precursor for hematopoietic and endothelial cells. Development 1998;125:725–732.
11. Shalaby F, Rossant J, Yamaguchi TP, et al. Failure of blood-island formation and vasculogenesis in Flk-1-deficient mice. Nature 1995;376:62–66.
12. Ferrara N, Carver-Moore K, Chen H, et al. Heterozygous embryonic lethality induced by targeted inactivation of the VEGF gene. Nature 1996;380:439–442.
13. Asahara T, Murohara T, Sullivan A, et al. Isolation of putative progenitor endothelial cells for angiogenesis. Science 1997;275:964–967.
14. Gunsilius E, Petzer AL, Duba HC, Kahler CM, Gastl G. Circulating endothelial cells after transplantation. Lancet 2001;357:1449–1450.
15. Rafii S, Lyden D. Therapeutic stem and progenitor cell transplantation for organ vascularization and regeneration. Nat Med 2003;9:702–712.

16. Peichev M, Naiyer AJ, Pereira D, et al. Expression of VEGFR-2 and AC133 by circulating human CD34(+) cells identifies a population of functional endothelial precursors. Blood 2000;95:952–958.
17. Gehling UM, Ergun S, Schumacher U, et al. In vitro differentiation of endothelial cells from AC133-positive progenitor cells. Blood 2000;95:3106–3112.
18. Shi Q, Rafii S, Wu MH, et al. Evidence for circulating bone marrow-derived endothelial cells. Blood 1998;92:362–367.
19. Harraz M, Jiao C, Hanlon HD, Hartley RS, Schatteman GC. CD34- blood-derived human endothelial cell progenitors. Stem Cells 2001;19:304–312.
20. Bertolini F, Paul S, Mancuso P, et al. Maximum Tolerable Dose and Low-Dose Metronomic Chemotherapy Have Opposite Effects on the Mobilization and Viability of Circulating Endothelial Progenitor Cells. Cancer Res 2003;63:4342–4346.
21. Asahara T, Masuda H, Takahashi T, et al. Bone marrow origin of endothelial progenitor cells responsible for postnatal vasculogenesis in physiological and pathological neovascularization. Circ Res 1999;85:221–228.
22. Davidoff AM, Ng CY, Brown P, et al. Bone marrow-derived cells contribute to tumor neovasculature and, when modified to express an angiogenesis inhibitor, can restrict tumor growth in mice. Clin Cancer Res 2001;7:2870–2879.
23. Garcia-Barros M, Paris F, Cordon-Cardo C, et al. Tumor response to radiotherapy regulated by endothelial cell apoptosis. Science 2003;300:1155–1159.
24. Lyden D, Hattori K, Dias S, et al. Impaired recruitment of bone-marrow-derived endothelial and hematopoietic precursor cells blocks tumor angiogenesis and growth. Nat Med 2001;7:1194–1201.
25. Gothert JR, Gustin SE, van Eekelen JA, et al. Genetically tagging endothelial cells in vivo: bone marrow-derived cells do not contribute to tumor endothelium. Blood 2004;104: 1769–1777.
26. De Palma M, Venneri MA, Roca C, Naldini L. Targeting exogenous genes to tumor angiogenesis by transplantation of genetically modified hematopoietic stem cells. Nat Med 2003;9:789–795.
27. Rajantie I, Ilmonen M, Alminaite A, Ozerdem U, Alitalo K, Salven P. Adult bone marrow-derived cells recruited during angiogenesis comprise precursors for periendothelial vascular mural cells. Blood 2004;104:2084–2086.
28. Gill M, Dias S, Hattori K, et al. Vascular trauma induces rapid but transient mobilization of VEGFR2(+)AC133(+) endothelial precursor cells. Circ Res 2001;88:167–174.
29. Kim KJ, Li B, Winer J, et al. Inhibition of vascular endothelial growth factor-induced angiogenesis suppresses tumour growth in vivo. Nature 1993;362:841–844.
30. Heissig B, Hattori K, Dias S, et al. Recruitment of stem and progenitor cells from the bone marrow niche requires MMP-9 mediated release of kit-ligand. Cell 2002;109:625–637.
31. Carmeliet P, Moons L, Luttun A, et al. Synergism between vascular endothelial growth factor and placental growth factor contributes to angiogenesis and plasma extravasation in pathological conditions. Nat Med 2001;7:5575–5583.
32. Millauer B, Shawver LK, Plate KH, Risau W, Ullrich A. Glioblastoma growth inhibited in vivo by a dominant-negative Flk-1 mutant. Nature 1994;367:576–579.
33. Lyden D, Hattori K, Dias S, et al. Impaired recruitment of bone-marrow-derived endothelial and hematopoietic precursor cells blocks tumor angiogenesis and growth. Nat Med 2001;7:1194–1201.
34. Luttun A, Tjwa M, Moons L, et al. Revascularization of ischemic tissues by PlGF treatment, and inhibition of tumor angiogenesis, arthritis and atherosclerosis by anti-Flt1. Nat Med 2002;8:831–840.

35. Kryczek I, Lange A, Mottram P, et al. CXCL12 and vascular endothelial growth factor synergistically induce neoangiogenesis in human ovarian cancers. Cancer Res 2005;65:465–472.
36. Carmeliet P, Ferreira V, Breier G, et al. Abnormal blood vessel development and lethality in embryos lacking a single VEGF allele. Nature 1996;380:435–439.
37. Shalaby F, Ho J, Stanford WL, et al. A requirement for Flk1 in primitive and definitive hematopoiesis and vasculogenesis. Cell 1997;89:981–990.
38. Gerber HP, McMurtrey A, Kowalski J, et al. Vascular endothelial growth factor regulates endothelial cell survival through the phosphatidylinositol 3′-kinase/Akt signal transduction pathway. Requirement for Flk-1/KDR activation. J Biol Chem 1998;273: 30,336–30,343.
39. Kim I, Kim HG, So JN, Kim JH, Kwak HJ, Koh GY. Angiopoietin-1 regulates endothelial cell survival through the phosphatidylinositol 3′-Kinase/Akt signal transduction pathway. Circ Res 2000;86:24–29.
40. Datta SR, Katsov A, Hu L, et al. 14-3-3 proteins and survival kinases cooperate to inactivate BAD by BH3 domain phosphorylation. Mol Cell 2000;6:41–51.
41. Dimmeler S, Haendeler J, Nehls M, Zeiher AM. Suppression of apoptosis by nitric oxide via inhibition of interleukin-1beta-converting enzyme (ICE)-like and cysteine protease protein (CPP)-32-like proteases. J Exp Med 1997;185:601–607.
42. Butzal M, Loges S, Schweizer M, et al. Rapamycin inhibits proliferation and differentiation of human endothelial progenitor cells in vitro. Exp Cell Res 2004;300:65–71.
43. Gerber HP, Dixit V, Ferrara N. Vascular endothelial growth factor induces expression of the antiapoptotic proteins Bcl-2 and A1 in vascular endothelial cells. J Biol Chem 1998;273: 13,313–13,316.
44. O'Connor DS, Schechner JS, Adida C, et al. Control of apoptosis during angiogenesis by survivin expression in endothelial cells. Am J Pathol 2000;156:393–398.
45. Tran J, Rak J, Sheehan C, et al. Marked induction of the IAP family antiapoptotic proteins survivin and XIAP by VEGF in vascular endothelial cells. Biochem Biophys Res Commun 1999;264:781–788.
46. Wang S, Sorenson CM, Sheibani N. Attenuation of retinal vascular development and neovascularization during oxygen-induced ischemic retinopathy in Bcl-2-/- mice. Dev Biol 2005;279:205–219.
47. Gratton JP, Morales-Ruiz M, Kureishi Y, Fulton D, Walsh K, Sessa WC. Akt down-regulation of p38 signaling provides a novel mechanism of vascular endothelial growth factor-mediated cytoprotection in endothelial cells. J Biol Chem 2001;276:30,359–30,365.
48. Carmeliet P, Lampugnani MG, Moons L, et al. Targeted deficiency or cytosolic truncation of the VE-cadherin gene in mice impairs VEGF-mediated endothelial survival and angiogenesis. Cell 1999;98:147–157.
49. Gao C, Sun W, Christofidou-Solomidou M, et al. PECAM-1 functions as a specific and potent inhibitor of mitochondrial-dependent apoptosis. Blood 2003;102:169–179.
50. Malyankar UM, Scatena M, Suchland KL, Yun TJ, Clark EA, Giachelli CM. Osteoprotegerin is an alpha vbeta 3-induced, NF-kappa B-dependent survival factor for endothelial cells. J Biol Chem 2000;275:20,959–20,962.
51. Stupack DG, Cheresh DA. Apoptotic cues from the extracellular matrix: regulators of angiogenesis. Oncogene 2003;22:9022–9029.
52. Benjamin LE, Hemo I, Keshet E. A plasticity window for blood vessel remodelling is defined by pericyte coverage of the preformed endothelial network and is regulated by PDGF-B and VEGF. Development 1998;125:1591–1598.
53. Murasawa S, Llevadot J, Silver M, Isner JM, Losordo DW, Asahara T. Constitutive human telomerase reverse transcriptase expression enhances regenerative properties of endothelial progenitor cells. Circulation 2002;106:1133–1139.

54. Dernbach E, Urbich C, Brandes RP, Hofmann WK, Zeiher AM, Dimmeler S. Antioxidative stress-associated genes in circulating progenitor cells: evidence for enhanced resistance against oxidative stress. Blood 2004;104:3591–3597.
55. Munshi N, Fernandis AZ, Cherla RP, Park IW, Ganju RK. Lipopolysaccharide-induced apoptosis of endothelial cells and its inhibition by vascular endothelial growth factor. J Immunol 2002;168:5860–5866.
56. Dhanabal M, Ramchandran R, Waterman MJ, et al. Endostatin induces endothelial cell apoptosis. J Biol Chem 1999;274:11,721–11,776.
57. Bannerman DD, Tupper JC, Ricketts WA, Bennett CF, Winn RK, Harlan JM. A constitutive cytoprotective pathway protects endothelial cells from lipopolysaccharide-induced apoptosis. J Biol Chem 2001;276:14,924–14,932.
58. Hanahan D, Folkman J. Patterns and emerging mechanisms of the angiogenic switch during tumorigenesis. Cell 1996;86:353–364.
59. Levenberg S, Golub JS, Amit M, Itskovitz-Eldor J, Langer R. Endothelial cells derived from human embryonic stem cells. Proc Natl Acad Sci USA 2002,99:4391–4396.
60. Wei J, Blum S, Unger M, et al. Embryonic endothelial progenitor cells armed with a suicide gene target hypoxic lung metastases after intravenous delivery. Cancer Cell 2004;5:477–488.
61. Ingram DA, Mead LE, Moore DB, Woodard W, Fenoglio A, Yoder MC. Vessel wall derived endothelial cells rapidly proliferate because they contain a complete hierarchy of endothelial progenitor cells. Blood 2004;104:2752–2760.
62. Murohara T, Ikeda H, Duan J, et al. Transplanted cord blood-derived endothelial precursor cells augment postnatal neovascularization. J Clin Invest 2000;105:1527–1536.
63. Reyes M, Dudek A, Jahagirdar B, Koodie L, Marker PH, Verfaillie CM. Origin of endothelial progenitors in human postnatal bone marrow. J Clin Invest 2002;109:337–346.
64. Asahara T, Takahashi T, Masuda H, et al. VEGF contributes to postnatal neovascularization by mobilizing bone marrow-derived endothelial progenitor cells. EMBO J 1999;18:3964–3972.
65. Dimmeler S, Aicher A, Vasa M, et al. HMG-CoA reductase inhibitors (statins) increase endothelial progenitor cells via the PI 3-kinase/Akt pathway. J Clin Invest 2001;108: 391–397.
66. Lin Y, Weisdorf DJ, Solovey A, Hebbel RP. Origins of circulating endothelial cells and endothelial outgrowth from blood. J Clin Invest 2000;105:71–77.
67. Ferrari N, Glod J, Lee J, Kobiler D, Fine HA. Bone marrow-derived, endothelial progenitor-like cells as angiogenesis-selective gene-targeting vectors. Gene Ther 2003;10: 647–656.
68. Freeman SM, Abboud CN, Whartenby KA, et al. The "bystander effect": tumor regression when a fraction of the tumor mass is genetically modified. Cancer Res 1993;53: 5274–5283.
69. Gomez-Navarro J, Contreras JL, Arafat W, et al. Genetically modified CD34+ cells as cellular vehicles for gene delivery into areas of angiogenesis in a rhesus model. Gene Ther 2000;7:43–52.
70. Arafat WO, Casado E, Wang M, et al. Genetically modified CD34+ cells exert a cytotoxic bystander effect on human endothelial and cancer cells. Clin Cancer Res 2000;6: 4442–4448.
71. Scappaticci FA. Mechanisms and future directions for angiogenesis-based cancer therapies. J Clin Oncol 2002;20:3906–3927.
72. Chester J, Ruchatz A, Gough M, et al. Tumor antigen-specific induction of transcriptionally targeted retroviral vectors from chimeric immune receptor-modified T cells. Nat Biotechnol 2002;20:256–263.
73. Jevremovic D, Gulati R, Hennig I, et al. Use of blood outgrowth endothelial cells as virus-producing vectors for gene delivery to tumors. Am J Physiol Heart Circ Physiol 2004;287: H494–H500.

74. Reynolds BA, Weiss S. Generation of neurons and astrocytes from isolated cells of the adult mammalian central nervous system. Science 1992;255:1707–1710.
75. Richards LJ, Kilpatrick TJ, Bartlett PF. De novo generation of neuronal cells from the adult mouse brain. Proc Natl Acad Sci USA 1992;89:8591–8595.
76. Lois C, Alvarez-Buylla A. Proliferating subventricular zone cells in the adult mammalian forebrain can differentiate into neurons and glia. Proc Natl Acad Sci USA 1993;90: 2074–2077.
77. Gage FH, Coates PW, Palmer TD, et al. Survival and differentiation of adult neuronal progenitor cells transplanted to the adult brain. Proc Natl Acad Sci USA 1995;92: 11,879–11,883.
78. Palmer TD, Ray J, Gage FH. FGF-2-responsive neuronal progenitors reside in proliferative and quiescent regions of the adult rodent brain. Mol Cell Neurosci 1995;6:474–486.
79. Shihabuddin LS, Horner PJ, Ray J, Gage FH. Adult spinal cord stem cells generate neurons after transplantation in the adult dentate gyrus. J Neurosci 2000;20:8727–8735.
80. Palmer TD, Markakis EA, Willhoite AR, Safar F, Gage FH. Fibroblast growth factor-2 activates a latent neurogenic program in neural stem cells from diverse regions of the adult CNS. J Neurosci 1999;19:8487–8497.
81. Kornack DR, Rakic P. Continuation of neurogenesis in the hippocampus of the adult macaque monkey. Proc Natl Acad Sci USA 1999;96:5768–5773.
82. Nunes MC, Roy NS, Keyoung HM, et al. Identification and isolation of multipotential neural progenitor cells from the subcortical white matter of the adult human brain. Nat Med 2003; 9:439–447.
83. Kornack DR, Rakic P. The generation, migration, and differentiation of olfactory neurons in the adult primate brain. Proc Natl Acad Sci USA 2001;98:4752–4757.
84. Uchida N, Buck DW, He D, et al. Direct isolation of human central nervous system stem cells. Proc Natl Acad Sci USA 2000;97:14,720–14,725.
85. Kelly S, Bliss TM, Shah AK, et al. Transplanted human fetal neural stem cells survive, migrate, and differentiate in ischemic rat cerebral cortex. Proc Natl Acad Sci USA 2004;101:11,839–11,844.
86. Aboody KS, Brown A, Rainov NG, et al. Neural stem cells display extensive tropism for pathology in adult brain: evidence from intracranial gliomas. Proc Natl Acad Sci USA 2000;97:12,846–12,851.
87. Ehtesham M, Kabos P, Gutierrez MA, et al. Induction of glioblastoma apoptosis using neural stem cell-mediated delivery of tumor necrosis factor-related apoptosis-inducing ligand. Cancer Res 2002;62:7170–7174.
88. Arnhold S, Hilgers M, Lenartz D, et al. Neural precursor cells as carriers for a gene therapeutical approach in tumor therapy. Cell Transplant 2003;12:827–837.
89. Lee J, Elkahloun AG, Messina SA, et al. Cellular and genetic characterization of human adult bone marrow-derived neural stem-like cells: a potential antiglioma cellular vector. Cancer Res 2003;63:8877–8889.
90. Brown AB, Yang W, Schmidt NO, et al. Intravascular delivery of neural stem cell lines to target intracranial and extracranial tumors of neural and non-neural origin. Hum Gene Ther 2003;14:1777–1785.
91. Tang Y, Shah K, Messerli SM, Snyder E, Breakefield X, Weissleder R. In vivo tracking of neural progenitor cell migration to glioblastomas. Hum Gene Ther 2003;14:1247–1254.
92. Anderson SA, Glod J, Arbab AS, et al. Noninvasive MR imaging of magnetically labeled stem cells to directly identify neovasculature in a glioma model. Blood 2005;105:420–425.
93. Imitola J, Raddassi K, Park KI, et al. Directed migration of neural stem cells to sites of CNS injury by the stromal cell-derived factor 1alpha/CXC chemokine receptor 4 pathway. Proc Natl Acad Sci USA 2004;101:18,117–18,122.

94. Fricker RA, Carpenter MK, Winkler C, Greco C, Gates MA, Bjorklund A. Site-specific migration and neuronal differentiation of human neural progenitor cells after transplantation in the adult rat brain. J Neurosci 1999;19:5990–6005.
95. Brustle O, Choudhary K, Karram K, et al. Chimeric brains generated by intraventricular transplantation of fetal human brain cells into embryonic rats. Nat Biotechnol 1998; 16:1040–1044.
96. Flax JD, Aurora S, Yang C, et al. Engraftable human neural stem cells respond to developmental cues, replace neurons, and express foreign genes. Nat Biotechnol 1998;16: 1033–1039.
97. Chu K, Kim M, Jeong SW, Kim SU, Yoon BW. Human neural stem cells can migrate, differentiate, and integrate after intravenous transplantation in adult rats with transient forebrain ischemia. Neurosci Lett 2003;343:129–133.
98. Roy NS, Wang S, Jiang L, et al. In vitro neurogenesis by progenitor cells isolated from the adult human hippocampus. Nat Med 2000;6:271–277.
99. Brustle O, Spiro AC, Karram K, Choudhary K, Okabe S, McKay RD. In vitro-generated neural precursors participate in mammalian brain development. Proc Natl Acad Sci USA 1997;94:14,809–14,814.
100. Kim JH, Auerbach JM, Rodriguez-Gomez JA, et al. Dopamine neurons derived from embryonic stem cells function in an animal model of Parkinson's disease. Nature 2002; 418:50–56.
101. Kuo HC, Pau KY, Yeoman RR, Mitalipov SM, Okano H, Wolf DP. Differentiation of monkey embryonic stem cells into neural lineages. Biol Reprod 2003;68:1727–1735.
102. Kawasaki H, Suemori H, Mizuseki K, et al. Generation of dopaminergic neurons and pigmented epithelia from primate ES cells by stromal cell-derived inducing activity. Proc Natl Acad Sci USA 2002;99:1580–1585.
103. Ben-Hur T, Idelson M, Khaner H, et al. Transplantation of human embryonic stem cell-derived neural progenitors improves behavioral deficit in Parkinsonian rats. Stem Cells 2004;22:1246–1255.
104. Snyder EY, Deitcher DL, Walsh C, Arnold-Aldea S, Hartwieg EA, Cepko CL. Multipotent neural cell lines can engraft and participate in development of mouse cerebellum. Cell 1992;68:33–51.
105. Renfranz PJ, Cunningham MG, McKay RD. Region-specific differentiation of the hippocampal stem cell line HiB5 upon implantation into the developing mammalian brain. Cell 1991;66:713–729.
106. Villa A, Snyder EY, Vescovi A, Martinez-Serrano A. Establishment and properties of a growth factor-dependent, perpetual neural stem cell line from the human CNS. Exp Neurol 2000;161:67–84.
107. Villa A, Navarro-Galve B, Bueno C, Franco S, Blasco MA, Martinez-Serrano A. Long-term molecular and cellular stability of human neural stem cell lines. Exp Cell Res 2004;294: 559–570.
108. Roy NS, Nakano T, Keyoung HM, et al. Telomerase immortalization of neuronally restricted progenitor cells derived from the human fetal spinal cord. Nat Biotechnol 2004;22:297–305.
109. Brazelton TR, Rossi FM, Keshet GI, Blau HM. From marrow to brain: expression of neuronal phenotypes in adult mice. Science 2000;290:1775–1779.
110. Mezey E, Chandross KJ, Harta G, Maki RA, McKercher SR. Turning blood into brain: cells bearing neuronal antigens generated in vivo from bone marrow. Science 2000;290: 1779–1782.
111. Sanchez-Ramos J, Song S, Cardozo-Pelaez F, et al. Adult bone marrow stromal cells differentiate into neural cells in vitro. Exp Neurol 2000;164:247–256.
112. Wagers AJ, Sherwood RI, Christensen JL, Weissman IL. Little evidence for developmental plasticity of adult hematopoietic stem cells. Science 2002;297:2256–2259.

113. Soares S, Sotelo C. Adult neural stem cells from the mouse subventricular zone are limited in migratory ability compared to progenitor cells of similar origin. Neuroscience 2004;128: 807–817.
114. Kuida K, Zheng TS, Na S, et al. Decreased apoptosis in the brain and premature lethality in CPP32-deficient mice. Nature 1996;384:368–372.
115. Cecconi F, Alvarez-Bolado G, Meyer BI, Roth KA, Gruss P. Apaf1 (CED-4 homolog) regulates programmed cell death in mammalian development. Cell 1998;94:727–737.
116. Kuida K, Haydar TF, Kuan CY, et al. Reduced apoptosis and cytochrome c-mediated caspase activation in mice lacking caspase 9. Cell 1998;94:325–337.
117. Yoshida H, Kong YY, Yoshida R, et al. Apaf1 is required for mitochondrial pathways of apoptosis and brain development. Cell 1998;94:739–750.
118. Roth KA, Kuan C, Haydar TF, et al. Epistatic and independent functions of caspase-3 and Bcl-X(L) in developmental programmed cell death. Proc Natl Acad Sci USA 2000;97: 466–471.
119. Yang X, Klein R, Tian X, Cheng HT, Kopan R, Shen J. Notch activation induces apoptosis in neural progenitor cells through a p53-dependent pathway. Dev Biol 2004;269:81–94.
120. Grandbarbe L, Bouissac J, Rand M, Hrabe de Angelis M, Artavanis-Tsakonas S, Mohier E. Delta-Notch signaling controls the generation of neurons/glia from neural stem cells in a stepwise process. Development 2003;130:1391–1402.
121. Lutolf S, Radtke F, Aguet M, Suter U, Taylor V. Notch1 is required for neuronal and glial differentiation in the cerebellum. Development 2002;129:373–385.
122. Ricci-Vitiani L, Pedini F, Mollinari C, et al. Absence of caspase 8 and high expression of PED protect primitive neural cells from cell death. J Exp Med 2004;200:1257–1266.
123. Bieberich E, MacKinnon S, Silva J, Noggle S, Condie BG. Regulation of cell death in mitotic neural progenitor cells by asymmetric distribution of prostate apoptosis response 4 (PAR-4) and simultaneous elevation of endogenous ceramide. J Cell Biol 2003;162:469–479.
124. Bieberich E, Silva J, Wang G, Krishnamurthy K, Condie BG. Selective apoptosis of pluripotent mouse and human stem cells by novel ceramide analogues prevents teratoma formation and enriches for neural precursors in ES cell-derived neural transplants. J Cell Biol 2004;167:723–734.
125. D'Sa-Eipper C, Leonard JR, Putcha G, et al. DNA damage-induced neural precursor cell apoptosis requires p53 and caspase 9 but neither Bax nor caspase 3. Development 2001;128:137–146.
126. Sun W, Winseck A, Vinsant S, Park OH, Kim H, Oppenheim RW. Programmed cell death of adult-generated hippocampal neurons is mediated by the proapoptotic gene Bax. J Neurosci 2004;24:11,205–11,213.
127. Lindsten T, Golden JA, Zong WX, Minarcik J, Harris MH, Thompson CB. The proapoptotic activities of Bax and Bak limit the size of the neural stem cell pool. J Neurosci 2003;23:11,112–11,119.
128. Li L, Liu F, Salmonsen RA, et al. PTEN in neural precursor cells: regulation of migration, apoptosis, and proliferation. Mol Cell Neurosci 2002;20:21–29.
129. Mizumatsu S, Monje ML, Morhardt DR, Rola R, Palmer TD, Fike JR. Extreme sensitivity of adult neurogenesis to low doses of X-irradiation. Cancer Res 2003;63:4021–4027.
130. Limoli CL, Giedzinski E, Rola R, Otsuka S, Palmer TD, Fike JR. Radiation response of neural precursor cells: linking cellular sensitivity to cell cycle checkpoints, apoptosis and oxidative stress. Radiat Res 2004;161:17–27.
131. Hori J, Ng TF, Shatos M, Klassen H, Streilein JW, Young MJ. Neural progenitor cells lack immunogenicity and resist destruction as allografts. Stem Cells 2003;21:405–416.
132. Reid R, Mar EC, Huang ES, Topal MD. Insertion and extension of acyclic, dideoxy, and ara nucleotides by herpesviridae, human alpha and human beta polymerases. A unique

inhibition mechanism for 9-(1,3-dihydroxy-2-propoxymethyl)guanine triphosphate. J Biol Chem 1988;263:3898–3904.

133. Beltinger C, Fulda S, Kammertoens T, Meyer E, Uckert W, Debatin KM. Herpes simplex virus thymidine kinase/ganciclovir-induced apoptosis involves ligand-independent death receptor aggregation and activation of caspases. Proc Natl Acad Sci USA 1999;96: 8699–8704.
134. Beltinger C, Fulda S, Kammertoens T, Uckert W, Debatin KM. Mitochondrial amplification of death signals determines thymidine kinase/ganciclovir-triggered activation of apoptosis. Cancer Res 2000;60:3212–3217.
135. Beltinger C, Fulda S, Walczak H, Debatin KM. TRAIL enhances thymidine kinase/ ganciclovir gene therapy of neuroblastoma cells. Cancer Gene Ther 2002;9:372–381.
136. Immonen A, Vapalahti M, Tyynela K, et al. AdvHSV-tk gene therapy with intravenous ganciclovir improves survival in human malignant glioma: a randomised, controlled study. Mol Ther 2004;10:967–972.
137. Ram Z, Culver KW, Oshiro EM, et al. Therapy of malignant brain tumors by intratumoral implantation of retroviral vector-producing cells. Nat Med 1997;3:1354–1361.
138. Uhl M, Weiler M, Wick W, Jacobs AH, Weller M, Herrlinger U. Migratory neural stem cells for improved thymidine kinase-based gene therapy of malignant gliomas. Biochem Biophys Res Commun 2005;328:125–129.
139. Huber BE, Austin EA, Richards CA, Davis ST, Good SS. Metabolism of 5-fluorocytosine to 5-fluorouracil in human colorectal tumor cells transduced with the cytosine deaminase gene: significant antitumor effects when only a small percentage of tumor cells express cytosine deaminase. Proc Natl Acad Sci USA 1994;91:8302–8306.
140. Barresi V, Belluardo N, Sipione S, Mudo G, Cattaneo E, Condorelli DF. Transplantation of prodrug-converting neural progenitor cells for brain tumor therapy. Cancer Gene Ther 2003;10:396–402.
141. Benedetti S, Pirola B, Pollo B, et al. Gene therapy of experimental brain tumors using neural progenitor cells. Nat Med 2000;6:447–450.
142. Ehtesham M, Kabos P, Kabosova A, Neuman T, Black KL, Yu JS. The use of interleukin 12-secreting neural stem cells for the treatment of intracranial glioma. Cancer Res 2002;62:5657–5663.
143. Niranjan A, Wolfe D, Fellows W, et al. Gene transfer to glial tumors using herpes simplex virus. Methods Mol Biol 2004;246:323–337.
144. Herrlinger U, Woiciechowski C, Sena-Esteves M, et al. Neural precursor cells for delivery of replication-conditional HSV-1 vectors to intracerebral gliomas. Mol Ther 2000;1: 347–357.
145. Staflin K, Honeth G, Kalliomaki S, Kjellman C, Edvardsen K, Lindvall M. Neural progenitor cell lines inhibit rat tumor growth in vivo. Cancer Res 2004;64:5347–5354.
146. Holland EC, Celestino J, Dai C, Schaefer L, Sawaya RE, Fuller GN. Combined activation of Ras and Akt in neural progenitors induces glioblastoma formation in mice. Nat Genet 2000;25:55–57.

13

Apoptosis and Cancer Therapy

Maurice Reimann and Clemens A. Schmitt

Summary

The observation that tumor development frequently is accompanied by apoptotic defects and the meanwhile accepted view that the vast majority of conventional anticancer agents primarily cause DNA damage to subsequently initiate the apoptotic machinery, have led to the conclusions that the capability to undergo apoptosis must contribute to the outcome of cancer therapy and that mutations compromising apoptosis might confer chemoresistance. However, experimental results from cell-culture-based systems and conflicting data from the clinic have prompted reasonable doubts about a critical role of apoptosis in cancer therapy. This chapter highlights the technical difficulties in properly assessing the impact of treatment-induced apoptosis, discusses the value of more complex model systems, and underscores the relevance of apoptosis by providing insights from mouse models harboring tumors with genetically defined lesions and clinical observations based on novel proapoptotic compounds.

Key Words: Apoptosis; chemotherapy; mouse models; resistance; senescence; targeted therapies.

1. Introduction

About one hundred years ago, Paul Ehrlich expressed his dream of "magic bullets"—therapeutics that would selectively eliminate harmful microbes or tumor cells *(1)*. In the 1940s, the first chemotherapeutic compound active against lymphomas, nitrogen mustard, was discovered as a byproduct of studies interested in elucidating the biological effects of mustard gas, a chemical warfare agent used during World War I *(2)*. During the following decades, empirical screens rather than rational efforts of drug development, added novel compounds to the growing arsenal of anticancer agents. Given their inherent toxicity to non-neoplastic cells, these drugs are conceptually far away from Ehrlich's "magic bullet" vision that anticipated the era of truly targeted therapies. Despite their different pharmacological properties and interference with a variety of molecular targets, most of the conventional anticancer compounds exert their cytotoxic action by producing direct or indirect damage of genomic DNA. Ironically, neither the role of DNA as the prime target of chemotherapeutic drug action, nor the discovery that genes are composed of DNA, nor the causal role of mutant DNA in oncogenic activation or tumor suppressor inactivation was appreciated in the early days of chemotherapy *(3)*. Even when DNA was recognized as the critical drug-damaged structure, it was believed for long periods of time that the extent of DNA damage delivered to fast growing cancer cells must be invariably lethal *per se*. Accordingly, the clinical problem of

From: *Apoptosis, Cell Signaling, and Human Diseases: Molecular Mechanisms, Volume 1*
Edited by R. Srivastava © Humana Press Inc., Totowa, NJ

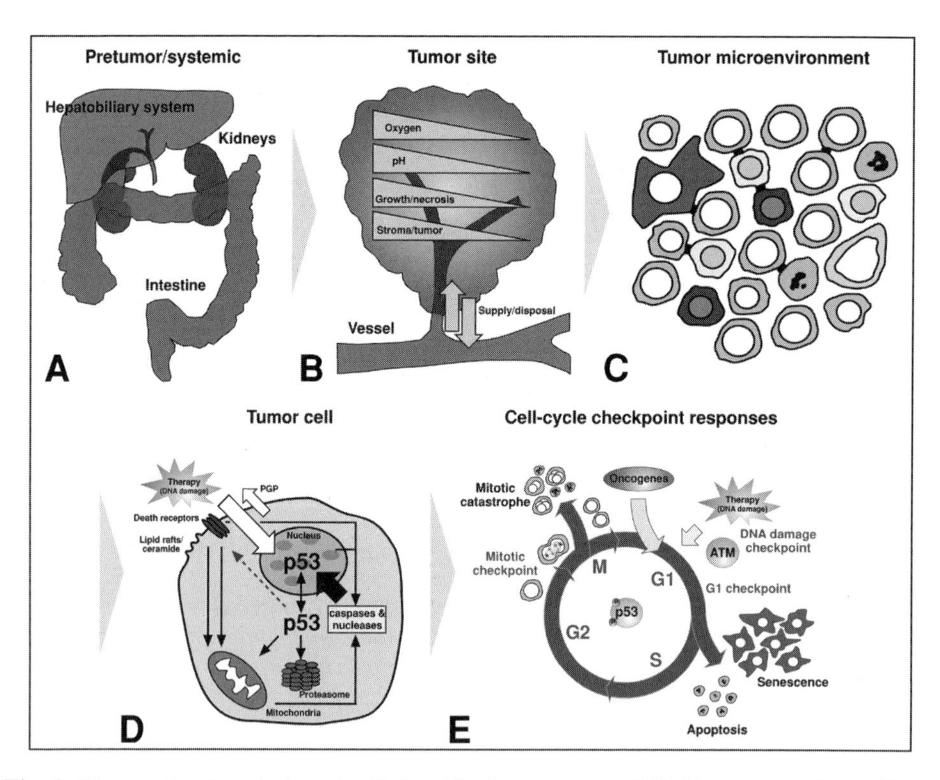

Fig. 1. The road to drug-induced cellular effector programs. **(A)** Upon systemic application, DNA damaging anticancer agents undergo pharmacological modification, for example hepatic toxificiation, and metabolization with subsequent excretion via the hepatobiliary system, the intestine and the kidneys. **(B)** Depending on the vascularization of the tumor, a fraction of the active drug is delivered to the tumor site, where it will reach individual parts of the tumor to varying degrees. In addition, gradients in oxygen supply and pH, as well as inhomogeneities between tumor and stromal cells and also between proliferating and necrotic areas will influence local drug activity. **(C)** The tumor microenvironment modulates drug action in many ways. Besides the local density of capillary vessels important for drug supply and metabolite disposal, nonneoplastic bystander cells secrete growth factor-like cytokines or directly interact via ligand-receptor mechanisms with tumor cells and alter their apoptotic threshold. Moreover, apoptotic bodies and senescent tumor cells not yet cleared may also impact on the susceptibility of tumor cells to the cytotoxic effects of anticancer agents. **(D)** If they are not instantly eliminated by cellular efflux pumps such as the P-glycoprotein (PGP) most anticancer agents, either in a direct or indirect manner, cause DNA damage. The prime consequence of DNA damage is the initiation of a p53-mediated mitochondrial apoptotic cascade ultimately activating caspases and nucleases. Primarily via changes in the cell membrane, via binding of specific ligands, or secondarily via crosstalk from the intrinsic apoptotic cascade, the extrinsic death-receptor-mediated pathway may augment proapoptotic signaling. **(E)** Cell-cycle checkpoints, sensitized by activated oncogenes, will translate the ATM-sensed DNA damage into p53-mediated pro-apoptotic signals, if the apoptotic machinery is available. As an alternative response, DNA damage may provoke in apoptotically incompetent cells a terminal cell-cycle arrest, termed cellular senescence. Some cells might continue to cycle for another few nuclear divisions, before entering mitotic catastrophe, an attenuated form of apoptosis often executed out of mitosis.

chemoresistance was inferred to mechanisms that may blunt drug action prior to the occurrence of DNA damage, such as accelerated systemic drug metabolization or induced expression of membrane-inserted drug efflux pumps *(4,5)*.

Following up on the discovery of apoptosis as a form of genetically encoded, programmed cell death *(6)*, and the observation that DNA damage may act as an inducer of apoptosis *(7)*, it was subsequently demonstrated that biochemically unrelated classes of DNA damaging anticancer agents may initiate apoptosis as a unique downstream effector cascade *(8)*. In turn, genetic alterations such as overexpression of the strictly anti-apoptotic molecule Bcl2 were shown to produce multidrug resistance to a variety of anticancer compounds in short-term assays in vitro *(9)*. Hence, programmed cell death was inaugurated as a novel concept of anticancer drug action based on a common post-damage drug effector mechanism, and genetic defects in this program became candidates for causing chemoresistance. However, despite the attractiveness of this model, the actual contribution of apoptosis to the outcome of cancer therapy still remains unclear.

2. Apoptosis and the Outcome of Therapy

Meanwhile, there is an enormous body of scientific work related to the field of "apoptosis and therapy." Unfortunately, imprecise use of either term makes it particularly difficult to judge the actual impact of apoptosis on the outcome of therapy.

Apoptosis is a complex sequential process of genetically determined self-destruction that ultimately leads to the activation of proteases with certain substrate specificities, the caspases, and nucleases that produce membrane blebs, degrade DNA into nucleosome-sized fragments and condensate cellular compartments *(10)*. The remainders, so called apoptotic bodies, are subject to phagocytosis devoid of a harmful inflammatory response. Paralleling the increasing molecular understanding of the processes involved in programmed cell death, numerous aberrant forms of apoptosis, termed atypical, caspase-independent apoptosis *(11)*, attenuated programmed death as a consequence of mitotic catastrophe *(12)*, autophagy *(13,14)*, or the possibility of programmed necrosis *(15)* have been described during the past few years. As a consequence, proper assessment of apoptotic features by experimental procedures has become an issue of debate, and many reports claiming to study apoptotic cell death simply measure viability as a surrogate marker.

Chemotherapeutic agents applied to a cancer cell in a Petri dish and to a cancer patient may not necessarily produce comparable biological effects. Surprisingly, many studies based on the short-term exposure of cancer cells to a cytotoxic compound in culture mutually assume that their findings establish a biologically meaningful approximation of a patient's outcome to therapy. In fact, only a very limited number of studies addressed the critical question to what extent the apoptotic sensitivity of primary tumor cells measured in a short-term in vitro assay may accurately predict the long-term response of the corresponding cancer patient. Remarkably, a substantial fraction of those studies failed to unveil a correlation between in vitro and in vivo responses, even when less stringent parameters such as cell growth or viability instead of apoptotic death were chosen *(16,17)*. Comparing short-term in vitro survival of primary samples obtained from patients diagnosed with acute myeloid leukemia with their respective clinical outcome, Möllgård and colleagues reported a significantly lower fraction of surviving cells following exposure to daunorubicin in vitro in the group of patients that later achieved a complete remission after induction therapy, whereas a similar effect was missed in response to cytarabin, the probably most effective compound against this entity in the clinic *(18)*. Patients displaying in vitro sensitivity to daunorubicin also showed a tendency towards better overall survival, whereas, paradoxically, patients who were in vitro sensitive to cytarabin suffered from a significantly reduced overall survival. The work by Rödel and colleagues is one of the sparse reports where the apoptotic capacity of a pretreatment biopsy had predictive value for the outcome of the same patient group: high-level spontaneous apoptosis assayed *in situ* in rectal cancer specimens obtained prior to neoadjuvant radiochemotherapy was associated with a significantly improved recurrence-free survival period *(19)*. Given the poor correlation between in vitro chemosensitivity and patient outcome irrespective of apoptosis or mere cell viability as the primary readout, one may conclude that drug-inducible cell death simply is not a critical component of chemotherapeutic effector mechanisms in vivo, or, as an alternative explanation, that short-term assays fail to sufficiently mimic the complexity of drug action in a cancer patient.

3. Apoptosis and Clonogenic Survival After Drug Exposure In Vitro

Clonogenic survival assays are considered the gold standard to assess the interference of cytotoxic therapies with the long-term growth capacity of cancer cells in vitro, because they are less biased by the complicated overlap of dying and regrowing subpopulations *(20)*. Although initiation of classic programmed cell death will force susceptible tumor cells to die within hours, attenuated forms of apoptosis may result in delayed death that can be missed by short-term viability assays, but will reduce clonogenic survival. In turn, drug-inducible arrest programs may affect clonogenic survival, because proliferation-stalled cells cannot contribute to colony formation, but may give rise to a relapse tumor when cells re-enter the cycle much later *(21)*. Moreover, defined genetic defects in arrest-controlling checkpoints, for example in the CDK inhibitor *p21*, have been shown not to impact on clonogenic survival after γ-irradiation, although the shift from an arrest response to more apoptotic death produced a significantly higher proportion of remissions in vivo *(22)*. Likewise, the comparison of cell populations differing in the expression status of the strictly antiapoptotic *bcl2* gene failed to unveil a significant difference in clonogenic outgrowth based on the apoptotic defect *(23–25)*, suggesting a drug-inducible arrest as an alternative outcome in the presence of an apoptotic block. In addition, two other mechanisms may contribute to the inability of clonogenic assays to accurately assess the role of apoptotic defects on long-term outcome: first, to score proliferating colonies that emerged from single cells plated at extremely low densities is highly artificial and based on strong selection for mutations that facilitate growth under this extreme condition *(25)*, and second, factors of the tumor microenvironment influencing growth in vivo might be completely missed in the Petri dish environment of a standard clonogenic assay in vitro.

4. Apoptosis and the Role of the Tumor Microenvironment

Indeed, components of the tumor microenvironment have been shown to profoundly alter clonogenic outgrowth in vitro *(25–27)*. Physiological life and death decisions, for example in the germinal center of a lymph follicle, not only depend on cell autonomous determinants, but also on prosurvival and pro-apoptotic factors of specific compartments or "niches," and even tumor cells that harbor severe apoptotic defects are not entirely refractory to extracellular survival or death signals. Importantly, environmental signals may select for tumor cell mutations providing a growth advantage, or may "epigenetically" adjust gene activities without provoking mutations (e.g., CD154/CD40 signaling to activate the NF-κB survival cascade *[28]*), hereby generating a form of nonstructural apoptotic resistance that relies on the presence of anti-apoptotic factors. Furthermore, there is increasing evidence that complex interactions between extracellular signals and cell-autonomous death mediators may be important for setting individual apoptotic thresholds: CD95 (or Fas) death receptor-triggered translocation of acid sphingomyelinase to the outer plasma membrane releases the extracellularly oriented sphingolipid ceramide, which, in turn, is essential for clustering of CD95 into lipid rafts and subsequent CD95-mediated cell death *(29)*, and the process of C95 redistribution into lipid rafts can be augmented by DNA damaging anticancer agents such as cisplatin *(30)*. Nevertheless, the potential crosstalk between the p53-controlled intrinsic or mitochondrial cell death pathway primarily engaged by DNA damaging compounds and death receptor mediated apoptosis still remains an issue of debate *(31–33)*. However, there is good evidence that the combination therapy of conventional anticancer agents and proapoptotic death receptor ligands such as the tumor necrosis factor-related apoptosis-inducing ligand (TRAIL) may overcome resistance to apoptosis *(34,35)*. Moreover, anoikis (i.e., apoptosis resulting from disruption of fibronectin-mediated cell anchorage) has been shown in an in vivo model of acute myelogenous leukemia to synergize with classic DNA damage-induced apoptosis *(36)*. Hence, cell-autonomous and nonautonomous functions control the susceptibility of a tumor cell to undergo apoptosis in response to therapeutic stimuli.

5. Apoptosis is Important for Treatment Outcome In Vivo

Despite the concerns about the validity of clonogenic assays to predict drug responses in vivo and to serve as an appropriate model system to assess the role of apoptosis as a therapeutic effector program, there is convincing evidence from more physiological models that apoptosis indeed is important for the outcome of cancer therapy.

Matched pairs of primary, Myc-induced mouse lymphomas infected with either a mock control or stably expressing Bcl2 displayed significant differences in their responsiveness to cyclophosphamide, an alkylating anticancer agent widely used in the clinic, when growing as systemic transplant tumors at their natural tumor sites in immunocompetent mice *(25,37)*. Moreover, long-term observation of treated mice harboring Bcl2 overexpressing vs control lymphomas unveiled a profound difference in their overall survival, demonstrating that an anti-apoptotic lesion may produce chemoresistance in vivo that was not detected by standard clonogenic assays in vitro *(38)*. Other transgenic cancer models have been exploited to study treatment responses of tumors with defined genetic lesions as well, and demonstrated inferior outcome for syngenic tumors with *p53* or *INK4a/ARF* deletions or activating *Akt* mutations *(39,40)*, but the impact of treatment-induced apoptosis or genetic defects in apoptosis-controlling genes such as *p53* on long-term responses remained uncertain when investigated in different mouse models *(41–43)*. Besides the fact that the power of transgenic mouse models to study treatment responses based on defined genetic lesions is still underrated, most of the mouse models used so far to analyze drug action and outcome in vivo are only correlative in nature. At least, overexpression of wild-type p53 in xenograft models underscored its potential to improve drug responses, albeit with no clear discrimination which of the p53-controlled effector functions actually contributed to treatment outcome *(44,45)*. In clinical samples of chronic lymphocytic leukemia (CLL), pretreatment with alkylating agents has been shown to correlate significantly with the occurrence of *p53* mutations *(46)*, and *p53* inactivation, in turn, correlates with impaired overall survival *(47)*. Although some clinical investigations failed—possibly for technical reasons—to identify a positive correlation between p53 status and treatment outcome, many clinical trials successfully linked *p53* mutations and poor outcome throughout various entities (*see* ref. *48* and references therein).

An intrinsic obstacle to scientifically address the potentially causative role of apoptosis in treatment outcome is the difficulty to generate in a biologically meaningful way isogenic primary tumors that only differ in the status of a strictly apoptosis-conferring genetic moiety relevant to drug-induced apoptosis. Matched pairs of primary tumor samples with and without a gene deletion can be generated by homologous recombination, stable expression of a small interfering RNA, a dominant-negative activity or introduction of an antiapoptotic candidate gene. Although these pairs can be used to directly assess the role of the respective moiety in treatment responses, the "addition" or "subtraction" of a genetic activity to an already established cancer cell certainly is different from a mutation acquired during tumor development. In turn, genetic lesions such as nonconditional "knockout" deletions provided during tumorigenesis may produce malignancies that indeed form in dependency on this defect, but the corresponding control tumors may not necessarily reflect a true biological "match" anymore.

6. Provoking Apoptosis in Apoptotically Compromised Tumors

Apoptotic defects are a hallmark of cancer cells *(49)*, because successful transformation of normal cells by mitogenic oncogenes such as Myc or oncogenic Ras requires the ablation of cell-cycle checkpoint-mediated cellular failsafe mechanisms such as apoptosis (Fig. 1E). Oncogenic signaling is sensed in primary cells by the ARF/p53 axis, which, in turn, is targeted by inactivating mutations in many cancer entities *(50–52)*. Importantly, drug-induced DNA damage is sensed and transduced into an apoptotic signal that merges with the antioncogenic signaling machinery at the level of p53. Hence, oncogene-provoked and DNA damage-induced cell death utilize pathways that, at least in part, overlap, and mutations acquired during tumor formation, for example in the apoptotic regulator genes *bax*, *bcl2*, or *Apaf-1*, may already co-select for resistance to DNA damaging therapies prior to any drug encounter *(53–55)*.

However, the broad prevalence of apoptotic defects throughout different cancer entities is not mutually linked to clinical resistance to first line anticancer therapies. In fact, most cancer patients, even when suffering from one of the typically "chemo-insensitive" solid tumor entities, primarily respond to DNA damaging therapies and achieve clinical remissions, or, at least, a substantial reduction of their tumor burden. Of note, this clinical observation is not necessarily a lethal blow for the "apoptotic concept" which considers programmed cell death a relevant component of drug action and treatment outcome. Although apoptosis ultimately reflects the activation of a relatively small set of effector caspases and nucleases, genetic defects in these downstream players seem to be much less frequent as compared with mutations in mediators that regulate apoptosis—and other cellular functions probably important for tumor development and possibly DNA damage signaling—more upstream. Whereas these lesions, for example at the level of p53 or the protein kinase Akt, can have profound impact on tumor development and clearly promote chemoresistance in suitable preclinical models *(39,40)*, dose and combinatorial effects of a multiple-agent chemotherapy may initially suffice to clinically override a genetic block, but apoptotically even more severely compromised cells will rapidly re-emerge and produce a clinical relapse that is now overtly refractory not only to the initial, but also to biochemically unrelated second- and third-line therapies *(56)*. Necrotic death, although undoubtedly provoked by anticancer agents to some extent, cannot account for the acquisition of more robustly resistant tumor cells during subsequent courses of chemotherapy. Thus, apoptosis signaling most likely reflects a well-controlled and reinforced network of interconnected mediators and modulators rather than much more vulnerable linear and unique pathways. Whereas subtle defects in response to activated oncogenes may still license tumor development, they might not fully block treatment-induced apoptosis as a consequence of supra-physiological DNA damage broadly activating multiple pro-apoptotic effector pathways. In turn, tumor-promoting and mutually resistance-conferring defects in apoptotic cellular failsafe programs can be highly specific for a certain oncogene-triggered pathway, thereby precluding tumors driven by different oncogenes to benefit from a lesion-specific resensitizing therapy *(40)*. Furthermore, the view of apoptosis as a "network concept" also integrates the common observation of multiple apoptotic defects not only in an individual tumor specimen as a consequence of tumor heterogeneity but in a single tumor cell. Interestingly, despite their, in theory, mutually exclusive action in the same linear death pathway, pro-apoptotic mediators such as Bax and p53 can be found co-mutated, and, as such, were reported to produce a more thoroughly chemoresistant phenotype *(57)*. Given the increasing acquisition of apoptotic defects during tumor progression and anticancer therapy, it is an attractive hypothesis that cumulative apoptotic insufficiency ultimately creates the molecular basis for clinically relevant chemoresistance.

7. Apoptogenic Therapy in Apoptosis Resistant Tumors

Using DNA damaging anticancer agents against an apoptosis-refractory tumor might be even more detrimental than just ineffective. Whereas alternative regimens based on empirically less effective compounds still produce DNA damage but little cytotoxicity, they may promote mutability and genomic instability in the surviving tumor. Patients suffering from high-frequency microsatellite instability-driven colon cancer that is based on defects in the mismatch-repair system reportedly have, in contrast with patients with sporadic, nonmismatch repair-deficiency-driven colon cancer, an inferior prognosis when exposed to adjuvant chemotherapy as compared to no treatment, implying that an otherwise apoptosis-inducing DNA damaging therapy can be harmful *per se* and may result in reduced patient survival *(58)*. Of note, a thorough investigation of a potential mutagenic impact of DNA damaging therapies on heavily pretreated, apoptotically resistant tumors is very difficult to conduct in clinical trials, because these patients have rapidly progressive diseases for numerous reasons, and this analysis would be only correlative at best. Unfortunately, to our knowledge, there are little data from preclinical models available that addressed the role of treatment-induced DNA damage in apoptosis-refractory tumors with respect to genomic instability, tumor progression, overall survival, and, on the other side, potentially beneficial, alternative effector programs. Indeed, as demonstrated in a transgenic mouse model, tumors that already harbor an apoptotic block can acquire mutations of genes unrelated to apoptosis during additional courses of chemotherapy, and these mutations may now interfere with nonapoptotic drug effector programs *(38)*.

8. Senescence, Mitotic Catastrophe, and Other Programmed Cellular Responses to Therapy

It should be stressed that apoptosis is not the only programmed cellular response to anticancer therapies. Aberrant types of apoptosis (i.e., forms of programmed cell death with atypical morphological features or delayed kinetics) including mitotic catastrophe as an attenuated form of apoptosis frequently observed in response to antimicrotubule agents in p53-inactivated or spindle checkpoint-deficient cells that may endoreduplicate and become multinucleated before they eventually undergo programmed cell death *(59,60)*, have already been discussed. The occurrence and functional relevance of these aberrant presentations of apoptosis once again underscore the view that distinct defects in checkpoint control or apoptotic execution may not necessarily prevent the ultimate outcome, but may modify its timing and biological way of execution. Importantly, severe DNA damage not only precludes a cell from further cycling by initiating programmed cell death, but may alternatively provoke an acutely inducible form of a terminal cell-cycle arrest, termed premature senescence. Various chemotherapeutic drugs have been shown to execute cellular senescence in vitro and in vivo *(38,61,62)*, and data from a preclinical mouse model provided evidence that drug-induced senescence in apoptotically compromised tumors can improve the overall outcome to therapy *(38)*. Whereas assessing the definite contribution of drug-inducible senescence and other programmed cellular responses to tumor biology and long-term prognosis requires further investigation, it is clear that nonapoptotic drug effector programs not only may interfere with cytotoxicity assays in vitro, but could also explain why so many studies failed to demonstrate a strong correlation between apoptotic capability and long-term outcome in vivo. The actual contribution of different components of drug action depends on many factors and cellular preconditions (i.e., cellular senescence may come into play as a backup mechanism when apoptosis is blocked). Finally, singular defects in crucial drug effector programs may not directly translate into clinical resistance, but very subtle differences in the rate of growth-eliminated cells in a certain time frame might very well determine a tumor's likelihood to reemerge as a more aggressive and more resistant relapse *(56,63)*.

9. Provoking Apoptosis in Nonneoplastic Cells

The clinical relevance of drug-inducible apoptosis depends on the therapeutic index (i.e., the margin of tumor cell death that can be achieved by a certain dose of a compound without unacceptably harming the normal cell compartment). The underlying biological mechanisms that determine the different sensitivities to anticancer agents in malignant compared with nonneoplastic cells are not fully elucidated yet. Although the proliferation rate reflecting the requirement for accurate DNA replication and the presentation at certain cell-cycle phases in which targets of specific chemotherapeutic agents might be particularly active undoubtedly play a role for drug efficacy, other factors seem to be important as well. Normal cells with stem cell-like properties, such as hematopoietic progenitor cells and self-renewing cells in the intestinal crypts and at the hair follicles, are prime targets for chemotherapeutic side effects, but can more easily recover from DNA damage as a result of their rather arrest- and repair-driven response compared with an apoptosis- or senescence-oriented response executed in cancer cells. Interestingly, pretreamtent of mice with a small compound temporarily inhibiting p53 function was shown to decrease the extent of typical chemotherapy-related side effects such as weight loss and intestinal crypt cell apoptosis *(64)*. Despite equal amounts of toxicity delivered to the cells, and irrespective of the possibility that cells with a high division rate may accumulate more unrepaired DNA damage, acutely transformed cells have been shown to be substantially more sensitive to apoptosis *(65)*. This oncogene-generated gain-of-sensitivity cannot be entirely attributed to altered growth characteristics, and relies on the different utilization of growth-independent regulators such as caspases and cathepsins in apoptosis or senescence-like effector programs *(66,67)*. Moreover, oncogenes may also sensitize to receptor-mediated cell death, as reported for the Fas-initiated death program as a component of Myc-provoked apoptosis *(68)*.

Furthermore, drug-induced apoptosis in non-neoplastic cells might still correspond with the genetics of the adjacent tumor cells. Although p53-deficient tumors are *per se* more resistant to environmental pro-apoptotic stimuli *(69)*, mutant p53 may promote neo-angiogenesis via enhanced secretion of factors such as the vascular endothelial growth factor (VEGF), and the rich network of tumor vessels, in turn, might protect the tumor from lethal shortage in supplies when some endothelial cells are codamaged during chemotherapy *(70)*. Interestingly, not only p53 inactivation in the neoplastic cells may increase their apoptotic resistance to the consequences of endothelial damage *(71)*, but cancer cells and their surrounding vasculature may even share the same resistance-conferring genetic defects if derived from a common precursor cell—an attractive, currently highly debated concept *(72)*. Accordingly, inherited genetic preconditions to evade apoptosis in normal cells for example in patients suffering from Li-Fraumeni syndrome resulting from germline *p53* mutations or suffering from ataxia telangiectasia caused by defects in the *ATM* gene may impact on treatment outcome, but clinical data cannot distinguish treatment effects on non-neoplastic from tumor cells *(73,74)*

Taken together, the spatial and temporal effects of DNA damaging therapies at natural tumor sites in vivo are complicated to dissect and reflect a combination of direct drug action in normal and tumor cells as well as complex interactions between different cellular compartments and secreted factors. Despite an intrinsically enhanced apoptotic propensity in cancer compared to normal cells it is clear that the scientific evaluation of the actual contribution of apoptosis—and other programmed responses—to the outcome of DNA damaging anticancer therapies requires model systems that can adequately mimic the patient scenario.

10. Therapy Induced Apoptosis and Solid Cancers

Although DNA damaging therapies are widely used in the clinic against virtually all kinds of disseminated cancers or in adjuvant scenarios of many tumor entities, their long-term efficacy in solid cancer settings with few exceptions such as germ cell tumors, is vastly reduced. Whereas many hematological malignancies are potentially curable by polychemotherapy, metastasized carcinomas, in principle, are not. Two different explanations could account for this clinically challenging problem: there is no signaling network available in epithelial cancer cells that could respond to DNA damage with some kind of acute or delayed programmed cell death, or doses applicable without producing unacceptable side effects are too low to activate an apoptotic response in these cells. To date, there is no definite answer to the underlying mechanisms of resistance in solid tumors. On one side, there are conflicting reports on the prognostic role of apoptotic defects: for example, specific *p53* mutations were shown to correlate with *de novo* resistance to doxorubicin in breast cancer patients *(75)*, whereas low-level Bcl2 expression was associated with short overall survival in this entity *(76)*. On the other side, high-dose chemotherapy (i.e., a relative dose escalation made possible by autologous stem cell support or as part of the conditioning regimen in allogeneic stem cell transplantation protocols) has mostly failed to improve outcome in patients with metastasized solid tumors, although other factors may interfere with outcome, especially in allogeneic transplant settings. Clinical trials based on refined strategies are required to reach more definite answers about the susceptibility to apoptosis under intensified dose regimens, which must not exceed the toxicity thresholds of nonhematopoietic organs *(77,78)*. Unlike lymphoid cells with their inherent propensity to commit suicide during the selection of self- vs non-self-reactivity in the germinal center, or to expand in response to a certain antigen and to later regress by controlled apoptosis, most postembryonic epithelial cells do not regulate their cell number by programmed cell death, but, for example, by shedding mucosal cells into the intestinal lumen. Experimental data suggest that epithelial tumor cells die with much slower kinetics than cells from hematological malignancies, and, therefore, may rely much more on the efficacy of their DNA repair machinery, before they might undergo an attenuated form of apoptosis *(20)*. Although carcinomas apparently select for mutations in pro-apoptotic genes during tumor development, these defects are hardly sufficient to explain the resistant behavior in response to therapy. We favor the explanation that oncogenic activation indeed provokes apoptotic defects required for tumor formation, albeit with rather modest interference with drug-related DNA damage-sensing and apoptosis-executing pathways. Although first line chemotherapies often achieve remissions against metastasized solid cancers, this potential is quickly lost in subsequent therapies, and additional DNA damage instead promotes the progression via mutability. Thus, DNA damage might be an inadequate trigger for cell death in epithelial cells, whereas other pro-apoptotic stimuli with different target specificity could have therapeutic potential.

11. Pro-Apoptotic Agents in the Clinic

The frustrating results of conventional anticancer agents in therapeutic regimens against solid tumors have prompted the development of pathogenesis-driven, lesion-specific compounds. Following the empirical strategies employed over decades for classic DNA damaging drugs, the unselected use of novel lesion-based, pro-apoptotic compounds in tumor entities irrespective of a matching genetic defect failed to produce encouraging results in various entities *(79–81)*. For example, the orally active tyrosine kinase inhibitor gefitinib targeting the epidermal growth factor receptor (EGFR) was tested alone or in combination with standard chemotherapy in non-small cell lung cancer patients, where it was of no therapeutic benefit *(82–85)*. However, when patients were substratified by their molecular EGFR status, mutations in the EGFR that activate antiapoptotic pathways were found to predict responsiveness to gefitinib *(86–88)*, and responses, at least in part, are based on the induction of apoptosis *(89)*. Interestingly, different mutations in the EGFR have now been described in patients that had acquired resistance to gefitinib after an initial response to the compound *(90)*. Likewise, the clinical responsiveness to the proapoptotic monoclonal antibody trastuzumab, targeting the human EGFR HER2 in breast cancer patients, was dramatically improved when patients were selected for gene amplification or overexpression of the trastuzumab target by immunohistochemistry or fluorescence *in situ* hybridization of their tumor specimens *(91–94)*. Moreover, many of the novel mechanism-based but not lesion-based agents such as histone deaceteylase inhibitors, DNA demethylating agents, or immunomodulatory thalidomide analogs interfere with complex cellular processes which may impact, in part, on programmed cell death as well. For example, the proteasome inhibitor bortezomib has pleiotropic effects on angiogenesis and tumor growth, but also blocks NF-κB prosurvival signaling *(95)*. Hence, proapoptotic agents can improve the clinical outcome of patients suffering from solid tumors if the compounds target cellular pathways that appear to be deregulated as components of the cell type specific transformation process.

Possibly as a result of their inherently higher apoptotic susceptibility, hematological malignancies seem to be particularly susceptible entities for proapoptotic targeted therapies. Evidence from early clinical trials suggests effectiveness of a *bcl2* antisense molecule as a single agent and in combination with conventional chemotherapy in patients from various groups of hematological malignancies *(96)*. Given the prominent role of Bcl2 family member-governed deregulation of apoptosis in the mitochondrial pathway acquired during tumor formation or in response to chemotherapy, proapoptotic moieties that counter this mechanism could be of enormous therapeutic potential, especially in hematological malignancies with explicit translocations involving the *bcl2* gene. Moreover, a peptide-based stapled form of the BH3 death domain derived from the proapoptotic Bcl2 family member BID exerted antileukemic activity in a preclinical model *in vivo (97)*. A similar approach based on peptide–peptide interactions has been employed to functionally disrupt another central regulator of cell survival in hematological malignancies: the Bcl6 oncoprotein, frequently activated in human lymphomas *(98)*, can suppress p53 expression in the germinal center *(99)*. Peptides that disrupt Bcl6 binding to corepressors have now been shown to cause apoptosis and cell-cycle arrest in Bcl6-positive B cell lymphomas *(100)* and might soon become a novel treatment option in the clinic.

12. Conclusions

Determining the role of apoptosis for the outcome of cancer therapy has been difficult. Compared with the complexity of the treatment response at a natural tumor site *in vivo*, most experimental settings fail to sufficiently mimic the tumor microenvironment and may miss components of drug action critical for long-term responses and, ultimately, for outcome in vivo. Tractable mouse models harboring primary tumors with defined genetic lesions and clinical evidence produced by the use of highly specific targeted therapies are expected to provide deeper and biologically more relevant insights than conventional cell culture-based approaches. Importantly, whereas apoptosis as a programmed response to DNA damage-inducing therapies clearly interferes with treatment outcome, other programmed effector mechanisms such as premature senescence may also contribute to the overall response, because most of the currently available compounds—conventional DNA damaging chemotherapy as well as lesion-specific small molecules—cannot exclusively execute apoptosis. Likewise, most of the novel agents targeting dysfunctional processes rather than specific molecular lesions will, via a broad spectrum of effector functions, ultimately interfere with the apoptosis machinery as well. Components capable of blocking pathogenesis-related, tumor-essential prosurvival pathways contribute to clinical remissions by executing apoptosis, and this effect is not limited to hematological malignancies, but occurs in solid tumors as well. Future investigations in appropriate preclinical models and probably soon in clinical trials will elucidate whether cotargeting tumor-sensitized apoptotic signal cascades such as the TRAIL receptor-mediated death cascade may be of additional therapeutic benefit. In light of the growing arsenal of cancer-specific, apoptosis-related, and dysfunction-targeting small compounds and antibodies, the broadly DNA damaging drugs with their imprecise and devastating action will expectedly loose their clinical value as prime choice in the treatment of disseminated tumors.

References

1. Ehrlich P. [The partial function of cells. (Nobel Prize address given on 11 December 1908 at Stockholm)]. Int Arch Allergy Appl Immunol 1954;5:67–86.
2. Gilman A, Philips FS. The biological actions and therapeutic implications of the beta-chloroethyl amines and sulfides. Science 1946;103:409–415.
3. Avery OT, MacLeod CM, McCarty M. Studies on the chemical nature of the substance inducing transformation of pneumococcal types. J Exp Med 1944;79:137–158.
4. Chen G, Teicher BA, Frei E, 3rd. Biochemical characterization of in vivo alkylating agent resistance of a murine EMT-6 mammary carcinoma. Implication for systemic involvement in the resistance phenotype. Cancer Biochem Biophys 1998;16:139–155.
5. Shen DW, Fojo A, Chin JE, et al. Human multidrug-resistant cell lines: increased mdr1 expression can precede gene amplification. Science 1986;232:643–645.
6. Kerr JF, Wyllie AH, Currie AR. Apoptosis: a basic biological phenomenon with wide-ranging implications in tissue kinetics. Br J Cancer 1972;26:239–257.
7. Sellins KS, Cohen JJ. Gene induction by gamma-irradiation leads to DNA fragmentation in lymphocytes. J Immunol 1987;139:3199–3206.
8. Barry MA, Behnke CA, Eastman A. Activation of programmed cell death (apoptosis) by cisplatin, other anticancer drugs, toxins and hyperthermia. Biochem Pharmacol 1990;40: 2353–2362.
9. Miyashita T, Reed JC. bcl-2 gene transfer increases relative resistance of S49.1 and WEHI7.2 lymphoid cells to cell death and DNA fragmentation induced by glucocorticoids and multiple chemotherapeutic drugs. Cancer Res 1992;52:5407–5411.
10. Okada H, Mak TW. Pathways of apoptotic and non-apoptotic death in tumour cells. Nat Rev Cancer 2004;4:592–603.
11. Woo M, Hakem R, Soengas MS, et al. Essential contribution of caspase 3/CPP32 to apoptosis and its associated nuclear changes. Genes Dev 1998;12:806–819.
12. Merritt AJ, Allen TD, Potten CS, Hickman JA. Apoptosis in small intestinal epithelial from p53-null mice: evidence for a delayed, p53-independent G2/M-associated cell death after gamma-irradiation. Oncogene 1997;14:2759–2766.

13. Bursch W, Hochegger K, Torok L, Marian B, Ellinger A, Hermann RS. Autophagic and apoptotic types of programmed cell death exhibit different fates of cytoskeletal filaments. J Cell Sci 2000; 113(Pt 7):1189–1198.
14. Levine B, Klionsky DJ. Development by self-digestion: molecular mechanisms and biological functions of autophagy. Dev Cell 2004;6:463–477.
15. Proskuryakov SY, Konoplyannikov AG, Gabai VL. Necrosis: a specific form of programmed cell death? Exp Cell Res 2003;283:1–16.
16. Nara N, Suzuki T, Nagata K, Yamashita Y, Murohashi I, Adachi Y. Relationship between the in vitro sensitivity to cytosine arabinoside of blast progenitors and the outcome of treatment in acute myeloblastic leukaemia patients. Br J Haematol 1988;70:187–191.
17. Phillips RM, Bibby MC, Double JA. A critical appraisal of the predictive value of in vitro chemosensitivity assays. J Natl Cancer Inst 1990;82:1457–1468.
18. Mollgard L, Tidefelt U, Sundman-Engberg B, Lofgren C, Paul C. In vitro chemosensitivity testing in acute non lymphocytic leukemia using the bioluminescence ATP assay. Leuk Res 2000;24:445–452.
19. Rodel C, Grabenbauer GG, Papadopoulos T, et al. Apoptosis as a cellular predictor for histopathologic response to neoadjuvant radiochemotherapy in patients with rectal cancer. Int J Radiat Oncol Biol Phys 2002;52:294–303.
20. Brown JM, Wouters BG. Apoptosis, p53, and tumor cell sensitivity to anticancer agents. Cancer Res 1999;9:1391–1399.
21. Di Leonardo A, Linke SP, Clarkin K, Wahl GM. DNA damage triggers a prolonged p53-dependent G1 arrest and long-term induction of Cip1 in normal human fibroblasts. Genes Dev 1994;8:2540–2551.
22. Waldman T, Zhang Y, Dillehay L, et al. Cell-cycle arrest versus cell death in cancer therapy. Nat Med 1997;3:1034–1036.
23. Yin DX, Schimke RT. BCL-2 expression delays drug-induced apoptosis but does not increase clonogenic survival after drug treatment in HeLa cells. Cancer Res 1995; 55:4922–4928.
24. Lock RB, Stribinskiene L. Dual modes of death induced by etoposide in human epithelial tumor cells allow Bcl-2 to inhibit apoptosis without affecting clonogenic survival. Cancer Res 1996;56:4006–4012.
25. Schmitt CA, Rosenthal CT, Lowe SW. Genetic analysis of chemoresistance in primary murine lymphomas. Nat Med 2000;6:1029–1035.
26. Walker A, Taylor ST, Hickman JA, Dive C. Germinal center-derived signals act with Bcl-2 to decrease apoptosis and increase clonogenicity of drug-treated human B lymphoma cells. Cancer Res 1997;57:1939–1945.
27. Taylor ST, Hickman JA, Dive C. Epigenetic determinants of resistance to etoposide regulation of Bcl-X(L) and Bax by tumor microenvironmental factors. J Natl Cancer Inst 2000;92:18–23.
28. Francis DA, Karras JG, Ke XY, Sen R, Rothstein TL. Induction of the transcription factors NF-kappa B, AP-1 and NF-AT during B cell stimulation through the CD40 receptor. Int Immunol 1995;7:151–161.
29. Grassme H, Jekle A, Riehle A, et al. CD95 signaling via ceramide-rich membrane rafts. J Biol Chem 2001;276:20,589–20,596.
30. Lacour S, Hammann A, Grazide S, et al. Cisplatin-induced CD95 redistribution into membrane lipid rafts of HT29 human colon cancer cells. Cancer Res 2004;64:3593–3598.
31. Muller M, Wilder S, Bannasch D, et al. p53 activates the CD95 (APO-1/Fas) gene in response to DNA damage by anticancer drugs. J Exp Med 1998;188:2033–2045.
32. Sheikh MS, Burns TF, Huang Y, et al. p53-dependent and -independent regulation of the death receptor KILLER/DR5 gene expression in response to genotoxic stress and tumor necrosis factor alpha. Cancer Res 1998;58:1593–1598.

33. Fulda S, Debatin KM. Exploiting death receptor signaling pathways for tumor therapy. Biochim Biophys Acta 2004;1705:27–41.
34. Belka C, Schmid B, Marini P, et al. Sensitization of resistant lymphoma cells to irradiation-induced apoptosis by the death ligand TRAIL. Oncogene 2001;20:2190–2196.
35. Kim MR, Lee JY, Park MT, et al. Ionizing radiation can overcome resistance to TRAIL in TRAIL-resistant cancer cells. FEBS Lett 2001;505:179–184.
36. Matsunaga T, Takemoto N, Sato T, et al. Interaction between leukemic-cell VLA-4 and stromal fibronectin is a decisive factor for minimal residual disease of acute myelogenous leukemia. Nat Med 2003;9:1158–1165.
37. Schmitt CA, Lowe SW. Bcl-2 mediates chemoresistance in matched pairs of primary Eμ-myc lymphomas in vivo. Blood Cells Mol Dis 2001;27:206–216.
38. Schmitt CA, Fridman JS, Yang M, et al. A senescence program controlled by p53 and p16INK4a contributes to the outcome of cancer therapy. Cell 2002;109:335–346.
39. Schmitt CA, McCurrach ME, de Stanchina E, Wallace-Brodeur RR, Lowe SW. INK4a/ARF mutations accelerate lymphomagenesis and promote chemoresistance by disabling p53. Genes Dev 1999;13:2670–2677.
40. Wendel HG, De Stanchina E, Fridman JS, et al. Survival signalling by Akt and eIF4E in oncogenesis and cancer therapy. Nature 2004;428:332–337.
41. Lallemand-Breitenbach V, Guillemin MC, Janin A, et al. Retinoic acid and arsenic synergize to eradicate leukemic cells in a mouse model of acute promyelocytic leukemia. J Exp Med 1999;189:1043–1052.
42. Bearss DJ, Subler MA, Hundley JE, Troyer DA, Salinas RA, Windle JJ. Genetic determinants of response to chemotherapy in transgenic mouse mammary and salivary tumors. Oncogene 2000;19:1114–1122.
43. Petit T, Bearss DJ, Troyer DA, Munoz RM, Windle JJ. p53-independent response to cisplatin and oxaliplatin in MMTV-ras mouse salivary tumors. Mol Cancer Ther 2003;2: 165–171.
44. Miyake H, Hara I, Hara S, Arakawa S, Kamidono S. Synergistic chemosensitization and inhibition of tumor growth and metastasis by adenovirus-mediated P53 gene transfer in human bladder cancer model. Urology 2000;56:332–336.
45. Kigawa J, Sato S, Shimada M, Kanamori Y, Itamochi H, Terakawa N. Effect of p53 gene transfer and cisplatin in a peritonitis carcinomatosa model with p53-deficient ovarian cancer cells. Gynecol Oncol 2002;84:210–215.
46. Sturm I, Bosanquet AG, Hermann S, Guner D, Dorken B, Daniel PT. Mutation of p53 and consecutive selective drug resistance in B-CLL occurs as a consequence of prior DNA-damaging chemotherapy. Cell Death Differ 2003;10:477–484.
47. Dohner H, Fischer K, Bentz M, et al. p53 gene deletion predicts for poor survival and non-response to therapy with purine analogs in chronic B-cell leukemias. Blood 1995;85: 1580–1589.
48. Schmitt CA, Lowe SW. Apoptosis and therapy. J Pathol 1999;187:127–137.
49. Hanahan D, Weinberg RA. The hallmarks of cancer. Cell 2000;100:57–70.
50. Palmero I, Pantoja C, Serrano M. p19ARF links the tumour suppressor p53 to Ras. Nature 1998;395:125–126.
51. Zindy F, Eischen CM, Randle DH, et al. Myc signaling via the ARF tumor suppressor regulates p53-dependent apoptosis and immortalization. Genes Dev 1998;12:2424–2433.
52. Schmitt CA. Senescence, apoptosis and therapy - cutting the lifelines of cancer. Nat Rev Cancer 2003;3:286–295.
53. Kitada S, Takayama S, De Riel K, Tanaka S, Reed JC. Reversal of chemoresistance of lymphoma cells by antisense-mediated reduction of bcl-2 gene expression. Antisense Res Dev 1994;4:71–79.

54. Meijerink JP, Mensink EJ, Wang K, et al. Hematopoietic malignancies demonstrate loss-of-function mutations of BAX. Blood 1998;91:2991–2997.
55. Soengas MS, Capodieci P, Polsky D, et al. Inactivation of the apoptosis effector Apaf-1 in malignant melanoma. Nature 2001;409:207–211.
56. Schmitt CA, Wallace-Brodeur RR, Rosenthal CT, McCurrach ME, Lowe SW. DNA damage responses and chemosensitivity in the Eμ-myc mouse lymphoma model. Cold Spring Harbor Symp. Quant. Biol. 2000;LXV:499–510.
57. Mrozek A, Petrowsky H, Sturm I, et al. Combined p53/Bax mutation results in extremely poor prognosis in gastric carcinoma with low microsatellite instability. Cell Death Differ 2003;10:461–467.
58. Ribic CM, Sargent DJ, Moore MJ, et al. Tumor microsatellite-instability status as a predictor of benefit from fluorouracil-based adjuvant chemotherapy for colon cancer. N Engl J Med 2003;349:247–257.
59. Wahl AF, Donaldson KL, Fairchild C, et al. Loss of normal p53 function confers sensitization to Taxol by increasing G2/M arrest and apoptosis. Nat Med 1996;2:72–79.
60. Nitta M, Kobayashi O, Honda S, et al. Spindle checkpoint function is required for mitotic catastrophe induced by DNA-damaging agents. Oncogene 2004;23:6548–6558.
61. Chang BD, Broude EV, Dokmanovic M, et al. A senescence-like phenotype distinguishes tumor cells that undergo terminal proliferation arrest after exposure to anticancer agents. Cancer Res 1999;59:3761–3767.
62. te Poele RH, Okorokov AL, Jardine L, Cummings J, Joel SP. DNA damage is able to induce senescence in tumor cells in vitro and in vivo. Cancer Res 2002;62:1876–1883.
63. Lee S, Schmitt CA. Chemotherapy response and resistance. Curr Opin Genet Dev 2003; 13:90–96.
64. Komarov PG, Komarova EA, Kondratov RV, et al. A chemical inhibitor of p53 that protects mice from the side effects of cancer therapy. Science 1999;285:1733–1737.
65. Fearnhead HO, McCurrach ME, O'Neill J, Zhang K, Lowe SW, Lazebnik YA. Oncogene-dependent apoptosis in extracts from drug-resistant cells. Genes Dev 1997;11:1266–1276.
66. Nahle Z, Polakoff J, Davulari RV, et al. Direct coupling of the cell cycle and cell death machinery by E2F. Nat Cell Biol 2002;4(11):859–864.
67. Fehrenbacher N, Gyrd-Hansen M, Poulsen B, et al. Sensitization to the lysosomal cell death pathway upon immortalization and transformation. Cancer Res 2004;64:5301–5310.
68. Hueber AO, Zornig M, Lyon D, Suda T, Nagata S, Evan GI. Requirement for the CD95 receptor-ligand pathway in c-Myc-induced apoptosis [see comments]. Science 1997;278: 1305–1309.
69. Graeber TG, Osmanian C, Jacks T, et al. Hypoxia-mediated selection of cells with diminished apoptotic potential in solid tumours [see comments]. Nature 1996;379:88–91.
70. Narendran A, Ganjavi H, Morson N, et al. Mutant p53 in bone marrow stromal cells increases VEGF expression and supports leukemia cell growth. Exp Hematol 2003;31:693–701.
71. Yu JL, Rak JW, Coomber BL, Hicklin DJ, Kerbel RS. Effect of p53 status on tumor response to antiangiogenic therapy. Science 2002;295:1526–1528.
72. Streubel B, Chott A, Huber D, et al. Lymphoma-specific genetic aberrations in microvascular endothelial cells in B-cell lymphomas. N Engl J Med 2004;351:250–259.
73. Yin KJ, Chen SD, Lee JM, Xu J, Hsu CY. ATM gene regulates oxygen-glucose deprivation-induced nuclear factor-kappaB DNA-binding activity and downstream apoptotic cascade in mouse cerebrovascular endothelial cells. Stroke 2002;33:2471–2477.
74. Sandoval C, Swift M. Hodgkin disease in ataxia-telangiectasia patients with poor outcomes. Med Pediatr Oncol 2003;40:162–166.
75. Aas T, Borresen AL, Geisler S, et al. Specific P53 mutations are associated with de novo resistance to doxorubicin in breast cancer patients. Nat Med 1996;2:811–814.

76. Sjostrom J, Blomqvist C, von Boguslawski K, et al. The predictive value of bcl-2, bax, bcl-xL, bag-1, fas, and fasL for chemotherapy response in advanced breast cancer. Clin Cancer Res 2002;8:811–816.
77. Armitage JO. High-dose chemotherapy and autologous hematopoietic stem cell transplantation: the lymphoma experience and its potential relevance to solid tumors. Oncology 2000;58:198–206.
78. Renga M, Pedrazzoli P, Siena S. Present results and perspectives of allogeneic non-myeloablative hematopoietic stem cell transplantation for treatment of human solid tumors. Ann Oncol 2003;14:1177–1184.
79. Gewirtz AM. Oligonucleotide therapeutics: clothing the emperor. Curr Opin Mol Ther 1999;1:297–306.
80. Morris MJ, Tong WP, Cordon-Cardo C, et al. Phase I trial of BCL-2 antisense oligonucleotide (G3139) administered by continuous intravenous infusion in patients with advanced cancer. Clin Cancer Res 2002;8:679–683.
81. Marshall JL, Eisenberg SG, Johnson MD, et al. A phase II trial of ISIS 3521 in patients with metastatic colorectal cancer. Clin Colorectal Cancer 2004;4:268–274.
82. Cappuzzo F, Gregorc V, Rossi E, et al. Gefitinib in pretreated non-small-cell lung cancer (NSCLC): analysis of efficacy and correlation with HER2 and epidermal growth factor receptor expression in locally advanced or metastatic NSCLC. J Clin Oncol 2003;21:2658–2663.
83. Kris MG, Natale RB, Herbst RS, et al. Efficacy of gefitinib, an inhibitor of the epidermal growth factor receptor tyrosine kinase, in symptomatic patients with non-small cell lung cancer: a randomized trial. Jama 2003;290:2149–2158.
84. Giaccone G, Herbst RS, Manegold C, et al. Gefitinib in combination with gemcitabine and cisplatin in advanced non-small-cell lung cancer: a phase III trial—INTACT 1. J Clin Oncol 2004;22:777–784.
85. Herbst RS, Giaccone G, Schiller JH, et al. Gefitinib in combination with paclitaxel and carboplatin in advanced non-small-cell lung cancer: a phase III trial—INTACT 2. J Clin Oncol 2004;22:785–794.
86. Lynch TJ, Bell DW, Sordella R, et al. Activating mutations in the epidermal growth factor receptor underlying responsiveness of non-small-cell lung cancer to gefitinib. N Engl J Med 2004;350:2129–2139.
87. Paez JG, Janne PA, Lee JC, et al. EGFR mutations in lung cancer: correlation with clinical response to gefitinib therapy. Science 2004;304:1497–1500.
88. Sordella R, Bell DW, Haber DA, Settleman J. Gefitinib-sensitizing EGFR mutations in lung cancer activate anti-apoptotic pathways. Science 2004;305:1163–1167.
89. Tracy S, Mukohara T, Hansen M, Meyerson M, Johnson BE, Janne PA. Gefitinib induces apoptosis in the EGFRL858R non-small-cell lung cancer cell line H3255. Cancer Res 2004;64:7241–7244.
90. Kobayashi S, Boggon TJ, Dayaram T, et al. EGFR mutation and resistance of non-small-cell lung cancer to gefitinib. N Engl J Med 2005;352:786–792.
91. Pegram MD, Lipton A, Hayes DF, et al. Phase II study of receptor-enhanced chemosensitivity using recombinant humanized anti-p185HER2/neu monoclonal antibody plus cisplatin in patients with HER2/neu-overexpressing metastatic breast cancer refractory to chemotherapy treatment. J Clin Oncol 1998;16:2659–2671.
92. Kunisue H, Kurebayashi J, Otsuki T, et al. Anti-HER2 antibody enhances the growth inhibitory effect of anti-oestrogen on breast cancer cells expressing both oestrogen receptors and HER2. Br J Cancer 2000;82:46–51.
93. Seidman AD, Fornier MN, Esteva FJ, et al. Weekly trastuzumab and paclitaxel therapy for metastatic breast cancer with analysis of efficacy by HER2 immunophenotype and gene amplification. J Clin Oncol 2001;19:2587–2595.

94. Tedesco KL, Thor AD, Johnson DH, et al. Docetaxel combined with trastuzumab is an active regimen in HER-2 3+ overexpressing and fluorescent in situ hybridization-positive metastatic breast cancer: a multi-institutional phase II trial. J Clin Oncol 2004;22:1071–1077.
95. Sunwoo JB, Chen Z, Dong G, et al. Novel proteasome inhibitor PS-341 inhibits activation of nuclear factor-kappa B, cell survival, tumor growth, and angiogenesis in squamous cell carcinoma. Clin Cancer Res 2001;7:1419–1428.
96. Chanan-Khan A. Bcl-2 antisense therapy in hematologic malignancies. Curr Opin Oncol 2004;16:581–585.
97. Walensky LD, Kung AL, Escher I, et al. Activation of apoptosis in vivo by a hydrocarbon-stapled BH3 helix. Science 2004;305:1466–1470.
98. Pasqualucci L, Migliazza A, Basso K, Houldsworth J, Chaganti RS, Dalla-Favera R. Mutations of the BCL6 proto-oncogene disrupt its negative autoregulation in diffuse large B-cell lymphoma. Blood 2003;101:2914–2923.
99. Phan RT, Dalla-Favera R. The BCL6 proto-oncogene suppresses p53 expression in germinal-centre B cells. Nature 2004;432:635–639.
100. Polo JM, Dell'Oso T, Ranuncolo SM, et al. Specific peptide interference reveals BCL6 transcriptional and oncogenic mechanisms in B-cell lymphoma cells. Nat Med 2004;10: 1329–1035.

14

Coupling Apoptosis and Cell Division Control in Cancer

The Survivin Paradigm

Dario C. Altieri

Summary

The interface between cell proliferation and cell death is thought to function as a pivotal crossroad essential to the preservation of normal homeostasis and to eliminate dangerous cells before they divide. Survivin is a prototype molecule at this crossroad, intercalated in protection against mitochondrial cell death and orchestrating various aspects of cell division. Dramatically exploited in cancer and an unfavorable gene signature for disease outcome, the survivin pathway provides tangible opportunities for targeted, rational cancer therapy.

Key Words: Survivin; apoptosis; cancer; IAP; stress response.

1. Introduction—Apoptotic Pathways

Apoptosis is generally defined as a genetic program of cellular suicide with unique morphologic characteristics *(1)*. Essential during embryonic and fetal development *(2)*, and for maintaining the homeostasis of differentiated tissues *(3)*, apoptotic mechanisms are commonly subverted in cancer, where increased cell survival and/or reduced cell death are invariable molecular traits of transformed cells *(4)*.

Among the regulators of apoptosis, Bcl-2 proteins *(5)* act at the mitochondria to decrease (antiapoptotic) or enhance (proapoptotic) permeability transition, typically by regulating cytochrome *c* release *(3)*. Conversely, members of the inhibitors of apoptosis (IAP) gene family have been implicated in the regulation of a downstream step in cell survival, by preventing the activation of caspases or by inhibiting the catalytic activity of the mature enzymes. Other mechanisms of IAP-mediated cytoprotection have been also demonstrated, including ubiquitin-dependent destruction of bound caspases or mitochondrial proapoptogenic proteins (i.e., Smac) *(6)*. The molecular signature of IAP is the presence of 1 to 3 copies of an approximate70-amino acid zinc finger fold designated baculovirus IAP repeat (BIR), which is conserved in related molecules from yeast to humans. Certain IAPs also contain a caspase-recruitment domain (CARD), a really interesting new gene (RING) finger, a ubiquitin-conjugating domain, and a nucleotide binding P-loop motif *(6)*. The anticaspase activity of IAPs is evolutionary conserved, and *Drosophila* IAP-like proteins are essential regulators of cell survival in flies *(7)*.

1.1. Survivin Structure Function

At 16.5 kDa, survivin is the smallest mammalian member of the IAP gene family *(6)*. Structurally, it contains a single BIR, an approximate 70-amino acid zinc finger fold that

From: *Apoptosis, Cell Signaling, and Human Diseases: Molecular Mechanisms, Volume 1*
Edited by R. Srivastava © Humana Press Inc., Totowa, NJ

is the hallmark of all IAPs, and an extended –COOH terminus α-helical coiled-coil, but no other identifiable domain *(8)* typically found in IAPs. Based on X-ray crystallography of the human *(9,10)* or mouse *(11)* protein, survivin is a stable homodimer in solution with the –COOH terminus α-helices protruding from the core dimer. A single copy *survivin* gene located on chromosome 17q25 (human), or 11E2 (mouse), gives rise to four alternatively spliced survivin transcripts. In addition to wild-type survivin (142 amino acids), three survivin isoforms are generated by insertion of an alternative exon 2 (survivin-2B, 165 amino acids), removal of exon 3 (survivin-ΔEx-3, 137 amino acids) *(14)*, or acquisition of an in-frame stop codon in intron 2 (Survivin-2α) *(15)*. In survivin-ΔEx-3, the splicing event introduces a frame shift that generates a unique –COOH terminus of potential functional significance *(16)*.

A unique property of survivin is a sharp cell cycle-dependent expression at mitosis. This is largely, but not exclusively, controlled at the level of gene transcription and involves canonical CDE/CHR boxes in the survivin promoter *(17,18)*, acting as potential G1-repressor elements. Among the post-translational modifications that affect survivin levels, interest has recently focused on the control of protein stability. Survivin is a relatively short-lived protein ($t^1/_2$ = 30 min), and polyubiquitylation followed by proteasomal destruction has been shown to contribute to cell-cycle periodicity by keeping survivin levels low at interphase *(19)*. In addition, mitotic phosphorylation of survivin on Thr34 by p34cdc2–cyclin B1 has been associated with increased survivin stability at metaphase *(20)*.

There are also examples of modulation of survivin expression independently of cell-cycle progression. For $CD34^+$ bone-marrow-derived stem cells, stimulation with hematopoietic cytokines resulted in survivin expression in the absence of cyclin D and with hypophosphorylated Rb *(21)*, and vascular remodeling by angiopoietin-1 (Ang-1) upregulated survivin in nonproliferating endothelial cells *(22,23)*. The search for signaling intermediates that control survivin expression has generated interesting findings. One common denominator of this pathway is the requirement for PI3-kinase/Akt activation, a general antiapoptotic signal *(24)* that has been implicated in up-regulation of survivin induced by granulocyte macrophage-colony stimulating factor (GM-CSF), CSF, Ang-1, and fibronectin-dependent cell adhesion *(21,22,25,26)*. In addition, modulation of survivin expression has been observed in endothelial cells stimulated with nonmitogenic concentrations of interleukin (IL)-11 *(27)*, engagement of vascular endothelial (VE)-cadherin expression, and function *(28)*, or angiotensin II stimulation *(29)*. Another critical transcriptional regulator of survivin expression in pleural effusion lymphoma cell sustained by vascular endothelial growth factor (VEGF), IL-6, or IL-10 *(30)* was identified as activated signal transducers and activators of transcription (STAT)3. This is potentially important for the known role of STAT3 in cytoprotection *(31)*, which may have direct consequences for oncogenic transformation *(32)*.

2. Role of Survivin in Cell Division

Upon expression at mitosis, survivin localizes to various components of the mitotic apparatus, including centrosomes, microtubules of the metaphase and anaphase spindle, and the remnants of the mitotic apparatus—midbodies *(12,33–36)*. A direct association between survivin and polymerized tubulin has been demonstrated in vitro *(12)*, potentially involving the –COOH terminus α-helices. A subcellular pool of survivin has also been localized to kinetochores of metaphase chromosomes *(35,37)*, and may comprise up to about 20% of total cellular survivin *(36)*. On the other hand, survivin is largely excluded from the nucleus *(36)* through a CRM1-dependent mechanism that may involve the survivin –COOH terminus *(38)*. Survivin is also a mitotic phosphoprotein. In addition to phosphorylation on Thr34 by the main mitotic kinase p34cdc2, phosphorylation of survivin on Thr117 by Aurora B kinase has been also reported *(39)* as potentially important in regulating the topography and/or association of the chromosomal passenger complex (discussed later).

The complex localization of survivin to the mitotic apparatus reflects an essential function at cell division. This first surfaced in targeting experiments using antisense or dominant-negative mutants, which resulted in a dual phenotype of apoptosis (discussed later) *and* aberrant mitotic progression, with supernumerary centrosomes, multipolar mitotic spindles, failed cytokinesis, and multinucleation *(33)*. Parallel experiments targeting survivin-like molecules in yeast *(40)* or *Caenorhabditis elegans (41)* also revealed lethal cell division defects with failure to assemble a cleavage furrow and inability to complete cytokinesis. The phenotype of survivin knockout mice is consistent with an essential role of this pathway at mitosis *(42)*. Beginning at embryonic day (E) 2.5, homozygous deletion of the *survivin* gene resulted in catastrophic defects of microtubule assembly, with absence of mitotic spindles, complete failure of cell division and multinucleation with 100% lethality by E3.5 to 4.5 *(42)*. An identical phenotype was independently confirmed on three different mouse genetic backgrounds *(43)*. Therefore, one model that has been put forward is that survivin functions in an essential and evolutionary conserved step in late stage cell division (i.e., cytokinesis, potentially involving cleavage furrow formation *[42,44])*. In support of this, survivin was shown to form a complex with molecules comprising the so-called "chromosomal passenger complex," including Aurora B kinase and INCENP on kinetochores and the anaphase central spindle *(37,45)*, and to enhance the activity of Aurora B kinase in both mammalian cells *(46)* and *Xenopus laevii (47)*. However, recent data suggest that survivin may play a broader role at various aspects of cell division. This point became apparent in antibody microinjection studies, which revealed a composite phenotype of prolonged metaphase arrest, occasionally followed by apoptosis, formation of multipolar mitotic spindles, and defects in chromosome attachments *(48,49)*. By immunofluorescence, cells microinjected with an antibody to survivin exhibited flattened and abortive mitotic spindles severely depleted of microtubules *(49)*. Conversely, retroviral or adenoviral expression of survivin restored spindle stability and dynamics against microtubule poisons in tumor cells *(49)* as well as endothelial cells *(50)*. A critical role of survivin in microtubule function and spindle formation was also demonstrated using RNAi ablation of survivin expression in nontransformed human fibroblasts *(51)*. This is consistent with a number of other investigations using survivin knockdown by RNA interference. In these experiments, loss of survivin expression was associated with inability to complete mitosis, stable prometaphase arrest, defective function of the spindle assembly checkpoint, and aberrant spindle formation *(52–54)*. Altogether, these data point to a more general role of survivin at cell division, and not exclusively limited to cytokinesis.

This possibility has recently gained further credibility from a more in-depth analysis of the chromosomal passenger complex. Different from earlier suggestions that implicated the chromosomal passenger complex solely in kinetochore attachment and cytokinesis, recent evidence has uncovered a key role of these molecules in spindle formation via enhanced microtubule stabilization *(55,56)*. Whether the proposed association of survivin with Aurora B kinase is directly involved in this mechanism remains to be seen, but it is clear that a large subcellular pool of mitotic survivin is not in complex with the chromosomal passenger complex *(36)*. This suggests that the chromosomal passenger complex and survivin may exert potentially independent roles in microtubule function and mitotic spindle assembly. Such a model is consistent with the observation that microtubule- and kinetochore-associated survivin are differentially regulated during the cell cycle, exhibit nonoverlapping patterns of phosphorylation, and are recognized by monoclonal antibodies (MAbs) in a mutually exclusive fashion *(36)*. As an alternative possibility, it cannot be ruled out that the differently spliced survivin isoforms could mediate independent functions at cell division *(14)*. This remains a possibility that requires formal testing when antibodies capable of discriminating between the various endogenous survivin isoforms become available. However, wild-type survivin is clearly the overwhelming form in all cell types tested thus far *(14)*, and, at least in over-expression experiments, survivin-2B does not associate with the mitotic apparatus, and survivin-ΔEx-3 has actually been localized to the nucleus potentially through a bipartite nuclear localization signal in the new –COOH terminus *(16,38)*.

In sum, it appears likely that the various survivin pools and/or alternatively spliced variants of survivin provide a continuum of functions throughout mitosis. This may involve regulation of centrosomal activity, the assembly of a competent bipolar mitotic spindles, the control of kinetochore-microtubule attachment, and cytokinesis. This model is also satisfactory if one considers the biology of Aurora kinases, which, like survivin, are found frequently over-expressed in human cancers, and may participate in oncogenic transformation. Altogether, the combined over-expression of both survivin and Aurora in cancer may obliterate the surveillance mechanism of the spindle assembly checkpoint, allowing cells with spindle defects, aberrant chromosome congression, or misaligned/unattached kinetochores to proceed through cell division, thus enhancing tumor-associated aneuploidy.

3. Role of Survivin in Apoptosis Inhibition

The question as to whether survivin had also a distinct and separable role in apoptosis inhibition, similarly to other mammalian IAPs *(6)*, or whether its function is limited to cell division, as with IAPs in yeast and *C. elegans (40,41)*, has been amply investigated in recent years. With the expansion of experimental work on survivin, three lines of experimental evidence clearly support the notion that survivin has an independent and separable role in apoptosis inhibition, in vitro and in vivo.

First, over-expression of survivin has been associated with inhibition of cell death initiated via the extrinsic or intrinsic apoptotic pathways *(57–60)*. Second, transgenic expression of survivin resulted in apoptosis inhibition in vivo *(61)*, and livers isolated from heterozygous survivin$^{+/-}$ animals exhibited exaggerated apoptosis in response to suboptimal ligation of Fas *(43)*. Third, molecular antagonists of survivin, including antisense, ribozymes, siRNA sequences, or dominant-negative mutants resulted in caspase-dependent cell death, enhancement of apoptotic stimuli, and anticancer activity, in vivo *(57–60)*. The apoptosis inhibitory function of survivin is evolutionary conserved, and a survivin-like molecule in *Drosophila, Deterin*, functions interchangeably with survivin to block apoptosis *(62)* in mammalian or insect cells *(34)*.

Although the notion that survivin inhibits apoptosis is well established, the more mechanistic aspects of this pathway have long remained elusive. As with other IAPs *(6)*, a physical interaction between survivin and initiator or effector caspases has been reported by several groups *(17,63,64)*, and a physical proximity between survivin and effector caspase-3 has been seen in vivo as well *(33)*. However, with the exception of two published studies *(65,66)*, this did not seem to translate in meaningful inhibition of caspase activity, in vitro *(9,64,67)*. In addition, the crystal structure of survivin does not suggest the presence of a "hook-and-sinker" region that mediates caspase binding in other IAPs *(9)*, and the residues in the linker upstream BIR2 that in other IAPs (i.e., X-linked mammalian inhibitor of apoptosis protein [XIAP]) dock to caspase-3 *(68)*, are not present in survivin. One can take the view that this does not automatically rule out that survivin may still function as an inhibitor of effector caspases similarly to other IAPs *(6)*, and arguments to explain the lack of activity in cell-free systems may include folding requirements of the recombinant protein *(66)*, or the need for additional protein partners, in vivo (discussed later).

On the other hand, evidence is accumulating that survivin may play a more selective role than other IAPs in antagonizing mitochondrial-dependent apoptosis. Overexpression of survivin is more efficient at blocking mitochondrial- but not death-receptor-induced apoptosis *(61)*, a complex between survivin and the upstream mitochondrial initiator caspase-9 has been demonstrated in vivo *(69)*, and survivin has been shown to associate with Smac/DIABLO *(70–73)*, a mitochondrial-released apoptogenic protein that relieves the inhibitory effect of IAPs on caspase activation *(68)*. Moreover, cell death induced by molecular antagonists of survivin or by heterozygous reduction of survivin levels *(43)* has the characteristics of mitochondrial-dependent apoptosis with cytochrome *c* release *(74)*, caspase-9 activation *(69)*, and involvement of the apoptosome components *(20)*, caspase-9 and Apaf-1 *(75)*. Collectively, these data suggest that survivin, differently from other IAPs that inhibit initiator or effector caspases through their independent BIRs *(68)*, may selectively target the multimolecular process of caspase-9 activation.

Although an attractive candidate, this pathway does not seem to be centered on the recognition of Smac/DIABLO by survivin *(76)*. Although this complex has been recently characterized in detail *(73)*, its role in regulating survivin-dependent cytoprotection has not been firmly demonstrated *(70,72)*. Conversely, recent experimental evidence provided unexpected clues to explain the pathway of survivin cytoprotection and its relevance to tumor progression, in vivo. In subcellular fractionation experiments, a novel and abundant pool of survivin was recently localized to mitochondria, and further mapped biochemically to the intermitochondrial membrane space *(77)*. Mitochondrial survivin was shown to function as a highly dynamic subcellular pool, enriched in response to cellular stress and promptly discharged in the cytosol following exposure to a variety of cell death agonists *(77)*. In turn, mitochondrially-released survivin was competent to inhibit the upstream initiation of mitochondrial apoptosis, prevent processing and activation of apoptosome-associated caspase-9, and enhance tumor growth in immunocompromised animals *(77)*. Conversely, survivin that could not be transported to mitochondria had no role in apoptosis inhibition and did not provide any advantage for accelerated tumor growth, in vivo. From these studies *(78)*, it is clear that the function of survivin in apoptosis inhibition is separable from its second role at mitosis, and is specifically mediated by a discrete subcellular (mitochondrial) pool of the molecule. In broader terms, these data also demonstrate that mitochondrial permeability transition results not only in the release of proapoptogenic mediators but also of at least one apoptosis inhibitor, potentially ideally positioned to counter-balance the apoptosome initiation of cell death *(78)*. Further studies are needed to clarify why survivin must travel to mitochondria and be dynamically discharged in the cytosol in order to become competent to inhibit apoptosis. However, given the complexity and functional relevance of survivin posttranslational modifications, it seems plausible to hypothesize that mitochondrial or cytosolic survivin undergo different and mutually exclusive posttranslational modifications that either enable or prevent its role in blocking apoptosis. Another critical aspect of the pathway is that mitochondrially released survivin acts in the cytosol to prevent apoptosis. This may reflect the mechanistic requirement of functional cofactors needed to perform as a high affinity apoptosis inhibitor. In this context, survivin has been shown to associate with cytosolic proteins, including HBXIP *(79)*, and another IAP family protein, XIAP *(80)*, and that formation of these complexes results in enhanced survivin anti-apoptotic activity.

A critical role of survivin in mitochondrial apoptosis is also consistent with a similar function ascribed to one of its splice variant, survivin-ΔEx3. It was recently reported that survivin-ΔEx3 contains a mitochondrial localization signal and a BH2 domain in the new, alternatively spliced –COOH terminus *(81)*, and that a survivin-ΔEx3-like molecule in the herpes simplex virus (HSV) genome localized to mitochondria and inhibited apoptosis by associating with Bcl-2 and by suppressing caspase-3 activity through its BIR *(81)*. Whether survivin-ΔEx3 behaves similarly to its viral counterpart remains to be seen, but its potential import to mitochondria would make it ideally suited to act, alone or by dimerizing with wild type survivin, as an upstream regulator of mitochondrial-dependent apoptosis.

4. Translational Targeting of the Survivin Pathway in Cancer

One of the most significant features of survivin is its differential expression in cancer vs normal tissues *(8)*. Reminiscent of "onco-fetal" antigens, survivin is strongly expressed in embryonic and fetal organs *(17,82)*, undetectable or found at very low levels in most terminally differentiated normal tissues *(8)*, and dramatically over-expressed in most human cancers surveyed *(83)*. The selective over-expression of survivin in cancer appears to reflect a global deregulation of *survivin* gene transcription in transformed cells. This is consistent with the cancer-specific transcription of the *survivin* promoter *(84)*, and the several oncogenic pathways that converge to up-regulate *survivin* gene expression in transformed cells. These include growth factor receptor signaling *(85,86)*, STAT activation *(30)*, PI3 kinase/Akt signaling *(87)*, oncogene (Ras) expression *(88)*, and loss of tumor suppressor molecules, p53 *(89,90)*, adenomatous polyposis coli (APC) *(91,92)*, and promyelocytic leukemia (PML) *(93)*.

Several retrospective studies have demonstrated that survivin expression in cancer is frequently associated with unfavorable disease outcome, abbreviated overall survival, increased rates of recurrences, resistance to therapy, and reduced apoptotic index, in vivo *(83)*. Accordingly, profiling studies identified survivin as a "risk-associated" gene signature for unfavorable outcome in breast cancer *(94)*, large cell non-Hodgkin's lymphoma *(95)*, and colorectal cancer *(96,97)*. Importantly, survivin was incorporated as one of the critical 16 genes predictors of recurrences in tamoxifen-treated node-negative breast cancer patients *(98)*.

Because of its differential expression in cancer and critical roles in mitotic progression and cell viability, survivin has been vigorously pursued as a new rational target for cancer treatment *(83)*. Three approaches have shown promise in both preclinical and early clinical testing. First, survivin has been used for novel cancer vaccination protocols *(99)*. Several studies have demonstrated that T-cells mount a vigorous cytolytic and antibody response to survivin peptides, in vitro and in vivo *(100–106)*. Accordingly, HLA Class I-restricted cytolytic T-cells against survivin peptides are found in cancer patients *(103,105)*, and exert strong antitumor activity when tested in preclinical models *(107–109)*. Similar promising results were obtained with a survivin-directed DNA cancer vaccine that was shown to exert anticancer activity in tumor models in vivo by both affecting the tumor cell population as well as survivin-expressing angiogenic endothelial cells *(110)*. Several phase I trials using targeted cancer immunotherapy with survivin are currently ongoing. Available published data are encouraging *(111–113)*, and demonstrate that cancer immunotherapy with survivin generates antigen-specific immunological response in a high proportion of treated patients and, importantly, this occurs without appreciable toxicity for normal tissues and organs. Altogether, these observations warrant further expansion of survivin cancer immunotherapy approaches in larger phase II trials.

Secondly, molecular antagonists of survivin, including antisense, ribozymes, siRNA, or dominant negative mutants have shown reproducible antitumor efficacy in vitro and in a variety of tumor models, in vivo. These have been consistently associated with enhanced tumor cell apoptosis, suppression of cell proliferation, and collapse of tumor-associated angiogenesis *(97,114–119)*. With respect to advanced clinical development, a phase I trial using survivin antisense oligonucleotides has recently been completed.

Thirdly, molecular *(74)* or pharmacological *(20)* antagonists of survivin phosphorylation on Thr34 have also shown promising preclinical results in suppression of tumor growth in mice, and enhancement of taxane-based chemotherapy *(20)*. In particular, sequential inhibition of p34cdc2 kinase activity by small molecule antagonists following mitotic arrest imposed by taxanes resulted in loss of survivin levels, massive induction of apoptosis, and inhibition of tumor growth in mouse models *(20,120)*.

A novel aspect of survivin cytoprotection in cancer has recently emerged with its link to the molecular chaperone Hsp90 *(121)*. Heat shock proteins (Hsps), in particular Hsp90 and Hsp70 are molecular chaperones that survey protein folding quality control, and play an essential role in the cellular stress response in cancer *(122)*, thus enabling transformed cells to constantly adapt to unfavorable environmental challenges *(123)*. Binding to Hsp90 preserves survivin levels in cells and disruption of this interaction results in proteasomal degradation of survivin and a dual phenotype of spontaneous apoptosis and mitotic defects *(121)*. Although at a much earlier stage of development compared to other approaches, selective disruption of the survivin-Hsp90 interaction could be envisioned as an alternative strategy to destabilize survivin levels in tumor cells, leading to suppression of cell proliferation and initiation of mitochondrial apoptosis *(121)*.

5. Conclusions

Despite its relatively recent discovery in 1997, survivin has attracted considerable attention from several viewpoints of biomedical sciences. In particular, its dramatic overexpression in virtually every human cancer coupled with its dual function in cell proliferation and cell survival and its clinical link to unfavorable disease outcome have made survivin an attractive target for rational cancer therapy. Importantly, survivin appears essential for tumor-cell maintenance and molecular or pharmacologic interference with the survivin pathway have been consistently associated with tumor cell death and strong anticancer activity, either alone or in combination with standard cytotoxics in several preclinical models. Importantly, no evidence of tumor resistance for survivin-based therapeutic strategies has been reported. Combined with the differential expression of this pathway in tumors as opposed to normal tissues, this inspires confidence that survivin-directed therapeutics may provide a novel class of targeted agents for rational cancer treatment in humans.

Acknowledgments

This work was supported by NIH grants HL54131, CA78810 and CA90917.

References

1. Kerr JF, Wyllie AH, Currie AR. Apoptosis: a basic biological phenomenon with wide-ranging implications in tissue kinetics. Br J Cancer 1972;26:239–257.
2. Meier P, Finch A, Evan GI. Apoptosis in development. Nature 2000;407:796–801.
3. Hengartner MO. The biochemistry of apoptosis. Nature 2000;407:770–776.
4. Hanahan D, Weinberg RA. The hallmarks of cancer. Cell 2000;100:57–70.
5. Cory S, Adams JM. The Bcl2 family: regulators of the cellular life-or-death switch. Nat Rev Cancer 2002;2:647–656.
6. Salvesen GS, Duckett CS. Apoptosis: IAP proteins: blocking the road to death's door. Nat Rev Mol Cell Biol 2002;3:401–410.
7. Yoo SJ, Huh JR, Muro I, et al. Hid, Rpr and Grim negatively regulate DIAP1 levels through distinct mechanisms. Nat Cell Biol 2002;4:416–424.
8. Ambrosini G, Adida C, Altieri DC. A novel anti-apoptosis gene, survivin, expressed in cancer and lymphoma. Nat Med 1997;3:917–921.
9. Verdecia MA, Huang H, Dutil E, Kaiser DA, Hunter T, Noel JP. Structure of the human anti-apoptotic protein survivin reveals a dimeric arrangement. Nat Struct Biol 2000;7: 602–608.
10. Chantalat L, Skoufias D, Kleman JP, Jung B, Dideberg O, Margolis RL. Crystal structure of human survivin reveals a bow tie-shaped dimer with two unusual alpha-helical extensions. Mol Cell 2000;6:183–189.
11. Muchmore SW, Chen J, Jakob C, et al. Crystal structure and mutagenic analysis of the inhibitor-of-apoptosis protein survivin. Mol Cell 2000;6:173–182.
12. Li F, Ambrosini G, Chu EY, et al. Control of apoptosis and mitotic spindle checkpoint by survivin. Nature 1998;396:580–584.
13. Li F, Altieri DC. The cancer antiapoptosis mouse survivin gene: characterization of locus and transcriptional requirements of basal and cell cycle-dependent expression. Cancer Res 1999;59:3143–3151.
14. Mahotka C, Wenzel M, Springer E, Gabbert HE, Gerharz CD. Survivin-deltaEx3 and survivin-2B: two novel splice variants of the apoptosis inhibitor survivin with different anti-apoptotic properties. Cancer Res 1999;59:6097–6102.
15. Caldas H, Honsey LE, Altura RA. Survivin 2alpha: a novel Survivin splice variant expressed in human malignancies. Mol Cancer 2005;4:11.
16. Mahotka C, Liebmann J, Wenzel M, et al. Differential subcellular localization of functionally divergent survivin splice variants. Cell Death Differ 2002;9:1334–1342.

17. Kobayashi K, Hatano M, Otaki M, Ogasawara T, Tokuhisa T. Expression of a murine homologue of the inhibitor of apoptosis protein is related to cell proliferation. Proc Natl Acad Sci USA 1999;96:1457–1462.
18. Li F, Altieri DC. Transcriptional analysis of human survivin gene expression. Biochem. J. 344 Pt 1999;2:305–311.
19. Zhao J, Tenev T, Martins LM, Downward J, Lemoine NR. The ubiquitin-proteasome pathway regulates survivin degradation in a cell cycle-dependent manner. J. Cell Sci. 113 Pt 2000; 23:4363–4371.
20. O'Connor DS, Wall NR, Porter AC, Altieri DC. A p34(cdc2) survival checkpoint in cancer. Cancer Cell 2002;2:43–54.
21. Fukuda S, Foster RG, Porter SB, Pelus LM. The antiapoptosis protein survivin is associated with cell cycle entry of normal cord blood CD34(+) cells and modulates cell cycle and proliferation of mouse hematopoietic progenitor cells. Blood 2002;100:2463–2471.
22. Papapetropoulos A, Fulton D, Mahboubi K, et al. Angiopoietin-1 inhibits endothelial cell apoptosis via the Akt/survivin pathway. J Biol Chem 2000;275:9102–9105.
23. Harfouche R, Hassessian HM, Guo Y, et al. Mechanisms which mediate the antiapoptotic effects of angiopoietin-1 on endothelial cells. Microvasc Res 2002;64:135–147.
24. Datta SR, Brunet A, Greenberg ME. Cellular survival: a play in three Akts. Genes Dev 1999;13:2905–2927.
25. Carter BZ, Milella M, Altieri DC, Andreeff M. Cytokine-regulated expression of survivin in myeloid leukemia. Blood 2001;97:2784–2790.
26. Fornaro M, Plescia J, Chheang S, et al. Fibronectin Protects Prostate Cancer Cells from Tumor Necrosis Factor-a-induced Apoptosis via the AKT/Survivin Pathway. J Biol Chem 2003;278:50,402–50,411.
27. Mahboubi K, Li F, Plescia J, et al. Interleukin-11 up-regulates survivin expression in endothelial cells through a signal transducer and activator of transcription-3 pathway. Lab Invest 2001;81:327–334.
28. Iurlaro M, Demontis F, Corada M, et al. VE-Cadherin Expression and Clustering Maintain Low Levels of Survivin in Endothelial Cells. Am J Pathol 2004;165:181–189.
29. Ohashi H, Takagi H, Oh H, et al. Phosphatidylinositol 3-Kinase/Akt Regulates Angiotensin II-Induced Inhibition of Apoptosis in Microvascular Endothelial Cells by Governing Survivin Expression and Suppression of Caspase-3 Activity. Circ Res 2004;94:785–793.
30. Aoki Y, Feldman GM, Tosato G. Inhibition of STAT3 signaling induces apoptosis and decreases survivin expression in primary effusion lymphoma. Blood 2003;101:1535–1542.
31. Shen Y, Devgan G, Darnell JE, Jr, Bromberg JF. Constitutively activated Stat3 protects fibroblasts from serum withdrawal and UV-induced apoptosis and antagonizes the proapoptotic effects of activated Stat1. Proc Natl Acad Sci USA 2001;98:1543–1548.
32. Turkson J, Jove R. STAT proteins: novel molecular targets for cancer drug discovery. Oncogene 2000;19:6613–6626.
33. Li F, Ackermann EJ, Bennett CF, Rothermel AL, et al. Pleiotropic cell-division defects and apoptosis induced by interference with survivin function. Nat Cell Biol 1999;1:461–466.
34. Jiang X, Wilford C, Duensing S, Munger K, Jones G, Jones D. Participation of Survivin in mitotic and apoptotic activities of normal and tumor-derived cells. J Cell Biochem 2001;83:342–354.
35. Skoufias DA, Mollinari C, Lacroix FB, Margolis RL. Human survivin is a kinetochore-associated passenger protein. J Cell Biol 2000;151:1575–1582.
36. Fortugno P, Wall NR, Giodini A, et al. Survivin exists in immunochemically distinct subcellular pools and is involved in spindle microtubule function. J Cell Sci 2002;115: 575–585.

37. Wheatley SP, Carvalho A, Vagnarelli P, Earnshaw WC. INCENP is required for proper targeting of Survivin to the centromeres and the anaphase spindle during mitosis. Curr Biol 2001;11:886–890.
38. Rodriguez JA, Span SW, Ferreira CG, Kruyt FA, Giaccone G. CRM1-mediated nuclear export determines the cytoplasmic localization of the antiapoptotic protein Survivin. Exp Cell Res 2002;275:44–53.
39. Wheatley SP, Henzing AJ, Dodson H, Khaled W, Earnshaw WC. Aurora-B Phosphorylation in Vitro Identifies a Residue of Survivin That Is Essential for Its Localization and Binding to Inner Centromere Protein (INCENP) in Vivo. J Biol Chem 2004;279:5655–5660.
40. Uren AG, Beilharz T, O'Connell MJ, et al. Role for yeast inhibitor of apoptosis (IAP)-like proteins in cell division. Proc Natl Acad Sci USA 1999;96:10,170–10,175.
41. Fraser AG, James C, Evan GI, Hengartner MO. Caenorhabditis elegans inhibitor of apoptosis protein (IAP) homologue BIR-1 plays a conserved role in cytokinesis. Curr Biol 1999;9:292–301.
42. Uren AG, Wong L, Pakusch M, et al. Survivin and the inner centromere protein INCENP show similar cell- cycle localization and gene knockout phenotype. Curr Biol 2000;10: 1319–1328.
43. Conway EM, Pollefeyt S, Steiner-Mosonyi M, et al. Deficiency of survivin in transgenic mice exacerbates Fas-induced apoptosis via mitochondrial pathways. Gastroenterology 2002;123:619–631.
44. Adams RR, Carmena M, Earnshaw WC. Chromosomal passengers and the (aurora) ABCs of mitosis. Trends Cell Biol 2001;11:49–54.
45. Speliotes EK, Uren A, Vaux D, Horvitz HR. The survivin-like C. elegans BIR-1 protein acts with the Aurora-like kinase AIR-2 to affect chromosomes and the spindle midzone. Mol. Cell 2000;6:211–223.
46. Chen J, Jin S, Tahir SK, et al. Survivin enhances Aurora-B kinase activity and localizes Aurora-B in human cells. J Biol Chem 2003;278:486–490.
47. Bolton MA, Lan W, Powers SE, et al. Aurora B Kinase Exists in a Complex with Survivin and INCENP and Its Kinase Activity Is Stimulated by Survivin Binding and Phosphorylation. Mol Biol Cell 2002;13:3064–3077.
48. Kallio MJ, Nieminen M, Eriksson JE. Human inhibitor of apoptosis protein (IAP) survivin participates in regulation of chromosome segregation and mitotic exit. Faseb J 2001;15:2721–2723.
49. Giodini A, Kallio M, Wall NR, et al. Regulation of microtubule stability and mitotic progression by survivin. Cancer Res. 2002;62:2462–2467.
50. Tran J, Master Z, Yu JL, Rak J, Dumont DJ, and Kerbel RS. A role for survivin in chemoresistance of endothelial cells mediated by VEGF. Proc Natl Acad Sci USA 2002;99: 4349–4354.
51. Yang D, Welm A, Bishop JM. Cell division and cell survival in the absence of survivin. Proc Natl Acad Sci USA 2004;101:15,100–15,105.
52. Lens SM, Wolthuis RM, Klompmaker R, et al. Survivin is required for a sustained spindle checkpoint arrest in response to lack of tension. Embo J 2003;22:2934–2947.
53. Carvalho A, Carmena M, Sambade C, Earnshaw WC, Wheatley SP. Survivin is required for stable checkpoint activation in taxol-treated HeLa cells. J Cell Sci 2003;116:2987–2998.
54. Beltrami E, Plescia J, Wilkinson JC, Duckett CS, Altieri DC. Acute ablation of survivin uncovers p53-dependent mitotic checkpoint functions and control of mitochondrial apoptosis. J Biol Chem 2004;279:2077–2084.
55. Gassmann R, Carvalho A, Henzing AJ, et al. Borealin: a novel chromosomal passenger required for stability of the bipolar mitotic spindle. J Cell Biol 2004;166:179–191.

56. Sampath SC, Ohi R, Leismann O, Salic A, Pozniakovski A, Funabiki H. The chromosomal passenger complex is required for chromatin-induced microtubule stabilization and spindle assembly. Cell 2004;118:187–202.
57. Altieri DC. Survivin in apoptosis control and cell cycle regulation in cancer. Prog Cell Cycle Res 2003;5:447–452.
58. Castedo M, Perfettini JL, Roumier T, Andreau K, Medema R, Kroemer G. Cell death by mitotic catastrophe: a molecular definition. Oncogene 2004;23:2825–2837.
59. Li F. Role of survivin and its splice variants in tumorigenesis. Br J Cancer 2005;92: 212–216.
60. Okada H, Mak TW. Pathways of apoptotic and non-apoptotic death in tumour cells. Nat Rev Cancer 2004;4:592–603.
61. Grossman D, Kim PJ, Blanc-Brude OP, et al. Transgenic expression of survivin in keratinocytes counteracts UVB-induced apoptosis and cooperates with loss of p53. J Clin Invest 2001;108:991–999.
62. Jones G, Jones D, Zhou L, Steller H, Chu Y. Deterin, a new inhibitor of apoptosis from Drosophila melanogaster. J. Biol. Chem. 2000;275:22,157–22,165.
63. Tamm I, Wang Y, Sausville E, et al. IAP-family protein survivin inhibits caspase activity and apoptosis induced by Fas (CD95), Bax, caspases, and anticancer drugs. Cancer Res 1998;58:5315–5320.
64. Kasof GM, Gomes BC. Livin, a novel inhibitor-of-apoptosis (IAP) family member. J. Biol. Chem. 2000;276:3238–3246.
65. Conway EM, Pollefeyt S, Cornelissen J, et al. Three differentially expressed survivin cDNA variants encode proteins with distinct antiapoptotic functions. Blood 2000;95: 1435–1442.
66. Shin S, Sung, BJ, Cho YS, et al. An anti-apoptotic protein human survivin is a direct inhibitor of caspase-3 and -7. Biochemistry 2001;40:1117–1123.
67. Banks DP, Plescia J, Altieri DC, et al. Survivin does not inhibit caspase-3 activity. Blood 2000;96:4002–4003.
68. Shi Y. Mechanisms of caspase activation and inhibition during apoptosis. Mol Cell 2002; 9:459–470.
69. O'Connor DS, Grossman D, Plescia J, et al. Regulation of apoptosis at cell division by $p34^{cdc2}$ phosphorylation of survivin. Proc. Natl. Acad. Sci. USA 2000;97:13,103–13,107.
70. McNeish IA, Lopes R, Bell SJ, et al. Survivin interacts with Smac/DIABLO in ovarian carcinoma cells but is redundant in Smac-mediated apoptosis. Exp Cell Res 2005;302:69–82.
71. Song Z, Liu S, He H, et al. A Single Amino Acid Change (Asp 53->Ala53) Converts Survivin from anti-apoptotic to pro-apoptotic. Mol Biol Cell 2004;15:1287–1296.
72. Song Z, Yao X, Wu M. Direct interaction between survivin and Smac/DIABLO is essential for the anti-apoptotic activity of survivin during taxol-induced apoptosis. J Biol Chem 2003;278:23,130–23,140.
73. Sun C, Nettesheim D, Liu Z, Olejniczak ET. Solution structure of human survivin and its binding interface with Smac/Diablo. Biochemistry 2005;44:11–17.
74. Mesri M, Wall NR, Li J, Kim RW, Altieri DC. Cancer gene therapy using a survivin mutant adenovirus. J Clin Invest 2001;108:981–990.
75. Wang X. The expanding role of mitochondria in apoptosis. Genes Dev 2001;15: 2922–2933.
76. Du C, Fang M, Li Y, Li L, Wang X. Smac, a mitochondrial protein that promotes cytochrome c-dependent caspase activation by eliminating IAP inhibition. Cell 2000;102:33–42.
77. Dohi T, Beltrami E, Wall NR, Plescia J, Altieri DC. Mitochondrial survivin inhibits apoptosis and promotes tumorigenesis. J Clin Invest 2004;114:1117–1127.

78. Dohi T, Altieri DC. Mitochondrial dynamics of survivin and "four dimensional" control of tumor cell apoptosis. Cell Cycle 2005;4:21–23.
79. Marusawa H, Matsuzawa S, Welsh K, et al. HBXIP functions as a cofactor of survivin in apoptosis suppression. Embo J 2003;22:2729–2740.
80. Dohi T, Okada K, Xia F, et al. An IAP-IAP complex inhibits apoptosis. J Biol Chem 2004;279:34,087–34,090.
81. Wang HW, Sharp TV, Koumi A, Koentges G, and Boshoff C. Characterization of an anti-apoptotic glycoprotein encoded by Kaposi's sarcoma-associated herpesvirus which resembles a spliced variant of human survivin. Embo J 2002;21:2602–2615.
82. Adida C, Crotty PL, McGrath J, Berrebi D, Diebold J, Altieri DC. Developmentally regulated expression of the novel cancer anti-apoptosis gene survivin in human and mouse differentiation. Am J Pathol 1998;152:43–49.
83. Altieri DC. Validating survivin as a cancer therapeutic target. Nat Rev Cancer 2003;3: 46–54.
84. Bao R, Connolly DC, Murphy M, et al. Activation of cancer-specific gene expression by the survivin promoter. J Natl Cancer Inst 2002;94:522–528.
85. O'Connor DS, Schechner JS, Adida C, et al. Control of apoptosis during angiogenesis by survivin expression in endothelial cells. Am J Pathol 2000;156:393–398.
86. Tran J, Rak J, Sheehan C, et al. Marked induction of the IAP family antiapoptotic proteins survivin and XIAP by VEGF in vascular endothelial cells. Biochem. Biophys. Res Commun 1999;264:781–788.
87. Dan HC, Jiang K, Coppola D, et al. Phosphatidylinositol-3-OH kinase/AKT and survivin pathways as critical targets for geranylgeranyltransferase I inhibitor-induced apoptosis. Oncogene 2004;23:706–715.
88. Sommer KW, Schamberger CJ, Schmidt GE, Sasgary S, Cerni C. Inhibitor of apoptosis protein (IAP) survivin is upregulated by oncogenic c-H-Ras. Oncogene 2003;22: 4266–4280.
89. Hoffman WH, Biade S, Zilfou JT, Chen J, Murphy M. Transcriptional repression of the anti-apoptotic survivin gene by wild type p53. J Biol Chem 2002;277:3247–3257.
90. Mirza A, McGuirk M, Hockenberry TN, et al. Human survivin is negatively regulated by wild-type p53 and participates in p53-dependent apoptotic pathway. Oncogene 2002;21:2613–2622.
91. Zhang T, Otevrel T, Gao Z, Ehrlich SM, Fields JZ, Boman BM. Evidence that APC regulates survivin expression: a possible mechanism contributing to the stem cell origin of colon cancer. Cancer Res 2001;61:8664–8667.
92. Kim PJ, Plescia J, Clevers H, Fearon ER, Altieri DC. Survivin and molecular pathogenesis of colorectal cancer. Lancet 2003;362:205–209.
93. Xu ZX, Zhao RX, Ding T, et al. Promyelocytic leukemia protein 4 induces apoptosis by inhibition of survivin expression. J Biol Chem 2004;279:1838–1844.
94. van't Veer LJ, Dai H, van de Vijver MJ, et al. Gene expression profiling predicts clinical outcome of breast cancer. Nature 2002;415:530–536.
95. Kuttler F, Valnet-Rabier MB, Angonin R, et al. Relationship between expression of genes involved in cell cycle control and apoptosis in diffuse large B cell lymphoma: a preferential survivin-cyclin B link. Leukemia 2002;16:726–735.
96. van de Wetering M, Sancho E, Vervweij C, et al. The b-catenin/TCF-4 complex imposes a crypt progenitor phenotype on colorectal cancer cells. Cell 2002;111:241–250.
97. Williams NS, Gaynor RB, Scoggin S, et al. Identification and validation of genes involved in the pathogenesis of colorectal cancer using cDNA microarrays and RNA interference. Clin Cancer Res 2003;9:931–946.

98. Paik S, Shak S, Tang G. et al. A multigene assay to predict recurrence of tamoxifen-treated, node-negative breast cancer. N Engl J Med 2004;351:2817–2826.
99. Andersen MH, Thor SP. Survivin—a universal tumor antigen. Histol Histopathol 2002;17: 669–675.
100. Rohayem J, Diestelkoetter P, Weigle B, et al. Antibody response to the tumor-associated inhibitor of apoptosis protein survivin in cancer patients. Cancer Res 2000;60: 1815–1817.
101. Yagihashi A, Asanuma K, Nakamura M, et al. Detection of anti-survivin antibody in gastrointestinal cancer patients. Clin Chem 2001;47:1729–1731.
102. Schmitz M, Diestelkoetter P, Weigle B, et al. Generation of survivin-specific CD8+ T effector cells by dendritic cells pulsed with protein or selected peptides. Cancer Res 2000;60: 4845–4849.
103. Andersen MH, Pedersen LO, Becker JC, Straten PT. Identification of a cytotoxic T lymphocyte response to the apoptosis inhibitor protein survivin in cancer patients. Cancer Res 2001;61:869–872.
104. Hirohashi Y, Torigoe T, Maeda A, et al. An HLA-A24-restricted Cytotoxic T Lymphocyte Epitope of a Tumor-associated Protein, Survivin. Clin Cancer Res 2002;8:1731–1739.
105. Schmidt SM, Schag K, Muller MR, et al. Survivin is a shared tumor-associated antigen expressed in a broad variety of malignancies and recognized by specific cytotoxic T cells. Blood 2003;102:571–576.
106. Idenoue S, Hirohashi Y, Torigoe T, et al. A potent immunogenic general cancer vaccine that targets survivin, an inhibitor of apoptosis proteins. Clin Cancer Res 2005;11:1474–1482.
107. Casati C, Dalerba P, Rivoltini L, et al. The apoptosis inhibitor protein survivin induces tumor-specific CD8+ and CD4+ T cells in colorectal cancer patients. Cancer Res 2003;63: 4507–4515.
108. Pisarev V, Yu B, Salup R, Sherman S, Altieri DC, Gabrilovich DI. Full-Length Dominant-Negative Survivin for Cancer Immunotherapy. Clin Cancer Res 2003;9:6523–6533.
109. Siegel S, Wagner A, Schmitz N, Zeis M. Induction of antitumour immunity using survivin peptide-pulsed dendritic cells in a murine lymphoma model. Br J Haematol 2003;122: 911–914.
110. Xiang R, Mizutani N, Luo Y., Chiodoni C, et al. A DNA vaccine targeting survivin combines apoptosis with suppression of angiogenesis in lung tumor eradication. Cancer Res 2005;65:553–561.
111. Hirschowitz EA, Foody T, Kryscio R, Dickson L, Sturgill J, Yannelli J. Autologous dendritic cell vaccines for non-small-cell lung cancer. J Clin Oncol 2004;22:2808–2815.
112. Otto K, Andersen MH, Eggert A, et al. Lack of toxicity of therapy-induced T cell responses against the universal tumour antigen survivin. Vaccine 2005;23:884–889.
113. Tsuruma T, Hata F, Torigoe T, et al. Phase I clinical study of anti-apoptosis protein, survivin-derived peptide vaccine therapy for patients with advanced or recurrent colorectal cancer. J Transl Med 2004;2:19.
114. Kanwar JR, Shen WP, Kanwar RK, Berg RW, Krissansen GW. Effects of survivin antagonists on growth of established tumors and b7-1 immunogene therapy. J Natl Cancer Inst 2001; 93:1541–1552.
115. Grossman D, Kim PJ, Schechner JS, Altieri DC. Inhibition of melanoma tumor growth in vivo by survivin targeting. Proc. Natl. Acad. Sci. USA 2001;98:635–640.
116. Yamamoto T, Manome Y, Nakamura M, Tanigawa N. Downregulation of survivin expression by induction of the effector cell protease receptor-1 reduces tumor growth potential and results in an increased sensitivity to anticancer agents in human colon cancer. Eur J Cancer 2002;38:2316–2324.

117. Blanc-Brude OP, Mesri M, Wall NR, Plescia J, Dohi T, Altieri DC. Therapeutic targeting of the survivin pathway in cancer: initiation of mitochondrial apoptosis and suppression of tumor-associated angiogenesis. Clin Cancer Res 2003;9:2683–2692.
118. Pennati M, Binda M, Colella G, et al. Ribozyme-mediated inhibition of survivin expression increases spontaneous and drug-induced apoptosis and decreases the tumorigenic potential of human prostate cancer cells. Oncogene 2004;23:386–394.
119. Ansell SM, Arendt BK, Grote DM, et al. Inhibition of survivin expression suppresses the growth of aggressive non-Hodgkin's lymphoma. Leukemia 2004;18:616–623.
120. Wall NR, O'Connor DS, Plescia J, Pommier Y, Altieri DC. Suppression of survivin phosphorylation on Thr34 by flavopiridol enhances tumor cell apoptosis. Cancer Res 62003;3:230–235.
121. Fortugno P, Beltrami E, Plescia J, et al. Regulation of survivin function by Hsp90. Proc Natl Acad Sci USA 2003;100:13,791–13,796.
122. Nollen EAA, Morimoto RI. Chaperoning signaling pathways: molecular chaperones as stress-sensing "heat shock" proteins. J Cell Sci 2002;115:2809–2816.
123. Neckers L. Hsp90 inhibitors as novel cancer chemotherapeutic agents. Trends Mol Med 2002;8:S55–S61.

15

Clinical Significance of Histone Deacetylase Inhibitors in Cancer

Sharmila Shankar and Rakesh K. Srivastava

Summary

Chromatin remodeling agents modulate gene expression in tumor cells. Acetylation and deacetylation are catalyzed by specific enzyme families, histone acetyltransferases (HATs) and deacetylases (HDACs), respectively. Aberrant acetylation of histone and nonhistone proteins has been linked to malignant diseases. HDAC inhibitors bear great potential as new drugs because of their ability to modulate transcription, induce differentiation and apoptosis, and inhibit angiogenesis. Furthermore, HDAC inhibitors also enhance the activity of other cancer therapeutics such as tumor necrosis factor-related apoptosis-inducing ligand (TRAIL), chemotherapeutic drugs, and radiotherapy. Some of the HDAC inhibitors are currently under clinical investigations. This chapter reviews the chemistry and the biology of HDACs and HDAC inhibitors, laying particular emphasis on those agents which have potentials for cancer therapy.

Key Words: HDAC inhibitors; SAHA; MS-275; TSA; TRAIL; apoptosis; caspase.

1. Introduction

Epigenic modifications, mainly DNA methylation and acetylation, are recognized as additional mechanisms contributing to the malignant phenotype *(1,2)*. Acetylation and deacetylation of histones plays a role in the regulation of gene expression *(3)*. Histone acetylation is a reversible process whereby histone acetyltransferase (HAT) transfers the acetyl moiety from acetyl coenzyme A to the lysine, and histone deacetylase (HDAC) removes the acetyl groups, re-establishing the positive charge in the histones. HATs and HDACs have recently been shown to regulate cell proliferation, differentiation, and apoptosis in various hematological and solid malignancies *(4,5)*. Altered HAT or HDAC activity is associated with cancer by changing the expression pattern of selected genes *(6,7)*. Hyperacetylation of histones H3 and H4 correlates with gene activation, whereas deacetylation mediates eukaryotic chromatin condensation and gene expression silencing *(8,9)*. Recently, new roles of histone acetylation have been uncovered, not only in transcription but also in DNA replication, repair, and heterochromatin formation *(10)*.

2. Histone Deacetylases

There are at least 12 human HDAC enzymes, HDAC1-11 *(11–15)* and HDAC-A *(16)*. Based on the structural properties, HDACs can be divided into three classes *(17)*. Class I members (HDAC 1, 2, 3, 8, and 11) are transcriptional corepressors homologous to yeast RPD3 and have a single deacetylase domain at the N termini and diversified C-terminal regions *(18)*. Class II members (HDAC 4, 5, 6, 7, 9, and 10) have domains

From: *Apoptosis, Cell Signaling, and Human Diseases: Molecular Mechanisms, Volume 1*
Edited by R. Srivastava © Humana Press Inc., Totowa, NJ

similar to yeast HDA1 with a deacetylase domain at a more C-terminal position *(19)*. In addition, HDAC 6 contains a second N-terminal deacetylase domain, which can function independently of its C-terminal counterpart. Class III histone deacetylases are distinct from class I and II HDACs and are homologous of the yeast silent information regulator 2 (Sir2). All of these HDACs apparently exist in the cell as subunits of multiprotein complexes. Class II HDACs have been shown to translocate from the cytoplasm to the nucleus in response to external stimuli, whereas class I HDACs are constitutively nuclear and play important roles in dynamic gene regulation *(20)*.

Sir2, a heterochromatin component in yeast, silences transcription at silent mating loci, telomerese, and ribosomal DNA (rDNA), and this also suppresses recombination in rDNA. Earlier experiments have shown that the overexpression of Sir2 in yeast induced the global deacetylation of histones, indicating that Sir2 was a histone deacetylase *(21)*. Later, it was shown that *cobB*, a bacterial homolog of Sir2, had ribosyltransferase activity, leading to experiments showing that Sir2 was also able to transfer ADP-ribose from NAD *(22)*. Subsequently, it was confirmed that Sir2 was an NAD-dependent histone deacetylase *(23)*. The ADP-ribosylation of an acetylated lysine residue is an intermediate state of the enzymatic reaction catalyzed by Sir2. Only class III enzymes use NAD as a cofactor. Thus, these enzymes are known as NAD-dependent histone deacetylases.

Recently, Sir2 has attracted much attention, because it is related to longevity *(24)*. The overexpression of Sir2 extends the lifespan of budding yeast, whereas its knockout shortens the lifespan by about 50% *(25)*. Sir2 (also known as sirtuins) is conserved from bacteria to humans. In the nematodes, the gene most homologs to yeast Sir2 gene is sir-2.1. A duplication containing the *sir-2.1* gene confers a lifespan that is extended by up to 50% *(26)*. The mammalian homologs consist of seven members, SIRT1–SIRT7. In mammalian cells, SIRT1 down-regulates stress-induced p53 and FOXO pathways for apoptosis, thus favoring survival under stress. In the absence of applied stress, SIRT1 silencing induces growth arrest and/or apoptosis in human epithelial cancer cells *(27)*. In contrast, normal human epithelial cells and normal human diploid fibroblasts seem to be refractory to SIRT1 silencing. Further studies have revealed that the SIRT1-regulated pathway is independent of p53, Bax, and caspase-2. Alternatively, SIRT1 may suppress apoptosis downstream from these apoptotic factors. FOXO4 (but not FOXO3) is required as proapoptotic mediator. Caspase-3 and caspase-7 act as downstream executioners of SIRT1/FOXO4-regulated apoptosis. These data suggest that SIRT1 as a novel target for selective killing of cancer vs noncancer epithelial cells. Upregulation of SIRT1 may be a double-edged sword that both promotes survival of aging cells and increases cancer risk in mammals.

Histones are part of the core proteins of nucleosomes. The recruitment of HATs and HDACs plays an important role in proliferation, differentiation and apoptosis *(4,5)*. Altered HAT or HDAC activity is associated with the development of cancer by changing the expression of several genes *(6,7)*. Hyperacetylation of the N-terminal tails of histones H3 and H4 correlates with gene activation, whereas deacetylation mediates transcriptional repression *(8)*. Treatment of malignant cells with HDAC inhibitors regulates only a small number (1–2%) of genes, as examined by DNA microarray studies *(28)*. HDAC1 interacts directly with other transcription repressors, including all three of the pocket proteins, Rb, p107, p130, and YY1. HDAC1 causes transcription repression by locally deacetylating histones, leading to a compact nucleosomal structure that prevents transcription factors from accessing DNA to promote transcription. Furthermore, HDAC1 knockout mice were embryonic lethal, possibly resulting from a proliferative defect upon unrestricted expressions of the cell cycle inhibitors $p21^{WAF1/CIP1}$ and $p27^{KIP1}$ *(29)*. Overexpression of histone deacetylase I confers resistance to sodium butyrate-mediated apoptosis in melanoma cells through a p53-mediated pathway *(30)*. We and others have shown that inhibition of HDAC activity induces apoptosis in various types of cancer *(31–35)*.

Stability of HDACs is an important factor in determining the biological activity. HDAC4 is unusually unstable, with a half-life of less than 8 h *(36)*. Consistent with the instability of HDAC4 protein, its mRNA was also highly unstable (with a half-life of less than 4 h). The exposure of cells to ultraviolet irradiation resulted in the degradation of HDAC4. This degradation was not dependent on proteasome or CRM1-mediated export activity but instead was caspase-dependent and was detectable in diverse human cancer lines. Of two potential caspase consensus motifs in HDAC4, both lying within a region containing proline, glutamic acid-, serine-, and threonine-rich (PEST) sequences, Asp-289 as the prime cleavage site was identified by site-directed mutagenesis *(36)*. Notably, this residue is not conserved among other class IIa members, HDAC5, -7, and -9. Finally, the induced expression of caspase-cleavable HDAC4 led to markedly increased apoptosis. These results therefore link the regulation of HDAC4 protein stability to caspases, enzymes that are important for controlling cell death and differentiation.

HDACs are important regulators of gene expression as part of transcriptional corepressor complexes. Caspases can repress the activity of the myocyte enhancer factor (MEF)2C transcription factor by regulating HDAC4 processing *(37)*. Cleavage of HDAC4 occurs at Asp 289 and disjoins the carboxy-terminal fragment, localized into the cytoplasm, from the amino-terminal fragment, which accumulates into the nucleus. In the nucleus, the caspase-generated fragment of HDAC4 is able to trigger cytochrome *c* release from mitochondria and cell death in a caspase-9-dependent manner. The caspase-cleaved amino-terminal fragment of HDAC4 acts as a strong repressor of the transcription factor MEF2C, independently from the HDAC domain. Removal of amino acids 166–289 from the caspase-cleaved fragment of HDAC4 abrogates its ability to repress MEF2 transcription and to induce cell death.

3. Histone Deacetylase Inhibitors

In transcriptionally active chromatin, histones generally are hyperacetylated and, conversely, hypoacetylated histones are coincident with silenced chromatin. Revived interest in these enzymatic pathways and how they modulate eukaryotic transcription has led to the identification of multiple cofactors whose complex interplay with HDAC affects gene expression. Concurrent with these discoveries, screening of natural product sources yielded new small molecules that were subsequently identified as potent inhibitors of HDAC. Whereas predominantly identified using antiproliferative assays, the biological activity of these new HDAC inhibitors also encompasses significant antiprotozoal, antifungal, phytotoxic, and antiviral applications. During the past decade, a number of HDAC inhibitors have been shown to induce growth arrest, differentiation, and/or apoptosis in caner cells *(31,33,38–40)*, and inhibit tumor growth in various xenograft models *(41–48)*. Recently, attention has been focused on the ability of HDAC inhibitors to induce perturbations in cell cycle regulatory proteins (e.g., p21[/WAF1/CIP1]), down regulation of survival signaling pathways (e.g., Raf/MAPkinase/ERK), and disruption of cellular redox state (e.g., reactive oxygen species [ROS]). Therefore, HDAC inhibitors are considered as candidate drugs in cancer therapy *(49–51)*. Five classes of HDAC inhibitors have been characterized and include benzamides (i.e., MS-275); hydroxamic acids (i.e., suberolanilide hydroxamic acid and trichostatin A); short-chain fatty acids (i.e., sodium butyrate and phenylbutyrate); cyclic tetrapeptide containing a 2-amino-8-oxo-9, 10-epoxy-decanoyl moiety (i.e., trapoxin A); and cyclic peptides without the 2-amino-8-oxo-9, 10-epoxy-decanoyl moiety (i.e., FK228). A wide variety of HDAC inhibitors of both natural and synthetic origin has been reported. Except depsispeptide/FK228, natural HDACs (trichostatin [TSA]), depudecin, trapoxins, apicidins) as well as sodium butyrate, phenylbutyrate and suberoyl anilide hydroxamic acid (SAHA), while effective in vivo, are inefficient due to instability and low retention. Subsequently, synthetic analogs isolated from screening libraries (oxamflatin, scriptaid) were discovered as having a common structure with TSA and SAHA: a hydroxamic acid zinc-binding group linked via a spacer (5 or 6 CH2) to a hydrophobic group. Design of a second generation of HDACs was based upon these data affording potent HDACs such as LAQ824 and PDX101 currently under phase I clinical trials. Simultaneously, synthetic benzamide-containing HDAC inhibitors were reported and two of them, MS-275 and CI-994, have reached phase II and I clinical trials, respectively.

3.1. Benzamides

Several benzamides have been found to inhibit HDAC activity in the low micromolar range. A 2′-hydroxy or amino function seems to be essential for the optimum activity *(52)*. The 2′-amino compound MS-275 is the first HDAC inhibitor discovered with oral anticancer activity in an animal model. It appears to be nontoxic in several animal models. Pretreatment of human leukemic cells with MS-275 significantly enhances the abrogative capacity of an established nucleoside analogue, fludarabine *(53)*. The study indicates that apart from promoting acetylation of histones and regulation of genes involved in differentiation and apoptosis, MS-275 also induces multiple perturbations in signal transduction, survival, and cell-cycle regulatory pathways that increase the fludarabine-mediated cell death. MS-275, acetyldinaline, and CI-994 are in clinical trials for the treatment of cancers *(54,55)*.

3.2. Trichostatin A and Analogs

Trichostatin A (TSA) from *Streptomyces hygroscopicus* was initially identified as an antifungal agent *(56)*, but later discovered that it induces histone hyperacetylation and terminal differentiation *(57)*. It acts as a noncompetitive inhibitor of HDAC by mimicking the lysine substrate as well as chelating a zinc atom crucial for enzymatic activity. TSA inhibits proliferation, causes cell-cycle arrest, and induces differentiation and/or apoptosis in numerous models of leukemia, lymphoma, multiple myeloma, and solid tumors *(33,54,58–69)*. TSA is also effective in xenograft models *(70,71)*. TSA attenuated the development of allergic airway inflammation by decreasing expression of the Th2 cytokines, interleukin (IL)-4, and IL-5, and IgE, which resulted from reduced T-cell infiltration, suggesting that HDAC inhibition may attenuate the development of asthma by a T-cell suppressive effect *(72)*. Other analogs of TSA such as oxamflatin, scriptaid, and amide derivatives have been reported to have anticancer activity *(54,73–75)*. Scriptaid induces reticulocytosis and human gamma-globin synthesis *(76)*, suggesting a potential for scriptaid as a treatment option for sickle cell disease.

3.3. Cyclic Tetrapeptides

Hydrophobic cyclotetrapeptides contain common amino acid (S)-2-Amino-9,10-epoxy-8-xodecanoic acid (L-Aoe) and have been reported to inhibit HDACs *(77,78)*. The epoxyketone was first thought to be essential for activity, as reduction or nucleophoilic attack resulted in inactivation of compounds *(77,78)*. Trapoxin A, a microbially derived cyclotetrapeptide, is an irreversible inhibitor in the low nanomolar range *(77)*. Trapoxin inhibits histone deacetylation in vivo and causes mammalian cells to arrest in the cell cycle *(11)*. On the other hand, related HC toxin (host-selective toxin of *Cochliobolus carbonum*) inhibits maize enzyme activity reversibly *(78)*. K-trap (an analogous of trapoxin A) inhibited HDAC1 activity. A novel fungal metabolite, apicidin [cyclo(N-*O*-methyl-L-tryptophanyl-L-isoleucinyl-D-pipecolinyl-L-2-amino-8-oxo-decanoyl)], exhibits potent, broad spectrum antiprotozoal activity in vitro against Apicomplexan parasites *(79)*. Apicidin's antiparasitic activity appears to result from low nanomolar inhibition of HDAC, which induces hyperacetylation of histones in treated parasites. Because apicidin and apicidin A possess only a ketone functional group and are active in the low nanomolar concentrations, it appears that the presence of the epoxy group is not essential for activity. Apicidin induces differentiation, cell-cycle arrest, and apoptosis, and inhibits metastasis and angiogenesis in several cancer models *(38,80–87)*. A number of derivatives such as 9-acyloxyapicidins or 9-hydroxy have been prepared and are under investigation. Trapoxin analogs that combine cyclotetrapeptide and hydroxamic acid moieties have been prepared. The inhibitors of quinolone analogs and the hydroxamic acid analogs of apicidin yielded promising results *(88,89)*. Depudecin, a natural epoxide derivative isolated from the fungus *A. brassicicola*, induces hyperacetylation of histones and morphological reversion in v-ras transformed NIH 3T3 cells *(90)*.

3.4. Tetrapeptide Analogs

Simple analogs of cyclic tetrapeptides that contains suberic acid linkers and hydroxamate, instead of epoxyketone or ketone functional group, inhibit HDAC activity *(91)*. The structurally related hybrid polar compounds (HPC) was shown to induce differentiation in a wide variety of transformed cells *(92)*. The first representative was hexamethylene bisacetamide (HMBA) which induced differentiation of transformed cells in millimolar range *(93)*. HMBA regulates genes that control G1-to-S phase transition, leading to G1 arrest and inhibition of DNA synthesis. Among the inducer-mediated changes, suppression of cyclin-dependent kinase cdk4, which may be required for phosphorylation of the retinoblastoma protein pRB and perhaps p107, is critical in the pathway of terminal differentiation. HMBA induces an increase in the level of $p21^{WAF1/CIP1}$ which inhibits cyclin-dependent kinase activity and, in turn, may cause cells to arrest in G1. p107 complexes with transcription factor E2F, which may alter E2F-dependent gene transcription. HMBA has also been shown to induce differentiation of neoplastic cells in patients. Furthermore, a second generation of hybrid polar compounds have been synthesized which are up to 1000-fold more potent than HMBA on a molar basis as inducers of murine erythroleukemia (MEL) cells and other transformed cells in vitro. Second-generation HPCs such as suberoylanilide hydroxamic acid (SAHA), suberic bishydroxamic acid (SBHA), and m-carboxycinnamic acid bishydroxamide (CBHA) inhibited HDAC activity and induced cancer cell differentiation *(69,94)*. SAHA has been shown to act as a chemopreventive agent in mammary tumors in the rat *(95)* and inhibited the growth of established tumors *(41,95,96)*. Polyaminohydroxamic acids (PAHAs) represent an important new chemical class of HDAC inhibitors and appear to be more specific than SAHA, TSA, and MS-275, because it is selectively directed to chromatin and associated histones by the positively charged polyamine side chain. Several other analogs of hydroxamic acids are being developed *(91,97)*.

3.5. Depsipeptide and Thiols

The depsipeptide (FR901228/FK228) induces differentiation, growth arrest, and apoptosis in the low-nanomolar range *(98–109)*. It also inhibits metastasis and angiogenesis *(108–110)*. Depsipeptide is also very promising antitumor agent against osteosarcoma, inducing apoptosis by the activation of the Fas/FasL system *(111)*. Depsipeptide induced apoptosis in multiple myeloma U266 and RPMI 8226 cell lines in a time- and dose-dependent fashion, and in primary patient myeloma cells *(100)*. Depsipeptide promotes histone acetylation, gene transcription, apoptosis and its activity is enhanced by DNA methyltransferase inhibitors in AML1/ETO-positive leukemic cells *(100)*. Preclinical studies with depsipeptide in chronic lymphocytic leukemia (CLL) and acute myeloid leukemia (AML) have demonstrated that it effectively induces apoptosis at concentrations at which HDAC inhibition occurs. A dose-dependent increase in H3 and H4 histone acetylation was noted in depsipeptide-treated AML1/ETO-positive Kasumi-1 cells and blasts from a patient with t(8;21) AML *(105)*. A phase 1 and pharmacodynamic study of depsipeptide in CLL and AML have yielded promising results *(112)*.

Opening of the disulfide bridge leads to a thiol that may be able to enter the active site and complex the zinc ion. In this regard, garlic constituents and their metabolites, such as diallylsulfide and allylmercaptan, inhibited HDAC activity. Diallyl disulfide caused increased acetylation of H3 and H4 histones in DS19 mouse erythroleukemic cells and K562 human leukemic cells *(113)*, suggesting that differentiation in erythroleukemic cells by diallyl disulfide and allyl mercaptan may be mediated through induction of histone acetylation. Acetylation was also induced in rat hepatoma and human breast cancer cells by diallyl disulfide or its metabolite, allyl mercaptan. Diallyl disulfide increased histone acetylation and $p21^{WAF1/CIP1}$ expression in human colon tumor cell lines *(114)*.

3.6. Psammaplins

Psammaplins, isolated from a marine sponge *Pseudoceratina purpurea,* inhibited histone deacetylase and DNA methyltransferase activities *(115)*. Psammaplin A (PsA) contains an α-oximatoamide functional group, which inhibits the HDAC activity at the catalytic site. The disulfide group is also an essential feature for HDAC inhibition. PsA showed a potent cytotoxicity against several cancer and endothelial cells *(116–123)*. PsA-induced cytotoxicity may correlate with its inhibition on DNA replication *(116)*. Furthermore, PsA was found to inhibit mammalian aminopeptidase N (APN) that plays a key role in tumor cell invasion and angiogenesis *(117)*. Interestingly, the antiproliferative effect of PsA was dependent on the cellular amount of APN expression. PsA suppressed the invasion and tube formation of endothelial cells stimulated by basic fibroblast growth factor. These findings suggest that PsA can be developed as antiangiogenic and anticancer agent. Several synthetic analogs of PsA are being developed.

4. Intracellular Mechanisms of HDAC Inhibitors

Recent studies have shown that HDAC inhibitors induce cell-cycle arrest, differentiation, and apoptosis in vitro and in vivo *(33,34,44,61,107,111,124–134)*. The mechanisms by which these inhibitors induce cell cycle arrest, differentiation, and apoptosis appear to involve multiple genes. In addition to inducing growth arrest and apoptosis, they also inhibit metastasis and angiogenesis *(68,87,107,135–138)*.

Molecular inhibition of the ErbB signaling pathway represents a promising cancer treatment strategy. Preclinical studies suggest that enhancement of antitumor activity can be achieved by maximizing ErbB signaling inhibition. It has been demonstrated that HDAC inhibitors decreased transcript expression of ErbB1 (epidermal growth factor receptor) and ErbB2 in DU145 (prostate) and ErbB2 in SKBr3 (breast) cancer cell lines *(139)*. HDAC inhibitors also inhibited caveolin-1 and HIF-α (hypoxia inducible factor 1-alpha), and upregulated gelsolin, p19 (INK4D) and Nur77 expressions in DU145 cells *(139)*. Enhanced proliferative inhibition, apoptosis induction, and signaling inhibition were demonstrated when combining HDAC inhibition with ErbB blockade. These results suggest that used cooperatively, anti-ErbB agents and HDAC inhibitors may offer a promising strategy of dual targeted therapy. The beneficial interaction of these agents may not derive solely from modulation of ErbB expression, but may result from effects on other oncogenic processes including angiogenesis, invasion and cell-cycle kinetics.

4.1. Effects of HDAC Inhibitors on Cell Cycle

During the cell-division cycle, chromosomal DNA must initially be precisely duplicated and then correctly segregated to daughter cells. Cell-cycle control of transcription seems to be a universal feature of proliferating cells, although relatively little is known about its biological significance and conservation between organisms.

Given the key role of cell-cycle integrity in tumor suppression and cancer therapy, a lot of attention has focused on the ability of HDAC inhibitors to alter the levels of cell-cycle regulatory proteins. HDAC inhibitors induce growth arrest at both the G1 and G2/M phases of cell cycle and induce differentiation and/or apoptosis of various types of tumor cell lines *(33,34,45,69,102,126,127,130,132,140–145)*. HDAC inhibitors induced both $p21^{WAF1/CIP1}$ and $p27^{KIP1}$ at protein levels, and caused hypophosphorylation of Rb *(33,69,132,146)*. Other cell-cycle inhibitors that participate in the proliferative arrest elicited by HDAC inhibitors are $p15^{INK4b}$, $p18^{INK4c}$, and $p19^{INK4d}$ *(147,148)*. Moreover, positive regulators of proliferation, such as cyclins D1 and D2, cMyc, or c-Src, are downregulated by HDAC inhibitors *(149–154)*. P53 is activated both by inhibitors of HDACs class I/II, as well as, by inhibitors of the Sir2 family *(155–158)*. Transcription factor Sp1 regulates $p21^{WAF1/CIP1}$ expression in a p53-independent fashion *(85,106,159,160)*. $p21^{WAF1/CIP1}$ expression is also transcriptionally regulated by p53 *(161)*.

4.2. Effects of HDAC Inhibitors on Apoptosis

HDAC inhibitors induce apoptosis in breast, prostate, lung and thyroid carcinoma, and leukemia and multiple myeloma *(18,31,33,45,130,146,162–169)*. In addition to the induction of death receptors by HDAC inhibitors, the regulation of Bcl-2 family members is also important for inducing sensitivity *(31,33,100,170–173)*. We and others have shown that HDAC inhibitors selectively induce proapoptotic members such as Bax, Bak, Noxa, Bim, and PUMA and inhibit antiapoptotic Mcl-1, Bcl-X_L and Bcl-2 expression. Bcl-2 family members mainly exert their apoptotic effects by acting at the level of mitochondria and play crucial role in cancer development *(174)*. HDAC inhibitors cleave poly(ADP-ribose) polymerase (PARP) and caspase-8, -9, -3, -7, and -2. Transfection of Bcl-2 cDNA partially suppressed SAHA-induced cell death. HDAC inhibitors can also induce TRAIL, suggesting the activation of death receptor pathway without the requirement of exogenous TRAIL. Thus, HDAC inhibitors can induce apoptosis by linking both death receptor and mitochondrial pathways of apoptosis.

Dysregulation in apoptosis has been associated with the development of cancer *(175)*. Recent studies have shown the involvement of mitochondria in many apoptotic signaling pathways *(176,177)*. Members of the Bcl-2 family of proteins that regulate apoptotic signaling through mitochondria are key regulators of apoptosis in mammalian development, and their deregulation is associated with disease, particularly cancer *(178,179)*. There are three classes of Bcl-2 family members: apoptosis promoters (e.g., Bax and Bak); apoptosis inhibitors (e.g., Bcl-2, Bcl-X_L, and adenoviral E1B 19K); and the BH3-only Bcl-2 family members (e.g., Bid, Puma, Noxa, Bad, and Nbk/Bik), that contain the BH3 interaction domain that act as apoptosis promoters and inhibitors *(180)*. Therefore, the regulation of Bcl-2 family members by HDAC inhibitor may play important roles on apoptosis by inducing a death activity or by antagonizing a survival activity. Inactivation of both Bax and Bak was required for tumor growth and was selected for in vivo during tumorigenesis *(181,182)*. $Bax^{-/-}$ and $Bak^{-/-}$ DKO MEFs were resistant to death signaling pathway, indicating that they are the required downstream components of mitochondrial signaling pathways *(177)*. The induction of Bim by HDAC inhibitors in leukemia cells is similar to other study where withdrawal of cytokines from survival factor-dependent lymphocyte cell lines results in an up-regulation of Bim expression, concomitant with the induction of apoptotic program. Bim has been implicated in modulating lymphocyte homeostasis in immune cells. $Bim^{-/-}$ mice succumb to autoimmune kidney disease, accumulation of lymphoid and myeloid cells and perturbed T-cell development *(183,184)*.

The inhibitor of apoptosis (IAP) proteins serve as key regulators of caspase activity *(185)*. IAPs (cIAP-1, cIAP-2, NIAP, Livin/ML-IAP, survivin, and XIAP) protect cells against apoptosis by acting as caspase inhibitors *(185)*. IAPs bind to and directly inhibit caspase-3, -7, and -9 *(185,186)*. IAP proteins are regulated by interactions with the mitochondrial proteins (e.g., Smac/DIABLO), which may be released into the cytosol upon apoptotic stimulation and through IAP sequestration results in elevated caspase activity *(187,188)*. Some IAP proteins are also regulated by proteolysis via the ubiquitin-proteasome pathway and caspase-dependent cleavage of XIAP in cells undergoing apoptosis. The inhibition of XIAP, cIAP1, and cIAP2 expressions by HDAC inhibitors may contribute in sensitization of cells to TRAIL. In this context, we have shown that TRAIL inhibits the expression of IAPs in breast and prostate cancer cells *(31,69)*. The combination of HDAC inhibitors and TRAIL may further inhibit the expression of some of the IAPs and contribute to the synergistic induction of apoptosis by these agents.

HDAC inhibitors activate the p53 molecule through acetylation of 320 and 373 lysine residues, upregulate PIG3 and NOXA and induce apoptosis in cancer cells expressing wild and pseudo-wild type p53 genes *(189)*.

SAHA induced polyploidy in human colon cancer cell line HCT116 and human breast cancer cell lines, MCF-7, MDA-MB-231, and MBA-MD-468, but not in normal human embryonic fibroblast SW-38 and normal mouse embryonic fibroblasts *(190)*. The polyploid cells lost the capacity for proliferation and committed to senescence. The induction of polyploidy was enhanced in HCT116 $p21^{WAF1-/-}$ or HCT116 $p^{53-/-}$ cells than in wild-type HCT116. The development of senescence of SAHA-induced polyploidy cells was similar in all colon cell lines *(191)*. The present findings indicate that the HDAC inhibitor could exert antitumor effects by inducing polyploidy, and this effect is more marked in transformed cells with nonfunctioning $p21^{WAF1/CIP1}$ or p53 genes.

In chronic myelocytic leukemia (CML) the activity of the Bcr-Abl tyrosine kinase is known to activate a number of molecular mechanisms, which inhibit apoptosis *(191,192)*. SAHA markedly decreases protein expression levels of Bcr-Abl and c-Myc and HDAC3 in CML, suggesting that SAHA exerts its biological activity by inhibiting survival pathway *(191)*. Differential expression of histone deacetylase has been reported in various cancer. To explore the mechanisms of disease-specific HDAC activity in acute myeloid leukaemia (AML), the expression of the HDAC in primary AML blasts and in four control cell types, namely $CD34^+$ progenitors from umbilical cord, either quiescent or cycling (postculture), cycling $CD34^+$ progenitors from GCSF-stimulated adult donors and peripheral blood mononuclear cells (PBMCs) was characterized. Only SIRT1 was consistently overexpressed in AML samples compared with all controls, while HDAC6 was overexpressed relative to adult, but not neonatal cells *(193)*. HDAC5 and SIRT4 were consistently underexpressed. HDAC inhibitors (valproate, butyrate, TSA, SAHA) caused hyperacetylation of histones in AML blasts and cell lines *(193)*. Such treatment also modulated the pattern of HDAC expression, with strong induction of HDAC11 in all myeloid cells tested, and lesser, more selective, induction of HDAC9 and SIRT4. The distinct pattern of HDAC expression in AML and its response to HDAC inhibitors is of relevance to the development of HDAC inhibitor-based therapeutic strategies and may contribute to observed patterns of clinical response and development of drug resistance.

4.3. Effects of HDAC Inhibitors on Angiogenesis

Tumor growth requires the development of new vessels that sprout from pre-existing normal vessels in a process known as "angiogenesis" *(194)*. These new vessels arise from local capillaries, arteries, and veins in response to the release of soluble growth factors from the tumor mass, enabling these tumors to grow beyond the diffusion-limited size of approx 2-mm diameter. Tumor growth and metastasis depend upon the development of a neovasculature in and around the tumor *(194–199)*. Angiogenesis is regulated by the balance between stimulatory (e.g., basic fibroblast growth factor [bFGF], IL-8, matrix metalloproteinase [MMP]-2, MMP-9, tumor growth factor [TGF]β1, vascular endothelial growth factor [VEGF]) and inhibitory (e.g., angiostatin,

IL-10, interferon) factors released by the tumor and its environment *(196,200)*. For example, overexpression of bFGF *(201,202)* and VEGF *(203–205)* has been found in the tissue, serum, and urine of patients with bladder cancer and has been associated with cancer progression, suggesting a direct involvement of these proteins in angiogenesis.

HDAC inhibitors also inhibit endothelial cell proliferation and angiogenesis by downregulating angiogenesis-related gene expression *(44,68,87,107–109,135–139, 206–219)*. TSA has been reported to inhibit angiogenesis both in vitro and in vivo *(68)*. TSA inhibits hypoxia-induced production of the angiogenic mediator VEGF by tumor cells and also inhibits directly endothelial cell migration and proliferation *(68)*. These data suggest that HDAC inhibitors can inhibit tumor growth by inhibiting angiogenesis, and their effects on angiogenesis can be further enhanced in the presence of TRAIL.

The combination of adenoviral vector carrying wild type p53 (Ad-p53) gene therapy with sodium butyrate resulted in a complete regression of xenografted human gastric tumor (KATO-III) cells in nude mice *(44)*. Tumors treated with the combination showed higher numbers of terminal dUTP nick end labeling (TUNEL) positive cells and lower CD34 staining than those treated with a single modality *(44)*. This was further supported by the finding that BAI-1 (brain specific angiogenesis inihibitor-1), an inhibitor of vascularization, was induced by sodium butyrate treatment in cells transfected with Ad-p53 *(44)*. These data suggest that HDAC inhibitors can be combined with p53 gene therapy for the treatment of cancer. The finding that HDAC appears to be a critical regulator of angiogenesis in addition to tumor cell growth will heighten interest in the development of HDAC inhibitors as potential antiangiogenic and anticancer drugs.

5. HDAC Inhibitors and TRAIL/Apo-2L

We and other have shown that several HDAC inhibitors can enhance the apoptosis-inducing potential of TRAIL and sensitize TRAIL-resistant breast, prostate, and lung cancer cells, and malignant mesothelioma, leukemia, and myeloma cells *(31,33,98,162, 220–224)*. The sensitization of TRAIL resistant cells appears to result from downregulation of the antiapoptotic protein Bcl-2, Bcl-X_L, Mcl-1, and upregulation of proapoptotic genes Bax, Bak, TRAIL, Fas, FasL, DR4, and DR5, and activation of caspases. HDAC inhibitors upregulate proapoptotic genes only in cancer cells but not in normal cells *(225,226)*. Sodium butyrate and TSA enhanced TRAIL-mediated apoptosis to a greater extent than depsipeptide, MS-275 and oxamflatin *(98)*. Both sodium butyrate and TSA treatment also increased mRNA and surface expression of TRAIL-R2/DR5 that was dependent on the transcription factor Sp1, thus providing a possible mechanism behind the increased sensitivity to TRAIL. These results show that sensitivity to HDAC inhibitors in cancer cells is a property of the fully transformed phenotype and depends on activation of a specific death pathway. Because HDAC inhibitors sensitize TRAIL-resistant cancer cells to undergo apoptosis by TRAIL, they appear to be promising candidates for combination chemotherapy.

Several studies have demonstrated the engagement of mitochondria during activation of death receptor pathway *(69,227–229)*. Crosstalk between the death-receptor (extrinsic) and mitochondrial (intrinsic) pathways requires caspase-8/-10 dependent cleavage of Bid *(31,33,69,227)*. tBid activates Bax and Bak to release cytochrome *c* and other mitochondrial proteins *(176,230)*. Because HDAC inhibitors induced cleavage of Bid, the truncated Bid may trigger activation of mitochondria in the absence of ligand

TRAIL. We have shown that the pan-caspase inhibitor z-VAD-fmk completely inhibited TRAIL-induced apoptosis in the presence of HDAC inhibitor *(31,33,69)*. The caspase-8 inhibitor z-IETD and DN-FADD completely inhibited the synergistic interaction between HDAC inhibitor and TRAIL. Furthermore, in the presence of HDAC inhibitors, TRAIL induced caspase-3 and caspase-9 activation and caused cleavage of their substrate PARP. Antiapoptotic proteins Bcl-2 and Bcl-X_L inhibit HDAC inhibitors and/or TRAIL-induced apoptosis by blocking cytochrome *c* release. The phosphorylation deficient mutant of Bcl-2 and Bcl-X_L also inhibited HDAC inhibitors and/or TRAIL-induced apoptosis. In cell-intrinsic pathway of apoptosis, mitochondria amplify the apoptotic signals leading to activation of caspase-9 *(177)*. Caspase-9 in turn activates downstream caspases and the cleavage of apoptotic substrates that finally kill cells. The synergistic effects of HDAC inhibitors and TRAIL on apoptosis occur through activation of downstream caspase-3, which can be activated by both extrinsic and intrinsic pathways *(31,33,69)*.

The sensitization of cancer cells to HDAC inhibitors appears to be p53 independent. We have recently shown that chemotherapeutic drugs *(231)* or irradiation *(232,233)* can sensitize breast and prostate cancer cells by upregulating death receptors DR4 and/or DR5 in cells harboring wild-type (MCF-7) and mutated (MDA-MB-231, and MDA-MB-468) p53. Recent studies have shown that HDAC inhibitors induce apoptosis in leukemia in a p53-independent manner but not in normal hematopoietic progenitors *(220,225)*. Other transcription factors such as NFκB and SP1 have been shown to regulate the expression of death receptors *(234–237)*.

Treatment of nude mice with HDAC inhibitors resulted in acetylation of histone H3 and H4, and down-regulation of hypoxia-inducible factor 1-α and VEGF expression in tumor cells. Furthermore, control mice demonstrating increased rate of tumor growth had increased numbers of CD31-positive or von Willebrand Factor (vWF)-positive blood vessels, and increased circulating vascular VEGFR2-positive endothelial cells compared with HDAC inhibitor and/or TRAIL treated mice. Sequential treatments of athymic nude mice with HDAC inhibitors followed by TRAIL cause a synergistic apoptotic response through activation of caspase-3 and caspase-7, which is accompanied by regression of tumor growth, inhibition of angiogenesis and enhancement of survival of xenografted nude mice. Together with our previous studies showing that cancer chemotherapeutic drugs and irradiation up-regulate DR4 and/or DR5 expression, thereby enhancing TRAIL-induced apoptosis in vivo *(231,233,238,239)*, these studies demonstrate the antitumor interactions of HDAC inhibitors with the TRAIL-death receptor pathway. Similarly, several recent studies including ours have demonstrated the additive or synergistic effects of HDAC inhibitors and TRAIL on apoptosis in vitro *(31,33,69,162,170,171,220–224)*. The ability of HDAC inhibitors to sensitize cancer cells to TRAIL suggests that HDAC inhibitors can reduce the minimal effective dose or side effects of TRAIL. Thus, these data provide the framework for clinical evaluation of HDAC inhibitors and TRAIL for the treatment of human cancer.

6. HDAC Inhibitors and Irradiation

HDAC inhibitors have been shown to radiosensitize breast, prostate, and glioma cells lines *(132,240,241)*. TSA has been shown to radiosensitize human glioblastoma U373MG and U87MG cell lines in a dose- and time-dependent manner *(240)*. Valproic acid enhanced the radiosensitivity of brain tumor SF539 and U251 cell lines in vitro and U251 xenografts in vivo, which correlated with the induction of histone hyperacetylation *(242)*. Similarly, MS-275 can enhance radiosensitivity of DU145 prostate carcinoma and U251 glioma cells suggesting that this effect may involve an inhibition of DNA repair *(241)*. The combination of HDAC inhibitors with irradiation may be a useful for the treatment of cancer and merit further investigation. Given the limited efficacy of standard treatments for patients with cancer, these data provide support for clinical trials integrating HDAC inhibitor with radiation therapy.

Caspase-2 and caspase-3 cleave HDAC4 in vitro and caspase-3 is critical for HDAC4 cleavage in vivo during UV-induced apoptosis *(37)*. After ultraviolet irradiation, GFP-HDAC4 translocates into the nucleus coincidentally/immediately before the retraction response, but clearly before nuclear fragmentation. Together, these data indicate that caspases could specifically modulate gene repression and apoptosis through the proteolytic processing of HDAC4. Among molecular cell cycle-targeted drugs currently in the pipeline for testing in early-phase clinical trials, HDAC inhibitors may have therapeutic potential as radiosensitizers.

7. HDAC Inhibitors and Chemotherapeutic Drugs

Chemotherapeutic treatment with combinations of drugs is front-line therapy for many types of cancer. Combining drugs which target different signaling pathways often lessens adverse side effects while increasing the efficacy of treatment and reducing patient morbidity. It has recently been shown that HDAC inhibitors facilitate the cytotoxic effectiveness of the topoisomerase I inhibitor camptothecin in the killing of tumor cells *(243)*. HDAC inhibitors (SAHA and sodium butyrate) interacted synergistically with camptothecin in inducing apoptosis of breast and lung cancer cell lines. Experiments have shown that cells arrested in G2-M by camptothecin were most sensitive to subsequent addition of HDAC inhibitor. In camptothecin-arrested cells, sodium butyrate decreases cyclin B levels, as well as the levels of the antiapoptotic proteins XIAP and survivin. Overall, these findings suggest that reducing the levels of these critical antiapoptotic factors may increase the efficacy of camptothecin in the clinical setting if given in a sequence that does not prevent or inhibit tumor cell progression through the S phase.

MS-275 also synergistically interacted with fludarabine in inducing apoptosis of lymphoid and myeloid human leukemia cells *(53)*. Prior exposure of Jurkat lymphoblastic leukemia cells to MS-275 increased mitochondrial injury, caspase activation, and apoptosis in response to fludarabine, resulting in highly synergistic antileukemic interactions and loss of clonogenic survival. Simultaneous exposure to MS-275 and fludarabine also led to synergistic effects, but these were not as pronounced as observed with sequential treatment. Similar interactions were noted in the case of (1) other human leukemia cell lines (e.g., U937, CCRF-CEM); (2) other HDAC inhibitors (e.g., sodium butyrate); and (3) other nucleoside analogues (e.g., 1-β-D-arabinofuranosylcytosine, gemcitabine). Potentiation of fludarabine-induced apoptosis by MS-275 was associated with acetylation of histones H3 and H4, down-regulation of the antiapoptotic proteins XIAP and Mcl-1, enhanced cytosolic release of proapoptotic mitochondrial proteins (e.g., cytochrome *c*, Smac/DIABLO, and AIF), and caspase activation. These events were accompanied by the caspase-dependent down-regulation of $p27^{/KIP1}$, cyclins A, E, and D1, and cleavage and diminished phosphorylation of retinoblastoma protein. Prior exposure to MS-275 attenuated fludarabine-mediated activation of MEK1/2, extracellular signal-regulated kinase, and Akt, and enhanced c-Jun NH(2)-terminal kinase phosphorylation; furthermore, inducible expression of constitutively active MEK1/2 or Akt significantly diminished MS-275/fludarabine-induced lethality. Combined exposure of cells to MS-275 and fludarabine was associated with a significant increase in generation of ROS; moreover, both the increase in ROS and apoptosis were largely attenuated by coadministration of the free radical scavenger L-*N*-acetylcysteine. Finally, prior administration of MS-275 markedly potentiated fludarabine-mediated generation of the proapoptotic lipid second messenger ceramide. Taken together, these findings indicate that MS-275 induces multiple perturbations in signal transduction, survival, and cell cycle regulatory pathways that lower the threshold for fludarabine-mediated mitochondrial injury and apoptosis in human leukemia cells.

A synergistic interaction of retinoic acid and CBHA was shown in a mouse model of neuroblastoma. DNA hypomethylating agents have been found to have synergistic effects with HDAC inhibitors. The combination of TSA with azacytidine caused a dramatic potentiation in the activation of silenced genes *(244–246)*. Depsipeptide and TSA induced apoptosis in human lung cancer cells, and HDAC inhibitor-induced apoptosis was greatly enhanced in the presence of the DNA methyltransferase inhibitor, 5-aza-2′-deoxycytidine (DAC), suggesting the DNA methylation status plays an important role on the effectiveness of HDAC inhibitors *(247)*. Furthermore, HDAC inhibitors enhanced paclitaxel-induced cell death in ovarian cancer cell lines independent of p53 status *(248)*. Similarly, commonly used anticancer drugs doxorubicin and decitabine have been reported to have synergistic effects with HDAC inhibitors *(249,250)*. Thus, the combination of anticancer drugs with other epigenetic therapies provides potentially safer therapeutic options.

8. HDAC Inhibitors and Chemopreventive Drugs

In recent years, the use of naturally occurring chemopreventive agents have attracted many investigators because of their nontoxic effects. The preclinical data on selected chemopreventive agents have been very promising. Sulforaphane (SFN), an isothiocyanate first isolated from broccoli, exhibits chemopreventive properties in prostate cancer cells through mechanisms that are poorly understood. SFN inhibits HDAC activity in colon and prostate cancer cells *(251)*. The inhibition of HDAC was accompanied by an increase in acetylated histones. SFN caused enhanced interaction of acetylated histone H4 with the promoter region of the $p^{21/WAF1/CIP1}$ gene and the bax gene. A corresponding 1.5- to 2-fold increase was seen for $p^{21/WAF1/CIP1}$ and Bax protein expression, consistent with previous studies using HDAC inhibitors such as trichostatin A. SFN induced cell-cycle arrest and apoptosis through caspase activation. These findings provide new insight into the mechanisms of SFN action in benign prostate hyperplasia, and they suggest a novel approach to chemoprotection and chemotherapy of prostate cancer through the inhibition of HDAC.

9. Clinical Trials With HDAC Inhibitors

Initial clinical trials indicate that HDAC inhibitors from several different structural classes are very well tolerated and exhibit clinical activity against a variety of human malignancies; however, the molecular basis for their anticancer selectivity remains largely unknown. HDAC inhibitors have also shown preclinical promise when combined with other therapeutic agents, and innovative drug delivery strategies, including liposome encapsulation, may further enhance their clinical development and anticancer potential. An improved understanding of the mechanistic role of specific HDACs in human tumorigenesis, as well as the identification of more specific HDAC inhibitors, will likely accelerate the clinical development and broaden the future scope and utility of HDAC inhibitors for cancer treatment.

Several HDAC inhibitors (SAHA, MS-275, CI-994 and depsipeptide) are currently undergoing clinical trials *(252–254)*. HDAC inhibitors represent a relatively new group of targeted anticancer compounds, which are showing significant promise as agents with activity against a broad spectrum of neoplasms, at doses that are well tolerated by cancer patients. SAHA is one of the HDAC inhibitors most advanced in development. It is in phase I and II clinical trials for patients with both hematologic and solid tumors *(252)*. Phase I/II clinical trials on depsipeptide have shown low toxicity and evidence of antitumor activity; on the other hand, this compound has potential for synergism with radiotherapy, chemotherapy, and biologicals. Second generation HDAC inhibitors such as LAQ824 and PDX101 are currently under phase I clinical trials. Simultaneously, synthetic benzamide-containing HDACs were reported and two of them, MS-275 and CI-994, have reached phase II and I clinical trials, respectively.

10. Conclusions

The histone deacetylase has been considered a target molecule for cancer therapy. The inhibition of HDAC activity by a specific inhibitor induces growth arrest, differentiation, and apoptosis of transformed or several cancer cells. The discovery and development of specific HDAC inhibitors are helpful for cancer therapy, and decipher the molecular mode of action for HDAC.

Given that HDAC inhibitors upregulate proapoptotic members TRAIL-R1/DR4, TRAIL-R2/DR5, Bax, Bak, PUMA, and NOXA and downregulate antiapoptotic Bcl-2, Bcl-X_L and Mcl-1 proteins, it is possible that sensitization of cancer cells to chemotherapy, irradiation, or TRAIL by HDAC inhibitors may occur at various stages of apoptotic pathways. Furthermore, the ability of HDAC inhibitors to inhibit angiogenesis may further affect tumor growth by regulating angiogenesis-related signaling pathways. Preliminary studies in animal models have revealed a relatively high tumor selectivity of HDAC inhibitors, strengthening their promising potential in cancer chemotherapy. Some of these inhibitors are in a clinical trial at phase I or phase II. Furthermore, the combination of HDAC inhibitors with commonly used anticancer drugs, irradiation or TRAIL will be useful for cancer therapy.

Acknowledgment

This work was supported by grants from the Susan G. Komen Breast Cancer Foundation.

References

1. Plass C. Cancer epigenomics. Hum Mol Genet 2002;11:2479–2488.
2. Jones PA. DNA methylation and cancer. Oncogene 2002;21:5358–5360.

3. Grunstein M. Histone acetylation in chromatin structure and transcription. Nature 1997;389:349–352.
4. Glass CK, Rosenfeld MG. The coregulator exchange in transcriptional functions of nuclear receptors. Genes Dev 2000;14:121–141.
5. Kouzarides T. Histone acetylases and deacetylases in cell proliferation. Curr Opin Genet Dev 1999;9:40–48.
6. Grignani F, De Matteis S, Nervi C, et al. Fusion proteins of the retinoic acid receptor-alpha recruit histone deacetylase in promyelocytic leukaemia. Nature 1998;391:815–818.
7. Lin RJ, Nagy L, Inoue S, Shao W, Miller WH, Jr, Evans RM. Role of the histone deacetylase complex in acute promyelocytic leukaemia. Nature 1998;391:811–814.
8. Strahl BD, Allis CD. The language of covalent histone modifications. Nature 2000;403: 41–45.
9. Johnstone RW, Licht JD. Histone deacetylase inhibitors in cancer therapy: is transcription the primary target? Cancer Cell 2003;4:13–18.
10. Kurdistani SK, Grunstein M. Histone acetylation and deacetylation in yeast. Nat Rev Mol Cell Biol 2003;4:276–284.
11. Taunton J, Hassig CA, Schreiber SL. A mammalian histone deacetylase related to the yeast transcriptional regulator Rpd3p. Science 1996;272:408–411.
12. Yang WM, Inouye C, Zeng Y, Bearss D, Seto E. Transcriptional repression by YY1 is mediated by interaction with a mammalian homolog of the yeast global regulator RPD3. Proc Natl Acad Sci USA 1996;93:12,845–12,850.
13. Emiliani S, Fischle W, Van Lint C, Al-Abed Y, Verdin E. Characterization of a human RPD3 ortholog, HDAC3. Proc Natl Acad Sci USA 1998;95:2795–2800.
14. Grozinger CM, Hassig CA, Schreiber SL. Three proteins define a class of human histone deacetylases related to yeast Hda1p. Proc Natl Acad Sci USA 1999;96:4868–4873.
15. Gao L, Cueto MA, Asselbergs F, Atadja P. Cloning and functional characterization of HDAC11, a novel member of the human histone deacetylase family. J Biol Chem 2002; 277:25,748–25,755.
16. Fischle W, Emiliani S, Hendzel MJ, et al. A new family of human histone deacetylases related to Saccharomyces cerevisiae HDA1p. J Biol Chem 1999;274:11,713–11,720.
17. Gray SG, Ekstrom TJ. The human histone deacetylase family. Exp Cell Res 2001; 262:75–83.
18. de Ruijter AJ, van Gennip AH, Caron HN, Kemp S, van Kuilenburg AB. Histone deacetylases (HDACs): characterization of the classical HDAC family. Biochem J 2003;370: 737–749.
19. Verdin E, Dequiedt F, Kasler HG. Class II histone deacetylases: versatile regulators. Trends Genet 2003;19:286–293.
20. McKinsey TA, Olson EN. Toward transcriptional therapies for the failing heart: chemical screens to modulate genes. J Clin Invest 2005;115:538–546.
21. Braunstein M, Rose AB, Holmes SG, Allis CD, Broach JR. Transcriptional silencing in yeast is associated with reduced nucleosome acetylation. Genes Dev 1993;7:592–604.
22. Frye RA. Characterization of five human cDNAs with homology to the yeast SIR2 gene: Sir2-like proteins (sirtuins) metabolize NAD and may have protein ADP-ribosyltransferase activity. Biochem Biophys Res Commun 1999;260:273–279.
23. Imai S, Armstrong CM, Kaeberlein M, Guarente L. Transcriptional silencing and longevity protein Sir2 is an NAD-dependent histone deacetylase. Nature 2000;403:795–800.
24. Bordone L, Guarente L. Calorie restriction, SIRT1 and metabolism: understanding longevity. Nat Rev Mol Cell Biol 2005;6:298–305.
25. Kaeberlein M, McVey M, Guarente L. The SIR2/3/4 complex and SIR2 alone promote longevity in Saccharomyces cerevisiae by two different mechanisms. Genes Dev 1999;13:2570–2580.

26. Tissenbaum HA, Guarente L. Increased dosage of a sir-2 gene extends lifespan in Caenorhabditis elegans. Nature 2001;410:227–230.
27. Ford J, Jiang M, Milner J. Cancer-specific functions of SIRT1 enable human epithelial cancer cell growth and survival. Cancer Res 2005;65:10,457–10,463.
28. Van Lint C, Emiliani S, Verdin E. The expression of a small fraction of cellular genes is changed in response to histone hyperacetylation. Gene Expr 1996;5:245–253.
29. Lagger G, O'Carroll D, Rembold M, et al. Essential function of histone deacetylase 1 in proliferation control and CDK inhibitor repression. EMBO J 2002;21:2672–2681.
30. Bandyopadhyay D, Mishra A, Medrano EE. Overexpression of histone deacetylase 1 confers resistance to sodium butyrate-mediated apoptosis in melanoma cells through a p53-mediated pathway. Cancer Res 2004;64:7706–7710.
31. Singh TR, Shankar S, Srivastava RK. HDAC inhibitors enhance the apoptosis-inducing potential of TRAIL in breast carcinoma. Oncogene 2005;24:4609–4623.
32. Marks PA, Miller T, Richon VM. Histone deacetylases. Curr Opin Pharmacol 2003;3: 344–351.
33. Fandy TE, Shankar S, Ross DD, Sausville E, Srivastava RK. Interactive effects of HDAC inhibitors and TRAIL on apoptosis are associated with changes in mitochondrial functions and expressions of cell cycle regulatory genes in multiple myeloma. Neoplasia 2005;7:646–657.
34. Fang JY. Histone deacetylase inhibitors, anticancerous mechanism and therapy for gastrointestinal cancers. J Gastroenterol Hepatol 2005;20:988–994.
35. Rosato RR, Wang Z, Gopalkrishnan RV, Fisher PB, Grant S. Evidence of a functional role for the cyclin-dependent kinase-inhibitor p21WAF1/CIP1/MDA6 in promoting differentiation and preventing mitochondrial dysfunction and apoptosis induced by sodium butyrate in human myelomonocytic leukemia cells (U937). Int J Oncol 2001;19:181–191.
36. Liu F, Dowling M, Yang XJ, Kao GD. Caspase-mediated specific cleavage of human histone deacetylase 4. J Biol Chem 2004;279:34,537–34,546.
37. Paroni G, Mizzau M, Henderson C, Del Sal G, Schneider C, Brancolini C. Caspase-dependent regulation of histone deacetylase 4 nuclear-cytoplasmic shuttling promotes apoptosis. Mol Biol Cell 2004;15:2804–2818.
38. Kwon SH, Ahn SH, Kim YK, et al. Apicidin, a histone deacetylase inhibitor, induces apoptosis and Fas/Fas ligand expression in human acute promyelocytic leukemia cells. J Biol Chem 2002;277:2073–2080.
39. Marks PA, Richon VM, Miller T, Kelly WK. Histone deacetylase inhibitors. Adv Cancer Res 2004;91:137–168.
40. Boyle GM, Martyn AC, Parsons PG. Histone deacetylase inhibitors and malignant melanoma. Pigment Cell Res 2005;18:160–166.
41. Butler LM, Agus DB, Scher HI, et al. Suberoylanilide hydroxamic acid, an inhibitor of histone deacetylase, suppresses the growth of prostate cancer cells in vitro and in vivo. Cancer Res 2000;60:5165–5170.
42. Tang XX, Robinson ME, Riceberg JS, et al. Favorable neuroblastoma genes and molecular therapeutics of neuroblastoma. Clin Cancer Res 2004;10:5837–5844.
43. Zhang Y, Adachi M, Zhao X, Kawamura R, Imai K. Histone deacetylase inhibitors FK228, N-(2-aminophenyl)-4-[N-(pyridin-3-yl-methoxycarbonyl)amino- methyl]benzamide and m-carboxycinnamic acid bis-hydroxamide augment radiation-induced cell death in gastrointestinal adenocarcinoma cells. Int J Cancer 2004;110:301–308.
44. Takimoto R, Kato J, Terui T, et al. Augmentation of Antitumor Effects of p53 Gene Therapy by Combination with HDAC Inhibitor. Cancer Biol Ther 2005;4:421–428.
45. Sakajiri S, Kumagai T, Kawamata N, Saitoh T, Said JW, Koeffler HP. Histone deacetylase inhibitors profoundly decrease proliferation of human lymphoid cancer cell lines. Exp Hematol 2005;33:53–61.

46. Bordin M, D'Atri F, Guillemot L, Citi S. Histone deacetylase inhibitors up-regulate the expression of tight junction proteins. Mol Cancer Res 2004;2:692–701.
47. Shao Y, Gao Z, Marks PA, Jiang X. Apoptotic and autophagic cell death induced by histone deacetylase inhibitors. Proc Natl Acad Sci USA 2004;101:18,030–18,035.
48. Park JH, Jung Y, Kim TY, et al. Class I histone deacetylase-selective novel synthetic inhibitors potently inhibit human tumor proliferation. Clin Cancer Res 2004;10: 5271–5281.
49. Marks PA, Richon VM, Breslow R, Rifkind RA. Histone deacetylase inhibitors as new cancer drugs. Curr Opin Oncol 2001;13:477–483.
50. McLaughlin F, La Thangue NB. Histone deacetylase inhibitors open new doors in cancer therapy. Biochem Pharmacol 2004;68:1139–1144.
51. Johnstone RW. Histone-deacetylase inhibitors: novel drugs for the treatment of cancer. Nat Rev Drug Discov 2002;1:287–299.
52. Suzuki T, Ando T, Tsuchiya K, et al. Synthesis and histone deacetylase inhibitory activity of new benzamide derivatives. J Med Chem 1999;42:3001–3003.
53. Maggio SC, Rosato RR, Kramer LB, et al. The histone deacetylase inhibitor MS-275 interacts synergistically with fludarabine to induce apoptosis in human leukemia cells. Cancer Res 2004;64:2590–2600.
54. Monneret C. Histone deacetylase inhibitors. Eur J Med Chem 2005;40:1–13.
55. Ryan QC, Headlee D, Acharya M, et al. Phase I and pharmacokinetic study of MS-275, a histone deacetylase inhibitor, in patients with advanced and refractory solid tumors or lymphoma. J Clin Oncol 2005;23:3912–3922.
56. Tsuji N, Kobayashi M, Nagashima K, Wakisaka Y, Koizumi K. A new antifungal antibiotic, trichostatin. J Antibiot (Tokyo) 1976;29:1–6.
57. Yoshida M, Kijima M, Akita M, Beppu T. Potent and specific inhibition of mammalian histone deacetylase both in vivo and in vitro by trichostatin A. J Biol Chem 1990;265: 17,174–17,179.
58. Taghiyev AF, Guseva NV, Sturm MT, Rokhlin OW, Cohen MB. Trichostatin A (TSA) Sensitizes the Human Prostatic Cancer Cell Line DU145 to Death Receptor Ligands Treatment. Cancer Biol Ther 2005;4:382–390.
59. Tsatsoulis A. The role of apoptosis in thyroid disease. Minerva Med 2002;93:169–180.
60. Inoue H, Shiraki K, Ohmori S, et al. Histone deacetylase inhibitors sensitize human colonic adenocarcinoma cell lines to TNF-related apoptosis inducing ligand-mediated apoptosis. Int J Mol Med 2002;9:521–525.
61. Fronsdal K, Saatcioglu F. Histone deacetylase inhibitors differentially mediate apoptosis in prostate cancer cells. Prostate 2005;62:299–306.
62. Toth KF, Knoch TA, Wachsmuth M, et al. Trichostatin A-induced histone acetylation causes decondensation of interphase chromatin. J Cell Sci 2004;117:4277–4287.
63. Vanhaecke T, Henkens T, Kass GE, Rogiers V. Effect of the histone deacetylase inhibitor trichostatin A on spontaneous apoptosis in various types of adult rat hepatocyte cultures. Biochem Pharmacol 2004;68:753–760.
64. Vanhaecke T, Papeleu P, Elaut G, Rogiers V. Trichostatin A-like hydroxamate histone deacetylase inhibitors as therapeutic agents: toxicological point of view. Curr Med Chem 2004;11:1629–1643.
65. Wang ZM, Hu J, Zhou D, Xu ZY, Panasci LC, Chen ZP. Trichostatin A inhibits proliferation and induces expression of p21WAF and p27 in human brain tumor cell lines. Ai Zheng 2002;21:1100–1105.
66. Yamashita Y, Shimada M, Harimoto N, et al. Histone deacetylase inhibitor trichostatin A induces cell-cycle arrest/apoptosis and hepatocyte differentiation in human hepatoma cells. Int J Cancer 2003;103:572–576.

67. Blagosklonny MV, Robey R, Sackett DL, et al. Histone deacetylase inhibitors all induce p21 but differentially cause tubulin acetylation, mitotic arrest, and cytotoxicity. Mol Cancer Ther 2002;1:937–941.
68. Williams RJ, Trichostatin A. an inhibitor of histone deacetylase, inhibits hypoxia-induced angiogenesis. Expert Opin Investig Drugs 2001;10:1571–1573.
69. Shankar S, Singh TR, Fandy TE, Luetrakul T, Ross DD, Srivastava RK. Interactive effects of histone deacetylase inhibitors and TRAIL on apoptosis in human leukemia cells: Involvement of both death receptor and mitochondrial pathways. Int J Mol Med 2005;16:1125–1138.
70. Touma SE, Goldberg JS, Moench P, et al. Retinoic acid and the histone deacetylase inhibitor trichostatin a inhibit the proliferation of human renal cell carcinoma in a xenograft tumor model. Clin Cancer Res 2005;11:3558–3566.
71. Canes D, Chiang GJ, Billmeyer BR, et al. Histone deacetylase inhibitors upregulate plakoglobin expression in bladder carcinoma cells and display antineoplastic activity in vitro and in vivo. Int J Cancer 2005;113:841–848.
72. Choi JH, Oh SW, Kang MS, Kwon HJ, Oh GT, Kim DY. Trichostatin A attenuates airway inflammation in mouse asthma model. Clin Exp Allergy 2005;35:89–96.
73. Jung M, Brosch G, Kolle D, Scherf H, Gerhauser C, Loidl P. Amide analogues of trichostatin A as inhibitors of histone deacetylase and inducers of terminal cell differentiation. J Med Chem 1999;42:4669–4679.
74. Kim YB, Lee KH, Sugita K, Yoshida M, Horinouchi S. Oxamflatin is a novel antitumor compound that inhibits mammalian histone deacetylase. Oncogene 1999;18:2461–2470.
75. Su GH, Sohn TA, Ryu B, Kern SE. A novel histone deacetylase inhibitor identified by high-throughput transcriptional screening of a compound library. Cancer Res 2000;60:3137–3142.
76. Johnson J, Hunter R, McElveen R, Qian XH, Baliga BS, Pace BS. Fetal hemoglobin induction by the histone deacetylase inhibitor, scriptaid. Cell Mol Biol (Noisy-le-grand) 2005; 51:229–238.
77. Kijima M, Yoshida M, Sugita K, Horinouchi S, Beppu T. Trapoxin, an antitumor cyclic tetrapeptide, is an irreversible inhibitor of mammalian histone deacetylase. J Biol Chem 1993;268:22,429–22,435.
78. Brosch G, Ransom R, Lechner T, Walton JD, Loidl P. Inhibition of maize histone deacetylases by HC toxin, the host-selective toxin of Cochliobolus carbonum. Plant Cell 1995;7:1941–1950.
79. Darkin-Rattray SJ, Gurnett AM, Myers RW, et al. Apicidin: a novel antiprotozoal agent that inhibits parasite histone deacetylase. Proc Natl Acad Sci USA 1996;93:13,143–13,147.
80. Kim JS, Jeung HK, Cheong JW, et al. Apicidin potentiates the imatinib-induced apoptosis of Bcr-Abl-positive human leukaemia cells by enhancing the activation of mitochondria-dependent caspase cascades. Br J Haematol 2004;124:166–178.
81. Cheong JW, Chong SY, Kim JY, et al. Induction of apoptosis by apicidin, a histone deacetylase inhibitor, via the activation of mitochondria-dependent caspase cascades in human Bcr-Abl-positive leukemia cells. Clin Cancer Res 2003;9:5018–5027.
82. Kouraklis G, Theocharis S. Histone deacetylase inhibitors and anticancer therapy. Curr Med Chem Anti-Canc Agents 2002;2:477–484.
83. Hong J, Ishihara K, Yamaki K, et al. Apicidin, a histone deacetylase inhibitor, induces differentiation of HL-60 cells. Cancer Lett 2003;189:197–206.
84. Han JW, Ahn SH, Park SH, et al. Apicidin, a histone deacetylase inhibitor, inhibits proliferation of tumor cells via induction of p21WAF1/Cip1 and gelsolin. Cancer Res 2000;60: 6068–6074.
85. Han JW, Ahn SH, Kim YK, et al. Activation of p21(WAF1/Cip1) transcription through Sp1 sites by histone deacetylase inhibitor apicidin: involvement of protein kinase C. J Biol Chem 2001;276:42,084–42,090.

86. Kim JS, Lee S, Lee T, Lee YW, Trepel JB. Transcriptional activation of p21(WAF1/CIP1) by apicidin, a novel histone deacetylase inhibitor. Biochem Biophys Res Commun 2001; 281:866–871.
87. Kim SH, Ahn S, Han JW, et al. Apicidin is a histone deacetylase inhibitor with anti-invasive and anti-angiogenic potentials. Biochem Biophys Res Commun 2004;315:964–970.
88. Meinke PT, Colletti SL, Doss G, et al. Synthesis of apicidin-derived quinolone derivatives: parasite-selective histone deacetylase inhibitors and antiproliferative agents. J Med Chem 2000;43:4919–4922.
89. Meinke PT, Liberator P. Histone deacetylase: a target for antiproliferative and antiprotozoal agents. Curr Med Chem 2001;8:211–235.
90. Kwon HJ, Owa T, Hassig CA. Shimada J, Schreiber SL. Depudecin induces morphological reversion of transformed fibroblasts via the inhibition of histone deacetylase. Proc Natl Acad Sci USA 1998;95:3356–3361.
91. Hoffmann K, Brosch G, Loidl P, Jung M. First non-radioactive assay for in vitro screening of histone deacetylase inhibitors. Pharmazie 2000;55:601–606.
92. Marks PA, Richon VM, Rifkind RA. Cell cycle regulatory proteins are targets for induced differentiation of transformed cells: Molecular and clinical studies employing hybrid polar compounds. Int J Hematol 1996;63:1–17.
93. Marks PA, Rifkind RA. Hexamethylene bisacetamide-induced differentiation of transformed cells: molecular and cellular effects and therapeutic application. Int J Cell Cloning 1988;6:230–240.
94. Richon VM, Emiliani S, Verdin E, et al. A class of hybrid polar inducers of transformed cell differentiation inhibits histone deacetylases. Proc Natl Acad Sci USA 1998;95:3003–3007.
95. Cohen LA, Amin S, Marks PA, Rifkind RA, Desai D, Richon VM. Chemoprevention of carcinogen-induced mammary tumorigenesis by the hybrid polar cytodifferentiation agent, suberanilohydroxamic acid (SAHA). Anticancer Res 1999;19:4999–5005.
96. Chinnaiyan P, Vallabhaneni G, Armstrong E, Huang SM, Harari PM. Modulation of radiation response by histone deacetylase inhibition. Int J Radiat Oncol Biol Phys 2005;62:223–229.
97. Qiu L, Burgess A, Fairlie DP, Leonard H, Parsons PG, Gabrielli BG. Histone deacetylase inhibitors trigger a G2 checkpoint in normal cells that is defective in tumor cells. Mol Biol Cell 2000;11:2069–2083.
98. Vanoosten RL, Moore JM, Karacay B, Griffith TS. Histone Deacetylase Inhibitors Modulate Renal Cell Carcinoma Sensitivity to TRAIL/Apo-2L-induced Apoptosis by Enhancing TRAIL-R2 Expression. Cancer Bio Ther 2005;4:1104–1112.
99. Doi S, Soda H, Oka M, et al. The histone deacetylase inhibitor FR901228 induces caspase-dependent apoptosis via the mitochondrial pathway in small cell lung cancer cells. Mol Cancer Ther 2004;3:1397–1402.
100. Khan SB, Maududi T, Barton K, Ayers J, Alkan S. Analysis of histone deacetylase inhibitor, depsipeptide (FR901228), effect on multiple myeloma. Br J Haematol 2004;125:156–161.
101. Sawa H, Murakami H, Kumagai M, et al. Histone deacetylase inhibitor, FK228, induces apoptosis and suppresses cell proliferation of human glioblastoma cells in vitro and in vivo. Acta Neuropathol (Berl) 2004;107:523–531.
102. Sato N, Ohta T, Kitagawa H, et al. FR901228, a novel histone deacetylase inhibitor, induces cell cycle arrest and subsequent apoptosis in refractory human pancreatic cancer cells. Int J Oncol 2004;24:679–685.
103. Klisovic DD, Katz SE, Effron D, et al. Depsipeptide (FR901228) inhibits proliferation and induces apoptosis in primary and metastatic human uveal melanoma cell lines. Invest Ophthalmol Vis Sci 2003;44:2390–2398.
104. Aron JL, Parthun MR, Marcucci G, et al. Depsipeptide (FR901228) induces histone acetylation and inhibition of histone deacetylase in chronic lymphocytic leukemia cells concurrent

with activation of caspase 8-mediated apoptosis and down-regulation of c-FLIP protein. Blood 2003;102:652–658.
105. Klisovic MI, Maghraby EA, Parthun MR, et al. Depsipeptide (FR 901228) promotes histone acetylation, gene transcription, apoptosis and its activity is enhanced by DNA methyltransferase inhibitors in AML1/ETO-positive leukemic cells. Leukemia 2003;17:350–358.
106. Sasakawa Y, Naoe Y, Inoue T, et al. Effects of FK228, a novel histone deacetylase inhibitor, on human lymphoma U-937 cells in vitro and in vivo. Biochem Pharmacol 2002;64: 1079–1090.
107. Sasakawa Y, Naoe Y, Noto T, et al. Antitumor efficacy of FK228, a novel histone deacetylase inhibitor, depends on the effect on expression of angiogenesis factors. Biochem Pharmacol 2003;66:897–906.
108. Mie Lee Y, Kim SH, Kim HS, et al. Inhibition of hypoxia-induced angiogenesis by FK228, a specific histone deacetylase inhibitor, via suppression of HIF-1alpha activity. Biochem Biophys Res Commun 2003;300:241–246.
109. Kwon HJ, Kim MS, Kim MJ, Nakajima H, Kim KW. Histone deacetylase inhibitor FK228 inhibits tumor angiogenesis. Int J Cancer 2002;97:290–296.
110. Klisovic DD, Klisovic MI, Effron D, Liu S, Marcucci G, Katz SE. Depsipeptide inhibits migration of primary and metastatic uveal melanoma cell lines in vitro: a potential strategy for uveal melanoma. Melanoma Res 2005;15:147–153.
111. Imai T, Adachi S, Nishijo K, et al. FR901228 induces tumor regression associated with induction of Fas ligand and activation of Fas signaling in human osteosarcoma cells. Oncogene 2003;22:9231–9242.
112. Byrd JC, Marcucci G, Parthun MR, et al. A phase 1 and pharmacodynamic study of depsipeptide (FK228) in chronic lymphocytic leukemia and acute myeloid leukemia. Blood 2005;105:959–967.
113. Lea MA, Randolph VM, Patel M. Increased acetylation of histones induced by diallyl disulfide and structurally related molecules. Int J Oncol 1999;15:347–352.
114. Druesne N, Pagniez A, Mayeur C, et al. Diallyl disulfide (DADS) increases histone acetylation and p21(waf1/cip1) expression in human colon tumor cell lines. Carcinogenesis 2004;25:1227–1236.
115. Pina IC, Gautschi JT, Wang GY, et al. Psammaplins from the sponge Pseudoceratina purpurea: inhibition of both histone deacetylase and DNA methyltransferase. J Org Chem 2003;68:3866–3873.
116. Jiang Y, Ahn EY, Ryu SH, et al. Cytotoxicity of psammaplin A from a two-sponge association may correlate with the inhibition of DNA replication. BMC Cancer 2004;4:70.
117. Shim JS, Lee HS, Shin J, Kwon HJ, Psammaplin A. A marine natural product, inhibits aminopeptidase N and suppresses angiogenesis in vitro. Cancer Lett 2004;203:163–169.
118. Park Y, Liu Y, Hong J, et al. New bromotyrosine derivatives from an association of two sponges, Jaspis wondoensis and Poecillastra wondoensis. J Nat Prod 2003;66:1495–1498.
119. Nicolaou KC, Hughes R, Pfefferkorn JA, Barluenga S. Optimization and mechanistic studies of psammaplin A type antibacterial agents active against methicillin-resistant Staphylococcus aureus (MRSA). Chemistry 2001;7:4296–4310.
120. Pham NB, Butler MS, Quinn RJ. Isolation of psammaplin A 11′-sulfate and bisaprasin 11′-sulfate from the marine sponge Aplysinella rhax. J Nat Prod 2000;63:393–395.
121. Kim D, Lee IS, Jung JH, Lee CO, Choi SU. Psammaplin A, a natural phenolic compound, has inhibitory effect on human topoisomerase II and is cytotoxic to cancer cells. Anticancer Res 1999;19:4085–4090.
122. Kim D, Lee IS, Jung JH, Yang SI. Psammaplin A, a natural bromotyrosine derivative from a sponge, possesses the antibacterial activity against methicillin-resistant Staphylococcus aureus and the DNA gyrase-inhibitory activity. Arch Pharm Res 1999;22:25–29.

123. Jung JH, Sim CJ, Lee CO. Cytotoxic compounds from a two-sponge association. J Nat Prod 1995;58:1722–1726.
124. Yoshida M, Hoshikawa Y, Koseki K, Mori K, Beppu T. Structural specificity for biological activity of trichostatin A, a specific inhibitor of mammalian cell cycle with potent differentiation-inducing activity in Friend leukemia cells. J Antibiot (Tokyo) 1990;43:1101–1106.
125. Yoshida M, Shimazu T, Matsuyama A. Protein deacetylases: enzymes with functional diversity as novel therapeutic targets. Prog Cell Cycle Res 2003;5:269–278.
126. Strait KA, Warnick CT, Ford CD, Dabbas B, Hammond EH, Ilstrup SJ. Histone deacetylase inhibitors induce G2-checkpoint arrest and apoptosis in cisplatinum-resistant ovarian cancer cells associated with overexpression of the Bcl-2-related protein Bad. Mol Cancer Ther 2005;4:603–611.
127. Marks PA, Jiang X. Histone deacetylase inhibitors in programmed cell death and cancer therapy. Cell Cycle 2005;4:549–551.
128. Mai A, Massa S, Rotili D, et al. Histone deacetylation in epigenetics: an attractive target for anticancer therapy. Med Res Rev 2005;25:261–309.
129. Henderson C, Brancolini C. Apoptotic pathways activated by histone deacetylase inhibitors: implications for the drug-resistant phenotype. Drug Resist Updat 2003;6: 247–256.
130. Donadelli M, Costanzo C, Faggioli L, et al. Trichostatin A, an inhibitor of histone deacetylases, strongly suppresses growth of pancreatic adenocarcinoma cells. Mol Carcinog 2003;38:59–69.
131. Marks P, Rifkind RA, Richon VM, Breslow R, Miller T, Kelly WK. Histone deacetylases and cancer: causes and therapies. Nat Rev Cancer 2001;1:194–202.
132. Nome RV, Bratland A, Harman G, Fodstad O, Andersson Y, Ree AH. Cell cycle checkpoint signaling involved in histone deacetylase inhibition and radiation-induced cell death. Mol Cancer Ther 2005;4:1231–1238.
133. Fenic I, Sonnack V, Failing K, Bergmann M, Steger K. In vivo effects of histone-deacetylase inhibitor trichostatin-A on murine spermatogenesis. J Androl 2004;25:811–818.
134. Hu J, Colburn NH. Histone deacetylase inhibition down-regulates cyclin D1 transcription by inhibiting nuclear factor-kappaB/p65 DNA binding. Mol Cancer Res 2005;3:100–109.
135. Kim MS, Kwon HJ, Lee YM, et al. Histone deacetylases induce angiogenesis by negative regulation of tumor suppressor genes. Nat Med 2001;7:437–443.
136. Zgouras D, Becker U, Loitsch S, Stein J. Modulation of angiogenesis-related protein synthesis by valproic acid. Biochem Biophys Res Commun 2004;316:693–697.
137. Sawa H, Murakami H, Ohshima Y, et al. Histone deacetylase inhibitors such as sodium butyrate and trichostatin A inhibit vascular endothelial growth factor (VEGF) secretion from human glioblastoma cells. Brain Tumor Pathol 2002;19:77–81.
138. Deroanne CF, Bonjean K, Servotte S, et al. Histone deacetylases inhibitors as anti-angiogenic agents altering vascular endothelial growth factor signaling. Oncogene 2002;21:427–436.
139. Chinnaiyan P, Varambally S, Tomlins SA, et al. Enhancing the antitumor activity of ErbB blockade with histone deacetylase (HDAC) inhibition. Int J Cancer 2005;118:1041–1050.
140. Duan H, Heckman CA, Boxer LM. Histone deacetylase inhibitors down-regulate bcl-2 expression and induce apoptosis in t(14;18) lymphomas. Mol Cell Biol 2005;25: 1608–1619.
141. Myzak MC, Karplus PA, Chung FL, Dashwood RH. A novel mechanism of chemoprotection by sulforaphane: inhibition of histone deacetylase. Cancer Res 2004;64:5767–5774.
142. Acharya MR, Figg WD. Histone deacetylase inhibitor enhances the anti-leukemic activity of an established nucleoside analogue. Cancer Biol Ther 2004;3:719–720.
143. Rosato RR, Almenara JA, Grant S. The histone deacetylase inhibitor MS-275 promotes differentiation or apoptosis in human leukemia cells through a process regulated by

generation of reactive oxygen species and induction of p21CIP1/WAF1 1. Cancer Res 2003;63: 3637–3645.
144. Lavelle D, Chen YH, Hankewych M, DeSimone J. Histone deacetylase inhibitors increase p21(WAF1) and induce apoptosis of human myeloma cell lines independent of decreased IL-6 receptor expression. Am J Hematol 2001;68:170–178.
145. Rocchi P, Tonelli R, Camerin C, et al. p21Waf1/Cip1 is a common target induced by short-chain fatty acid HDAC inhibitors (valproic acid, tributyrin and sodium butyrate) in neuroblastoma cells. Oncol Rep 2005;13:1139–1144.
146. Mitsiades CS, Poulaki V, McMullan C, et al. Novel histone deacetylase inhibitors in the treatment of thyroid cancer. Clin Cancer Res 2005;11:3958–3965.
147. Yokota T, Matsuzaki Y, Miyazawa K, Zindy F, Roussel MF, Sakai T. Histone deacetylase inhibitors activate INK4d gene through Sp1 site in its promoter. Oncogene 2004;23: 5340–5349.
148. Hitomi T, Matsuzaki Y, Yokota T, Takaoka Y, Sakai T. p15(INK4b) in HDAC inhibitor-induced growth arrest. FEBS Lett 2003;554:347–350.
149. Dehm SM, Hilton TL, Wang EH, Bonham K. SRC proximal and core promoter elements dictate TAF1 dependence and transcriptional repression by histone deacetylase inhibitors. Mol Cell Biol 2004;24:2296–2307.
150. Heruth DP, Zirnstein GW, Bradley JF, Rothberg PG. Sodium butyrate causes an increase in the block to transcriptional elongation in the c-myc gene in SW837 rectal carcinoma cells. J Biol Chem 1993;268:20,466–20,472.
151. Lallemand F, Courilleau D, Sabbah M, Redeuilh G, Mester J. Direct inhibition of the expression of cyclin D1 gene by sodium butyrate. Biochem Biophys Res Commun 1996; 229:163–169.
152. Souleimani A, Asselin C. Regulation of c-myc expression by sodium butyrate in the colon carcinoma cell line Caco-2. FEBS Lett 1993;326:45–50.
153. Takai N, Desmond JC, Kumagai T, et al. Histone deacetylase inhibitors have a profound antigrowth activity in endometrial cancer cells. Clin Cancer Res 2004;10:1141–1149.
154. Vaziri C, Stice L, Faller DV. Butyrate-induced G1 arrest results from p21-independent disruption of retinoblastoma protein-mediated signals. Cell Growth Differ 1998;9:465–474.
155. Juan LJ, Shia WJ, Chen MH, et al. Histone deacetylases specifically down-regulate p53-dependent gene activation. J Biol Chem 2000;275:20,436–20,443.
156. Luo J, Nikolaev AY, Imai S. et al. Negative control of p53 by Sir2alpha promotes cell survival under stress. Cell 2001;107:137–148.
157. Luo J, Su F, Chen D, Shiloh A, Gu W. Deacetylation of p53 modulates its effect on cell growth and apoptosis. Nature 2000;408:377–381.
158. Vaziri H, Dessain SK, Ng Eaton E, et al. hSIR2(SIRT1) functions as an NAD-dependent p53 deacetylase. Cell 2001;107:149–159.
159. Savickiene J, Treigyte G, Pivoriunas A, Navakauskiene R, Magnusson KE. Sp1 and NF-kappaB transcription factor activity in the regulation of the p21 and FasL promoters during promyelocytic leukemia cell monocytic differentiation and its associated apoptosis. Ann N Y Acad Sci 2004;1030:569–577.
160. Varshochi R, Halim F, Sunters A, et al. ICI182,780 induces p21Waf1 gene transcription through releasing histone deacetylase 1 and estrogen receptor alpha from Sp1 sites to induce cell cycle arrest in MCF-7 breast cancer cell line. J Biol Chem 2005;280: 3185–3196.
161. Parker SB, Eichele G, Zhang P, et al. p53-independent expression of p21Cip1 in muscle and other terminally differentiating cells. Science 1995;267:1024–1027.
162. Rosato RR, Almenara JA, Dai Y, Grant S. Simultaneous activation of the intrinsic and extrinsic pathways by histone deacetylase (HDAC) inhibitors and tumor necrosis factor-related

apoptosis-inducing ligand (TRAIL) synergistically induces mitochondrial damage and apoptosis in human leukemia cells. Mol Cancer Ther 2003;2:1273–1284.
163. Papeleu P, Vanhaecke T, Elaut G, et al. Differential effects of histone deacetylase inhibitors in tumor and normal cells-what is the toxicological relevance? Crit Rev Toxicol 2005;35:363–378.
164. Zhang Y, Jung M, Dritschilo A, Jung M. Enhancement of radiation sensitivity of human squamous carcinoma cells by histone deacetylase inhibitors. Radiat Res 2004;161:667–674.
165. Mori N, Matsuda T, Tadano M, et al. Apoptosis induced by the histone deacetylase inhibitor FR901228 in human T-cell leukemia virus type 1-infected T-cell lines and primary adult T-cell leukemia cells. J Virol 2004;78:4582–4590.
166. Kim DH, Kim M, Kwon HJ. Histone deacetylase in carcinogenesis and its inhibitors as anti-cancer agents. J Biochem Mol Biol 2003;36:110–119.
167. Vigushin DM, Coombes RC. Histone deacetylase inhibitors in cancer treatment. Anticancer Drugs 2002;13:1–13.
168. Amin HM, Saeed S, Alkan S. Histone deacetylase inhibitors induce caspase-dependent apoptosis and downregulation of daxx in acute promyelocytic leukaemia with t(15;17). Br J Haematol 2001;115:287–297.
169. Chen CS, Weng SC, Tseng PH, Lin HP, Chen CS. Histone Acetylation-independent Effect of Histone Deacetylase Inhibitors on Akt through the Reshuffling of Protein Phosphatase 1 Complexes. J Biol Chem 2005;280:38,879–38,887.
170. Neuzil J, Swettenham E, Gellert N. Sensitization of mesothelioma to TRAIL apoptosis by inhibition of histone deacetylase: role of Bcl-xL down-regulation. Biochem Biophys Res Commun 2004;314:186–191.
171. Zhang XD, Gillespie SK, Borrow JM, Hersey P. The histone deacetylase inhibitor suberic bishydroxamate: a potential sensitizer of melanoma to TNF-related apoptosis-inducing ligand (TRAIL) induced apoptosis. Biochem Pharmacol 2003;66:1537–1545.
172. Mitsiades N, Mitsiades CS, Richardson PG, et al. Molecular sequelae of histone deacetylase inhibition in human malignant B cells. Blood 2003;101:4055–4062.
173. Zhang XD, Gillespie SK, Borrow JM, Hersey P. The histone deacetylase inhibitor suberic bishydroxamate regulates the expression of multiple apoptotic mediators and induces mitochondria-dependent apoptosis of melanoma cells. Mol Cancer Ther 2004;3:425–435.
174. Green DR, Reed JC. Mitochondria and apoptosis. Science 1998;281:1309–1312.
175. Johnstone RW, Ruefli AA, Lowe SW. Apoptosis: a link between cancer genetics and chemotherapy. Cell 2002;108:153–164.
176. Wei MC, Lindsten T, Mootha VK, et al. tBID, a membrane-targeted death ligand, oligomerizes BAK to release cytochrome c. Genes Dev 2000;14:2060–2071.
177. Kandasamy K, Srinivasula SM, Alnemri ES, et al. Involvement of proapoptotic molecules Bax and Bak in tumor necrosis factor-related apoptosis-inducing ligand (TRAIL)-induced mitochondrial disruption and apoptosis: differential regulation of cytochrome c and Smac/DIABLO release. Cancer Res 2003;63:1712–1721.
178. Grimm S, Bauer MK, Baeuerle PA, Schulze-Osthoff K. Bcl-2 down-regulates the activity of transcription factor NF-kappaB induced upon apoptosis. J Cell Biol 1996;134:13–23.
179. Gross A, McDonnell JM, Korsmeyer SJ. BCL-2 family members and the mitochondria in apoptosis. Genes Dev 1999;13:1899–1911.
180. Gross A. BCL-2 proteins: regulators of the mitochondrial apoptotic program. IUBMB Life 2001;52:231–236.
181. Degenhardt K, Chen G, Lindsten T, White E. BAX and BAK mediate p53-independent suppression of tumorigenesis. Cancer Cell 2002;2:193–203.
182. Degenhardt K, Sundararajan R, Lindsten T, Thompson C, White E. Bax and Bak independently promote cytochrome C release from mitochondria. J Biol Chem 2002;277: 14,127–14,134.

183. Bouillet P, Purton JF, Godfrey DI, et al. BH3-only Bcl-2 family member Bim is required for apoptosis of autoreactive thymocytes. Nature 2002;415:922–926.
184. Bouillet P, Strasser A. BH3-only proteins - evolutionarily conserved proapoptotic Bcl-2 family members essential for initiating programmed cell death. J Cell Sci 2002;115: 1567–1574.
185. Deveraux QL, Reed JC. IAP family proteins—suppressors of apoptosis. Genes Dev 1999; 13:239–252.
186. Deveraux QL, Stennicke HR, Salvesen GS, Reed JC. Endogenous inhibitors of caspases. J Clin Immunol 1999;19:388–398.
187. Du C, Fang M, Li Y, Li L, Wang X. Smac, a mitochondrial protein that promotes cytochrome c-dependent caspase activation by eliminating IAP inhibition. Cell 2000; 102:33–42.
188. Verhagen AM, Ekert PG, Pakusch M, et al. Identification of DIABLO, a mammalian protein that promotes apoptosis by binding to and antagonizing IAP proteins. Cell 2000; 102:43–53
189. Terui T, Murakami K, Takimoto R, et al. Induction of PIG3 and NOXA through acetylation of p53 at 320 and 373 lysine residues as a mechanism for apoptotic cell death by histone deacetylase inhibitors. Cancer Res 2003;63:8948–8954.
190. Xu WS, Perez G, Ngo L, Gui CY, Marks PA. Induction of polyploidy by histone deacetylase inhibitor: a pathway for antitumor effects. Cancer Res 2005;65:7832–7839.
191. Xu Y, Voelter-Mahlknecht S, Mahlknecht U. The histone deacetylase inhibitor suberoylanilide hydroxamic acid down-regulates expression levels of Bcr-abl, c-Myc and HDAC3 in chronic myeloid leukemia cell lines. Int J Mol Med 2005;15:169–172.
192. Nimmanapalli R, Fuino L, Stobaugh C, Richon V, Bhalla K. Cotreatment with the histone deacetylase inhibitor suberoylanilide hydroxamic acid (SAHA) enhances imatinib-induced apoptosis of Bcr-Abl-positive human acute leukemia cells. Blood 2003;101:3236–3239.
193. Bradbury CA, Khanim FL, Hayden R, et al. Histone deacetylases in acute myeloid leukaemia show a distinctive pattern of expression that changes selectively in response to deacetylase inhibitors. Leukemia 2005;19:1751–1759.
194. Folkman J. Role of angiogenesis in tumor growth and metastasis. Semin Oncol 2002;29: 15–18.
195. Folkman J. Fundamental concepts of the angiogenic process. Curr Mol Med 2003; 3:643–651.
196. Folkman J. Angiogenesis and proteins of the hemostatic system. J Thromb Haemost 2003; 1:1681–1682.
197. Folkman J. Angiogenesis and apoptosis. Semin Cancer Biol 2003;13:159–167.
198. Folkman J, Kalluri R. Cancer without disease. Nature 2004;427:787.
199. Liotta LA, Steeg PS, Stetler-Stevenson WG. Cancer metastasis and angiogenesis: an imbalance of positive and negative regulation. Cell 1991;64:327–336.
200. Folkman J. Antiangiogenic activity of a matrix protein. Cancer Biol Ther 2003;2:53–54.
201. Ravery V, Jouanneau J, Gil Diez S, et al. Immunohistochemical detection of acidic fibroblast growth factor in bladder transitional cell carcinoma. Urol Res 1992;20:211–214.
202. Allen LE, Maher PA. Expression of basic fibroblast growth factor and its receptor in an invasive bladder carcinoma cell line. J Cell Physiol 1993;155:368–375.
203. Brown LF, Berse B, Jackman RW, et al. Increased expression of vascular permeability factor (vascular endothelial growth factor) and its receptors in kidney and bladder carcinomas. Am J Pathol 1993;143:1255–1262.
204. Brown LF, Berse B, Jackman RW, et al. Expression of vascular permeability factor (vascular endothelial growth factor) and its receptors in adenocarcinomas of the gastrointestinal tract. Cancer Res 1993;53:4727–4735.

205. O'Brien T, Cranston D, Fuggle S, Bicknell R, Harris AL. Different angiogenic pathways characterize superficial and invasive bladder cancer. Cancer Res 1995;55:510–513.
206. Caponigro F, Basile M, de Rosa V, Normanno N. New drugs in cancer therapy, National Tumor Institute, Naples, 17-18 June 2004. Anticancer Drugs 2005;16:211–221.
207. Nam NH, Parang K. Current targets for anticancer drug discovery. Curr Drug Targets 2003;4:159–179.
208. Wiedmann MW, Caca K. Molecularly targeted therapy for gastrointestinal cancer. Curr Cancer Drug Targets 2005;5:171–193.
209. Michaelis M, Suhan T, Michaelis UR, et al. Valproic acid induces extracellular signal-regulated kinase 1/2 activation and inhibits apoptosis in endothelial cells. Cell Death Differ 2006;13(3):446–453.
210. He GH, Helbing CC, Wagner MJ, Sensen CW, Riabowol K. Phylogenetic analysis of the ING family of PHD finger proteins. Mol Biol Evol 2005;22:104–116.
211. Murakami J, Asaumi J, Maki Y, et al. Effects of demethylating agent 5-aza-2(')-deoxycytidine and histone deacetylase inhibitor FR901228 on maspin gene expression in oral cancer cell lines. Oral Oncol 2004;40:597–603.
212. Michaelis M, Michaelis UR, Fleming I, et al. Valproic acid inhibits angiogenesis in vitro and in vivo. Mol Pharmacol 2004;65:520–527.
213. Bapna A, Vickerstaffe E, Warrington BH, Ladlow M, Fan TP, Ley SV. Polymer-assisted, multi-step solution phase synthesis and biological screening of histone deacetylase inhibitors. Org Biomol Chem 2004;2:611–620.
214. Momparler RL. Cancer epigenetics. Oncogene 2003;22:6479–6483.
215. Wang S, Yan-Neale Y, Fischer D, et al. Histone deacetylase 1 represses the small GTPase RhoB expression in human nonsmall lung carcinoma cell line. Oncogene 2003;22:6204–6213.
216. Liu LT, Chang HC, Chiang LC, Hung WC. Histone deacetylase inhibitor up-regulates RECK to inhibit MMP-2 activation and cancer cell invasion. Cancer Res 2003;63: 3069–3072.
217. Rossig L, Li H, Fisslthaler B. Inhibitors of histone deacetylation downregulate the expression of endothelial nitric oxide synthase and compromise endothelial cell function in vasorelaxation and angiogenesis. Circ Res 2002;91:837–844.
218. Pili R, Kruszewski MP, Hager BW, Lantz J, Carducci MA. Combination of phenylbutyrate and 13-cis retinoic acid inhibits prostate tumor growth and angiogenesis. Cancer Res 2001;61:1477–1485.
219. Qian DZ, Wang X, Kachhap SK, et al. The histone deacetylase inhibitor NVP-LAQ824 inhibits angiogenesis and has a greater antitumor effect in combination with the vascular endothelial growth factor receptor tyrosine kinase inhibitor PTK787/ZK222584. Cancer Res 2004;64:6626–6634.
220. Nebbioso A, Clarke N, Voltz E, et al. Tumor-selective action of HDAC inhibitors involves TRAIL induction in acute myeloid leukemia cells. Nat Med 2005;11:77–84.
221. Inoue S, MacFarlane M, Harper N, Wheat LM, Dyer MJ, Cohen GM. Histone deacetylase inhibitors potentiate TNF-related apoptosis-inducing ligand (TRAIL)-induced apoptosis in lymphoid malignancies. Cell Death Differ 2004;11 Suppl 2:S193–206.
222. Facchetti F, Previdi S, Ballarini M, Minucci S, Perego P, La Porta CA. Modulation of pro- and anti-apoptotic factors in human melanoma cells exposed to histone deacetylase inhibitors. Apoptosis 2004;9:573–582.
223. Shetty S, Graham BA, Brown JG, et al. Transcription factor NF-kappaB differentially regulates death receptor 5 expression involving histone deacetylase 1. Mol Cell Biol 2005; 25:5404–5416.
224. Goldsmith KC, Hogarty MD. Targeting programmed cell death pathways with experimental therapeutics: opportunities in high-risk neuroblastoma. Cancer Lett 2005;228:133–141.

225. Insinga A, Monestiroli S, Ronzoni S, et al. Inhibitors of histone deacetylases induce tumor-selective apoptosis through activation of the death receptor pathway. Nat Med 2005;11:71–76.
226. Insinga A, Minucci S, Pelicci PG. Mechanisms of selective anticancer action of histone deacetylase inhibitors. Cell Cycle 2005;4:741–743.
227. Suliman A, Lam A, Datta R, Srivastava RK. Intracellular mechanisms of TRAIL: apoptosis through mitochondrial-dependent and -independent pathways. Oncogene 2001;20: 2122–2133.
228. Sartorius U, Schmitz I, Krammer PH. Molecular mechanisms of death-receptor-mediated apoptosis. Chembiochem 2001;2:20–29.
229. Debatin KM, Krammer PH. Death receptors in chemotherapy and cancer. Oncogene 2004;23:2950–2966.
230. Luo X, Budihardjo I, Zou H, Slaughter C, Wang X. Bid, a Bcl2 interacting protein, mediates cytochrome c release from mitochondria in response to activation of cell surface death receptors. Cell 1998;94:481–490.
231. Singh TR, Shankar S, Chen X, Asim M, Srivastava RK. Synergistic interactions of chemotherapeutic drugs and tumor necrosis factor-related apoptosis-inducing ligand/Apo-2 ligand on apoptosis and on regression of breast carcinoma in vivo. Cancer Res 2003;63: 5390–5400.
232. Shankar S, Singh TR, Chen X, Thakkar H, Firnin J, Srivastava RK. The sequential treatment with ionizing radiation followed by TRAIL/Apo-2L reduces tumor growth and induces apoptosis of breast tumor xenografts in nude mice. Int J Oncol 2004;24:1133–1140.
233. Shankar S, Singh TR, Srivastava RK. Ionizing radiation enhances the therapeutic potential of TRAIL in prostate cancer in vitro and in vivo: Intracellular mechanisms. Prostate 2004;61:35–49.
234. Keane MM, Ettenberg SA, Nau MM, Russell EK, Lipkowitz S. Chemotherapy augments TRAIL-induced apoptosis in breast cell lines. Cancer Res 1999;59:734–741.
235. Nagane M, Pan G, Weddle JJ, Dixit VM, Cavenee WK, Huang HJ. Increased death receptor 5 expression by chemotherapeutic agents in human gliomas causes synergistic cytotoxicity with tumor necrosis factor-related apoptosis-inducing ligand in vitro and in vivo. Cancer Res 2000;60:847–853.
236. Ravi R, Bedi GC, Engstrom LW, et al. Regulation of death receptor expression and TRAIL/Apo2L-induced apoptosis by NF-kappaB. Nat Cell Biol 2001;3:409–416.
237. Chen X, Kandasamy K, Srivastava RK. Differential roles of RelA (p65) and c-Rel subunits of nuclear factor kappa B in tumor necrosis factor-related apoptosis-inducing ligand signaling. Cancer Res 2003;63:1059–1066.
238. Shankar S, Chen X, Srivastava RK. Effects of sequential treatments with chemotherapeutic drugs followed by TRAIL on prostate cancer in vitro and in vivo. Prostate 2005;62: 165–186.
239. Chinnaiyan AM, Prasad U, Shankar S. Combined effect of tumor necrosis factor-related apoptosis-inducing ligand and ionizing radiation in breast cancer therapy. Proc Natl Acad Sci USA 2000;97:1754–1759.
240. Kim JH, Shin JH, Kim IH. Susceptibility and radiosensitization of human glioblastoma cells to trichostatin A, a histone deacetylase inhibitor. Int J Radiat Oncol Biol Phys 2004;59:1174–1180.
241. Camphausen K, Burgan W, Cerra M, et al. Enhanced radiation-induced cell killing and prolongation of gammaH2AX foci expression by the histone deacetylase inhibitor MS-275. Cancer Res 2004;64:316–321.
242. Camphausen K, Cerna D, Scott T, et al. Enhancement of in vitro and in vivo tumor cell radiosensitivity by valproic acid. Int J Cancer 2005;114:380–386.

243. Bevins RL, Zimmer SG. It's about time: scheduling alters effect of histone deacetylase inhibitors on camptothecin-treated cells. Cancer Res 2005;65:6957–6966.
244. Chen WY, Bailey EC, McCune SL, Dong JY, Townes TM. Reactivation of silenced, virally transduced genes by inhibitors of histone deacetylase. Proc Natl Acad Sci USA 1997;94:5798–5803.
245. Baylin S, Bestor TH. Altered methylation patterns in cancer cell genomes: cause or consequence? Cancer Cell 2002;1:299–305.
246. Baylin SB, Esteller M, Rountree MR, Bachman KE, Schuebel K, Herman JG. Aberrant patterns of DNA methylation, chromatin formation and gene expression in cancer. Hum Mol Genet 2001;10:687–692.
247. Zhu WG, Lakshmanan RR, Beal MD, Otterson GA. DNA methyltransferase inhibition enhances apoptosis induced by histone deacetylase inhibitors. Cancer Res 2001;61:1327–1333.
248. Chobanian NH, Greenberg VL, Gass JM, et al. Histone deacetylase inhibitors enhance paclitaxel-induced cell death in ovarian cancer cell lines independent of p53 status. Anticancer Res 2004; 24:539–545.
249. Blagosklonny MV, Robey R, Bates S, Fojo T. Pretreatment with DNA-damaging agents permits selective killing of checkpoint-deficient cells by microtubule-active drugs. J Clin Invest 2000;105:533–539.
250. Gozzini A, Santini V. Butyrates and decitabine cooperate to induce histone acetylation and granulocytic maturation of t(8;21) acute myeloid leukemia blasts. Ann Hematol 2005;84(Suppl 13):1–7.
251. Myzak MC, Hardin K, Wang R, Dashwood RH, Ho E. Sulforaphane inhibits histone deacetylase activity in BPH-1, LnCaP, and PC-3 prostate epithelial cells. Carcinogenesis 2006;27(4):811–819.
252. Kelly WK, O'Connor OA, Krug LM, et al. Phase I study of an oral histone deacetylase inhibitor, suberoylanilide hydroxamic acid, in patients with advanced cancer. J Clin Oncol 2005;23:3923–3931.
253. Hess-Stumpp H. Histone deacetylase inhibitors and cancer: from cell biology to the clinic. Eur J Cell Biol 2005;84:109–121.
254. Blanchard F, Chipoy C. Histone deacetylase inhibitors: new drugs for the treatment of inflammatory diseases? Drug Discov Today 2005;10:197–204.

Index

A

ABL, *see BCR-ABL*; TEL-ABL
Acute myeloid leukemia (AML),
 cellular origin of human disease, 16
 histone deacetylase expression, 343
Akt,
 FOXO regulation, 107
 survival pathway activation, 101, 102
AML, *see* Acute myeloid leukemia
Angiogenesis,
 definition, 35
 endothelial progenitor cell roles, 37, 281, 282
 histone deacetylase inhibitor effects, 343, 344
 inhibitors,
  endorepellin, 52
  endostatin, 43, 45
  perlecan, 52
  tumstatin, 45–51
 c-Myc role, 154, 155
 nuclear factor-κB role, 242
 regulators,
  angiopoietins, 40
  ephrins, 41
  integrins, 42
  matrix metalloproteinases, 40, 41
  vascular endothelial growth factor, 37–40
  VE-cadherin, 42
 sphingosine-1-phosphate role, 191
 steps, 35–38
 tumor mechanisms, 42, 43
Angiopoietins, angiogenesis regulation, 40
Anoikis, metastasis and resistance, 79, 80
Antisense oligonucleotides, c-Myc targeting, 164, 165
APAF-1,
 metastasis and altered expression, 69
 c-Myc interactions, 149
Apicidin, histone deacetylase inhibition, 339
ARF,
 c-Myc induction, 150, 151
 p53-MDM2-ARF network and negative feedback loops, 223–225
Arrestin, angiogenesis inhibition, 47
Arsenic trioxide, BCR-ABL suppression, 22
ATM,
 cell cycle checkpoint control, 125
 DNA damage sensing, 120, 121
ATR,
 cell cycle checkpoint control, 123, 124, 126
 DNA damage sensing, 121

B

Bak, function, 100
Bax,
 function, 100
 metastasis and altered expression, 71, 72
Bcl-2,
 carcinogenesis role, 100, 101
 function, 100
 metastasis and altered expression, 70
 c-Myc repression, 150
 nuclear factor-κB regulation, 241
Bcl-xL,
 function, 100
 metastasis and altered expression, 71
 c-Myc repression, 150
 nuclear factor-κB regulation, 241
BCR-ABL,
 ABL function, 3, 4
 animal models, *see* chronic myelogenous leukemia
 cellular origin of human disease, 15, 16
 genotype–phenotype correlations, 4, 5
 healthy individuals, 10
 major breakpoint cluster region, 5
 signaling pathways, 7
 therapeutic targeting,

Gleevec,
acute leukemia and blast crisis trials, 20
historical perspective, 19, 20
Philadelphia chromosome cell effects, 21, 22
resistance mechanisms, 20, 21
protein suppression, 22
pyridopyrimidines, 22
rationale, 16–18
signaling molecule targeting, 23
transcription targeting, 22
tyrosine kinase inhibitor overview, 18, 19
transcript types, 7–9
transformation induction, 10, 11
BRCA1, cell cycle checkpoint control, 126

C

VE-Cadherin, angiogenesis regulation, 42
Canstatin, angiogenesis inhibition, 47, 48
Caspases,
apoptosis module, 220, 221
classification, 98
metastasis dysfunction,
inhibitor of apoptosis protein altered expression, 68
mutation and altered expression, 67, 68
CD95, *see* Fas
Cdc25A,
cell cycle checkpoint control, 124, 126
G1/S module, 219
Cell cycle checkpoints,
G1, 123
G1/S, 123, 124
mitotic checkpoint control, 126, 127
S, 125
transcriptional control nodes linking cell cycle and apoptosis, *see* Transcriptional control nodes
Checkpoint, *see* Cell cycle checkpoints
Chemoprevention therapy, histone deacetylase inhibitor effects on response, 347
Chemotherapy,
apoptosis and outcomes,
clinical importance, 307, 308
clonogenic survival in vitro, 306
overview, 305, 306
therapeutic induction of apoptosis
apoptotically compromised tumors, 308–310
clinical prospects, 312–314
non-neoplastic cells, 310, 311
solid cancers, 311, 312
tumor microenvironment effects, 306, 307
histone deacetylase inhibitor effects on response, 346, 347
historical perspective, 303
mitotic catastrophe induction, 310
senescence induction, 310
Chronic myelogenous leukemia (CML),
animal models,
grafts in immunodeficient mice, 12
knockout mice, 15
retroviral transduction, 13
syngeneic mice, 12
transgenic mice, 13, 15
gene mutations, 6, 7
natural history, 6
Philadelphia chromosome, *see BCR-ABL*
Clonogenic survival assays, chemotherapy and apoptosis, 306
CML, *see* Chronic myelogenous leukemia
Control nodes, *see* Transcriptional control nodes
Coronary artery stenosis, c-Myc therapeutic targeting, 165
COX-2, *see* Cyclooxygenase-2
Cyclooxygenase-2 (COX-2), nuclear factor-κB regulation, 242
Cytochrome c,
APAF-1 interactions, 149
prosurvival activity, 104

D

DAP kinases, *see* Death-associated protein kinases
DcR3, *see* Decoy receptor-3
Death-associated protein (DAP) kinases, metastasis and altered expression, 68, 69
Death-induced signaling complex (DISC), apoptosis module, 220
Decoy receptor-3 (DcR3), therapeutic targeting, 270
Depsipeptide, histone deacetylase inhibition, 340
Depudecin, histone deacetylase inhibition, 339
Diabetes,
epidemiology, 161
c-Myc apoptosis and b-cell failure, 161, 162
Diabetic nephropathy, tumstatin management, 50, 51

DISC, *see* Death-induced signaling complex
DNA methyltransferase, c-Myc complex, 142
DNA repair,
damage,
coupling to cell cycle checkpoints, 122—127
effects, 119, 120
mechanisms, 119
sensing, 120, 121
double-strand break repair, 121, 122
5-fluorouracil resistance, 266, 267
therapeutic targeting, 127, 129
ultraviolet-induced lesions, 122

E

E2F, transcriptional control node, 218, 219, 221, 222
ECM, *see* Extracellular matrix
EGFR, *see* Epidermal growth factor receptor
Endorepellin, angiogenesis inhibition, 52
Endostatin, angiogenesis inhibition, 43, 45
Endothelial progenitor cell (EPC),
angiogenesis, 37, 281, 282
apoptotic and antiapoptotic mechanisms, 283, 284
cellular vehicles for tumor gene therapy,
antiangiogenic genes, 288
cell sources, 285
clinical prospects, 288, 289
expansion in vitro, 285, 286
genetic manipulation, 286
packaging cells for viral vectors, 288
rationale, 286, 287
suicide genes, 287
functions, 280, 281
markers, 281
EPC, *see* Endothelial progenitor cell
Ephrins, angiogenesis regulation, 41
Epidermal growth factor receptor (EGFR), therapeutic targeting, 270, 271, 312, 313
ETV-ABL, *see* TEL-ABL
Extracellular matrix (ECM), angiogenesis, 36
Extrinsic apoptotic pathway,
5-fluorouracil induction, 269
metastasis and altered signaling, 73–75
overview, 97–100

F

Fas (CD95),
5-fluorouracil apoptosis induction of ligand, 269, 270
metastasis and resistance, 73–75
survival pathway activation, 106
Fibroblast growth factor, nuclear factor-κB induction, 242
FLIP,
c-Myc in downregulation, 149
nuclear factor-κB induction, 228
therapeutic targeting, 270
5-Fluorouracil (FU),
apoptosis induction, 268–270
mechanism of action,
DNA damage, 26
RNA misincorporation, 265, 266
thymidylate synthase inhibition, 264, 265
metabolism, 263, 264
resistance mechanisms,
DNA repair, 266, 267
drug activation/inactivation, 266
p53 deficiency, 267, 268
prosurvival signaling, 270, 271
thymidylate synthase expression, 266
resistance to pharmacologically-induced apoptosis, 81
FOXO3a, apoptosis regulation, 107, 108
FU, see 5-Fluorouracil

G

βGBP,
cell cycle regulation, 204, 205, 207
molecular phenotypes of apoptosis pathways,
molecular phenotype I, 207
molecular phenotype II, 207–209
receptors, 204, 205
therapeutic potential of recombinant protein, 209, 210, 212
Gene therapy,
apoptotic sensitivity restoration approaches in cancer, 85, 86
brain tumors, 289
cellular vehicles,
endothelial progenitor cells,
antiangiogenic genes, 288
cell sources, 285
clinical prospects, 288, 289
expansion in vitro, 285, 286
genetic manipulation, 286
packaging cells for viral vectors, 288
rationale, 286, 287
suicide genes, 287
neural progenitor cells,
antiangiogenic genes, 294
cell sources, 290, 291

clinical prospects, 294, 295
cytokine expression, 293
homing to brain tumors, 289, 290
immune response protection, 292
oncolytic virus delivery, 293, 294
suicide genes, 292, 293
TRAIL expression, 294
overview, 279, 280
c-Myc targeting, 164
Gleevec, *see* Imatinib mesylate

H

HDACs, *see* Histone deacetylases
Hexamethylene bisacetamide (HMBA), histone deacetylase inhibition, 339
Histone deacetylases (HDACs),
classification, 335, 336
corepressor complexes, 337
functional overview, 335
inhibitors,
angiogenesis response, 343, 344
benzamides, 338
cell cycle effects, 341
chemoprevention therapy response, 347
chemotherapy response, 346, 347
clinical trials, 347, 348
cyclic tetrapeptides, 339
depsipeptide, 340
inhibitor of apoptosis protein expression effects, 342, 343
mechanisms of action, 341
overview, 337, 338
psammaplin A, 340, 341
radiosensitization, 345, 346
tetrapeptide analogs, 339, 340
TRAIL response, 344, 345
trichostatin A, 338
nuclear factor-κB regulation, 248, 249
recruitment, 336, 337
Sir2 and longevity, 336
stability, 337
Histone H2AX, DNA damage sensing, 120, 121
HMBA, *see* Hexamethylene bisacetamide
Homologous recombination, double-strand break repair, 121, 122

I

IAPs, *see* Inhibitor of apoptosis proteins
IGF-1, *see* Insulin-like growth factor-1
Imatinib mesylate (Gleevec),
chronic myelogenous leukemia management,
acute leukemia and blast crisis trials, 20
Philadelphia chromosome cell effects, 21, 22
historical perspective, 19, 20
resistance mechanisms, 20, 21
Inhibitor of apoptosis proteins (IAPs), *see also* Survivin,
caspase interactions, 99
functional overview, 321
histone deacetylase inhibitor effects on expression, 342, 343
metastasis and altered expression, 68
nuclear factor-κB induction, 228
Insulin-like growth factor-1 (IGF-1), c-Myc apoptosis inhibition, 150
Integrins,
angiogenesis regulation, 42
tumstatin binding, 48
Intrinsic apoptotic pathway,
5-fluorouracil induction, 268, 269
metastasis and altered signaling, 69–73
overview, 97–100

J

JUNB, knockout mouse model of chronic myelogenous leukemia, 15

L

Lipocalins, functions, 105
LPA, *see* Lysophosphatidic acid
Lysophosphatidic acid (LPA),
cancer pathophysiology, 189–191
growth factor activity, 184–186
intracellular actions, 188
metabolism, 179, 180
receptors and signaling, 182–184
survival factor activity, 186–188

M

Mad,
Max complex, 146
c-Myc cell differentiation role, 146
MAPK, *see* Mitogen-activated protein kinase
Maspin, metastasis and altered expression, 72, 73
Matrix metalloproteinases (MMPs),
angiogenesis regulation, 40, 41
nuclear factor-κB regulation, 244
Max,
Mad complex, 146
c-Myc,
complex, 142, 145

therapeutic targeting, 165, 166
MDM2,
cell cycle checkpoint control, 123, 124
p53-MDM2-ARF network and negative feedback loops, 223–225
Metastasis,
apoptotic resistance,
diagnosis and treatment, 84, 85
mechanisms
anoikis resistance, 79, 80
caspase dysfunction, 66–69
extrinsic apoptotic pathway altered signaling, 73–75
intrinsic apoptotic pathway altered signaling, 69–73
transcriptional regulators, 76–79
overview, 64, 65
resistance to pharmacologically induced apoptosis
5-fluorouracil, 81
nitric oxide, 82–84
paclitaxel, 81, 82
sensitivity restoration approaches, 85, 86
assays, 65, 66
nuclear factor-κB role, 243, 244
stages, 63, 64
Microenvironment, *see* Tumor microenvironment
Mitogen-activated protein kinase (MAPK), survival pathway activation, 102, 103
Mitotic catastrophe, chemotherapy induction, 310
MMPs, *see* Matrix metalloproteinases
MRN complex, cell cycle checkpoint control, 125
MS-275, histone deacetylase inhibition, 338–340
c-Myc,
activation in tumors, 160
apoptosis evasion role in cancer, 152–160
developmental effects on activation effects, 158
diabetes role, 161, 162
function, 138
overexpression in cancer, 138, 153
prospects for study, 166, 167
regulation of activity,
activation, 140
apoptosis modulation, 146, 147, 149–152
cell cycle effects, 143–145
cell differentiation effects, 146
transcriptional activity, 142, 143
turnover, 140–142
therapeutic targeting, 138, 139, 162–166
transcriptional control node, 218, 219, 221, 222
transgenic mouse, 153–155
tumor regression following deactivation, 155, 156

N

NER, *see* Nucleotide excision repair
Neural progenitor cell (NPC),
antitumor effects, 294
apoptotic and antiapoptotic mechanisms, 291, 292
functions, 289
gene therapy vehicles,
antiangiogenic genes, 294
cell sources, 290, 291
clinical prospects, 294, 295
cytokine expression, 293
homing to brain tumors, 289, 290
immune response protection, 292
oncolytic virus delivery, 293, 294
suicide genes, 292, 293
TRAIL expression, 294
Neuropilin, vascular endothelial growth factor binding, 40
NF-κB, *see* Nuclear factor-κB
NGAL, apoptosis modulation, 105
Nitric oxide (NO), resistance to pharmacologically induced apoptosis, 82–84
NO, *see* Nitric oxide
NPC, *see* Neural progenitor cell
Nuclear factor-κB (NF-κB),
acetylation, 247–249
activation, 239, 240, 249
family of proteins, 239
5-fluorouracil resistance role, 271
history of study, 239
metastasis,
constitutive activation, 76
therapeutic targeting, 76, 77
oncogenesis roles,
angiogenesis role, 242
apoptosis inhibition, 241, 242
cell proliferation, 240, 241
metastasis role, 243, 244
phosphorylation, 246, 247
proteasome degradation, 244, 245
survival pathway activation, 101, 106, 108
transcriptional control node, 218, 227–229

Nucleotide excision repair (NER), ultraviolet-induced lesions, 122

O

Oncogenes, *see also* c-Myc,
tumor regression following deactivation, 155–158

P

p53,
cell cycle checkpoint control, 123, 124
5-fluorouracil resistance and deficiency, 267, 268
metastasis and dysfunction, 77
c-Myc activation, 150
therapeutic targeting, 129
transcriptional control node, 218, 222–227
Paclitaxel (Taxol), resistance to pharmacologically induced apoptosis, 81, 82
PD compounds,
chromic myelogenous leukemia trials, 22
mechanism of action, 129
Perlecan, angiogenesis inhibition, 52
Philadelphia chromosome, *see* BCR-ABL
Phosphatidylinositol 3-kinase (PI3K),
survival pathway activation, 101, 102
therapeutic targeting, 204
PI3K, *see* Phosphatidylinositol 3-kinase
Polyaminohydroxamic acids, histone deacetylase inhibition, 340
Psammaplin A, histone deacetylase inhibition, 340, 341

R

Radiation therapy, histone deacetylase inhibitors and radiosensitization, 345, 346
Raf, therapeutic targeting, 204
Reactive oxygen species (ROS),
apoptosis role, 99, 100, 108
c-Myc and accumulation, 147
RNA interference, c-Myc targeting, 165
ROS, *see* Reactive oxygen species

S

S1P, *see* Sphingosine-1-phosphate
SAHA, *see* Suberoylanilide hydroxamic acid
Senescence, chemotherapy induction, 310
SLUG, expression in chronic myelogenous leukemia, 7
Sphingosine-1-phosphate (S1P),
cancer pathophysiology, 189–191
growth factor activity, 184–186
intracellular actions, 189
metabolism, 180–182
receptors and signaling, 182–184
survival factor activity, 186–188
STATs, activation in cancer, 271
STI571, *see* Imatinib mesylate
Suberoylanilide hydroxamic acid (SAHA), histone deacetylase inhibition, 340, 343
Survivin,
function,
apoptosis inhibition, 324–326
cell division, 322–324
overview, 322
structure, 321, 322
therapeutic targeting, 327, 328
tumor expression and prognosis, 326, 327
Syndecans, nuclear factor-κB regulation, 243

T

Taxol, *see* Paclitaxel
TEL-ABL,
leukemias, 9
proteins, 9
TGF-β, *see* Transforming growth factor-β
Tiam1, c-Myc interactions, 152
TRAIL,
histone deacetylase inhibitor effects on response, 344, 345
metastasis and resistance, 75
survival pathway activation, 106
therapeutic targeting, 269, 270
Transcriptional control nodes,
cell cycle and apoptosis modules
apoptosis module, 220, 221
G1/S module, 219
É2F-Myc node,
regulation and targets, 221, 222
S phase coordination with apoptosis, 222
nuclear factor-κB node,
antiapoptotic, proapoptotic, and proliferative pathways, 228, 229
regulation and targets, 227
overview, 217–219
p53 node,
decision determinants in life and death, 226, 227
overview, 222, 223
p53-MDM2-ARF network and negative feedback loops, 223–225

positive feedback loops, 225, 226
prospects for study, 229, 230
Transforming growth factor-β (TGF-β), metastasis and dysfunction, 77–79
Trapoxin A, histone deacetylase inhibition, 339
Trichostatin A, histone deacetylase inhibition, 338
Tumor microenvironment, effects on chemotherapy-induced apoptosis, 306, 307
Tumstatin,
angiogenesis inhibition, 45–51
antitumor activity, 50
diabetic nephropathy management, 50, 51
endothelial cell protein synthesis inhibition, 49, 50
Goodpasture syndrome role, 46, 47
integrin binding, 48
peptide studies, 49
tum-5 domain, 48, 49
24p3, apoptosis modulation, 105, 108

U

UCN-01, mechanism of action, 127
uPA, *see* Urokinase-type plasminogen activator
Urokinase-type plasminogen activator (uPA), nuclear factor-κB regulation, 243, 244

V

Vascular endothelial growth factor (VEGF),
angiogenesis regulation, 37–40
c-Myc induction, 154, 155
nuclear factor-κB induction, 242
receptors,
neuropilin, 40
VEGFR-1, 39
VEGFR-2, 39, 40
splice variants, 39
therapeutic targeting, 52
types, 38, 39
VEGF, *see* Vascular endothelial growth factor